The Elements of Neutron Interaction Theory

The Elements of
NEUTRON INTERACTION THEORY

ANTHONY FODERARO

THE MIT PRESS
Cambridge, Massachusetts, and London, England

CONTENTS

PREFACE

If you publish a book, a parish curate accuses you of heresy, a college sophomore denounces you, an illiterate condemns you, the public derides you, your publisher renounces you, and your wine dealer cuts off your credit.

In the face of Voltaire's warning, why did I write this book? Indeed, why does anyone write any book? I submit that the answer lies in some form of overriding selfishness. Given the time and the tavern, I feel reasonably sure I could convince you, if it is not already evident to you, that the underlying motives of the author of any book are selfish ones: the desires for fame, fortune, or esteem in the eyes of one's peers or gods are perhaps the most widespread. I wrote this book in a selfish attempt to reduce the effort that I must expend to teach interaction theory. Although I enjoy lecturing, I dislike preparing those class notes which are necessary adjuncts to lectures in courses that have no textbooks. Had a suitable textbook been in existence or had I been able to convince someone else to write one, I should never have begun this one.

In writing this book I had in mind the background and the specific needs of the first-year graduate student in nuclear engineering. Among the first courses he should take are those dealing with the interactions of neutrons, photons, and charged particles with nuclei, atoms, and electrons. This subject matter is fundamental to his further course work in neutron and gamma-ray transport theory, reactor theory, radiation shielding, and radiation effects in materials. Unfortunately, most first-year graduate students in nuclear engineering have had neither advanced classical mechanics nor quantum mechanics, without which it is effectively impossible to meaningfully discuss interaction cross sections. A possible solution is to delay the interactions course for about two years so that the student might first take classical and quantum mechanics. But to delay for two years would not only prevent the M.S. candidate from taking the interactions course, and the courses which depend on it, but also would disrupt the Ph.D. candidate's nuclear engineering course sequence and, in effect, push much of the curriculum back two years. My solution to this dilemma is embodied in this book. I have attempted to write a textbook that can be read and understood by anyone who has obtained the equivalent of a bachelor's degree in physics, chemistry, or one of the engineering disciplines. I assume no mathematical background beyond differential equations and elementary vector analysis and no physics background beyond elementary modern physics.

Content and Purpose

This book consists of material pertinent to the understanding of neutron interactions in the energy range below 20 Mev, the range of interest in nuclear reactors. The first four chapters present those topics in classical and quantum mechanics which are fundamental to an understanding of the kinematics and dynamics of any nonrelativistic two-body collision between spinless particles. Chapter 5 covers the properties of neutrons and nuclei that influence their interactions. Chapter 6 generalizes the theory of the first five chapters to include interactions between particles with spin and culminates in the general theory of nuclear interactions. Each of the next five chapters is devoted to one of the principal neutron interactions of interest to nuclear engineers, from elastic scattering to fission. The final chapter deals with neutron interactions in which the motion and the binding of the target atoms are significant.

The aim of this book is to present, as simply as possible, those aspects of neutron interaction theory which follow directly from conservation laws and elementary quantum mechanics. References are included primarily for the purpose of further amplification. Most references are to books, rather than to technical articles, because the books are generally more clear and complete than the original articles and are more easily obtained by the student. A reference to an article is given only when the article has considerable intrinsic historical value or when the material in the article is not to be found in a standard reference book.

It is my hope that the student who has diligently read this book will be able to understand, with little additional effort, any of the hundreds of technical papers on neutron interactions and cross sections that are published each year. It is my further hope that he will have sufficient background to be able to apply known cross sections properly and be able to estimate unknown cross sections intelligently. The nuclear engineering

student who wishes to make his career in the area of experimental or theoretical cross-section analysis is strongly advised to take additional courses in quantum mechanics and nuclear theory.

To the Teacher

We have found that the entire contents of this book can be covered rather comfortably in about 75 fifty-minute class periods. Chapters 1 through 6 require a full semester of 45 class periods; Chaps. 7 through 12 require 30 class periods. In our interactions course, we devote the final 15 class periods of the second semester to gamma-ray and charged-particle interactions. The kinematics of these interactions are fully derived, but the cross sections are not derived, merely displayed.

The greater depth in treatment of neutron interactions compared with photon and charged-particle interactions can be justified on several bases: First, the average nuclear engineer spends considerably more of his time on problems involving neutron interactions. Second, neutron cross sections are much less well known than the others; thus the nuclear engineer must be conversant with the theory in order to estimate unknown neutron cross sections. On the other hand, the theories of photon and charged-particle interactions have been worked out in exhaustive detail in all important cases. The nuclear engineer need only know (1) what the principal interactions are, (2) their kinematics, and (3) where to obtain the known cross sections. These topics can be readily covered in 15 class periods.

Acknowledgments

It gives me great pleasure to acknowledge the help and cooperation I have received from many persons during the preparation of this book. I am very grateful to Professors John R. Lamarsh, Richard K. Osborn, and Thomas G. Williamson and to Dr. William L. Whittemore for suggesting numerous technical improvements. My sincere thanks are due to the following graduate students for their help: Joseph Cardito, Tara Chawla, James Cook, Theodore Ginsberg, Jacek Jedruch, Robert Kreahling, Kenneth Lindquist, Leo Mariani, William McCurdy, William Naughton, K. N. Prasad, Finn Skogen, Edmond Tourigny, and William Woolson. Very particular thanks are due Carolyn M. Lee and Marian C. Fox, Division of Technical Information Extension, who, as editors, contributed a vast number of stylistic improvements. I also wish to thank the Division of Technical Information of the United States Atomic Energy Commission for sponsoring this book. Without their financial aid to my graduate students and myself, this book would have been long delayed, perhaps never written. In this regard, I wish particularly to thank James D. Cape, John Inglima, and Joseph Gratton for their patience, understanding, and helpful suggestions. Finally I am greatly indebted to Mary Bell, Patricia Kidder, and Alex Giedroc for their aid in the physidal preparation of the manuscript and art work.

Anthony Foderaro

University Park
Pennsylvania

GENERAL FEATURES OF INTERACTIONS 1

Two bodies moving relative to one another are said to interact when they measurably influence one another's internal states or trajectories. An *interaction* is defined, in general, as a mutual or reciprocal action or influence. Operationally, however, it can only be defined in terms of measurable consequences. Hence, if the internal or external states of two bodies have not been measurably changed owing to each other's presence, they have not interacted.

The term *collision* designates an interaction that occurs continuously for only a comparatively short period of time. Two bodies that collide influence one another only during a finite or infinitesimal time period. Bodies, such as the earth and the sun, that are in continuous interaction through their gravitational force fields are not in collision.

Every collision can be conveniently divided into three time periods, before, during, and after, defined loosely as follows: Before the collision the particles are sufficiently far apart that they exert no influence, or negligible influence, on one another. During the collision the particles or their force fields are in contact. After the collision the resultant particles are sufficiently far apart that they exert only negligible influences on one another, or they have somehow merged to form a single resultant particle.

1.1 Elastic Scattering, Inelastic Scattering, and Reactions

The pair of particles that enters a collision is called the *initial constellation*. The group of one, two, or more particles that leaves the collision is called the *final constellation*.

A *scattering collision* is defined as a collision in which the final constellation is the same as the initial constellation or differs only in the degree of excitation of the internal states of its members. A scattering collision is *elastic* if the total kinetic energy is conserved, i.e., if the total kinetic energy of the particles in the final constellation equals the total kinetic energy of the particles in the initial constellation. Note, however, that particles which scatter elastically can, and indeed almost always do, interchange kinetic energy in the laboratory frame of reference. A common misconception is that in elastic scattering there is no change in the kinetic energy of either of the interacting particles. This is generally not so. Elastic scattering is defined as a collision in which there is no change in the sum of the kinetic energies of the interacting particles. If part of the kinetic energy of the initial constellation is expended in increasing the internal energy of one or both of the colliding particles, a scattering collision is called *inelastic*.

A *reaction* is a collision in which the final constellation differs from its initial constellation, either completely or in that one or both of the members of the final constellation emerge with an internal state different from the initial internal state.

If we designate the members of the initial constellation A and B and the members of the final constellation C, D, E, etc., and, if we use an asterisk to denote a particle in an excited state, then we can give symbolic examples of the types of collisions defined above:

Scattering collisions:
$A + B \rightarrow A + B$ (elastic scattering)
$A + B \rightarrow A + B^*$ (inelastic scattering)

Reactions:
$A + B \rightarrow C^*$ (one-particle final constellation)
$A + B \rightarrow C + D$ (two-particle final constellation)
$A + B \rightarrow C + D + E + \cdots$ (more-than-two-particle final constellation)
$A + B \rightarrow A + B^*$ (inelastic scattering)

We note that the terms *scattering collision* and *reaction* are not mutually exclusive; an inelastic-scattering collision is both a scattering collision and a reaction. But, the terms *elastic scattering* and *reaction* are mutually exclusive; a collision cannot be both an elastic-scattering collision and a reaction. This distinction is sometimes made more evident by calling reactions *nonelastic collisions*; thus all collisions are divided into two major classes: elastic collisions and nonelastic collisions.

The initial constellation of almost all the collisions we will encounter consists of *target particles* which are at rest in the laboratory system and *projectile particles* which are moving toward the target particles. The projectile particles considered in this book are neutrons; the target particles are atomic nuclei. Although our primary interest is in neutron interactions with nuclei, we will continue to use the general terms projectile particle and target particle whenever the theory is independent of the specific nature of the particles in the initial constellation.

The neutron energy range considered is that of primary interest to the nuclear engineer: from zero to approximately 18×10^6 electron volts (ev).* The upper limit of the energy range (approximately 18 Mev) is the energy of the most energetic fission neutrons. Interactions of neutrons having all energies below this upper limit are significant in reactor-core and shielding design and in radiation-effects studies.

1.2 Kinematics and Cross Sections

Let us now focus our attention upon two distinct aspects of any collision: the *kinematics* of the collision and the *cross sections* for the collision. By the kinematics of the collision, we mean the analysis of the interaction which is based solely on the general laws of conservation and which is not concerned with the forces between the particles.

Under the heading kinematics we seek answers to such questions as: Is this reaction possible; that is, does it violate any conservation law? What is the threshold energy of this reaction? If the incident particle elastically scatters, what is the relation between its initial energy, final energy, and angle of scattering? The answers to these and other questions, as we shall see, are independent of the specific forces that may exist between the interacting particles.

When considering the kinematics of an interaction, we do not ask such questions as: What are the relative probabilities of this and that interaction? What is the probability that a particle of type A will scatter between angles θ_1 and θ_2 when it elastically scatters off a particle of type B? Such questions as these, involving as they do the probabilities of various interactions, are questions that are completely answered by the appropriate cross sections for the interactions.

The cross section for a collision is a measure of the probability of the occurrence of the collision. The precise way in which the cross section is a measure of the probability is described in the next section. Unlike the kinematics of an interaction, the cross sections are dependent on the specific forces between the interacting particles.

* The electron volt is defined as the amount of kinetic energy an electron gains in falling through an electrostatic potential difference of one common volt. One electron volt equals 1.6×10^{-12} erg.

Dynamics is that branch of the science of mechanics which deals with motion. Dynamics itself has two branches, kinematics and kinetics. Kinematics is the study of motion itself, apart from causes. Kinetics is the study of the action of forces in producing or changing motion.

Classical and quantum kinematics when based solely on the laws of conservation of energy and linear momentum are identical; so all collisions in effect can be described by classical kinematics. Classical and quantum kinetics, on the other hand, are quite distinct. Cross sections, which depend on the specific forces between bodies, must be derived through the appropriate kinetic analysis: through classical kinetics for large bodies and through quantum kinetics for atomic and subatomic bodies.

1.3 Microscopic and Macroscopic Cross Sections

All interactions within the scope of this book are usually, and quite conveniently, characterized by cross sections. A cross section is a number proportional to the rate at which a particular interaction will occur under given conditions.

Microscopic Cross Sections

Consider a small region of some homogeneous material through which there are particles of type A moving with speed v_a relative to the stationary target particles. Let us concentrate on one type of interaction, the jth, with one particular type of target entity in the material, the type we call B. The jth interaction may be elastic scattering, inelastic scattering, or any of the possible reactions that can occur between particles A and B. Type A particles may be, for example, neutrons, photons, or alpha particles; type B particles may be, for example, atoms, nuclei, or electrons.

The *number densities* of particles A and B, that is, the numbers per unit volume will be designated $[A]$ and $[B]$, respectively.

Under the assumption that the region is sufficiently thin or that the target particles are sufficiently far apart relative to their sizes that there is no shadowing of one particle by another, it is clear that the number of j-interactions per unit volume per unit time can safely be taken as directly proportional to the number density of target particles,

$$I_j \propto [B],$$

and directly proportional to number density of moving particles,

$$I_j \propto [A].$$

It is obvious that the number of interactions that will occur in any time interval is directly proportional to the number of A particles that pass through the region in this time interval; hence, the interaction rate is directly proportional to the speed of the A particles,

$$I_j \propto v_a.$$

Since there are no other variables, we may write

$$I_j \propto v_a[A][B].$$

A constant of proportionality must be inserted to form an equation from this relation. The Greek letter σ has been favored historically for this purpose. The constant σ obviously depends on the type of A and B particles and on the type of interaction and, in general, may depend on the energy of the incoming particles. We indicate these dependencies by subscripts, superscripts, and arguments; thus

$$I_j = \sigma_{ab}^j(E_a)v_a[A][B]. \tag{1.1}$$

Now we note that

$$\sigma_{ab}^j = \frac{I_j}{v_a[A][B]}$$

has dimensions

$$\frac{(L^{-3}T^{-1})}{(LT^{-1})(L^{-3})(L^{-3})} = L^2.$$

Because the dimensions of the constant of proportionality, σ, are those of an area, σ is called a *cross section*. Specifically, $\sigma_{ab}^j(E_a)$ is the *microscopic cross section* of target entity B for the jth type interaction with particles A having energy E_a.

Microscopic neutron cross sections of nuclei and gamma-ray cross sections of atoms are usually of approximate order 10^{-24} cm^2. It has therefore proved convenient to define a unit of area, the *barn* (10^{-24} cm^2), to express microscopic cross sections for atomic and nuclear interactions.

Equation 1.1 is the defining equation for cross sections. A more common form of this equation is obtained by noting that the product $v_a[A]$ is the flux of particles of type A. If this flux is called Φ_a, then Equation 1.1 becomes

$$I_j = \sigma_{ab}^j\Phi_a[B]. \tag{1.2}$$

Microscopic cross sections as measures of the individual interactions between particles can be determined experimentally by direct application of Equation 1.2. If we know the number of target entities and the flux, then we can measure the rate of particular interactions and determine the microscopic cross section by division:

$$\sigma_{ab}^{j} = \frac{I_j}{[B]\Phi_a}.$$ (1.3)

The definition of σ is purely operational; we may imagine all cross sections as being measured by an idealized experimental arrangement such as that sketched in Figure 1.1. A uniform beam of monoenergetic, monodirectional A particles is incident on a thin foil of B particles. The A particles are sufficiently far apart that they do not interact with one another, only with the B particles. The foil of B particles is sufficiently thin that almost all the A particles pass through it without interaction; so the flux of A particles is a constant throughout the foil. Furthermore, the flux is sufficiently small that the number of interactions that will occur during the experiment is many orders of magnitude less than the number of B particles in the target; so the target composition does not vary significantly during the experiment.

An idealized detector, i.e., one capable of detecting the numbers and types of particles in the final constellation, their energies, and any other variables that may describe them, is located a great distance from the foil relative to the size of the foil and to the cross-sectional area of the beam.

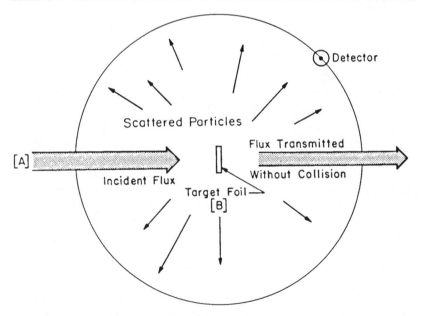

Figure 1.1—Idealized experimental arrangement for measuring cross sections.

By successively positioning the detector on the imaginary spherical shell centered on the foil, we can determine the total number of j interactions, N_j, that occur during an experiment of duration Δt. If the foil has cross-sectional area ΔA and thickness Δz, the number of interactions per unit volume of the foil per unit time is simply

$$I_j = \frac{N_j}{\Delta A \, \Delta z \, \Delta t},$$

and the cross section may be determined by direct application of Equation 1.3.

In the preceding discussion it was assumed that the B particles were at rest and that only the A particles were moving. Indeed, this is most often the situation. Usually the B particles are nuclei or atoms that, for all practical purposes, may be considered to be at rest in the laboratory, and usually the A particles are rapidly moving neutrons or gamma rays.

We will, however, also consider interactions in which both sets of particles are moving. In this instance the distinction between target particles and projectile particles no longer can be drawn, and the generalized equation for the interaction rate must be written

$$I_j = \sigma_{ab}^j v_{ab}[A][B], \tag{1.4}$$

where $[A]$ and $[B]$ are the number densities of A and B particles; v_{ab} is the *relative speed* between the A and B particles, i.e., the speed of approach of the A and B particles; and σ_{ab}^j is the microscopic cross section for the jth type interaction between A and B particles moving with relative speed v_{ab}. The cross section for the interaction will be a function of the *relative energy*, i.e., $\sigma_{ab}^j = \sigma_{ab}^j(E_{ab})$.

The derivation of Equation 1.4 follows precisely the arguments used in deriving Equation 1.1 except that the interaction rate is proportional to the relative speed between the A and B particles. This is most easily proved by considering the interaction as viewed from the coordinate system in which one of the beams of particles is at rest. If, for example, we choose a coordinate system in which the B particles are at rest, then the rate of interactions is proportional to the speed with which the A particles pass through the B particles. The speed is, of course, the relative speed v_{ab}.

The physical picture used in the derivation of Equation 1.4 is sketched in Figure 1.2. Two monodirectional beams of particles are shown approaching each other at angle α. The first beam contains $[A]$ particles per unit volume moving with vector velocity $\mathbf{v}_a$. The second beam contains $[B]$ particles per unit volume moving with vector velocity $\mathbf{v}_b$. The beams intersect, and collisions occur in the volume that is shown shaded in

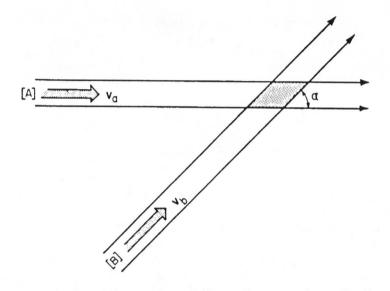

Figure 1.2—Intersection of two beams of particles.

Figure 1.2. As usual, I_j is the rate of primary interactions of type j per unit volume.

The A and B particles approach one another with a relative speed v_{ab}, which for nonrelativistic particles is given by the square root of the square of the velocity difference,

$$v_{ab} = [(\mathbf{v}_a - \mathbf{v}_b) \cdot (\mathbf{v}_a - \mathbf{v}_b)]^{1/2}$$

or

$$v_{ab} = (v_a^2 + v_b^2 - 2v_a v_b \cos \alpha)^{1/2}, \tag{1.5}$$

where α is the angle between the velocity vectors of the A and B particles and

$$\cos \alpha = \frac{(\mathbf{v}_a \cdot \mathbf{v}_b)}{v_a v_b}.$$

Equations 1.1 and 1.2, obviously special cases of Equation 1.4, are obtained by setting the velocity of the B particles equal to zero.

Macroscopic Cross Sections

The product of the microscopic cross section for an interaction and the number density of target particles is called the *macroscopic cross section* for the interaction. If the flux under consideration consists of neutrons, then historically the symbol for the macroscopic cross section is Σ; if the flux consists of photons, the symbol for the macroscopic cross section is μ. Thus Equation 1.2 may be rewritten, in terms of macroscopic cross sections, in the form

$$I_j = \Sigma_{ab}^j \Phi_a, \tag{1.6}$$

where $\Sigma_{ab}^j = \sigma_{ab}^j[B]$ for neutrons, and in the form

$$I_j = \mu_{ab}^j \Phi_a, \tag{1.7}$$

where $\mu_{ab}^j = \sigma_{ab}^j[B]$ for photons.

Since microscopic cross sections have dimensions of area and number densities have dimensions of reciprocal volume, the dimension of a macroscopic cross section is reciprocal length.

Upon rearranging Equation 1.6 in the form

$$\Sigma = \frac{I}{\Phi}, \tag{1.8}$$

it becomes apparent that the macroscopic cross section for an interaction is the number of interactions per unit volume per unit time per unit flux of incident particles.

1.4 Differential Cross Sections

Differential cross sections are employed to describe interactions in which the states of the particles in the final constellation have a continuous distribution in one or more variables.

Suppose in the final constellation of a collision that one of the variables, q, has a continuous range of values in some interval q_1 to q_2. The differential cross section, $\sigma(q)$, for this collision is defined as follows: $\sigma(q)\,dq$ is the cross section for the interaction in which the variable q obtains a value in the range q to $q + dq$.

The number of interactions per unit volume per unit time, designated $I(q)\,dq$, which results in a value between q and $q + dq$ for the variable q is given by an expression analogous to Equation 1.2,

$$I(q)\,dq = \sigma(q)\,dq\,\Phi_a[B]. \tag{1.9}$$

It is most important to note that the dimensions of the product $\sigma(q)\,dq$ are those of area; so the dimensions of $\sigma(q)$ itself are those of area per unit q dimension. Similarly, since the dimensions of $I(q)\,dq$ are interactions per unit volume per unit time, the dimensions of $I(q)$ itself are interactions per unit volume per unit time per unit q dimension.

The total cross section for the interaction is given by the integral of the differential cross section over all possible final values of q,

$$\sigma = \int_{q_1}^{q_2} \sigma(q)\, dq. \tag{1.10}$$

Change of Variables

If, as is often the case, the final constellation of the collision is characterized by a second variable h which is a function of the variable q, $h = h(q)$ such that for every value of h there is a single value of q, $q = q(h)$, then the differential cross section can be expressed in terms of the variable h by simple transformation of variables. Since it is identically true that $\sigma(h)\, dh \equiv \sigma(q)\, dq$, where dh is the interval of h corresponding to the interval dq of q,

$$\sigma(h)\, dh = \sigma(q) \left| \frac{dq}{dh} \right| dh. \tag{1.11}$$

If q and dq/dh on the right-hand side are expressed as explicit functions of h, then $\sigma(h)$ is an explicit function of h. The absolute value sign is placed around dq/dh to ensure that the cross section remains positive, as it physically must, even if increasing h results in decreasing q.

Elastic Scattering

The most common example of differential cross sections is the differential cross section for elastic scattering. In elastic scattering the initial and final constellations consist of the same particles. Projectile particles A merely change their directions of motion and their speeds in scattering from target particles B. Interest naturally centers on determining the cross section as a function of scattering angle.

The basic, idealized, experimental procedure is shown in Figure 1.3. A uniform, collimated, monoenergetic beam of A particles is projected at a foil of volume ΔV containing many identical B particles. The foil is sufficiently thin that the A particles have extremely low probability of scattering more than once. A detector of infinitesimal size is mounted on a lever arm centered at the foil. The length of the lever arm is much greater than the length or width of the foil. The detector can be positioned at any *scattering angle* $\tilde{\theta}$ and *azimuthal angle* $\tilde{\phi}$. [The tilde ($\sim$) designates angles measured in the laboratory.] The scattering angle is defined as the angle between the velocity vectors of the incident and scattered A particles. The azimuthal angle, for our purposes, may be defined with reference to any axis perpendicular to the incident-flux velocity vector.

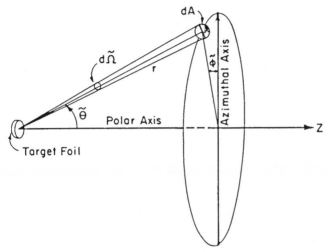

Figure 1.3—Geometry of differential scattering.

The detector subtends a solid angle at the foil given by

$$d\tilde{\Omega} = \frac{dA}{r^2}, \tag{1.12}$$

where dA is the projected area of the detector on the plane normal to the radius vector, $\mathbf{r}$. Since r is much greater than the largest dimension of the detector or the foil, the solid angle subtended by the detector is the same regardless of the point on the foil from which it is measured.

To summarize, the ideal experiment to measure a scattering cross section is characterized by the following conditions:

1. The projectile particles are monoenergetic and monodirectional and form a uniform beam.

2. The projectile particles do not interact with one another.

3. No projectile particle interacts more than once in the target.

4. The only forces that act are those between the projectile particles and the target particles; there are no external forces.

5. The detector is at a great distance from the target relative to the largest dimension of both the target and the detector.

The number of elastically scattered A particles that enters the detector per second will in general be a function of $\tilde{\theta}$ and $\tilde{\phi}$, which we call $dN_s(\tilde{\theta}, \tilde{\phi})$. It is clear from previous arguments that dN_s will be proportional to the incident flux Φ_a and the total number of target entities $[B]\,\Delta V$. It is also apparent that the number of particles that intersect the detector is directly proportional to the solid angle $d\tilde{\Omega}$ subtended by the detector. (Remember, the detector is infinitesimal so that only negligible variations might exist in the scattered flux as a function of angle over the dimensions of the detector.) Again, after converting our proportionality to an equation, we have

$$dN_s(\tilde{\theta}, \tilde{\phi}) = \sigma_s(\tilde{\theta}, \tilde{\phi}) \, d\tilde{\Omega} \, \Phi_a[B] \, \Delta V, \qquad (1.13)$$

where the constant of proportionality, $\sigma_s(\tilde{\theta}, \tilde{\phi})$, is called the *elastic-scattering differential cross section*. Note that, if $d\tilde{\Omega}$ is measured in steradians, as is common, the dimensions of $\sigma_s(\tilde{\theta}, \tilde{\phi})$ are those of area per steradian, e.g., cm²/steradian or barns/steradian.

If we define $dI_s(\tilde{\theta}, \tilde{\phi})$ as the number of scattered A particles entering the detector per second per unit volume of target, then, since

$$dI_s(\tilde{\theta}, \tilde{\phi}) = \frac{dN_s(\tilde{\theta}, \tilde{\phi})}{\Delta V},$$

we can rewrite Equation 1.13 in the more compact form

$$dI_s(\tilde{\theta}, \tilde{\phi}) = \sigma_s(\tilde{\theta}, \tilde{\phi}) d\tilde{\Omega} \, \Phi_a[B]. \qquad (1.14)$$

In the derivation of Equation 1.14, $d\tilde{\Omega}$ was introduced as the solid angle subtended by a detector having cross-sectional area dA, but $d\tilde{\Omega}$ could have been introduced as simply a differential solid angle in the spherical coordinate system having the target as center,

$$d\tilde{\Omega} = \sin \tilde{\theta} \, d\tilde{\theta} \, d\tilde{\phi}. \qquad (1.15)$$

In place of the scattering angle $\tilde{\theta}$, it will prove convenient to introduce a variable $\tilde{\mu}$ defined by the equation $\tilde{\mu} = \cos \tilde{\theta}$, in which case $d\tilde{\mu} = -\sin \tilde{\theta} \, d\tilde{\theta}$. We will drop the minus sign since it merely indicates the relative directions of variations of $\tilde{\mu}$ and $\cos \tilde{\theta}$ and assert that

$$d\tilde{\mu} = \sin \tilde{\theta} \, d\tilde{\theta}. \qquad (1.16)$$

Thus the differential solid angle in spherical coordinates can be written

$$d\tilde{\Omega} = d\tilde{\mu} \, d\tilde{\phi}.$$

When this expression is substituted into Equation 1.14,

$$dI_s(\tilde{\mu}, \tilde{\phi}) = \sigma_s(\tilde{\mu}, \tilde{\phi}) \, d\tilde{\mu} \, d\tilde{\phi} \, \Phi_a[B], \qquad (1.17)$$

where $\sigma_s(\tilde{\mu}, \tilde{\phi}) \, d\tilde{\mu} \, d\tilde{\phi}$ is the cross section for elastic scattering into the polar range $d\tilde{\mu}$ about $\tilde{\mu}$ and the azimuthal range $d\tilde{\phi}$ about $\tilde{\phi}$.

If the cross section is independent of azimuthal angle, Equation 1.17 may be directly integrated over the entire range of $\tilde{\phi}$ from 0 to 2π to yield

$$dI_s(\tilde{\mu}) = \sigma_s(\tilde{\mu})2\pi \, d\tilde{\mu} \, \Phi_a[B], \tag{1.18}$$

where $dI_s(\tilde{\mu})$ is the number of A particles scattered per second per unit volume of target foil into the range $d\tilde{\mu}$ about $\tilde{\mu}$ and $\sigma_s(\tilde{\mu})2\pi \, d\tilde{\mu}$ is the cross section for scattering into the solid-angle range $2\pi \, d\tilde{\mu}$ about $\tilde{\mu}$.

It is important to note that the flux of scattered particles must be independent of the azimuthal angle $\tilde{\phi}$ unless either the projectile particles or the target particles are polarized in some sense or other. This follows from the fact that the origin of the angle $\tilde{\phi}$ cannot be physically defined in the absence of some physical vector entity having a component perpendicular to the z-axis. This physical entity could be, for example, the intrinsic angular momentum of either the A or B particles provided this spin was not random but had a net component perpendicular to the z-axis. It could be an asymmetric potential shape or an external electric, magnetic, or gravitational field. The nuclear engineer rarely, if ever, encounters situations in which he must deal with polarized particles or external fields. Thus, unless specifically excepted, we shall consider only azimuthally independent interactions.

If the scattering is not only independent of the azimuthal angle $\tilde{\phi}$ but also is independent of the polar angle $\tilde{\theta}$, then the scattering is called *isotropic*. If the scattering is isotropic in the laboratory, the same flux of scattered particles will pass through the detector regardless of the angular orientation of the detector.

A clear picture of isotropic scattering can be drawn from a simple gedanken experiment. [*Gedanken* is the German word for *thought*. A gedanken experiment is one capable of being mentally envisioned and is employed primarily for its instructional value. Whether or not it can be carried out in the laboratory exactly as envisioned is irrelevant.] Imagine a monodirectional beam of A particles incident on a small target. Suppose there is a large spherical shell centered at the target which is so fashioned that every scattered A particle leaves a permanent record in the form of a dot on the sphere's surface as it passes through it. After a time sufficient to produce many scattered particles, the beam is stopped and the distribution of dots on the sphere's surface is examined (Figure 1.4). If the dots are distributed uniformly over the surface so that each element of equal area has the same number of dots, then the scattering was isotropic, i.e., the same number of particles were scattered through each unit solid angle regardless of the orientation of the solid angle.

It is important to understand that isotropic scattering does not imply that the same number of particles scatter through equal increments in $\tilde{\theta}$.

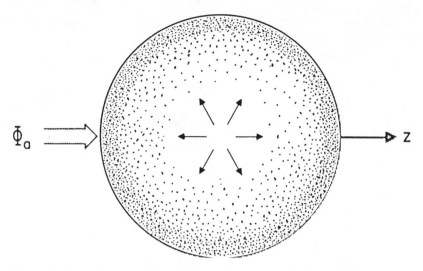

Figure 1.4—Isotropic scattering gedanken experiment.

For example, it is clear that far more particles scatter through the 10° increment in $\tilde{\theta}$ from 80° to 90° than from 0° to 10°. To prove this, just compare the numbers of dots in these two angular regions. The numbers of dots will be directly proportional to the areas of the regions on the sphere surface or to the solid angles subtended by the regions. To calculate the solid angle between any two angles $\tilde{\theta}_1$ and $\tilde{\theta}_2$, we need an expression for the differential solid angle in polar coordinates. In Figure 1.5, by definition the solid angle subtended by the differential segment of the spherical surface between $\tilde{\theta}$ and $\tilde{\theta} + d\tilde{\theta}$ is the ratio of the area of this segment to the square of the distance between the center of the sphere and the area segment,

$$d\tilde{\Omega} = \frac{dA}{r^2} = \frac{(r\,d\tilde{\theta})(2\pi r \sin \tilde{\theta})}{r^2} = 2\pi \sin \tilde{\theta}\,d\tilde{\theta}. \tag{1.19}$$

Alternatively, from the definition of $d\tilde{\mu}$, we may write

$$d\tilde{\Omega} = 2\pi\,d\tilde{\mu} \tag{1.20}$$

Therefore, the magnitude of the solid angle between $\tilde{\theta}_1$ and $\tilde{\theta}_2$ is given by

$$\tilde{\Omega}(\tilde{\theta}_1, \tilde{\theta}_2) = 2\pi \int_{\tilde{\theta}_1}^{\tilde{\theta}_2} \sin \tilde{\theta}\,d\tilde{\theta} = 2\pi(\cos \tilde{\theta}_1 - \cos \tilde{\theta}_2). \tag{1.21}$$

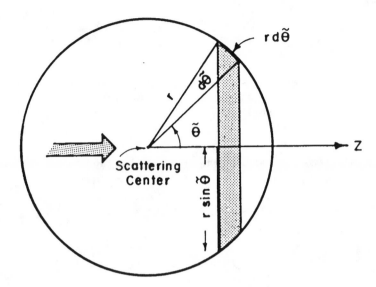

Figure 1.5—The differential solid angle in polar coordinates with no azimuthal dependence.

Thus, if $\tilde{\theta}_1 = 0$ and $\tilde{\theta}_2 = 10°$,

$$\tilde{\Omega}(0°, 10°) = 2\pi(\cos 0° - \cos 10°) = 2\pi(0.015),$$

whereas, if $\tilde{\theta}_1 = 80°$ and $\tilde{\theta}_2 = 90°$,

$$\tilde{\Omega}(80°, 90) = 2\pi(\cos 80° - \cos 90°) = 2\pi(0.174)$$

and we note that over 10 times more particles will isotropically scatter into the angular range 80° to 90° than into the range 0° to 10°.

Angular differential cross sections, such as those for elastic scattering and inelastic scattering, will be expressed in this book, as in almost all technical literature, in units of area per unit solid angle and not in units of area per unit angle. There are several reasons for this choice of units, the primary practical reason being that departures from isotropic scattering are much more easily seen in solid-angle units. For example, suppose we are dealing with isotropic scattering. The differential cross section in solid-angle units will be some constant C:

$$\sigma_s(\tilde{\theta}) = C \text{ barns/steradian.}$$

The cross section expressed in barns/radian can be obtained from the identity $\sigma_s(\tilde{\theta})\, d\tilde{\Omega} = \sigma_s'(\tilde{\theta})\, d\tilde{\theta}$ and the fact that $d\tilde{\Omega} = 2\pi \sin \tilde{\theta}\, d\tilde{\theta}$. Thus

$$\sigma_s'(\tilde{\theta}) = 2\pi C \sin \tilde{\theta} \text{ barns/radian.}$$

Graphs of these two cross sections are shown in Figure 1.6. Note that, when expressed in solid-angle units, the cross section for isotropic scattering is a constant; when expressed in angular units, it is no longer a constant. It is clear that small departures from isotropy are more easily detected from a graph of $\sigma_s(\tilde{\theta})$ than from a graph of $\sigma_s'(\tilde{\theta})$. Note also that the graph of $\sigma_s'(\tilde{\theta})$ conveys the false impression that a detector would find more scattered particles at $\pi/2$ than at any other angle; however, the graph of $\sigma_s(\tilde{\theta})$ quite properly indicates that the detector will find the same flux of scattered particles regardless of its angular orientation.

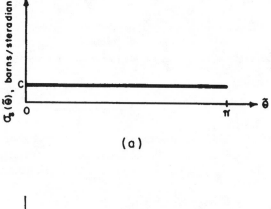

(a)

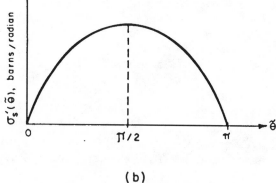

(b)

Figure 1.6—Isotropic differential scattering cross sections (a) in units of area per unit solid angle and (b) in units of area per unit angle.

1.5 Force-Field Collisions and Contact Collisions

Bodies in collision seem to exert two apparently different types of force on one another. *Contact forces* are the more common. A ball striking a bat or a wall, a bullet striking a target, and two billiard balls colliding are all examples of *contact collisions*. Collisions between macroscopic objects some distance from one another are also rather common, e.g., *force-field collisions* of electrified pith balls or of bar magnets suspended by strings. An apple dropping from a tree is in collision with the earth through their mutual gravitational forces while dropping and through contact forces when it reaches the ground.

The distinction between contact collisions and force-field collisions disappears when one considers the microscopic structure of macroscopic objects. In the region of apparent contact, there are billions of atoms exerting electromagnetic forces on one another, forces at a distance. In this sense bodies never touch, and all collision between bodies ultimately may be reduced to force-field collisions. Nevertheless, in practice the distinction between these two types of collisions is maintained simply because it is practically impossible to treat the simultaneous force-field collisions of billions of particles.

Collisions on the microscopic level between, for example, neutrons and nuclei or photons and electrons are force-field collisions and are analyzed in terms of the forces acting at a distance between the interacting particles. In some cases the forces are of extremely short range and are so localized that, in effect, they are what might be called contact forces. Thus the distinction between contact collisions and force-field collisions again becomes one of convenience rather than of absolute necessity.

If the force between two bodies is conservative, i.e., if no energy is expended in friction or other dissipative forces, then the force **F** between the two bodies can be expressed as the gradient of a scalar function V called the *potential energy*, *potential function*, or sometimes simply *potential*,

$$\mathbf{F} = -\nabla V. \tag{1.22}$$

The spatial point where the potential energy between two particles is zero is arbitrary; therefore we define the potential energy between two bodies as zero for infinite separation of the two bodies, so that the potential energy of a two-body system as a function of the distance of separation, r, of the centers of the two bodies will be the work which must be expended to bring the bodies from infinity to the distance r. Thus we define the potential at separation distance r by rewriting Equation 1.22 in the form

$$V(r) = -\int_{\infty}^{r} \mathbf{F} \cdot d\mathbf{s}, \tag{1.23}$$

where ds is an element of path length along the path of approach of the bodies.

If the force between the two bodies is repulsive, $\mathbf{F} \cdot ds$ will be negative for bodies being brought together; so the potential at any point will be positive. Conversely, if the force is attractive, the potential will be negative. In other words, energy must be expended to bring bodies together against

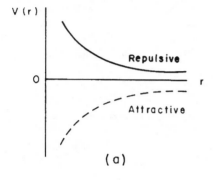

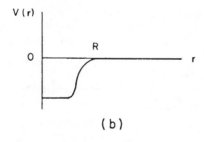

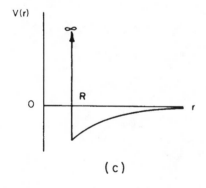

Figure 1.7—Examples of potential functions. (a) Inverse-r potential. (b) Attractive short-range potential. (c) Potential for massive impenetrable spheres.

their mutual forces of repulsion, but energy is gained when attracting bodies are brought together. The potential energy that is lost or gained in this way appears as kinetic energy of the interacting bodies; so the sum of kinetic plus potential energy remains constant.

Some common examples of potential functions are shown in Figure 1.7. Figure 1.7a depicts a repulsive inverse-r potential such as would exist between two bodies having electrostatic charges of the same sign. The mirror image of this potential would arise from an attractive inverse-r^2 force, such as the electrostatic force between bodies having unlike charges or the gravitational force between massive bodies.

A typical short-range potential such as might roughly describe the force between a neutron and a nucleus is shown in Figure 1.7b. No force acts until the particle centers are within a distance R of each other, then a strong attractive surface force acts, followed again by no force after the neutron enters the inner nucleus.

The contact collision between two massive, impenetrable, uncharged spheres may be described by a potential like that in Figure 1.7c. Only gravitational forces act until the spheres make contact; then the potential becomes infinite and remains so to $r = 0$. In this figure R represents the sum of the radii of the two spheres.

We shall ignore the influence of the external magnetic and gravitational fields of the earth in our analysis of collisions: these fields exert completely negligible influences on the trajectories of particles moving with speeds of 1000 m/sec and greater over distances of the order of tens of meters. All interactions of interest to us are between particles with relative speeds greater than 1000 m/sec. Thus, if we restrict ourselves to collisions in the laboratory where the distances between the source of the particle and the target and between the target and the detector are of the order of tens of meters or less, we need not consider external field effects, and the only forces of consequence are those between the interacting particles. There are no external forces acting on our particles, only internal forces between particles. Because our colliding particles form an isolated system, the analysis of their interactions is considerably simplified, as we shall see in subsequent chapters.

EXERCISES

1. Classify each of the following collisions in as many ways as possible under the headings scattering, elastic scattering, inelastic scattering, and reaction.

(a) $A + n \rightarrow B$ (b) $A + \gamma \rightarrow C + n$
(c) $D + n \rightarrow F + p$ (d) $A + n \rightarrow A^* + n$
(e) $A + n \rightarrow A + n$ (f) $C + n \rightarrow D + F + G$

2. Show that the elastic-scattering differential cross section in the case of a single scattering center may be written in the form

$$\sigma(\theta, \phi) = \frac{r^2 \times \text{scattered flux at } (r, \theta, \phi)}{\text{incident flux}}$$

3. Prove the following assertions: (a) Half of all particles that scatter isotropically scatter between angles 60° and 120°. (b) Most of the straws in a random pile of hay lie more nearly perpendicular to the line of vision than parallel to it (regardless of how you look at it).

4. Suppose all the people on the earth were distributed uniformly and walked randomly, blindfolded, over the entire surface. (Assume they can walk on water.) Estimate, to order of magnitude, the number of collisions between people that would occur per day. List all assumptions in your estimate.

5. Show, by order-of-magnitude calculation, that the trajectory of a neutron of energy greater than 0.01 ev is practically unaffected by the gravitational field of the earth over distances of the order of meters. Show, similarly, that the trajectories of electrons with energies of interest to nuclear engineers (say, $E > 10^5$ ev) are practically unaffected by either the gravitational or the magnetic fields of the earth over distances of the order of meters.

6. The differential scattering cross section for a particular collision is found to be of the form

$$\sigma_s(\mu) = (50 + 35\mu) \text{ barns/steradian.}$$

(a) Calculate the total scattering cross section. (b) Derive an expression for the number of particles that scatter between any two angles θ_1 and θ_2. (c) Derive an expression for the differential cross section $\sigma(E)$ based on the assumption that the energy of the scattered particles, E, depends on their initial energy E_0 and angle of scattering through the equation

$$E = E_0(\tfrac{1}{2} + \tfrac{1}{2}\mu).$$

(d) Calculate the total scattering cross section by integration of $\sigma(E)$, and compare it with the cross section calculated in Part (a).

7. Develop some examples of elastic collisions between macroscopic objects that would result in (a) azimuthally independent scattering and (b) azimuthally dependent scattering.

CLASSICAL COLLISIONS 2

In this chapter ordinary nonrelativistic mechanics, i.e., Newtonian mechanics, will be used to develop kinematic relations and cross sections for two-body collisions.

The kinematic relations, developed solely from the laws of conservation of energy and linear momentum, are equally applicable to all nonrelativistic collisions whether the colliding bodies are large or small. These equations, for example, will be valid for neutron collisions and for subatomic charged-particle collisions.

Since the cross sections derived in this chapter are based on Newtonian kinetics, they will be valid only for classical interactions, i.e., interactions between bodies sufficiently large and massive to obey the Newtonian approximation to quantum mechanics.

2.1 Conservation Laws

The laws of conservation of linear momentum and energy are the essential tools, and usually the only tools, required for the analysis of the kinematics of classical and quantum-mechanical interactions. Practically all the parameters whose values describe a particular interaction, except the cross sections for that interaction, can be determined by straightforward application of these two laws.

Conservation of Linear Momentum

The law of conservation of linear momentum of a system of particles follows directly from Newton's laws of motion. Consider an isolated system of particles, i.e., a system of particles on which no external forces act. The equation of motion for the ith particle is given by Newton's second law,

$$\sum_j \mathbf{F}_{ji} = \frac{d\mathbf{p}_i}{dt}, \tag{2.1}$$

where $\mathbf{p}_i$ is the momentum of the ith particle, $\mathbf{F}_{ji}$ is the force on the ith particle due to the jth particle, and, quite naturally, $\mathbf{F}_{ii} = 0$.

Summed over all particles, Equation 2.1 becomes

$$\sum_i \sum_j \mathbf{F}_{ji} = \frac{d}{dt} \sum_i \mathbf{p}_i.$$

However, since Newton's third law states that $\mathbf{F}_{ji} = -\mathbf{F}_{ij}$, the left-hand side reduces to zero. Therefore the time rate of change of the total linear momentum is zero, or, equivalently,

$$\sum_i \mathbf{p}_i \equiv \mathbf{P} = \text{constant in time.} \tag{2.2}$$

Equation 2.2 is the mathematical statement of the law of conservation of linear momentum.

Law of Conservation of Linear Momentum of a System of Particles: The total linear momentum of an isolated system of particles moving and interacting in any way remains constant.

Linear momentum is a vector quantity; thus Equation 2.2 is actually three equations, one for each component of total momentum in three-dimensional space. Each of the components is separately, i.e., independently, conserved.

$$\sum_i p_{ix} \equiv P_x = \text{constant,}$$

$$\sum_i p_{iy} \equiv P_y = \text{constant,} \tag{2.3}$$

$$\sum_i p_{iz} \equiv P_z = \text{constant.}$$

Conservation of Energy

The law of conservation of energy is a generalization of a vast number of observations extending back to the cannon-boring experiments of Count Rumford around 1798.

Law of Conservation of Energy of a System of Particles: The total energy of a system of isolated particles moving and interacting in any way remains constant.

Note that it is total energy, and not one particular type of energy, that is conserved. Three types of energy may be distinguished in a system of interacting particles: (1) kinetic energy relative to a given frame of reference: (2) potential energy relative to infinite separation of particles; and (3) internal energy, into which we might lump rest energies (mc^2), energies of internal motion, and rotational energies about fixed axes. The law of conservation of energy has nothing to say about the separate conservation, or lack of conservation, of each of these types of energy. It says only that the combined total of all types is conserved.

To illustrate this point, consider what is perhaps the simplest of all interactions: A mass m moving with speed v strikes and coalesces with a mass M that was at rest before the collision. This might describe, for example, a bullet striking and becoming embedded in a block of wood. The question immediately arises: What is the speed V of the block plus the embedded bullet after the collision? One is strongly tempted to write an equation based on conservation of energy in the form

$$\frac{mv^2}{2} = \frac{(M + m)V^2}{2},$$

but to do so is absolutely wrong. One has no a priori proof that kinetic energy is conserved in this reaction; as we shall soon discover, it is not. Instead we should use our knowledge that linear momentum is always conserved; thus

$$mv = (M + m)V. \tag{2.4}$$

Solving for the velocity after the collision, we obtain

$$V = \left(\frac{m}{M + m}\right)v, \tag{2.5}$$

which is the correct answer to the problem. Note that this speed differs from the speed we would have obtained from assuming (incorrectly) conservation of kinetic energy. The actual total kinetic energy of the system after the collision is

$$E = \frac{(M + m)V^2}{2} = \frac{1}{2}\frac{m^2}{(M + m)}v^2. \tag{2.6}$$

The total kinetic energy before the collision was $mv^2/2$; therefore the quantity of kinetic energy converted to other forms of energy as a result of the collision was

$$\Delta E = \frac{mv^2}{2} - E = \frac{1}{2}\frac{Mm}{(M + m)}v^2. \tag{2.7}$$

This elementary example provides the following note of caution: Always be sure you have included all relevant forms of energy when you write an equation based on the conservation of energy. Remember that *momentum is always conserved in the interaction of isolated particles whereas kinetic energy is conserved only in the elastic collision of isolated particles.* This observation is not as profound as it may seem, since an elastic collision is by definition a collision in which kinetic energy is conserved.

If the mass of the block, M, is much greater than the mass of the bullet, m, Equation 2.7 implies that almost all the initial kinetic energy is converted to other forms of energy. The forms into which it is converted depend on the details of the interaction and the structure, if any, of the interacting bodies. For example, when a bullet strikes a block of wood, part of the converted energy generally will go into rotation of the block and part into heat; the fraction that goes into rotational energy may be determined by application of the law of conservation of angular momentum, which is derived in the next section.

Conservation of Angular Momentum

The laws of conservation of linear momentum and energy provide general tools for the analysis of all interactions, they apply equally to classical and quantum-mechanical interactions, to nonrelativistic and, with proper definitions, relativistic interactions. The third conservation law, that for angular momentum, is valid in the form we shall derive here only for classical interactions. The quantum-mechanical expressions for angular momentum and its conservation are radically different in form from those we shall now investigate.

Once again, consider a system of particles on which no external forces act. These "particles" may be point particles or differential volumes of bodies whose masses extend over finite volumes; i.e., some of the particles under consideration might interact so strongly as to be considered the elements of a rigid body. Let ρ_i be the radius vector from the origin of some arbitrary fixed-coordinate system to the ith particle. The angular momentum of the ith particle is defined by the cross product $\rho_i \times p_i$, and the total angular momentum, L, of the system is defined by the summation over i,

$$L = \sum_i \rho_i \times p_i. \tag{2.8}$$

The time rate of change of L is given by

$$\dot{L} = \sum_i \rho_i \times \dot{p}_i + \sum_i \dot{\rho}_i \times p_i. \tag{2.9}$$

The second term on the right side of Equation 2.9 is zero because the momentum and the velocity of the particle are in the same direction. In the first term, $\dot{\mathbf{p}}_i$ can be replaced by its expression from Equation 2.1 to give

$$\dot{\mathbf{L}} = \sum_i \boldsymbol{\rho}_i \times \sum_j \mathbf{F}_{ji}. \tag{2.10}$$

The right-hand side of Equation 2.10 is a sum of terms of the form

$$(\boldsymbol{\rho}_i \times \mathbf{F}_{ji}) + (\boldsymbol{\rho}_j \times \mathbf{F}_{ij}) = (\boldsymbol{\rho}_i - \boldsymbol{\rho}_j) \times \mathbf{F}_{ji},$$

where the last step follows from Newton's third law. Since $(\boldsymbol{\rho}_i - \boldsymbol{\rho}_j)$, the vector between the ith and jth particles, and $\mathbf{F}_{ji}$, the force between the ith and jth particles, are parallel, their cross product is zero. (We have assumed that the mutual forces between the interacting bodies are central forces.) Therefore, the sum on the right-hand side of Equation 2.10 vanishes and $\dot{\mathbf{L}} = 0$ or

$$\mathbf{L} = \sum_i \boldsymbol{\rho}_i \times \mathbf{p}_i = \text{constant in time.} \tag{2.11}$$

This equation is the mathematical statement of the *Law of Conservation of Angular Momentum of a System of Particles: The total angular momentum of a system of isolated particles moving and interacting in any way remains constant provided the mutual forces are central forces.*

Note that this is a vector law; so all three components of total angular momentum L_x, L_y, and L_z are independently conserved. Note further that it is the total angular momentum that is conserved, not any particular type of angular momentum. Suppose, for example, that groups of the particles are rigidly fixed to one another. One can define the intrinsic angular momentum of the gth group by the equation

$$\mathbf{L}_g^{\text{in}} = I_g \boldsymbol{\omega}_g,$$

where I_g is the moment of inertia of the gth group about its axis of rotation and $\boldsymbol{\omega}_g$ is its angular velocity about this same axis. In addition, we can define an extrinsic angular momentum for the entire gth group relative to our fixed-coordinate system by the equation

$$\mathbf{L}_g^{\text{ex}} = \boldsymbol{\rho}_g \times \mathbf{p}_g,$$

where $\boldsymbol{\rho}_g$ is the vector from the origin of our coordinate system to the center of mass of the group and $\mathbf{p}_g$ is the total linear momentum of the group (given by the total mass of the group times the velocity of its center of mass relative to the fixed system). The total intrinsic and extrinsic angular momenta may be defined by summing over groups and individual point particles that are not members of any groups:

$$L^{in} = \sum_g L_g^{in},$$

$$L^{ex} = \sum_g L_g^{ex} + \sum_j L_j.$$

The law of conservation of angular momentum states that the total angular momentum, $L = L^{in} + L^{ex}$, is conserved. It says nothing about the possibility of the separate conservation of L^{in} and L^{ex}. In general, it may be expected that they will not be separately conserved. As an elementary example, when a bullet becomes embedded in a block of wood, all the extrinsic angular momentum of the bullet relative to the center of mass of the block is converted to intrinsic angular momentum in the spinning block–bullet system.

The derivations of the conservation laws of linear and angular momentum presented here depend on Newton's law of action and reaction. As long as we restrict ourselves to collisions between bodies that exert only central forces, we may freely use these laws; however, we must refrain from applying them blindly to interactions involving noncentral forces. Quite conveniently, most of the interactions of consequence to the nuclear engineer can be treated as central-force interactions.

2.2 Coordinate Systems

Of the unlimited number of possible coordinate systems from which a two-body collision might be viewed, three have been found particularly useful—the laboratory coordinate system, the center-of-momentum (or mass) coordinate system, and the relative coordinate system.

The laboratory coordinate system (L-system) is a system at rest with respect to a stationary observer in the laboratory.

The center-of-momentum coordinate system (C-system) is a system in which the total momentum of the colliding particles is zero. If the particles have nonzero rest mass, the center of momentum is moving with the same velocity in the laboratory as the center of mass. In such cases the C-system is usually called the center-of-mass coordinate system.

The relative coordinate system (R-system) is a system at rest with respect to one of the particles in the initial constellation. Generally, the target particle is chosen as the origin of the R-system.

Three systems of coordinates may appear excessive since, in principle, the physical consequences of any particular interaction are independent of the system from which it is viewed. The L-system is clearly the "natural" system since targets, whether foils or reactor cores, are at rest in the L-system and the particles generated and directed at the targets have known energies in the L-system. Why trouble ourselves with other coordinate systems? Another coordinate system becomes useful because the L-system is not the natural system to the particles in the initial constellation. One could argue that the particles have no knowledge of our stationary targets and measuring devices; that each of the particles sees the other rushing toward it while it remains stationary in its own R-system. So one or the other of the R-systems might be a more natural system from which to view the collision. The following questions then arise: Which of the two particles should preferentially be chosen as the origin of the R-system? If neither of the particles survives the collision, how can either R-system survive? These considerations seem to suggest that the collisions might very well be viewed from a system that is not at rest with respect to the laboratory or to either of the two particles, a system that occupies a more or less neutral position with respect to the two particles. There is one such system, the C-system, that has a unique and extremely useful property, a property that may be stated as follows:

The velocity of the center of mass of an isolated system of particles remains constant regardless of the motion or interaction of the particles comprising the system.

Let us prove this theorem for the two-particle system, the one of present interest.* Consider two massive† particles moving in three dimensions under the mutual influence of each other's forces. The equations of motion of the two particles are

$$m_1 \ddot{\rho}_1 = F_{21},$$

$$m_2 \ddot{\rho}_2 = F_{12}, \tag{2.12}$$

where ρ_1 and ρ_2 are the position vectors of particles 1 and 2 relative to the arbitrarily placed origin of the fixed L-system (see Figure 2.1).

The center of mass of particles 1 and 2 is defined as the point, determined by m_1 and m_2, such that the vectors r_1 and r_2 drawn from that point to masses m_1 and m_2, respectively, satisfy the condition

$$m_1 r_1 + m_2 r_2 \equiv 0. \tag{2.13}$$

The position vector to the center of mass, ρ_c, can be found readily by substituting $r_2 = \rho_2 - \rho_c$ and $r_1 = \rho_1 - \rho_c$ into Equation 2.13:

$$m_1 \rho_1 + m_2 \rho_2 = (m_1 + m_2)\rho_c. \tag{2.14}$$

---

*See any good book on classical mechanics, for example, Herbert Goldstein's *Classical Mechanics* (Reference 1 at the end of this chapter), for the general proof of this theorem.
†The word "massive" is used throughout this book in the technical sense of having nonzero rest mass.

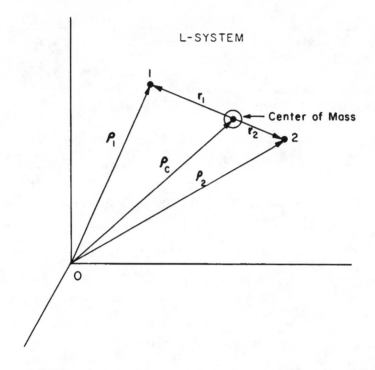

Figure 2.1—Two particles in the L-system.

Differentiating this equation twice with respect to time and substituting from Equations 2.12, we obtain

$$m_1\ddot{\boldsymbol{\rho}}_1 + m_2\ddot{\boldsymbol{\rho}}_2 = (m_1 + m_2)\ddot{\boldsymbol{\rho}}_c = \mathbf{F}_{21} + \mathbf{F}_{12}.$$

By Newton's third law, $\mathbf{F}_{21} = -\mathbf{F}_{12}$; therefore $\ddot{\boldsymbol{\rho}}_c = 0$ and

$$\dot{\boldsymbol{\rho}}_c = \text{constant in time.} \tag{2.15}$$

The velocity of the center of mass is indeed a constant in time, and the proof of the theorem is complete.

If the second of Equations 2.12 is subtracted from the first, we get

$$\ddot{\boldsymbol{\rho}}_1 - \ddot{\boldsymbol{\rho}}_2 = \frac{\mathbf{F}_{21}}{m_1} - \frac{\mathbf{F}_{12}}{m_2} = \left(\frac{m_1 + m_2}{\cdot\, m_1 m_2}\right)\mathbf{F}_{21} \tag{2.16}$$

since $-\mathbf{F}_{12} = \mathbf{F}_{21}$. An R-system originating at particle 2 can be set up by defining the position vector $\mathbf{r}$ of particle 1 relative to particle 2 as

$$\mathbf{r} \equiv \mathbf{r}_1 - \mathbf{r}_2 = \boldsymbol{\rho}_1 - \boldsymbol{\rho}_2. \tag{2.17}$$

Introducing the reduced mass (m) by the definition

$$m \equiv \frac{m_1 m_2}{m_1 + m_2},$$ (2.18)

Equation 2.16 can be written

$$m\ddot{\mathbf{r}} = \mathbf{F}_{21},$$ (2.19)

which is recognizable as Newton's second law of motion for a particle of mass m moving in the coordinate system in which particle 2 is at rest under the influence of the force exerted by particle 2 on particle 1.

In the sense of the equations developed above, the isolated two-particle system has been found equivalent to two one-particle systems. The center of mass moves with constant velocity in the laboratory system. One of the particles moves in the coordinate system in which the other particle is at rest. This movement is like that of a particle of reduced mass $m_1 m_2/(m_1 + m_2)$ acted on by a force equal to that exerted by the stationary particle on the moving particle.

The second of the two systems mentioned above is almost always called the center-of-mass coordinate system in technical literature. Unfortunately this name is somewhat misleading. The center of mass is not the origin of this system; the origin is one of the particles. The system is one of coordinates relative to the fixed particle, where r is the distance of one particle from the other.

Previously the question was posed: Why bother with coordinate systems other than the L-system? Hopefully it has become apparent that the answer is: Because the mathematical description of the collision is much more elementary in the C-system or the R-system than in the L-system. It has been demonstrated that the two-body problem in the L-system reduces to two more tractable one-body problems in the R-system. It will be shown shortly that the speeds of the particles measured in the C-system are not changed by an elastic collision. In general, the derivation of cross sections by either classical or quantum-mechanical procedures is considerably simplified in the R-system. Long experience has shown that usually the least complicated procedure for studying collisions is as follows:

1. Investigate the kinematics of the collision in the C-system where the mathematics is extremely simple, almost trivial.

2. Derive cross sections, usually by quantum mechanics, in the R-system where the mathematics is in general most tractable.

3. Transform the results obtained in the C- and R-systems back to the L-system.

4. Make comparisons between theory and experiment on the basis of L-system results. Use L-system results in the design of reactor cores, radiation shields, etc.

2.3 Kinematics of Collisions

The kinematics of nonrelativistic collisions will now be investigated using the conservation laws of linear momentum and energy as the sole tools. The reader may be pleasantly surprised by the quantity of information that can be obtained by application of these laws alone. The common procedure for investigating the kinematics of all collisions will be to derive the desired equations in the C-system and then transform the results back to the L-system. To carry out this transformation, one must find an expression for the velocity of the center of mass relative to the L-system. Since the velocity of the center of mass does not change during an interaction, one may most readily find its value by examination of the initial constellation before the interaction has taken place, i.e., before the forces between the particles begin to affect their trajectories.

The initial constellation consists of a target particle at rest in the laboratory and a projectile particle moving with constant velocity toward the target particle. The z-axis is defined as the direction of motion of the projectile particle before the collision. The subscript 1 is used to denote quantities connected with the projectile particle, quantities such as mass, speed, and energy. Similarly, the subscript 2 is associated with the target particle, and numerical subscripts greater than 2 are associated with the particle or particles in the final constellation. A tilde is placed over a symbol to denote a quantity measured relative to the L-system; symbols without tildes are understood to represent quantities measured relative to the C-system.

Figure 2.2 depicts the laboratory view of the constellation before the collision. The coordinate plane is the plane that contains velocity vector $\tilde{v}_1$ and particle 2. As usual, the z-axis has been taken to be parallel to the velocity vector $\tilde{v}_1$. The coordinate positions of particles 1 and 2 and of the center of mass (CM) are, respectively, $(\tilde{x}_1, \tilde{z}_1)$, $(\tilde{x}_2, \tilde{z}_2)$, and $(\tilde{x}_c, \tilde{z}_c)$. These positions are relative to an arbitrary origin of coordinates fixed in the laboratory.

Before the collision, only particle 1 and the center of mass are moving. The center of mass, indicated by an open circle, is pictured in Figure 2.2 as moving parallel to the z-axis with velocity of magnitude $\tilde{u}$. We will now demonstrate that the center of mass does indeed move parallel to the z-axis.

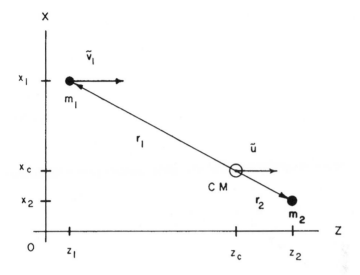

Figure 2.2—Two particles before collision in the L-system.

The position of the center of mass is defined as the point about which the moments of the masses vanish, that is,

$$m_1\mathbf{r}_1 + m_2\mathbf{r}_2 = 0, \tag{2.20}$$

where $\mathbf{r}_1$ and $\mathbf{r}_2$ are the vectors from the center of mass to particles 1 and 2. Taking the time derivative of Equation 2.20, one obtains

$$m_1\dot{\mathbf{r}}_1 + m_2\dot{\mathbf{r}}_2 = 0, \tag{2.21}$$

which implies that the center of mass is also the center of momentum. The x component of Equation 2.21 is

$$m_1(\dot{x}_1 - \dot{x}_c) + m_2(\dot{x}_2 - \dot{x}_c) = 0$$

and, since both $\dot{x}_1$ and $\dot{x}_2$ are zero,

$$(m_1 + m_2)\dot{x}_c = 0.$$

This relation can be satisfied for arbitrary m_1 and m_2 only if the x component of the velocity of the center of mass is zero, proving that the center of mass does move parallel to the z-axis.

To obtain an expression for the speed of the center of mass, let the z component of Equation 2.21 be written

$$m_1(\dot{z}_1 - \dot{z}_c) + m_2(\dot{z}_2 - \dot{z}_c) = 0.$$

We know that $\dot{z}_1 = \tilde{v}_1$, $\dot{z}_2 = 0$, and $\dot{z}_c$ is by definition $\tilde{u}$; therefore, upon rearranging the above equation, we obtain

$$\tilde{u} = \frac{m_1}{m_1 + m_2}\,\tilde{v}_1. \tag{2.22}$$

This is the speed of the center of mass, and of the C-system, as viewed from the laboratory. The velocity of the C-system is parallel to the z-axis, thus the components of $\tilde{\mathbf{u}}$ are given by

$$\tilde{\mathbf{u}} = (\tilde{u}_x, \tilde{u}_y, \tilde{u}_z) = \left(0, 0, \frac{m_1}{m_1 + m_2}\,\tilde{v}_1\right). \tag{2.23}$$

The transverse distance between centers of the two particles before the interaction is called the impact parameter of the collision and is denoted by the letter b. From Figure 2.2 we can see that b is given by

$$b = |x_2 - x_1|. \tag{2.24}$$

In classical elastic collisions, the impact parameter and the forces together determine the angle of scattering. In quantum-mechanical collisions, the impact parameter cannot be defined. Indeed, we will show that it is impossible to set up an experiment involving submicroscopic particles that have a fixed impact parameter and zero transverse velocity. As long as attention is restricted to kinematics based only on the laws of conservation of linear momentum and energy there need be no anxiety about specifying the impact parameter; it will not enter into the equations.

In the kinematic analyses to follow, the impact parameter, if one can be defined, will be taken as unspecified. For graphical clarity only, collisions will be pictured as though the impact parameter were zero. The reader should also be aware that the actual trajectories during the collision are not known and do not need to be known. The kinematics are completely independent of these trajectories. In effect, useful kinematic relations between what went in and what came out are derived without ever considering what went on during the collision. Of course, there is no way of predicting what will come out until one has calculated cross sections, which do indeed depend on what went on during the collision.

We will now consider the following examples of two-body kinematics in the order of increasing complexity:

1. One-body final constellation.
2. Two-body final constellation—elastic scattering.
3. Two-body final constellation—inelastic scattering.
4. Two-body final constellation—reaction.

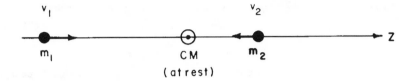

Figure 2.3—Two particles before collision in the center-of-mass system.

The initial constellations of all these collisions are identical. Before the collision an observer in the C-system will see both particles moving toward the center of mass (see Figure 2.3). The target particle, which is at rest in the L-system, will be seen moving toward the center of mass in the C-system with the speed of the center of mass in the L-system, that is,

$$v_2 = \tilde{u} = \frac{m_1}{m_1 + m_2}\,\tilde{v}_1. \tag{2.25}$$

The projectile particle, which has speed $\tilde{v}_1$ in the L-system, will have speed v_1 in the C-system given by

$$v_1 = \tilde{v}_1 - \tilde{u} = \frac{m_2}{m_1 + m_2}\,\tilde{v}_1. \tag{2.26}$$

Note that the relative speed of approach of the two particles is the same in the C- and L-systems:

$$v_1 + v_2 = \tilde{v}_1.$$

The total kinetic energy, however, is less in the C-system than in the L-system. In the C-system, the total kinetic energy, T, is

$$T = \frac{1}{2}m_1 v_1^2 + \frac{1}{2}m_2 v_2^2 = \frac{1}{2}\frac{m_1 m_2}{m_1 + m_2}\tilde{v}_1^2 = \frac{1}{2}m\tilde{v}_1^2, \tag{2.27}$$

whereas in the L-system the total kinetic energy is

$$\tilde{T} = \tfrac{1}{2}m_1\tilde{v}_1^2. \tag{2.28}$$

The difference between $\tilde{T}$ and T is the kinetic energy associated with the motion of the center of mass in the L-system,

$$\tilde{T}_c = \tilde{T} - T = \tfrac{1}{2}(m_1 + m_2)\tilde{u}^2. \tag{2.29}$$

Since the center of mass must continue to move with speed $\tilde{u}$ regardless of the interaction between the two particles, the energy associated with the center of mass, $\tilde{T}_c$, is unavailable; that is, it cannot be transformed into any other form of energy but must remain as the kinetic energy of the entire system of particles. It is the kinetic energy in the C-system, T, that is available for transformation into other forms. This is another reason why the C-system is the preferred system for kinematic analysis.

1. *One-body final constellation:* The single body formed by the fusing of the two bodies in the initial constellation sits at rest in the C-system with total internal energy $T + (m_1 + m_2)c^2$. In the laboratory system it is moving with velocity $\tilde{u}$, kinetic energy $\tilde{T}_c$, and the same total internal energy, $T + (m_1 + m_2)c^2$. Nothing more remains to be said about the kinematics of this most simple of all collisions.

2. *Two-body final constellation—elastic scattering:* Consider the problem of a projectile particle of mass m_1 and laboratory speed $\tilde{v}_1$ which elastically scatters through angle $\tilde{\theta}_1$ off a particle of mass m_2 which is at rest in the laboratory. Figure 2.4 shows the before and after views of this collision in the L-system.

After the collision, particle 1 moves off with speed $\tilde{v}_3$ at angle $\tilde{\theta}_1$ relative to its initial direction, and particle 2 moves off with speed $\tilde{v}_4$ at angle $\tilde{\theta}_2$ relative to its initial direction. The center of mass continues to move along the z direction with unchanged speed $\tilde{u}$. The scattering angles in the C-system for particles 1 and 2 are θ_1 and θ_2, respectively. It is assumed that $m_1, m_2, \tilde{v}_1$, and $\tilde{\theta}_1$ are given; thus one attempts to evaluate the unknowns. In particular, answers are sought to the following questions:

1. What is the kinetic energy (or speed) of particle 1 after the collision?

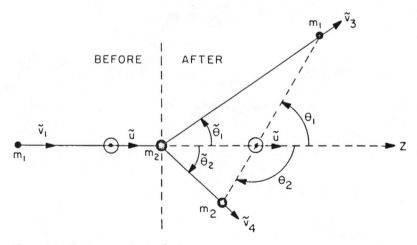

Figure 2.4—Elastic scattering in the L-system.

2. How much kinetic energy will particle 2 be given? (This question is trivial once we answer question 1.)

3. What is the relation between the laboratory and center-of-mass scattering angles?

4. What is the angle of scattering of particle 2?

5. What range of scattering angles is possible?

The answers will be found by first solving the problem in the C-system and then transforming the answers to the L-system. The full power and beauty of this method of procedure will become apparent after the following theorem has been proved.

The speeds of both particles as measured in the C-system are unchanged in an elastic collision.

In elastic scattering, the total kinetic energy of the bodies after the collision is equal to the total kinetic energy of the bodies before the collision. In the C-system this may be written

$$\tfrac{1}{2}m_1 v_3^2 + \tfrac{1}{2}m_2 v_4^2 = \tfrac{1}{2}m_1 v_1^2 + \tfrac{1}{2}m_2 v_2^2, \tag{2.30}$$

where v_1 and v_2 are the C-system speeds of particles 1 and 2 before the collision and v_3 and v_4 are the C-system speeds of particles 1 and 2 after the collision. One can solve Equation 2.30 for v_3 in terms of $\tilde{v}_1$ by replacing v_1 and v_2 with their expressions from Equations 2.26 and 2.25, respectively, and by replacing v_4 with its equivalent from the center-of-momentum expression

$$m_2 v_4 = m_1 v_3 \tag{2.31}$$

After making these substitutions and after a bit of elementary algebra, we find that

$$v_3 = \frac{m_2}{m_1 + m_2}\, \tilde{v}_1. \tag{2.32}$$

The right-hand side of this expression is v_1 in Equation 2.26; therefore it has been proved that

$$v_3 = v_1. \tag{2.33}$$

In a similar fashion one may prove that

$$v_4 = v_2. \tag{2.34}$$

Thus this remarkable and very useful theorem is proved. In other words, an observer sitting on the center of mass would see the two particles recede from the center of mass after an elastic collision with exactly the same speeds with which they approached. Note that "before" and "after" are used in the technical sense in which they were defined in Chapter 1, that

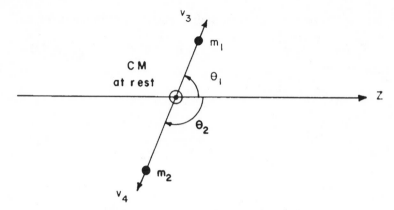

Figure 2.5—The center-of-mass system after elastic scattering.

is, before and after forces come in to play. This theorem does not state anything concerning the speeds of the particles during the collision. When cross sections are calculated, one is concerned about what went on during the collision, but for present purposes this is irrelevant.

Figure 2.3 depicts the initial constellation of the collision as seen in the center-of-mass system. Figure 2.5 depicts the final constellation after elastic scattering as seen in the center-of-mass system. The center of mass is still at rest, and the particles are receding from the center of mass with the same speeds with which they approached. Particle 1 has scattered through a center-of-mass angle θ_1, and particle 2 has scattered through a center-of-mass angle θ_2. Since the center of mass is at rest in the C-system, we immediately have the important relation

$$\theta_2 = \pi - \theta_1. \tag{2.35}$$

The kinematic problem has been completely solved in the C-system. The speeds of the particles are unchanged, and there is a determined relation between the angles of scattering. We must now transform these results to the L-system.

To transform velocities, we use the rather obvious fact that the vector velocity of any particle in the L-system is the sum of the vector velocity of that particle in the C-system plus the vector velocity of the C-system in the L-system: *

$$\tilde{\mathbf{v}} = \mathbf{v} + \tilde{\mathbf{u}}. \tag{2.36}$$

* The author hastens to point out that this "rather obvious fact," known as the Galilean transformation, is only an approximation to the rigorously correct Lorentz transformation. It is used here because it is quite accurate when applied to velocities well below that of light and because it is algebraically less complex than the Lorentz transformation.

The laboratory velocity of particle 1 after the collision is given by

$\tilde{\mathbf{v}}_3 = \mathbf{v}_3 + \tilde{\mathbf{u}}.$

These velocity vectors have as components along the z-axis

$$\tilde{v}_3 \cos \tilde{\theta}_1 = v_3 \cos \theta_1 + \tilde{u} \tag{2.37}$$

and as components perpendicular to the z-axis

$$\tilde{v}_3 \sin \tilde{\theta}_1 = v_3 \sin \theta_1. \tag{2.38}$$

After substituting v_1 for v_3 (Equation 2.33) and then substituting for v_1 and $\tilde{u}$ from Equations 2.26 and 2.22, respectively, we find that

$$\tilde{v}_3 \cos \tilde{\theta}_1 = \frac{m_2}{m_1 + m_2} \tilde{v}_1 \cos \theta_1 + \frac{m_1}{m_1 + m_2} \tilde{v}_1 \tag{2.39}$$

and

$$\tilde{v}_3 \sin \tilde{\theta}_1 = \frac{m_2}{m_1 + m_2} \tilde{v}_1 \sin \theta_1. \tag{2.40}$$

By squaring these two equations and adding them, we eliminate $\tilde{\theta}_1$ and obtain an equation for the final speed of particle 1,

$$\tilde{v}_3 = \tilde{v}_1 \left[\frac{m_1^2 + m_2^2 + 2m_1 m_2 \cos \theta_1}{(m_1 + m_2)^2} \right]^{1/2}. \tag{2.41}$$

We can find an expression for the final kinetic energy of particle 1 in the L-frame by squaring Equation 2.41,

$$\tilde{T}_3 = \tilde{T}_1 \left[\frac{m_1^2 + m_2^2 + 2m_1 m_2 \cos \theta_1}{(m_1 + m_2)^2} \right]. \tag{2.42}$$

These expressions relate the final velocity and final energy of particle 1 to the center-of-mass scattering angle θ_1. We can get a relation involving the laboratory scattering angle $\tilde{\theta}_1$ by rearranging Equation 2.37 to find an expression for $\cos \theta_1$, squaring this equation and Equation 2.40, and then adding. This eliminates θ_1, and we get

$$\tilde{v}_3 = \tilde{v}_1 \left\{ \frac{m_1}{m_1 + m_2} \cos \tilde{\theta}_1 + \left[\frac{m_2 - m_1}{m_1 + m_2} + \left(\frac{m_1}{m_1 + m_2} \right)^2 \cos^2 \tilde{\theta}_1 \right]^{1/2} \right\}. \tag{2.43}$$

An equation for the final kinetic energy of particle 1 may now be found by squaring Equation 2.43:

$$\tilde{T}_3 = \tilde{T}_1 \left\{ \frac{m_1}{m_1 + m_2} \cos \tilde{\theta}_1 + \left[\frac{m_2 - m_1}{m_1 + m_2} + \left(\frac{m_1}{m_1 + m_2} \right)^2 \cos^2 \tilde{\theta}_1 \right]^{1/2} \right\}^2.$$

(2.44)

Equations 2.43 and 2.44 are the explicit answers to the first question on page 34. They express the final kinetic energy and speed of particle 1 as a function of the known masses and the postulated scattering angle.

The second question is answered readily because kinetic energy is conserved in an elastic interaction; hence

$$\tilde{T}_4 = \tilde{T}_1 \quad \tilde{T}_3.$$

(2.45)

The third question is most readily answered by dividing Equation 2.39 by Equation 2.40 to eliminate the speeds,

$$\frac{\cos \tilde{\theta}_1}{\sin \tilde{\theta}_1} = \frac{\cos \theta_1}{\sin \theta_1} + \frac{m_1}{m_2 \sin \theta_1}$$

or, upon rearranging,

$$\tan \tilde{\theta}_1 = \frac{\sin \theta_1}{(m_1/m_2) + \cos \theta_1}.$$

(2.46)

A somewhat similar relation can be derived in identical fashion for particle 2,

$$\tan \tilde{\theta}_2 = \frac{\sin \theta_2}{1 + \cos \theta_2},$$

(2.47)

and, since $\theta_2 = \pi - \theta_1$, we may write this in terms of θ_1 as

$$\tan \tilde{\theta}_2 = \frac{\sin \theta_1}{1 - \cos \theta_1}.$$

(2.48)

The angle of scattering of particle 2 may be obtained from Equations 2.48 and 2.46; however, there is a much more direct procedure. Momentum conservation demands that

$$m_1 \tilde{v}_3 \sin \tilde{\theta}_1 = m_2 \tilde{v}_4 \sin \tilde{\theta}_2$$

(2.49)

or

$$\sin \tilde{\theta}_2 = \frac{m_1 \tilde{v}_3}{m_2 \tilde{v}_4} \sin \tilde{\theta}_1.$$

(2.50)

Since $\tilde{\theta}_1$ is given and $\tilde{v}_3$ and $\tilde{v}_4$ are known from Equations 2.43 and 2.45, Equation 2.50 gives the sine of the scattering angle of particle 2. With the additional fact, which will be proved shortly, that the target particle cannot scatter backward, Equation 2.50 gives the angle $\bar{\theta}_2$ and so answers the fourth question.

There will be various answers to the fifth question, depending on the relative masses of the two particles, as we shall see. However, let it first be stated that, regardless of the masses of the particles, the target particle can never scatter in any backward direction in the L-system, that is,

$$\bar{\theta}_2 \le \frac{\pi}{2}. \tag{2.51}$$

The proof is rather obvious. If particle 2 were to scatter in some backward direction, then particle 1 would have to scatter forward with a greater speed than its initial speed in order to conserve momentum. In this event the total kinetic energy of the constellation would be increased in violation of the definition of elastic scattering. Hence, particle 2 cannot scatter backward.

Examine now the range of possible values of the center-of-mass scattering angle of particle 1. We see by inspection of Equation 2.41 that the expression within the radical,

$$m_1^2 + m_2^2 + 2m_1 m_2 \cos \theta_1,$$

remains positive for any value of $\cos \theta_1$ between 1 and -1. Thus any value of the center-of-mass scattering angle is possible,

$$0 \le \theta_1 \le \pi. \tag{2.52}$$

The range of possible values for the laboratory scattering angle of particle 1 can be deduced from examination of the expression on the right-hand side of Equation 2.43. One demands on physical grounds that the speed $\bar{v}_3$ never become negative or imaginary. If the target particle is heavier than the projectile particle, the speed remains real and positive for any value of $\bar{\theta}_1$; so all values of $\bar{\theta}_1$ are possible

$$0 \le \bar{\theta}_1 \le \pi \qquad \text{(for } m_2 > m_1 \text{)}. \tag{2.53}$$

If the particle masses are equal, the angle $\bar{\theta}_1$ must be less than $\pi/2$, otherwise the speed becomes negative,

$$0 \le \bar{\theta}_1 \le \frac{\pi}{2} \qquad \text{(for } m_2 = m_1 \text{)}. \tag{2.54}$$

If the target particle is lighter than the projectile particle, the expression within the radical in Equation 2.43 will become negative unless

$$\cos \bar{\theta}_1 \le \left| \frac{m_1^2 - m_2^2}{m_1^2} \right|^{1/2}$$

so the angle of scattering is limited to the range

$$0 \le \tilde{\theta}_1 \le (\tilde{\theta}_1)_{max} \qquad \text{(for } m_1 > m_2\text{)}, \qquad (2.55)$$

where

$$(\tilde{\theta}_1)_{max} = \cos^{-1}\left(\frac{m_1^2 - m_2^2}{m_1^2}\right)^{1/2}. \qquad (2.56)$$

3. *Two-body final constellation—inelastic scattering:* As the third example of nonrelativistic kinematics, consider an interaction in which particle 1, moving with speed $\tilde{v}_1$ in the laboratory, scatters off particle 2, which is at rest before the collision. After the scattering, particle 1 moves off at angle $\tilde{\theta}_1$ with speed $\tilde{v}_3$. Particle 2 moves off at angle $\tilde{\theta}_2$ with speed $\tilde{v}_4$. It is assumed that mass is conserved but that kinetic energy is not. A part, E^*, of what was kinetic energy in the initial constellation appears as internal energy in the particles of the final constellation.

It is assumed that m_1, m_2, $\tilde{v}_1$, $\tilde{\theta}_1$, and E^* are given and that we wish to determine all other angles, speeds, and energies. In the C-system the kinematics are trivial since conservation of linear momentum demands that

$$m_1 v_3 = m_2 v_4 \qquad (2.57)$$

or, written in terms of the kinetic energies of the particles,

$$m_1 T_3 = m_2 T_4. \qquad (2.58)$$

Conservation of energy may be written

$$T_1 + T_2 = T_3 + T_4 + E^*. \qquad (2.59)$$

We can now determine the kinetic energies of both particles after the collision using Equations 2.58 and 2.59:

$$T_3 = \frac{m_2}{m_1 + m_2}(T_1 + T_2 - E^*), \qquad (2.60)$$

$$T_4 = \frac{m_1}{m_1 + m_2}(T_1 + T_2 - E^*), \qquad (2.61)$$

where the total available kinetic energy in the C-system, $T_1 + T_2$, may be obtained directly from Equation 2.27,

$$T_1 + T_2 = \tfrac{1}{2}m\tilde{v}_1^2. \qquad (2.62)$$

Note that either Equation 2.60 or 2.61 implies that the maximum possible excitation energy is limited to a value no greater than the total available kinetic energy, $T_1 + T_2$. Everything on the right-hand sides of the previous three equations is known; so the final kinetic energies in the C-system are known, and the problems are solved in the C-system. Note that these results are independent of angle. This is always the case in the C-system; it is one of the principal reasons why mathematics in the C-system is so simple. If a trusted observer could live in the C-system, there would be no problems at all. Since this is not feasible, these simple equations must be transformed back into the L-system.

To apply the velocity transformation formula, Equation 2.36, we must solve Equations 2.60 and 2.61 for the speeds of the particles,

$$v_3 = \left[2 \frac{m_2/m_1}{m_1 + m_2} (T_1 + T_2 - E^*) \right]^{1/2}, \tag{2.63}$$

and

$$v_4 = \left[2 \frac{m_1/m_2}{m_1 + m_2} (T_1 + T_2 - E^*) \right]^{1/2}. \tag{2.64}$$

The algebra from this point on is identical with that of the elastic-scattering transformation studied previously except that the quantity E^* is carried along. After algebraic steps equivalent to those in Equations 2.37 through 2.44, we arrive at an expression for the final laboratory kinetic energy of particle 1 in terms of known quantities,

$$\tilde{T}_3 = \tilde{T}_1 \left\{ \frac{m_1}{m_1 + m_2} \cos \tilde{\theta}_1 + \left[\frac{m_2 - m_1}{m_1 + m_2} \right. \right.$$
$$\left. \left. + \left(\frac{m_1}{m_1 + m_2} \right)^2 \cos^2 \tilde{\theta}_1 - \frac{m_2}{m_1 + m_2} \frac{E^*}{\tilde{T}_1} \right]^{1/2} \right\}^2. \tag{2.65}$$

This expression properly reduces to that for elastic scattering when the excitation energy E^* is zero. We could now derive a similar equation for $\tilde{T}_4$ and expressions that relate the scattering angles in the L-system to those in the C-system. However, because these expressions and all those we derived for elastic scattering are actually special cases of a more general one which will be derived presently, their derivations will not be given.

4. *Two-body final constellation—reaction:* Consider now the most general nonrelativistic collision having a two-body final constellation. Particle 1 moving with speed $\tilde{v}_1$ in the laboratory reacts with particle 2, which is at rest before the collision. Particles 3 and 4 leave the collision

with total excitation energy E^*. The total mass in the final constellation is not necessarily equal to the total mass in the initial constellation. The quantities m_1, m_2, m_3, m_4, $\tilde{v}_1$, E^*, and one of the angles are assumed to be known. Thus a determination of all other angles, speeds, and energies is the task at hand.

First it is necessary to go back and develop the transformation equations in their most general form. From the addition equation for velocity,

$$\tilde{\mathbf{v}} = \mathbf{v} + \bar{\mathbf{u}}, \tag{2.66}$$

we obtain the velocity components parallel to and perpendicular to the z-axis for one of the particles in the final constellation,

$$\tilde{v}_j \cos \tilde{\theta}_j = v_j \cos \theta_j + \bar{u}, \tag{2.67}$$

$$\tilde{v}_j \sin \tilde{\theta}_j = v_j \sin \theta_j, \tag{2.68}$$

where $\tilde{v}_j$ and v_j are the laboratory and center-of-mass speeds of particle j, one of the particles in the final constellation; $\tilde{\theta}_j$ and θ_j are the laboratory and center-of-mass angles of particle j; and $\bar{u}$ is the speed of the center of mass in the laboratory.

If the laboratory angle $\tilde{\theta}_j$ is known, we proceed as follows: Since both $\bar{u}$ and $\tilde{\theta}_j$ are known and since an expression for v_j can be found by examining the kinematics in the C-system, we algebraically eliminate θ_j between Equations 2.67 and 2.68 to obtain the final laboratory speed of particle j,

$$Speed: \quad \tilde{v}_j = \bar{u} \cos \tilde{\theta}_j + (\bar{u}^2 \cos^2 \tilde{\theta}_j + v_j^2 - \bar{u}^2)^{1/2}. \tag{2.69}$$

The laboratory kinetic energy can then be obtained from

$$Energy: \quad \tilde{T}_j = \tfrac{1}{2} m_j \tilde{v}_j^2, \tag{2.70}$$

where m_j is the mass of particle j. Now that $\tilde{v}_j$ is known, we can make use of Equations 2.67 and 2.68 to obtain the center-of-mass angle in terms of the laboratory angle

$$C\text{-}angle\ of\ j: \quad \tan \theta_j = \frac{\sin \tilde{\theta}_j}{\cos \tilde{\theta}_j - (\bar{u}/\tilde{v}_j)}. \tag{2.71}$$

The center-of-mass angle θ_k of the other particle (particle k) in the final constellation is $\pi - \theta_j$. Its laboratory angle is given by solution of Equations 2.67 and 2.68 with index j replaced by index k,

$$L\text{-}angle\ of\ k: \quad \tan \tilde{\theta}_k = \frac{\sin \theta_k}{\cos \theta_k + (\bar{u}/v_k)}. \tag{2.72}$$

If, instead of the laboratory angle $\tilde{\theta}_j$, the center-of-mass angle θ_j is known, then we can eliminate $\tilde{\theta}_j$ from Equations 2.67 and 2.68 to obtain

$$Speed\ of\ j:\quad \tilde{v}_j = (v_j^2 + \tilde{u}^2 + 2v_j\tilde{u}\cos\theta_j)^{1/2}. \qquad (2.73)$$

Then the kinetic energy is given by Equation 2.70 and the laboratory angle is given by solution of Equations 2.67 and 2.68

$$L\text{-}angle\ of\ j:\quad \tan\tilde{\theta}_j = \frac{\sin\theta_j}{\cos\theta_j + (\tilde{u}/v_j)}. \qquad (2.74)$$

Equations 2.69 through 2.74 represent the solution to the kinematics of any nonrelativistic two-body final constellation problem provided one can find an expression for v_j, the speed of particle j in the C-system. But one can always find v_j from the conservation laws in the C-system:

Conservaticn of Linear Momentum in C-system
Elastic scattering: $m_1v_3 = m_2v_4$ (2.75)
Inelastic scattering: $m_1v_3 = m_2v_4$ (2.76)
Reaction: $m_3v_3 = m_4v_4$ (2.77)

Conservation of Energy in C-system
Elastic scattering: $T_1 + T_2 = T_3 + T_4$ (2.78)
Inelastic scattering: $T_1 + T_2 = T_3 + T_4 + E^*$ (2.79)
Reaction:

$$T_1 + T_2 + (m_1 + m_2)c^2 = T_3 \quad T_4 + (m_3 + m_4)c^2 + E^* \qquad (2.80)$$

The unknown quantities in Equations 2.78, 2.79, and 2.80 are T_3 and T_4, but one of them can be eliminated from the appropriate conservation of momentum equation. For example, to find the center-of-mass speed of particle 3 after a reaction, we replace T_4 in Equation 2.80 by its expression from Equation 2.77, namely,

$$T_4 = \frac{1}{2}m_4v_4^2 = \frac{1}{2}\frac{m_3^2}{m_4}v_3^2, \qquad (2.81)$$

and we replace T_3 by $m_3v_3^2/2$, then solve Equation 2.80 for v_3:

$$v_3 = \left\{\frac{2m_4}{m_3(m_3 + m_4)}[T_1 \quad T_2 + (m_1 + m_2 - m_3 - m_4)c^2 - E^*]\right\}^{1/2}. \qquad (2.82)$$

If it is desired that the expression for v_3 be explicitly in terms of given quantities, merely replace the sum $T_1 + T_2$ by its known equivalent,

$m\tilde{v}_1^2/2$, and so obtain

$$v_3 = \left\{ \frac{2m_4}{m_3(m_3 + m_4)} \right.$$

$$\left. \times \left[\frac{m_1 m_2}{2(m_1 + m_2)} \tilde{v}_1^2 + (m_1 + m_2 - m_3 - m_4)c^2 - E^* \right] \right\}^{1/2} . (2.83)$$

Equation 2.83 is the general equation for v_3 for any collision. For a scattering collision, the term $(m_1 + m_2 - m_3 - m_4)$ is equal to zero. For an elastic-scattering collision, the term E^* is also equal to zero.

Substituting u_3, determined by dividing Equation 2.83 into Equation 2.69 or 2.73, we obtain the laboratory speed of particle 3. This speed can then be substituted into Equation 2.70 to find the laboratory kinetic energy and into Equation 2.71 or 2.74 to find the unknown angle.

2.4 Elastic-Scattering Cross Sections

We now leave the realm of kinematics and enter that of kinetics, the branch of science that deals with the action of forces in producing or changing motion. Our immediate objective is to calculate differential elastic-scattering cross sections. The discussion will be restricted to collisions between bodies that exert only central forces, i.e., forces whose vector directions lie along the line between the centers of the bodies. Furthermore, only forces that depend on the distance between the particles, and not, for example, on their velocities or angular momenta, are considered. It will be assumed that the force falls to zero at infinite separation. The methods of classical mechanics, not quantum mechanics, will be used; so the results can be expected to be applicable to large-particle scattering and not to the scattering of particles of atomic and subatomic dimensions.

The procedure will be to calculate the cross sections in the R-system, where the target particle is always at rest, then transform the cross sections back to the L-system. The advantage of this procedure is, as previously explained, that it reduces the two-body scattering problem to a one-body problem. The target particle is replaced, in effect, by a fixed center of force, and we need only investigate the motion of the reduced-mass projectile particle under the influence of this fixed center of force.

Imagine an infinite, uniform, monodirectional flux of monoenergetic, monotype particles incident on a single fixed center of force. The particles in this flux are so spaced in time that they do not interact with one another; each interacts only with the center of force. The word "uniform" means that a cross-sectional cut through the beam would reveal that the same

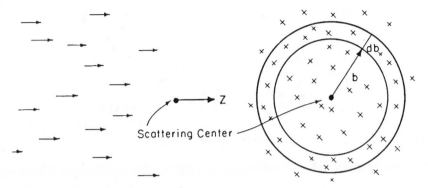

Figure 2.6—Side and cross-sectional views of flux incident on fixed scattering area.

flux of particles passes through each unit area. Thus the flux of particles having impact parameter in the range b to $b + db$ is simply proportional to the area in the impact-parameter range b to $b + db$ (see Figure 2.6). The flux of particles in the range b to $b + db$ is therefore

$$\Phi_0 2\pi b \, db, \qquad (2.84)$$

where Φ_0 is the incident flux per unit area, a constant.

The trajectory of any given particle depends on its impact parameter. All particles with a given impact parameter will scatter through the same angle. The angle with which we are concerned is the scattering angle in the R-system, θ_R. But it is very easy to prove that θ_R is, in fact, identical to the scattering angle of the projectile particle in the C-system, θ. The proof follows directly from inspection of Figure 2.7, which shows the laboratory view of the trajectories of both particles before, during, and

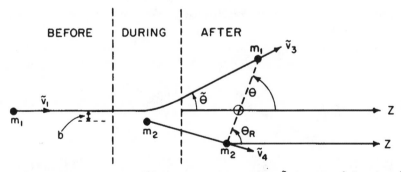

Figure 2.7—Laboratory view of scattering angles. L-system, $\tilde{\theta}$; C-system, θ; R-system, θ_R.

after a collision. Note that the angles θ_R and θ are both defined as the angle between the direction of the incident particle (the z-axis) and the direction of the target-to-projectile vector (or center of mass-to-projectile vector) after the collision. Henceforth the symbol θ shall be used for the scattering angle in both the R- and C-systems, although it is to be understood that cross sections will be calculated in the R-system.

The number of particles per second that scatter into angles between θ and $\theta + d\theta$ from a single fixed scattering center (in the R-system) is, by definition of the differential scattering cross section,

$$\Phi_0 \sigma(\theta) 2\pi \sin \theta \, d\theta. \tag{2.85}$$

But these particles were incident on the force center in some impact parameter range b to $b + db$. Since no particles were created or destroyed, the number incident per second in the range b to $b + db$ equals the number scattered per second into the range θ to $\theta + d\theta$, that is,

$$\Phi_0 2\pi b \, db = \Phi_0 \sigma(\theta) 2\pi \sin \theta \, d\theta \tag{2.86}$$

or.

$$\sigma(\theta) = \frac{b}{\sin \theta} \left| \frac{db}{d\theta} \right|. \tag{2.87}$$

The absolute value sign is introduced in Equation 2.87 because the cross section must remain positive even if the scattering angle decreases with increasing impact parameter.

Equation 2.87 was based on the implied assumption that corresponding to each scattering angle θ there is a unique impact parameter b. Certainly the laws of mechanics will demand that corresponding to each impact parameter b there is one, and only one, possible scattering angle θ,

$$\theta = \theta(b), \tag{2.88}$$

but this does not imply the converse relation. We have no a priori reason for assuming that b is a unique function of θ. To account for the possibility that several impact parameters may produce the same scattering angle, we generalize Equation 2.86 by summing over all impact areas that produce a given θ:

$$\Phi_0 \sum_i 2\pi b_i \, db_i = \Phi_0 \sigma(\theta) 2\pi \sin \theta \, d\theta, \tag{2.89}$$

so that

$$\sigma(\theta) = \sum_i \frac{b_i}{\sin \theta} \left| \frac{d\theta}{db} \right|_i^{-1} \tag{2.90}$$

Equation 2.90 is the basic equation from which all classical differential cross sections may be calculated. All that need be known to evaluate the right-hand side of this equation is an expression of the form $\theta = \theta(b)$. This expression may be an algebraic equation or simply a graph of θ vs. b. If it is an algebraic equation, the values of b_i corresponding to any θ may be found, and the values of $|d\theta/db|_i$ may be obtained by differentiation. If θ is known as a function of b through a plot of θ vs. b then the values of b_i and the derivatives corresponding to a given θ can be obtained directly from the graph. In either case the right-hand side of Equation 2.90 can be evaluated for all values of θ.

Thus we see that the problem of obtaining a classical differential cross section is essentially equivalent to the problem of determining the scattering angle for any value of the impact parameter.

Consider a particle of mass m_1 and laboratory speed $\tilde{v}_1$ incident with impact parameter b on a center of force fixed in the R-system. The force exerted by the fixed center on the moving particle is described by a potential function

$$V = V(r), \tag{2.91}$$

where r is the distance between particle centers. Consider a polar coordinate system set up with its origin on the center of force. Let $r(t)$ be the distance from the origin to the projectile particle at time t and $\beta(t)$ be the angular position of the projectile particle relative to the z-axis at time t (see Figure 2.8). The speed of the particle in the R-system both before and after the interaction will be $\tilde{v}_1$. During the interaction, however, the speed will change because of the action of the force. The speed at any time is given in polar coordinates by the well known expression

$$v(t) = [\dot{r}^2 + r^2\dot{\beta}^2]^{1/2}. \tag{2.92}$$

The law of conservation of energy demands that the total energy of the system before the interaction be equal to the total energy at any time during the interaction; hence, in the R-system

$$E = \tfrac{1}{2}m\tilde{v}_1^2 = \tfrac{1}{2}m(\dot{r}^2 + r^2\dot{\beta}^2) + V(r), \tag{2.93}$$

where $V(r)$ is the potential energy; m, the reduced mass; and E, the available energy. The law of conservation of angular momentum demands that the total angular momentum before and after the interaction be equal to the total angular momentum during the interaction; hence

$$L = m\tilde{v}_1 b = |\mathbf{r} \times m\mathbf{v}| = -mr^2\dot{\beta}. \tag{2.94}$$

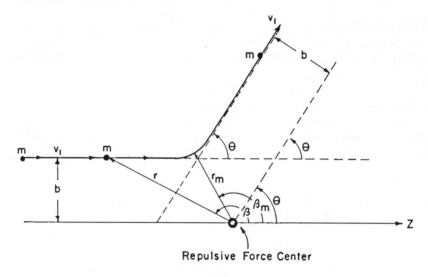

Repulsive Force Center

R-System

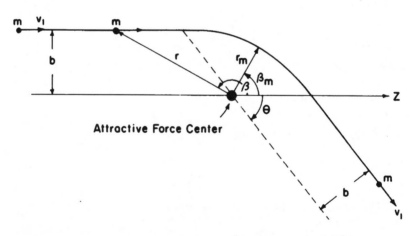

Attractive Force Center

Figure 2.8—Typical trajectories of particles under the influence of repulsive and attractive force centers.

The minus sign is attached to Equation 2.94 because β is decreasing so that $\dot{\beta}$ is negative (see Figure 2.8). Substituting $\dot{\beta}$ from Equation 2.94 into 2.93 and solving for $\dot{r}$, we obtain

$$\dot{r} = -\left\{\frac{2}{m}\left[E\left(1 - \frac{b^2}{r^2}\right) - V(r)\right]\right\}^{1/2}, \tag{2.95}$$

which may be written

$$dt = \frac{-dr}{\{2(E/m)[1 - b^2/r^2 \quad V(r)/E]\}^{1/2}}.$$

Now $d\beta$ may replace dt by the conservation of angular momentum equation, Equation 2.94, written in the form

$$d\beta = -\frac{\tilde{v}_1 b}{r^2} dt,$$

and since $\sqrt{2E/m} = \tilde{v}_1$,

$$d\beta = \frac{b\, dr}{r^2[1 - b^2/r^2 - V(r)/E]^{1/2}}. \qquad (2.96)$$

This equation may be integrated over the range of r from infinity to its minimum value, r_m, and over β from π to its value when $r = r_m$, namely, β_m (see Figure 2.8). The result may be written

$$\beta_m - \pi = -\left| \int_\infty^{r_m} \frac{b\, dr}{r^2[1 - b^2/r^2 - V(r)/E]^{1/2}} \right|. \qquad (2.97)$$

The absolute value signs are placed on the right-hand side of Equation 2.97 to avoid future difficulty with the algebraic sign of this integral. Since β_m is less than π, the right-hand side of Equation 2.97 is certainly negative; thus the negative sign is placed before the absolute value sign.

The scattering angle θ is readily obtained as a function of β_m from Figure 2.8. For repulsive forces

$$\theta = 2\beta_m - \pi, \qquad (2.98)$$

and for attractive forces

$$\theta = \pi - 2\beta_m. \qquad (2.99)$$

Note that the minimum value of β_m for a repulsive force is $\pi/2$ and that the maximum value of β_m for an attractive force is $\pi/2$; so θ always remains positive (as it certainly must because its range of definition extends only from 0 to π). Substituting the expression for β_m from Equation 2.97 into Equations 2.98 and 2.99, we obtain

$$\textit{Repulsive:} \quad \theta = \pi - 2b\left| \int_\infty^{r_m} \frac{dr}{r^2[1 - b^2/r^2 - V(r)/E]^{1/2}} \right|, \qquad (2.100)$$

$$\textit{Attractive:} \quad \theta = -\pi + 2b\left| \int_\infty^{r_m} \frac{dr}{r^2[1 - b^2/r^2 - V(r)/E]^{1/2}} \right| \qquad (2.101)$$

An expression for the turning point, r_m, may be obtained by use of the condition that $\dot{r} = 0$ at $r = r_m$, which becomes, by Equation 2.95,

$$E\left(1 - \frac{b^2}{r_m^2}\right) = V(r_m). \tag{2.102}$$

Equation 2.102 is an implicit equation for r_m; i.e., if we are given an expression for $V(r)$, r_m can be found by using Equation 2.102.

Equations 2.90, 2.100, 2.101, and 2.102 can be used to determine the classical differential elastic-scattering cross section for particles scattering off any potential, $V(r)$. Whether or not the form of the potential is such that the integrals in Equations 2.100 and 2.101 can be carried out in closed form is completely irrelevant. Integrals can always be evaluated graphically or numerically. The engineering student should be aware of the fact that only a vanishingly small fraction of all the problems he will encounter in his professional career can be solved in closed form. Those that cannot be solved in closed form nevertheless have solutions.

It is clear that, given b, one can determine θ from Equation 2.100 or 2.101 for any potential; therefore these equations can be employed to produce a plot of θ vs. b, from which $\sigma(\theta)$ can be determined by use of Equation 2.90.

Now that all classical scattering problems have been solved (solved, at least, in the hand-waving sense) let us determine the cross sections associated with some potentials which allow solutions in closed form.

Example 2.1   The Square Well
A simple potential, yet one of wide utility in neutron physics, is the spherical square-well, or more simply square-well, potential, which is defined by the conditions that

$$V(r) = \begin{cases} 0 & \text{(for } r > R), \\ -V_0 & \text{(for } r \leq R), \end{cases} \tag{2.103}$$

where V_0 is a positive constant (see Figure 2.9).

The kinetic energy of a particle moving toward this potential will, when the particle gets to a distance R from the center, suddenly increase from E to $E + V_0$ because of an impulse directed toward the center of the potential. Once inside the potential well, the particle will feel no force at all ($F = -dV/dr = 0$ inside the well); so it will move in a straight line across the well until it once again gets to $r = R$, where it will again feel an impulse directed toward the center. Once outside the well the particle moves in the R-system with the same speed $\tilde{v}_1$ and impact parameter b it had before the collision. Inside

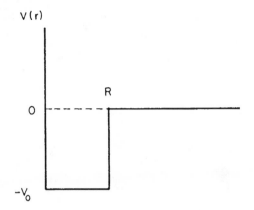

Figure 2.9—The square-well potential.

the well its speed was increased to

$$v_1' = \tilde{v}_1 \left(\frac{E + V_0}{E} \right)^{1/2}. \tag{2.104}$$

It is instructive to examine the trajectories of particles which encounter this potential at various impact parameters. The trajectories are remarkably easy to draw. They must be straight lines both within the well and outside it since no forces act in either place. To conserve angular momentum, the impact parameter within the well, b', must be related to the impact parameter outside the well, b, by Equation 2.105 or Equation 2.106:

$$m\tilde{v}_1 b = mv_1' b', \tag{2.105}$$

$$b' = \frac{\tilde{v}_1}{v_1'} b, \tag{2.106}$$

where the ratio of speeds is given by Equation 2.104.

Figure 2.10 shows a few of the trajectories of particles passing through a square well of depth $V_0 = 15E$. This particular depth was chosen so that the speed inside is exactly four times the speed outside, and therefore each impact parameter inside is one-fourth its value outside. It is seen that the larger the impact parameter, the larger the scattering angle; a particle with zero impact parameter is undeflected, and a particle with impact parameter equal to the radius of the well, R, suffers maximum deflection. Furthermore, the maximum scattering angle is less than π.

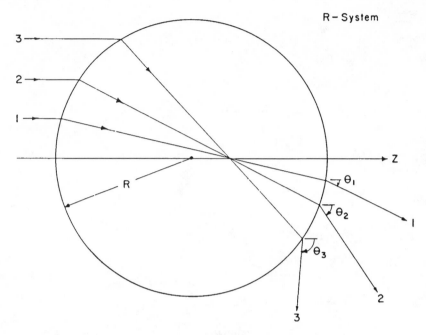

Figure 2.10—Trajectories of particles scattered by square-well potential with $V_0 = 15E$.

Figure 2.10 should remind you of optical ray tracing. You would obtain an identical figure if you used Snell's law to trace the paths of optical photons through a crystal ball with an index of refraction

$$n = \frac{v_1'}{\tilde{v}_1} = \left(\frac{E + V_0}{E}\right)^{1/2} \tag{2.107}$$

relative to the index of the outside medium. Ray tracing methods could be employed to find θ as a function of b for any potential, regardless of its complexity. The trajectories would be more or less complicated curves depending on the specific potential being investigated. The choice between ray tracing and numerical integration for otherwise intractable potentials is pretty much a matter of personal taste. Neither method is elegant, but both methods will accomplish the desired end of producing a curve of θ vs. b.

The square well is one of the limited class of potentials for which a closed-form solution for θ as a function of b can be obtained. Actually there are at least two ways in which this can be done. One can obtain $\theta(b)$ geometrically. (It should be clear that θ can be related to b and b', hence to b, E and V_0, simply by extending the lines of the initial and final trajectories shown in Figure 2.10 until they intersected and then employing elementary geometry on the resulting figure.) Such a procedure will not be used here. Instead, $\theta(b)$ will be found by the more general procedure of integrating Equation 2.101.

Step 1. The turning point r_m is given by Equation 2.102 with $V(r_m) = -V_0$:

$$E\left(1 - \frac{b^2}{r_m^2}\right) = -V_0$$

or

$$r_m = b\left(\frac{E}{E + V_0}\right)^{1/2}$$

and the equation for θ is obtained by substituting $V(r)$ into Equation 2.101:

$$\theta = -\pi + 2b\left|\int_\infty^R \frac{dr}{r^2(1 - b^2/r^2)^{1/2}} + \int_R^{r_m} \frac{dr}{r^2[(E + V_0)/E - b^2/r^2]^{1/2}}\right|.$$

Step 2. To work these integrals into recognizable forms, transform variables by defining x as $1/r$ so that $dx = -dr/r^2$. Then

$$\theta = -\pi + 2b\left|\int_0^{1/R} \frac{-dx}{(1 - b^2x^2)^{1/2}} + \int_{1/R}^{n/b} \frac{-dx}{(n^2 - b^2x^2)^{1/2}}\right|$$

where use has been made of the definitions

$$n \equiv \left(\frac{E + V_0}{E}\right)^{1/2}$$

and

$$x_m \equiv \frac{1}{r_m} = \frac{n}{b}.$$

Step 3. Using the integration formula

$$\int \frac{dx}{[A^2 - B^2x^2]^{1/2}} = \frac{1}{B}\sin^{-1}\left[\frac{B}{A}x\right],$$

evaluate the integrals. As a result,

$$\theta = 2\left[\sin^{-1}\left(\frac{b}{R}\right) - \sin^{-1}\left(\frac{b}{nR}\right)\right]. \tag{2.108}$$

Step 4. Set $b = R$ to find the maximum value of θ:

$$\theta_{max} = \pi - 2\sin^{-1}\left[\frac{E}{E + V_0}\right]^{1/2}. \tag{2.109}$$

This result confirms what might very well have been guessed on purely physical grounds, that the stronger the potential V_0, the larger the maximum scattering angle. An alternative statement, based on the optical analogy, would be that the greater the index of refraction of a crystal ball, the greater the angular spread of light which has passed through it.

Step 5. Solve Equation 2.108 explicitly for b as a function of θ. After some trigonometry and algebra, we find that

$$b^2 = \frac{R^2 \sin^2 (\theta/2)}{1 + 1/n^2 - (2/n)\cos (\theta/2)}. \tag{2.110}$$

Step 6. Differentiate Equation 2.110 with respect to θ and substitute the resulting expression for $b|db/d\theta|$ into the right-hand side of

$$\sigma(\theta) = \frac{b}{\sin \theta}\left|\frac{db}{d\theta}\right|$$

to produce the differential cross section for the rectangular well,

$$\sigma(\theta) = \frac{n^2 R^2}{4\cos (\theta/2)}\frac{[n - \cos (\theta/2)][n\cos (\theta/2) - 1]}{[1 + n^2 - 2n\cos (\theta/2)]^2}. \tag{2.111}$$

In this equation the permitted values of θ are restricted, as has been previously found, to lie in the range

$$0 \le \theta \le [\pi - 2\sin^{-1}(1/n)]. \tag{2.112}$$

This range of θ can easily be shown to be equivalent to the range of $\cos (\theta/2)$ given by

$$1 \ge \cos (\theta/2) \ge (1/n). \tag{2.113}$$

If the depth of the well, V_0, is much greater than the available kinetic energy of the incident particle, E, then n is much larger than unity, all scattering angles are possible, and Equation 2.111 reduces to

$$\sigma(\theta) = \frac{R^2}{4}, \tag{2.114}$$

which says that the scattering from a deep square well is independent of angle, i.e., isotropic, in the C-system. The total scattering cross section in this case is

$$\sigma = \int_{4\pi} \sigma(\theta) \, d\Omega = \int_0^\pi \left(\frac{R^2}{4} \right) 2\pi \sin \theta \, d\theta = \pi R^2, \tag{2.115}$$

which is just the expression for the geometric cross section of a sphere of radius R. It will be left as an exercise to prove by direct integration of Equation 2.111 that, regardless of the depth of the potential, the total scattering cross section for a classical square well is always πR^2.

If the well is shallow, n is near unity, and Equation 2.111 predicts that the scattering will be strongly peaked in the forward direction, as expected, since in the limit as the well depth approaches zero the force on any particle approaches zero so that the particle passes through the well without deflection.

Example 2.2   The Repulsive Inverse-r Potential
The potential corresponding to a repulsive force that varies as the inverse square of the distance between the particles is, in general,

$$V(r) = \frac{a}{r} \qquad \text{(with } a > 0\text{)}. \tag{2.116}$$

The inverse-r potential is plotted in Figure 1.7a. Unlike the square well potential, which is zero except for a finite region of space, the inverse-r potential extends to infinity. The reader could probably guess what total cross section will be found.

Figure 2.11 depicts some typical trajectories of particles scattered by the inverse-r potential. It is relatively easy to prove * from the equations of motion that these trajectories are hyperboles with foci lying on the center of force. Note the larger the impact parameter, the smaller the scattering angle. A particle with zero impact parameter must be scattered directly backward regardless of its initial energy since it encounters a repulsive force that approaches infinity at the origin. A particle with large impact parameter will feel only small force and so be practically undeflected.

The differential scattering cross section corresponding to this potential can be found by direct application of the general equations previously derived.

---

* See, for example, Goldstein's *Classical Mechanics*.

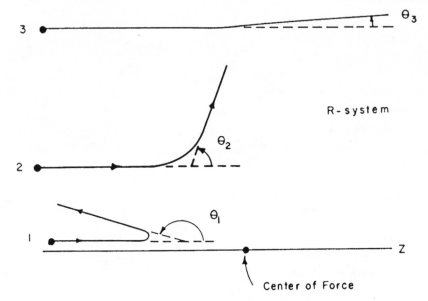

Figure 2.11—Typical trajectories of particles scattered by a repulsive inverse-r potential.

Step 1. Substitute $V(r)$ into Equation 2.100 and change variables from r to x by the transformation $x = 1/r$ so that

$$0 = \pi - 2b \left| \int_0^{x_m} \frac{dx}{[1 - (a/E)x - b^2x^2]^{1/2}} \right|. \tag{2.117}$$

Step 2. Integrate Equation 2.117, using the formula,

$$\int \frac{dx}{(A - Bx - Cx^2)^{1/2}} = \frac{1}{\sqrt{C}} \sin^{-1} \frac{2Cx + B}{(B^2 + 4AC)^{1/2}},$$

with the result that

$$\frac{\theta}{2} = \frac{\pi}{2} - \left| \sin^{-1} \frac{2b^2x + a/E}{[(a/E)^2 + 4b^2]^{1/2}} \right|_0^{x_m}. \tag{2.118}$$

Step 3. Substitute $V(r_m) = a/r_m$ into Equation 2.102 and solve the resultant quadratic for the turning point, r_m. Since r_m must be positive, choose the positive root of the quadratic, which is

$$r_m = \frac{1}{2} \left\{ \frac{a}{E} + \left[\left(\frac{a}{E} \right)^2 + 4b^2 \right]^{1/2} \right\}. \tag{2.119}$$

Step 4. Substitute $x_m = 1/r_m$ from Equation 2.119 into the upper limit of Equation 2.118. After appropriate algebraic and trigonometric manipulations, reduce the resultant equation to

$$\cot\frac{\theta}{2} = \frac{2E}{a}b. \tag{2.120}$$

Once again, as in the case of the square well, a closed form solution for θ as a function of b is obtained.

Step 5. Solve Equation 2.120 for $b(\theta)$ and $|db/d\theta|$ and substitute these expressions into Equation 2.87 to produce

$$\sigma(\theta) = \frac{b}{\sin\theta}\left|\frac{db}{d\theta}\right| = \frac{1}{2}\left(\frac{a}{2E}\right)^2 \frac{\cot(\theta/2)}{\sin\theta}\frac{1}{\sin^2(\theta/2)}.$$

Using the trigonometric identity $\sin\theta = 2\sin(\theta/2)\cos(\theta/2)$, we find

$$\sigma(\theta) = \frac{1}{16}\left(\frac{a}{E}\right)^2 \frac{1}{\sin^4(\theta/2)}. \tag{2.121}$$

This equation was first derived by Rutherford[*] in 1911 in a successful effort to explain the angular distribution of alpha particles scattered by matter in the form of thin foils. Rutherford assumed that the positive charge of the atom was concentrated in a tiny center, or nucleus, which was surrounded by a much larger diffuse sphere of negative charge. He further assumed that the observed scattering was caused by the repulsive electrostatic force between the nucleus and the alpha particles when they were in close proximity and that the force between the alpha particle and the surrounding negative charge could be neglected. The force between the alpha particle and the nucleus was assumed to be the Coulomb force $F = 2Ze^2/r^2$ where $2e$ is the charge of the alpha particle and Ze is the charge of the nucleus. Using these assumptions Rutherford derived an equation identical to Equation 2.121 with $a = 2Ze^2$. Rutherford's equation was experimentally verified by Geiger and Marsden[†] in 1913. In this classic experiment alpha particles were scattered off foils of silver and gold. Since the masses of these nuclei are much greater than the mass of the alpha particle, the center-of-mass angle and the laboratory angle are practically identical, and θ in Equation 2.121 can be replaced by the directly measurable laboratory angle $\tilde{\theta}$. Geiger and Marsden found that

[*] E. Rutherford, Scattering of α and β Particles by Matter and the Structure of the Atom, *Phil. Mag.*, 21: 669 (1911).

[†] H. Geiger and E. Marsden, The Laws of Deflection of α Particles Through Large Angles, *Phil. Mag.*, 25: 604 (1913).

Rutherford's equation was correct, within the limitations of their experiments, in every way. The number of scattered alpha particles was found to be directly proportional to Z^2 and inversely proportional to E^2 and $\sin^4(\theta/2)$, as predicted. Herein lies one of the great ironies of physics. Rutherford's equation, Equation 2.121, has no right to be correct since it is based entirely on Newtonian mechanics, which (we now know) is inapplicable to things as small as alpha particles and nuclei. It just happens that one gets exactly the same formula for the cross section corresponding to the inverse-r potential whether one derives it classically or quantum mechanically.

Just two years after Rutherford incorrectly derived the correct formula for alpha-particle scattering, Bohr succeeded in incorrectly deriving the correct formula for the energy levels of the hydrogen atom. Again a form of classical mechanics produced the same results as quantum mechanics for an inverse-r potential. It almost seems that nature conspired to hide the defects in classical mechanics.

The total scattering cross section for Rutherford scattering, obtained by integrating Equation 2.121 over all solid angles, is infinite. This should come as no surprise. The force field extends to infinity; so every incident particle, no matter how large its impact parameter, is scattered. In effect the particles are scattering off a sphere of infinite radius and hence of infinite cross-sectional area. This is not a peculiarity of the inverse-r potential; any potential that extends to infinity will give rise to an infinite classical cross section. It will be found, strangely enough, that the quantum-mechanical total cross section is finite for all potentials that decrease more rapidly than $1/r^2$ as r approaches infinity.

2.5 Transformation of Cross Sections to Laboratory Coordinates

In the previous section, some special cases of a general procedure for calculating classical differential elastic-scattering cross sections in the R-system were examined. It has been shown that one can always find an expression for, or at least a graph of, $\sigma(\theta)$, the differential cross section as a function of the center-of-mass angle. However, what is really needed for most calculations and for comparison with experiment is an expression for or a graph of $\sigma(\bar{\theta})$, the differential cross section as a function of the laboratory angle of the projectile particle after the scattering. The relation between $\sigma(\bar{\theta})$ and $\sigma(\theta)$ is readily derived from purely kinematic arguments and is valid regardless of the sizes of the colliding particles.

Let us draw an imaginary sphere about our scattering center, as in Figure 2.12, and focus our attention on those particles which scatter

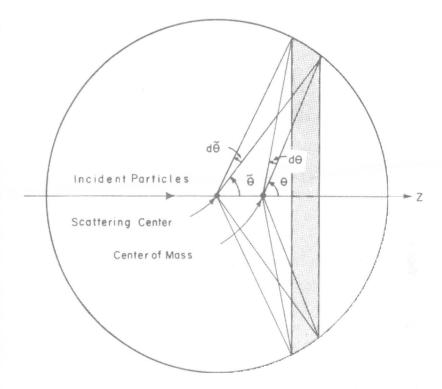

Figure 2.12—Solid angles in the L- and C-systems.

through a circular band on this sphere. This band subtends an angle between $\tilde{\theta}$ and $\tilde{\theta} + d\tilde{\theta}$ with respect to the scattering center and an angle between θ and $\theta + d\theta$ with respect to the position of the center of mass at the instant when the particles pass through the band.

The number of projectile particles that pass through this band per second per target particle is, by definition of the differential cross section in the laboratory,

$$d\tilde{N} = \sigma(\tilde{\theta})\,d\tilde{\Omega}$$

and, by definition of the differential cross section in the C-system,

$$dN = \sigma(\theta)\,d\Omega.$$

Regardless of which coordinate system is used to describe the direction of the particles, the same number pass through the band every second; thus

$$d\tilde{N} = dN,$$

and so

$$\sigma(\tilde{\theta}) = \sigma(\theta)\frac{d\Omega}{d\tilde{\Omega}} \tag{2.122}$$

or, since $d\Omega = 2\pi \sin \theta \, d\theta$ and $d\tilde{\Omega} = 2\pi \sin \tilde{\theta} \, d\tilde{\theta}$,

$$\sigma(\tilde{\theta}) = \frac{\sin \theta}{\sin \tilde{\theta}} \frac{d\theta}{d\tilde{\theta}} \sigma(\theta). \tag{2.123}$$

This is the transformation equation. It is still not very useful because the right-hand side is not yet an explicit function of θ alone. The quantities $\sin \tilde{\theta}$ and $d\theta/d\tilde{\theta}$ can be eliminated by use of the kinematic relation between θ and $\tilde{\theta}$ (Equation 2.46),

$$\tan \tilde{\theta} = \frac{\sin \theta}{(m_1/m_2) + \cos \theta}.$$

By taking differentials of Equation 2.46, we find

$$\frac{d\theta}{d\tilde{\theta}} = \frac{1}{\cos^2 \tilde{\theta}} \frac{[(m_1/m_2) + \cos \theta]^2}{[1 + (m_1/m_2)\cos \theta]}. \tag{2.124}$$

By making a geometric construction of the angle $\tilde{\theta}$ based on Equation 2.46, we get

$$\sin \tilde{\theta} = \frac{\sin \theta}{[1 + (m_1/m_2)^2 + 2(m_1/m_2)\cos \theta]^{1/2}} \tag{2.125a}$$

and

$$\cos \tilde{\theta} = \frac{[(m_1/m_2) + \cos \theta]}{[1 + (m_1/m_2)^2 + 2(m_1/m_2)\cos \theta]^{1/2}}. \tag{2.125b}$$

Then using Equations 2.124, 2.125a, and 2.125b, we obtain

$$\frac{\sin \theta \, d\theta}{\sin \tilde{\theta} \, d\tilde{\theta}} = \frac{[1 + (m_1/m_2)^2 + 2(m_1/m_2)\cos \theta]^{3/2}}{[1 + (m_1/m_2)\cos \theta]}$$

and so finally

$$\sigma(\tilde{\theta}) = \frac{[1 + (m_1/m_2)^2 + 2(m_1/m_2)\cos \theta]^{3/2}}{[1 + (m_1/m_2)\cos \theta]} \sigma(\theta). \tag{2.126}$$

Equation 2.126 is the expression by which an elastic-scattering cross section in the R-system can be transferred into the L-system. Please note that $\tilde{\theta}$ is not, in general, equal to θ. Given the center-of-mass cross section at C-angle θ, application of Equation 2.126 will produce the laboratory cross section at an L-angle $\tilde{\theta}$ given by Equation 2.46, that is, at angle

$$\tilde{\theta} = \tan^{-1}\left[\frac{\sin\theta}{(m_1/m_2) + \cos\theta}\right]. \qquad (2.127)$$

Some special cases of this transformation formula are of particular interest. If the target mass is much greater than the projectile mass, $m_2 \gg m_1$, then Equations 2.127 and 2.126 predict, as expected, that the laboratory and center-of-mass angles are approximately equal and that the laboratory and center-of-mass differential cross sections are also approximately equal,

$$\tilde{\theta} \simeq \theta \quad \text{and} \quad \sigma(\tilde{\theta}) \simeq \sigma(\theta) \qquad \text{(for } m_2 \gg m_1\text{)}.$$

If the masses of the projectile and target particles are equal, then Equation 2.127 reduces to $\tilde{\theta} = \theta/2$. Since the maximum C-angle for elastic scattering of equally massive particles is π, the maximum laboratory angle is $\pi/2$. This should not surprise anyone who has ever experimented with collisions between coins sliding on a smooth surface. Try to make a coin scatter backwards in a collision with an equally massive coin. Relatively few attempts will soon convince any observer of the impossibility of the feat. For equally massive particles Equation 2.126 can be reduced to the form

$$\sigma(\tilde{\theta}) = 4\cos\tilde{\theta}\,\sigma(\theta) \qquad \text{(for } m_1 = m_2\text{)},$$

which states that isotropic scattering of equally massive particles in the C-system will have a cosine-like distribution in the laboratory, peaked in the forward direction and zero beyond 90°.

We have been examining the transformation of differential elastic-scattering cross sections from the R-system, where they are calculated, back to the L-system. The transformation equations for inelastic scattering and for reactions can be derived in exactly the same way—by substituting the appropriate kinematic relation between θ and $\tilde{\theta}$ into Equation 2.123. These details will be left as exercises for the student.

Before leaving the subject of transformation, your attention is called to what is probably obvious. The total cross section for any interaction is the same in all coordinate systems; in particular,

$$\sigma = \int_{4\pi} \sigma(\theta)\,d\Omega = \int_{4\pi} \sigma(\tilde{\theta})\,d\tilde{\Omega}. \qquad (2.128)$$

This equation can be used to provide a convenient, necessary, but not sufficient, check on the validity of transformed cross sections.

2.6 The Centrifugal Potential

The reader's attention is now briefly directed to a very important aspect of the motion of particles under the influence of central forces. Newton's second law of motion for a particle of mass m_1 that is moving in the R-system whose origin is a particle of mass m_2 has been found to be of the form (see Equation 2.19)

$$m\ddot{\mathbf{r}} = \mathbf{F},$$

where m is the reduced mass, $m_1 m_2/(m_1 + m_2)$.

In polar coordinates (r, β), the r components of the acceleration and the force are $(\ddot{r} - r\dot{\beta}^2)$ and $(-dV/dr)$, respectively; so Equation 2.19 may be written

$$m(\ddot{r} - r\dot{\beta}^2) = -\frac{dV(r)}{dr}$$

or

$$m\ddot{r} = -\frac{dV(r)}{dr} + mr\dot{\beta}^2. \tag{2.129}$$

The law of conservation of angular momentum (see Equation 2.94),

$$L = mr^2\dot{\beta} = \text{constant},$$

provides an expression for $\dot{\beta}$ which, when substituted into Equation 2.129, yields

$$m\ddot{r} = -\frac{d}{dr}\left[V(r) + \frac{L^2}{2mr^2} \right]. \tag{2.130}$$

Equation 2.130 can now be interpreted as Newton's second law for a particle of mass m moving in one dimension under the action of an effective potential given by

$$V_{\text{eff}}(r) = V(r) + \frac{L^2}{2mr^2}. \tag{2.131}$$

This is a most remarkable result. It means that the radial position of the particle can be found as a function of time without ever considering the angular position β. The two-dimensional problem has been reduced to a one-dimensional problem for $r(t)$. But the particle moving in this one dimension moves as though it were in a potential consisting of the actual potential $V(r)$ plus a "centrifugal potential," $V_c(r)$, where

$$V_c(r) = \frac{L^2}{2mr^2}. \tag{2.132}$$

In classical mechanics the angular momentum, L, can have any value; its magnitude is determined by the linear momentum and impact parameter of the particle according to the equation $L = m\tilde{v}b$. As will be demonstrated later, in quantum mechanics the total angular momentum may have only certain values given by the quantum condition $L = [l(l + 1)]^{1/2}\hbar$, where l is an integer and $\hbar$ is a universal constant.

The centrifugal force, F_c, corresponding to the centrifugal potential is given by $-dV_c/dr$, or by

$$F_c = \frac{L^2}{mr^3}. \tag{2.133}$$

The name "centrifugal force" is appropriate since the force is indeed directed away from the origin.

Now consider a particle that is inside a potential well. The arguments about to be made are independent of the shape of the well, but for ease in visualization imagine that the well is a simple square well as in Figure 2.9. It is clear that the particle will never get out of the well if its kinetic energy is less than the depth of the well; that is, if $T' < V_0$, the particle is trapped. The symbol T' is used for the available kinetic energy of the particle in the well, i.e., $T' = m(v')^2/2$ where v' is its speed within the well with respect to the walls of the well. If the particle were to get out of the well, it would either have negative kinetic energy (whatever that means) or would be in violation of the law of conservation of energy. Hence, it is trapped.

Now suppose the particle has a kinetic energy greater than the depth of the well. One is tempted to jump to the conclusion, without reservation, that the particle can get out of the well. Actually, the particle may or may not be able to get out of the well depending on its angular momentum with respect to the center of the well. *The particle may be confined within the well because of the repulsive centrifugal potential.* How can a repulsive force confine a particle? Figure 2.13 shows a square-well potential, a centrifugal potential for $L > 0$, and the sum of the two potentials. It is the sum of the two potentials that affects the r position of the particle. Also shown on the figure is the kinetic energy of the particle in the well, deliberately chosen to be greater than V_0 but less than the total effective potential at the edge of the well. Now it is clear that to go beyond the edge of the well the particle would either have to have negative kinetic energy or be in violation of the law of conservation of energy. Hence, it is trapped. Notice that the particle not only is trapped within the well but also is further restricted to the region of r greater than r_0 where r_0 is the radius at which the total energy, potential plus kinetic, becomes zero.

The above may seem somewhat contrived and misleading if not downright mysterious. By introducing the terms "centrifugal potential"

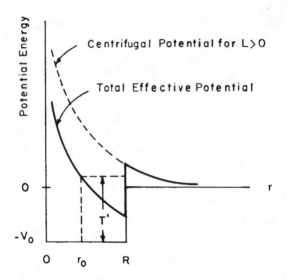

Figure 2.13—Total effective potential for square well.

and "centrifugal force," we have complicated what is, in truth, an almost trivial consequence of the law of conservation of angular momentum.

The easiest way to illustrate this is to draw a few trajectories within a square well. Suppose, for example, one chooses a square-well depth and a particle energy such that the speed of a particle within the well is four times the speed it would have if it got out of the well; that is

$$\frac{v'}{v} = \left(\frac{E + V_0}{E}\right)^{1/2} = 4.$$

Angular-momentum conservation demands that the ratio of outside impact parameters to inside impact parameters be 4 to 1:

$$\frac{b}{b'} = 4.$$

Figure 2.14 shows what happens when we attempt to draw a few trajectories within the well. Trajectory 1 is the path of a particle with small angular momentum. This trajectory can quite easily be continued outside the well along a path such that $b_1 = 4b_1'$. Trajectory 2 is the path of a particle with larger angular momentum. But there is no continuation of trajectory 2 outside the well, which obeys the condition that $b_2 = 4b_2'$; so this particle must be reflected from the surface and remain within the well with its same angular momentum and hence its same impact parameter. It's that simple. A particle with large angular momentum stays within the well because it cannot get out without losing angular momen-

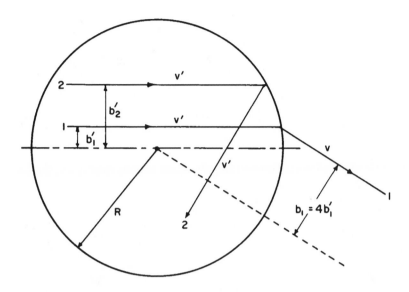

Figure 2.14—Trajectories of particles that leave and particles that are trapped within a square well for which $v'/v = 4$.

tum and thus violating the conservation law. If trajectory 2 is extended, one will find that this particle continues to reflect off the walls always staying within the radial region between b'_2 and R. It is easy to prove that b'_2 is the same as r_0, the minimum possible radial position of a particle with angular momentum $mv'b'_2$.

The reader will notice in Figure 2.13 that the centrifugal potential extends beyond $r = R$. It will now be shown that this potential, though repulsive, is never strong enough to prevent the penetration of an incident particle into the well. At the surface of the well, its magnitude is

$$V_c(R) = \frac{L^2}{2mR^2} = \frac{(m\tilde{v}b)^2}{2mR^2} = E\left(\frac{b}{R}\right)^2,$$

which is always less than the energy E of any incident particle that has an impact parameter small enough ($b < R$) to strike the well. Another way of saying this is: It is always possible to continue a trajectory from the outside through a potential well in such a way that angular momentum and energy are conserved. It will be left to the reader to prove that a particle, however, may be reflected from a potential barrier even though its kinetic energy is greater than the barrier height.

SELECTED REFERENCES *

1. Herbert Goldstein, *Classical Mechanics*, Addison-Wesley Publishing Company, Inc., Reading, Mass., 1950. This is a superbly executed textbook, one that should be on the desk of every nuclear engineer and physicist.
2. Robert B. Lindsay and Henry Margenau, *Foundations of Physics*, John Wiley & Sons, Inc., New York, 1936. The fundamental problems of physics are discussed from a philosophical point of view. Chapter I, "The Meaning of a Physical Theory," and Chapter III, "The Foundations of Mechanics," are particularly valuable references for the material in Chapters 2 and 3 of this textbook. Pick it up and discover, among other things, why $F = ma$ is not at all obvious.

EXERCISES

1. A bullet of mass m strikes and becomes embedded in a solid spherical block of wood of mass M and radius R, initially at rest. Assume that angular momentum is conserved, as well as linear momentum and energy, in the collision. Calculate (a) the final linear velocity of the center of mass of the block plus embedded bullet system and (b) the angular velocity of the system as a function of the impact parameter b.
2. Calculate and plot the laboratory scattering angle $\bar{\theta}$ vs. the center-of-mass scattering angle θ for neutrons elastically scattering off nuclei of masses 1, 2, 10, and 50 times that of a neutron. Briefly summarize the results.
3. For the $^{16}O(n,p)^{16}N$ reaction to proceed, the ^{16}O nucleus must be given 9.6 Mev of energy. How much laboratory kinetic energy must a neutron incident on a stationary ^{16}O nucleus have to produce this reaction; i.e., what is the threshold energy for this reaction?
4. A beam of 10^{12} monoenergetic neutrons per second is incident on a thin target containing 10^{24} nuclei of mass 3 times that of a neutron. Assume that the scattering is elastic and isotropic in the C-system and that the cross section in the C-system is 2 barns/steradian. Calculate and plot as a function of laboratory scattering angle (a) the number of scattered neutrons that would be detected per second by a spherical neutron detector 10 cm in diameter located 2 meters from the target and (b) the laboratory energies of the detected neutrons.
5. Plot the Rutherford differential elastic-scattering cross section in the C-system and in the L-system vs. angle for the elastic scattering of alpha particles off ^{12}C nuclei.
6. Derive the classical expression for the differential cross section for the scattering of particles of energy E off a potential barrier which has the form

$$V(r) = \begin{cases} 0 & \text{(for } r > R), \\ V_1 & \text{(for } r < R), \end{cases}$$

where V_1 is a positive constant. Consider the cases $E < V_1$ and $E > V_1$.
7. In the previous problem show that there are values of $E > V_1$ for which the particle is prevented from entering the region $r < R$ because of the centrifugal

---

*The number of selected references has deliberately been kept to an absolute minimum. It is hoped that the student will thereby be encouraged to spend at least a few hours in delightful exploration of each and every one of the listed references. Be assured that such time will be well employed. As Samuel Johnson once remarked, "Knowledge is of two kinds: We know a subject ourselves, or we know where we can find information upon it."

barrier. Derive and discuss the conditions under which a particle is excluded from the region $r < R$ because of the centrifugal barrier.

8. Estimate by the ray tracing method and draw the differential cross section for particles of energy $E = V_0$ incident on a spherical well of the form

$$V(r) = \begin{cases} 0 & \text{(for } r > R), \\ -V_0 & \text{(for } R/2 < r < R), \\ 2V_0 & \text{(for } r < R/2), \end{cases}$$

where V_0 is a positive constant.

9. Determine the classical differential elastic-scattering cross section for a particle incident on an inverse-cube attractive potential $(V(r) = -A/r^3$, where A is a positive constant). Is the total elastic cross section finite?

10. Derive equations for $\theta = \theta(b)$ that are similar to Equations 2.100 and 2.101 but are applicable to inelastic scattering. Clearly state any assumptions you must make.

11. Prove the following assertions: In an elastic scattering of a moving heavy particle by a stationary (in the lab) light particle, there is not a one to one correspondence between laboratory and center-of-mass scattering angles. Only one value of the laboratory angle corresponds to a given value of the center-of-mass angle, but the converse is not true. In fact, for every value of the laboratory scattering angle between zero and its maximum possible value, θ_m, there are two possible center-of-mass scattering angles.

12. The elastic scattering of neutrons of energy less than 10 Mev off ^{1}H nuclei is isotropic in the C-system. What is the mean scattering angle in the laboratory?

3 ELEMENTS OF QUANTUM MECHANICS

In 1924 a definite reorientation of physical theory occurred. Classical mechanics, which had reigned supreme for three hundred years, was shaken; quantum mechanics was born. That year marked the beginning of the most exciting decade in the history of natural philosophy. In rapid succession a number of outstanding mysteries in the observed behavior of microscopic matter were explained. Thus quantum mechanics, a new and powerful analytical tool, met with immediate and brilliant success when applied to many physical problems that previously had defied solution.

The key that unlocked the mystery of matter was forged by Louis Victor de Broglie who hypothesized, in his doctoral dissertation in 1924, that a particle with momentum p would act as a wave having wavelength λ, given by $\lambda = h/p$, where h is Planck's constant.* Experimental verification of de Broglie's astonishing postulate was not long in coming. In 1927 C. J. Davisson and L. H. Germer experimentally demonstrated that electrons exhibited wavelike interference patterns when scattered from a crystal and that their experimentally determined wavelengths were those predicted by the de Broglie relation. In 1928 T. H. Johnson diffracted slow hydrogen atoms off crystals to demonstrate that the de Broglie relation was valid for electrically neutral particles.

During the interim de Broglie formulated a mathematical theory, based on his postulate, that bore considerable resemblance to the classical theory of ray optics. This approach was very quickly superseded by two distinct, although equivalent, formulations of what has come to be called

---

* Numerical values of universal physical constants are listed in the Appendix at the end of the book.

quantum mechanics. These were the *wave mechanics* of Erwin Schrödinger and the *matrix mechanics* of Werner Heisenberg. These treatments are mathematically distinct; however, they are physically equivalent.

Note that quantum mechanics did not spring full grown from the minds of de Broglie, Schrödinger, and Heisenberg. It had its origins deep in the classical theories of mechanics and optics; in fact, a form of wave mechanics had been invented by Hamilton almost a hundred years before de Broglie put forth his germinal concept in quantum mechanics. In 1924 the time was ripe, and de Broglie's ideas were not overlooked. We can fully appreciate the attitudes and problems prevailing at that time by examining in some detail the classical formulation of mechanics. Particular attention should be directed to some striking analogies between classical mechanics and the ray and wave theories of light.

This chapter presents a development of classical mechanics aimed at implementing the plausibility and heritage of the Schrödinger theory. A postulatory structure of quantum mechanics is also introduced along with a rudimentary presentation of basic concepts, terminology, and illustrative examples.

3.1 Survey of Classical Mechanics

Of the numerous extant formulations of classical dynamics, those of Newton, Lagrange, and Hamilton have proved to have the widest applicability. Each of these formulations can be derived mathematically from either of the other two. Although each has a distinct mathematical form, all three describe the same physical reality, the motion of massive particles either in a free state of motion or under the influence of forces.

Newtonian Dynamics

The three famous laws of Isaac Newton were promulgated in the seventeenth century.

The first law states that a mass point not subject to external forces continues in a state of rest or straight line motion with constant velocity,

$$\ddot{\mathbf{r}} = 0. \tag{3.1}$$

The second law states that a point of mass m will experience an acceleration equal to the force acting on it divided by its inertial mass,

$$\mathbf{F} = m\mathbf{a}. \tag{3.2}$$

The third law states that, if a force is exerted by mass point 1 on mass point 2, an equal and opposite force is exerted by point 2 on point 1,

$$\mathbf{F}_{12} = -\mathbf{F}_{21}. \tag{3.3}$$

From the Newtonian laws we can derive the momentum conservation laws, as in Sec. 2.1. For a system of particles that exert central forces on one another but are not subject to external forces, these conservation laws take the form

Total linear momentum: $\dot{P} = 0,$ (3.4)

Total angular momentum: $\dot{L} = 0.$ (3.5)

The conservation law of total, kinetic plus potential, energy follows directly from the conservation law of linear momentum if, in addition to the forces being central, mass is conserved and if the potential energy does not depend explicitly on time. Written mathematically, this law is

Energy: $\dot{E} = \dfrac{d}{dt}(T + V) = 0.$ (3.6)

Lagrangian Dynamics

Classical dynamics was reformulated in the eighteenth century by Joseph Louis Lagrange, who introduced the concept of generalized coordinates and derived equations of motion that are valid in all inertial coordinate systems and that are often easier to apply than the equations derived from Newton's laws.

A system of N particles subject to k constraints will have $3N - k$ *degrees of freedom*. Lagrange suggested that the equations of motion of such a system should be written in terms of $3N - k = f$ generalized coordinates q_j, all of which are independent of one another. By first writing the old Euclidean coordinates as explicit functions of the independent generalized coordinates,

$$x_i = x_i(q_j),$$

$$y_i = y_i(q_j),$$

$$z_i = z_i(q_j),$$

where i has values from 1 to N and j has values from 1 to f, Lagrange derived* the equation of motion from Newton's laws. For a conservative system, i.e., a system in which the forces can be represented as gradients of potentials, Lagrange's equations have the form

$$\frac{d}{dt}\left(\frac{\partial L}{\partial \dot{q}_j}\right) - \frac{\partial L}{\partial q_j} = 0,$$ (3.7)

*The reader who has never seen this derivation and others that are not included in this section is urged to spend an hour or two with a good classical mechanics book, such as Goldstein's *Classical Mechanics* (Addison-Wesley Publishing Company, Reading, Mass., 1950) or *Classical Mechanics* by H. C. Corben and Philip Stehle (John Wiley & Sons, Inc., New York, 1960).

where j has values from 1 to f. The Lagrangian, L, is defined as the kinetic energy of the system minus the potential energy expressed as a function of the generalized coordinates and velocities,

$$L(q_j, \dot{q}_j) \equiv T(q_j, \dot{q}_j) - V(q_j), \tag{3.8}$$

where $j = 1, \ldots, f$. The system of f equations, Equation 3.7, is called *Lagrange's equations of motion*. The quantities

$$p_j = \frac{\partial L}{\partial \dot{q}_j} \tag{3.9}$$

are called the *generalized momenta*.

Solutions of Lagrange's equations are in the form of an expression for each q_j as a function of time. Since Lagrange's equations are second-order differential equations, there will be two arbitrary constants of integration for each differential equation. These integration constants are evaluated from the initial conditions on q_j and $\dot{q}_j$.

As a very elementary example of the use of Lagrange's equations, consider a particle of mass m moving in one dimension q under the action of a potential $V(q)$. In this case $\dot{q}$ is the velocity and $T = m\dot{q}^2/2$ is the kinetic energy. The Lagrangian, $L = T - V$, is

$$L = \tfrac{1}{2}m\dot{q}^2 - V(q),$$

and the single Lagrangian equation of motion becomes the usual Newtonian equation,

$$\frac{d}{dt}(m\dot{q}) + \frac{\partial V}{\partial q} = 0$$

or

$$m\ddot{q} = -\frac{\partial V}{\partial q}.$$

Why the quantities $\partial L/\partial \dot{q}_j$ are called generalized momenta seems apparent from this example, but do not be misled. In more complicated systems the so-called generalized momenta may bear little or no resemblance to linear or angular momenta.

Consider as a second example of the Lagrangian method the motion of a point particle of mass m in a central force field. Spherical coordinates with origin on the center of force are quite naturally chosen as generalized coordinates. Generalized coordinates r, θ, and ϕ are related to the rectangular coordinates by the expressions:

$$x = r \sin \theta \cos \phi,$$
$$y = r \sin \theta \sin \phi, \tag{3.10}$$
$$z = r \cos \theta.$$

The kinetic energy of the particle in rectangular coordinates is

$$T = \tfrac{1}{2}m(\dot{x}^2 + \dot{y}^2 + \dot{z}^2) \tag{3.11}$$

and, in generalized coordinates (from the transformation relations),

$$T = \tfrac{1}{2}m(\dot{r}^2 + r^2\dot{\theta}^2 + r^2 \sin^2 \theta \cdot \dot{\phi}^2). \tag{3.12}$$

Therefore, the Lagrangian function is

$$L = \tfrac{1}{2}m(\dot{r}^2 + r^2\dot{\theta}^2 + r^2 \sin^2 \theta \cdot \dot{\phi}^2) - V(r). \tag{3.13}$$

The Lagrangian equations (Equation 3.7) become, after substituting $q_1 = r, q_2 = \theta, q_3 = \phi$, and L from Equation 3.13,

$$\frac{d}{dt}(m\dot{r}) - mr(\dot{\theta}^2 + \sin^2 \theta \cdot \dot{\phi}^2) + \frac{dV}{dr} = 0, \tag{3.14}$$

$$\frac{d}{dt}(mr^2\dot{\theta}) - mr^2 \sin \theta \cos \theta \cdot \dot{\phi}^2 = 0, \tag{3.15}$$

$$\frac{d}{dt}(mr^2 \sin^2 \theta \cdot \dot{\phi}) = 0. \tag{3.16}$$

Equation 3.16 implies that

$$\dot{\phi} = \frac{C}{mr^2 \sin^2 \theta}, \tag{3.17}$$

where C is an arbitrary constant.

When Equation 3.17 is substituted into Equations 3.14 and 3.15, the resultant equations are independent of ϕ. This merely reflects the fact that central motion takes place in a plane. We may choose any plane for this motion; so, for simplicity, we usually choose the plane $\phi = 0$ (the x-z plane) in which case $C = 0$ and Equations 3.14 and 3.15 become

$$\frac{d}{dt}(m\dot{r}) - mr\dot{\theta}^2 + \frac{dV}{dr} = 0 \tag{3.18}$$

and

$$\frac{d}{dt}(mr^2\dot{\theta}) = 0. \tag{3.19}$$

Equation 3.19 is the statement of conservation of angular momentum in polar coordinates. (Compare it with Equation 2.94.) Equation 3.18 is identical to Equation 2.129, which was derived from Newton's second law.

The application of the Lagrangian method, after one has carefully selected a set of generalized coordinates and written the Lagrangian in terms of these coordinates, is more or less automatic. From this point of view the Lagrangian method provides the differential equations describing a particular motion in almost a "cookbook" fashion. No physical information is provided which could not have been obtained just as well from Newton's equations. However, much of the difficulty in applying Newton's laws lies in trying to determine the forces, particularly forces of constraint. The Lagrangian method eliminates this difficulty; forces, as such, do not appear in Lagrange's formalism, only energies.

Hamiltonian Dynamics

In the nineteenth century William Rowan Hamilton introduced yet another form of classical dynamics, a form which is now called, appropriately, *Hamiltonian dynamics*. For a conservative system of f degrees of freedom, Hamilton showed that Lagrange's f second-order differential equations of motion are equivalent to $2f$ simple first-order differential equations. These equations are

$$\dot{p}_j = -\frac{\partial H}{\partial q_j} \tag{3.20}$$

and

$$\dot{q}_j = \frac{\partial H}{\partial p_j}, \tag{3.21}$$

where j has values from 1 to f.

The Hamiltonian function, H, is the sum of the kinetic and potential energies of the system expressed as functions of generalized coordinates, momenta, and, possibly, time,

$$H = T + V. \tag{3.22}$$

Since $L = T - V$, this equation may be written $H = 2T - L^*$ where L^* is the Lagrangian written in terms of generalized momenta rather than generalized velocities. Since $2T = \sum_j p_j \dot{q}_j$, the Hamiltonian may be written

$$H(q_j, p_j, t) = \sum_j p_j \dot{q}_j - L^*(q_j, p_j, t). \tag{3.23}$$

Notice that the generalized velocities on the right-hand side of Equation 3.23 must be replaced by appropriate functions of coordinates and momenta before the function H is a proper Hamiltonian function.

The symbolic $2f$-dimensional space of the q_j and p_j coordinates is called the *phase space* of the dynamical problem. The f-dimensional space of the coordinates q_j is called the *configuration space*. Equations 3.20 and 3.21 are usually called the *canonical equations of motion*. Each pair of coordinates and momenta (q_j, p_j) that appear in symmetric positions in these equations is a pair of *canonically conjugate variables*.

We can illustrate these points and the Hamiltonian method of solution of dynamic problems by considering again the problem of the motion of a single mass point in a conservative central force field. To obtain the Hamiltonian function, we must first identify the generalized momenta. From Equations 3.9 and 3.13 we find

$$p_r = \frac{\partial L}{\partial \dot{r}} = m\dot{r},$$

$$p_\theta = \frac{\partial L}{\partial \dot{\theta}} = mr^2\dot{\theta}, \tag{3.24}$$

$$p_\phi = \frac{\partial L}{\partial \dot{\phi}} = mr^2 \sin^2\theta \cdot \dot{\phi}.$$

The generalized velocities $\dot{r}$, $\dot{\theta}$, and $\dot{\phi}$ can be replaced in the kinetic energy expression, Equation 3.12, by the generalized momenta from Equations 3.24. The Hamiltonian function then becomes

$$H = \frac{1}{2m}\left[p_r^2 + \frac{p_\theta^2}{r^2} + \frac{p_\phi^2}{r^2 \sin^2\theta}\right] \quad V(r). \tag{3.25}$$

Notice that the coordinate ϕ does not appear explicitly in the Hamiltonian. Such a coordinate is called *cyclic*. The generalized momentum corresponding to a cyclic coordinate is conserved since, by Hamilton's equation (Equation 3.20), if H is not a function of q_K, $\dot{p}_K = 0$ and p_K is a constant of the motion. Thus, simply by inspection of the Hamiltonian, we can say immediately that p_ϕ is a constant of the motion. With proper choice of initial conditions, namely, $\phi = 0$ at $t = 0$, we can completely eliminate the variable ϕ from the Hamiltonian since ϕ, $\dot{\phi}$, and p_ϕ are equal to 0. When H is substituted into the four remaining canonical equations, the resulting equations are:

$$\frac{d}{dt}(p_r) = \frac{p_\theta^2}{mr^3} - \frac{dV}{dr}, \tag{3.26}$$

$$\frac{d}{dt}(p_\theta) = 0, \tag{3.27}$$

$$\frac{d}{dt}(r) = \frac{p_r}{m}, \tag{3.28}$$

$$\frac{d}{dt}(\theta) = \frac{p_\theta}{mr^2}. \tag{3.29}$$

The last two equations are trivial, being merely the definitions p_r and p_θ. The first two equations are identical to Equations 3.18 and 3.19, the equations of motion obtained by the Lagrangian method.

Poisson Brackets

Let $a(q_j,p_j,t)$ and $b(q_j,p_j,t)$ be any two dynamical variables expressed, like the Hamiltonian, as functions of the generalized coordinates, momenta, and time. These variables could be, for example, coordinates, momenta, angular momenta, kinetic energies, Lagrangians, Hamiltonians, or any other functions of generalized coordinates, momenta, and time.

The *Poisson bracket* of the variables a and b is defined by the expression

$$(a,b) \equiv \sum_j \left(\frac{\partial a}{\partial q_j} \frac{\partial b}{\partial p_j} - \frac{\partial b}{\partial q_j} \frac{\partial a}{\partial p_j} \right). \tag{3.30}$$

Now the time rate of change of any variable a is given by

$$\frac{da}{dt} = \dot{a} = \sum_j \left(\frac{\partial a}{\partial q_j} \dot{q}_j + \frac{\partial a}{\partial p_j} \dot{p}_j \right) + \frac{\partial a}{\partial t}, \tag{3.31}$$

but, in view of the canonical equations, Equation 3.31 may be written

$$\dot{a} = \sum_j \left(\frac{\partial a}{\partial q_j} \frac{\partial H}{\partial p_j} - \frac{\partial a}{\partial p_j} \frac{\partial H}{\partial q_j} \right) + \frac{\partial a}{\partial t}, \tag{3.32}$$

or, from the definition of the Poisson bracket,

$$\dot{a} = (a, H) + \frac{\partial a}{\partial t}. \tag{3.33}$$

If a variable a does not contain the time explicitly, then

$$\dot{a} = (a, H). \tag{3.34}$$

In particular,

$$\dot{q}_j = (q_j, H) \tag{3.35}$$

and

$$\dot{p}_j = (p_j, H), \tag{3.36}$$

so the canonical equations of motion for conservative systems may be written

$$(q_j, H) = \frac{\partial H}{\partial p_j}, \tag{3.37}$$

$$(p_j, H) = -\frac{\partial H}{\partial q_j}. \tag{3.38}$$

It follows directly from the definition of the Poisson bracket that

$$(q_j, q_k) = 0, \tag{3.39}$$

$$(p_j, p_k) = 0, \tag{3.40}$$

and

$$(q_j, p_k) = \delta_{jk}, \tag{3.41}$$

where δ_{jk} is the *Kronecker delta* defined by

$$\delta_{jk} \equiv \begin{cases} 0 & (j \neq k), \\ 1 & (j = k). \end{cases} \tag{3.42}$$

Note that the Poisson bracket of two canonically conjugate variables does not vanish. This will be significant in quantum mechanics where the Poisson bracket is intimately related to the so-called commutator bracket of two quantum-mechanical operators.

Hamilton's Principle

In 1824 Hamilton postulated that the actual path of a mechanical system in configuration space will be that path along which S, the time integral of the Lagrangian, is stationary when compared with any other paths which begin and terminate at the same configuration points in the same time interval. The term "stationary" means either minimum or maximum. In the notation of the calculus of variations, Hamilton's principle may be written

$$\delta S = \delta \int_{t_1}^{t_2} L \, dt = 0, \tag{3.43}$$

where L is the Lagrangian, a function of the generalized coordinates and velocities or of coordinates, momenta, and, possibly, time. The function S is called the *Hamiltonian action*.

Hamilton's principle can be derived* from Lagrange's equations, Hamilton's canonical equations, D'Alembert's equations, or, ultimately, Newton's laws of motion. Conversely, Hamilton's principle can be used to derive Lagrange's, Hamilton's, or D'Alembert's equations. Classical mechanics is self-consistent despite the number of alternative formulations.

Hamilton's principle can be used to solve dynamical problems, but it is seldom used because it simply duplicates either the Lagrangian equations or the canonical equations of the system, depending on whether L is written in terms of generalized velocities or momenta. The primary application of Hamilton's principle today is in the formulation of new field theories of radiation and matter.

The Principle of Least Action

The *principle of least action* was rigorously formulated by Lagrange in the middle of the eighteenth century. It is widely called *Maupertius' principle* after Pierre de Maupertius, the French mathematician and astronomer who guessed at a roughly similar, though vague, universal principle a few years before Lagrange formulated the rigorously correct one.

The *action* of a system of N particles is defined as

$$A \equiv \int_{t_1}^{t_2} \sum_j p_j \dot{q}_j \, dt. \tag{3.44}$$

The principle of least action states that, in a system in which H is conserved ($\partial H/\partial t = 0$), the path of the system in configuration space between fixed initial and final points is that path along which A is stationary,

$$\delta A = \delta \int_{t_1}^{t_2} \sum_j p_j \dot{q}_j \, dt = 0. \tag{3.45}$$

Note that the principle states only that A is stationary, i.e., is either minimum or maximum. The principle of least action is misnamed; it should really be called the principle of stationary action but rarely is.

For a single particle moving in a conservative force field, the principle takes the form

$$\delta A = \delta \int_{t_1}^{t_2} mv^2 \, dt = \delta \int_{P_1}^{P_2} mv \, ds \tag{3.46}$$

*For the derivation see any textbook on classical mechanics.

or, finally,

$$\delta \int_{P_1}^{P_2} v \, ds = 0, \tag{3.47}$$

where P_1 and P_2 are the initial and final points of the trajectory in configuration space.

Equation 3.47 is strikingly similar to a principle of geometric optics formulated by Pierre de Fermat. *Fermat's principle* states that the path of a light ray through a medium between fixed initial and final points is that along which the integral of the index of refraction is stationary, or

$$\delta \int_{P_1}^{P_2} n \, ds = 0. \tag{3.48}$$

The index of refraction, n, is generally a function of position along the light ray just as the speed v in Equation 3.47 is generally a function of position along the trajectory.

The close resemblance between the principles of Maupertius and Fermat, one applying to particles and the other to light rays, led Hamilton to attempt to construct a wave theory of matter. Hamilton argued that the particle theory of mechanics might be a special case of a wave theory of mechanics just as the ray theory of geometric optics is a special case of the general wave theory of optics. Hamilton reasoned that the mechanical entity which was analogous to the optical amplitude was the Hamiltonian action S since the trajectories of particles are orthogonal to surfaces of equal action just as light rays are orthogonal to surfaces of equal phase. He wished to find a differential equation, in the form of a wave equation, for S. To understand the procedure he followed, we must examine yet another formulation of classical mechanics.

Hamilton–Jacobi Dynamics

It follows from Hamilton's canonical equations that the generalized momentum corresponding to a cyclic coordinate is a constant of the motion. The method of Hamilton and Jacobi consists in finding a canonical transformation such that all the new coordinates are cyclic. A canonical transformation is defined as a transformation of coordinates and momenta that leaves the canonical equations unchanged in form. The transformation equations will, in general, have the forms

$$\bar{q}_j = \bar{q}_j(q_j, p_j, t),$$

$$\bar{p}_j = \bar{p}_j(q_j, p_j, t), \tag{3.49}$$

$$\bar{H} = H + \frac{\partial S}{\partial t},$$

where the q_j and the p_j are the old coordinates and momenta and the $\bar{q}_j$ and $\bar{p}_j$ are the new coordinates and momenta, and where S is the *generating function* of the transformation. The requirement that this be a canonical transformation will be satisfied if there exists some function $\bar{H} = \bar{H}(\bar{q}_j,\bar{p}_j,t)$ such that the equations of motion in the new phase space are in the canonical form and the coordinates are cyclic. In other words, the function must satisfy the conditions

$$\dot{\bar{q}}_j = \frac{\partial \bar{H}}{\partial \bar{p}_j} = 0$$

and

$$\dot{\bar{p}}_j = -\frac{\partial \bar{H}}{\partial \bar{q}_j} = 0. \tag{3.50}$$

One of the possible ways* of accomplishing this transformation is to find a generating function $S = S(q_j,\bar{p}_j,t)$ such that

$$p_j = \frac{\partial S}{\partial q_j},$$

$$\bar{q}_j = \frac{\partial S}{\partial \bar{p}_j}. \tag{3.51}$$

It is clear from Equations 3.50 that the new variables $\bar{q}_j$ and $\bar{p}_j$ will indeed be constant in time if the new Hamiltonian function $\bar{H}$ is identically zero. This requirement is seen from the third of Equations 3.49 to be equivalent to

$$H\left(q_j,\frac{\partial S}{\partial q_j},t\right) + \frac{\partial S}{\partial t} = 0. \tag{3.52}$$

Equation 3.52 is the *Hamilton–Jacobi partial differential equation* for the unknown generating function S. We will now show that in a conservative system the function S is, in fact, the Hamiltonian action, i.e., the time integral of the Lagrangian function. From Equations 3.51,

$$\frac{dS}{dt} = \sum_j \frac{\partial S}{\partial q_j}\dot{q}_j + \frac{\partial S}{\partial t} \tag{3.53}$$

or

$$dS = \sum_j p_j\dot{q}_j\,dt + \frac{\partial S}{\partial t}\,dt = 2T\,dt + \frac{\partial S}{\partial t}\,dt. \tag{3.54}$$

*See Goldstein's *Classical Mechanics*, Chapter 8.

In a conservative system $H = E =$ a constant, and Equation 3.52 becomes

$$\frac{\partial S}{\partial t} = -E. \tag{3.55}$$

Equation 3.54 can be written

$$dS = (2T - E)\,dt = (T - V)\,dt = L\,dt$$

and, finally,

$$S = \int L\,dt,$$

which shows that in a conservative system the generating function that makes all coordinates cyclic is the Hamiltonian action.

As an example of the use of the Hamilton–Jacobi formalism, consider once again the motion of a point of mass m in a central force field described by a potential, $V(r)$. The Hamiltonian function is given by Equation 3.25,

$$H = \frac{1}{2m}\left[p_r^2 + \frac{p_\theta^2}{r^2} + \frac{p_\phi^2}{r^2 \sin^2 \theta} \right] + V(r).$$

The Hamiltonian function does not contain the time; therefore, the Hamilton–Jacobi partial differential equation (Equation 3.52) becomes, with the aid of Equation 3.55,

$$\left(\frac{\partial S}{\partial r}\right)^2 + \frac{1}{r^2}\left(\frac{\partial S}{\partial \theta}\right)^2 + \frac{1}{r^2 \sin^2 \theta}\left(\frac{\partial S}{\partial \phi}\right)^2 + 2m[V(r) - E] = 0.$$

Assuming separability in the form

$$S = S_r(r) + S_\theta(\theta) + S_\phi(\phi),$$

we obtain three ordinary differential equations,

$$\left(\frac{dS_\phi}{d\phi}\right) = \alpha_\phi, \text{ a constant,}$$

$$\left(\frac{dS_\theta}{d\theta}\right)^2 + \frac{\alpha_\phi^2}{\sin^2 \theta} = \alpha_\theta^2, \text{ a constant,}$$

$$\left(\frac{dS_r}{dr}\right)^2 + \frac{\alpha_\theta^2}{r^2} + 2m[V(r) - E] = 0,$$

which can be written, using Equation 3.51,

$$\frac{dS_\phi}{d\phi} = p_\phi = \alpha_\phi,$$

$$\frac{dS_\theta}{d\theta} = p_\theta = \left(\alpha_\theta^2 - \frac{\alpha_\phi^2}{\sin^2\theta}\right)^{1/2}, \tag{3.56}$$

$$\frac{dS_r}{dr} = p_r = \left\{2m[E - V(r)] - \frac{\alpha_\theta^2}{r^2}\right\}^{1/2}$$

The first of Equations 3.56 expresses the conservation of the component of angular momentum along the polar axis, the second expresses the conservation of total angular momentum, and the third expresses the conservation of energy. By proper choice of the initial conditions, namely, by choosing $p_\phi = \alpha_\phi = 0$, we obtain $p_\theta = \alpha_\theta = a$ constant. If $\tilde{v}_1$ denotes the approach velocity and b denotes the impact parameter, as in Chapter 2, then, since $\alpha_\theta^2 = p_\theta^2 = (m\tilde{v}_1 b)^2 = 2mEb^2$ and since $p_r = m\dot{r}$, the third of Equations 3.56 becomes

$$\dot{r} = \left\{\frac{2}{m}\left[E\left(1 - \frac{b^2}{r^2}\right) - V(r)\right]\right\}^{1/2}.$$

This expression is Equation 2.95, which was derived by invoking the conservation laws of energy and angular momentum.

Hamilton's Wave Equation

Consider the simple case of a single mass point moving with constant total energy E. In Cartesian coordinates, $S = S(x,y,z,t)$ represents a surface in space on which the value of S is a constant at each instant in time t. This surface is moving in space with velocity $u = u(x,y,z,t)$. However, on a surface of constant S,

$$\frac{dS}{dt} = 0 = \frac{\partial S}{\partial t} + \frac{\partial S}{\partial x}\frac{dx}{dt} + \frac{\partial S}{\partial y}\frac{dy}{dt} + \frac{\partial S}{\partial z}\frac{dz}{dt}$$

or

$$\frac{\partial S}{\partial t} + \mathbf{u}\cdot\nabla S = 0, \tag{3.57}$$

where $\mathbf{u}$ is the wave velocity, i.e., the velocity of propagation of surfaces of constant S. But Equation 3.55 states that

$$\frac{\partial S}{\partial t} = -E,$$

and, since

$$\nabla S = \int_{t_1}^{t_2} \nabla L \, dt = \int_{t_1}^{t_2} - \nabla V \, dt = \int_{t_1}^{t_2} m\dot{\mathbf{v}} \, dt = m\mathbf{v}, \qquad (3.58)$$

Equation 3.57 may be written

$E = m\mathbf{u} \cdot \mathbf{v}.$

Both **u** and **v** are perpendicular to the constant-S surfaces; therefore **u** and **v** are parallel and

$$E = mvu. \qquad (3.59)$$

Hence, the wave velocity is

$$u = \frac{E}{mv} = \frac{E}{\sqrt{2m(E - V)}}. \qquad (3.60)$$

If, into the Hamilton–Jacobi equation

$$H + \frac{\partial S}{\partial t} = 0,$$

we substitute

$$H = T + V = \frac{p^2}{2m} + V = \frac{1}{2m} |\nabla S|^2 + V,$$

we obtain

$$\frac{1}{2m} |\nabla S|^2 + V + \frac{\partial S}{\partial t} = 0. \qquad (3.61)$$

or, from Equations 3.55 and 3.60.

$$|\nabla S|^2 = 2m(E - V) = \frac{E^2}{u^2}.$$

Once again replacing E by its equivalent, from Equation 3.55, we obtain

$$|\nabla S|^2 = \frac{1}{u^2} \left(\frac{\partial S}{\partial t} \right)^2. \qquad (3.62)$$

Equation 3.62 is *Hamilton's wave equation.* It certainly does not look much like the wave equation of optics, which, for a single component

$Y(x,y,z,t)$ of the electromagnetic field vector, has the form

$$\nabla^2 Y = \frac{n^2}{c^2} \frac{\partial^2 Y}{\partial t^2}. \tag{3.63}$$

where n is the index of refraction and c is the velocity of light in vacuum. Under the conditions for which ray optics are valid, however, the optical wave equation reduces to a form identical to that of Hamilton's equation. To prove this, assume that Equation 3.63 has a solution of the form

$$Y = Ae^{i\phi}. \tag{3.64}$$

where the amplitude A is either constant or a very slowly varying function of position and where the phase ϕ is given by

$$\phi = 2\pi\left(\frac{s}{\lambda} - vt\right) = 2\pi v\left(\frac{s}{u} - t\right) = 2\pi v\left(\frac{sn}{c} - t\right), \tag{3.65}$$

where v is the frequency of the wave, λ is its wavelength, and s is the length of the optical path. At time $t = 0$, path length s equals 0, and the phase is taken to be zero. Substituting Equation 3.64 into Equation 3.63, we obtain an equation for the phase,

$$i\nabla^2\phi - |\nabla\phi|^2 = i\frac{n^2}{c^2}\frac{\partial^2\phi}{\partial t^2} - \frac{n^2}{c^2}\left(\frac{\partial\phi}{\partial t}\right)^2. \tag{3.66}$$

But $\partial^2\phi/\partial t^2 = 0$ by Equation 3.65, and, if it is assumed as a necessary condition for the validity of ray optics that n is changing very slowly with position so that the rays are nearly parallel, then $\nabla^2\phi = 0$ and Equation 3.66 reduces to

$$|\nabla\phi|^2 = \frac{1}{u^2}\left(\frac{\partial\phi}{\partial t}\right)^2. \tag{3.67}$$

The definition of the index of refraction,

$$n \equiv \frac{c}{u}, \tag{3.68}$$

was used in deriving the preceding equation.

Now Equation 3.67 is identical in form to Hamilton's wave equation for matter (Equation 3.62). But this form of the optical wave equation is only an approximation to the rigorously correct form (Equation 3.63). Is it not possible that Hamilton's wave equation is only an approximation to a rigorously correct wave equation for matter? This question was apparently not asked, at least not aloud, by Hamilton or his followers until de Broglie's time. But there was no real reason to ask this question in the nineteenth century; there were no known natural phenomena which violated classical mechanics, and no one had ever seen evidence that massive particles exhibited interference effects that would require a complete wave equation for their description.

3.2 Schrödinger's Equation

The key that was missing, as has been mentioned, was finally supplied by de Broglie's hypothesis some sixty years after Hamilton formulated his wave mechanics. In the interim several momentous developments had occurred in physics.

Around the turn of the twentieth century, Max Planck introduced the first quantum theory into physics. To derive the experimentally observed electromagnetic spectrum emerging from a heated black box, he had to assume that changes in action could only occur in integral multiples of a very small quantity h, which he determined to be of order 10^{-26} erg-sec.

A few years later, in 1905, Albert Einstein explained the photoelectric effect by endowing electromagnetic waves with a particle character. Specifically, he assumed that the momentum carried by an electromagnetic photon was given by

$$\text{Momentum} = \frac{h\nu}{c} = \frac{h}{\lambda}, \tag{3.69}$$

where h is Planck's constant.

Rutherford advanced his nuclear model of the atom in 1911. This model pictured the hydrogen atom as a single negative electron orbiting around a positively charged, much heavier nucleus. Classical electromagnetic theory predicts that in such a case the electron would continually radiate energy and very shortly spiral into the nucleus. This does not, in fact, happen; atoms enjoy long lives.

In 1913 Niels Bohr developed a semiclassical, semiquantum theory of the hydrogen atom that successfully explained some of the major features of the optical spectrum of the hydrogen atom. Bohr formulated the following postulates:

1. Atomic systems exist in certain stationary states of definite total energy E_j.

2. Transitions between the stationary states are accompanied by the absorption or emission of energy. The frequency of the emitted radiation is given by a form of Einstein's photoelectric relation,

$$\nu = \frac{E_2 - E_1}{h},$$

where E_1 and E_2 are the energies of the initial and final stationary states involved in a particular electron transition.

3. The stationary states are given by the condition

$$2\pi r(mv) = nh,$$

where mv is the momentum of the electron, r is the radius of its orbit, and n takes on the successive integer values $1, 2, 3, \ldots$.

In 1915 Wilson and Sommerfeld announced a new method of quantization, a generalization of Bohr's third postulate, obtained by equating the Maupertusian action of the orbital electron to an integral multiple of Planck's constant,

$$\int p \, dq = nh,$$

where $n = 1, 2, 3, \ldots$. This reduces to Bohr's third postulate for an electron in a circular orbit.

In Bohr's theory the atomic electron is pictured as moving in a stationary orbit (corresponding to a definite energy) in which the centrifugal force on the electron is balanced by the electrostatic attraction of the nucleus,

$$\frac{mv^2}{r} = \frac{e^2}{r^2}.$$

In the emission or absorption of energy, the atomic electron moves into a different circular orbit that is, again, a stationary state of lesser or greater energy.

Bohr's theory, now commonly referred to as the "old" quantum theory, quantitatively explains the frequency spectrum of hydrogen and hydrogen-like ions, such as He^+ and Li^{2+}. His theory does not, however, explain the entire spectrum of the next most complicated atom, helium. Furthermore the theory affords no means of predicting the intensities and polarizations of the spectrum lines.

Bohr's theory, which suffers from many defects, most of which were recognized and displayed by Bohr himself, was nevertheless an important stepping stone to the "new" quantum mechanics. Indeed, the new quantum mechanics was developed primarily in an effort to remove the deficiencies of the old quantum theory.

Light, which had been considered to be a form of wave motion ever since Young's interference experiments in the early 1800s, began to take on a particle aspect with Einstein's explanation of the photoelectric-effect in 1905. It only remained for de Broglie to complete the wave–particle symmetry of nature by endowing particles with wavelike characteristics.

De Broglie merely applied Einstein's relation,

$$\text{Momentum} = mv = \frac{h}{\lambda},$$ (3.70)

to particles and solved for λ; thus

$$\lambda = \frac{h}{mv}.$$ (3.71)

He next speculated that a material particle with momentum mv would act as a wave of wavelength $\lambda = h/mv$. The algebra was trivial, but the speculation profound.

Derivation of the Schrödinger Equation

A few years after de Broglie's hypothesis, Schrödinger completed the job Hamilton had begun over half a century before. Schrödinger noted that the generating function S in Hamilton's wave equation has the dimensions of an action, whereas the optical phase ϕ is a pure number. Schrödinger speculated that the connection between S and ϕ might be Planck's constant, which has the dimensions of action. For a conservative system therefore,

$$\phi = \frac{2\pi}{h} S = \frac{2\pi}{h}(S^* - Et),$$ (3.72)

where S^*, the reduced action function, is independent of time t. The 2π is required because E/h is the frequency v from Einstein's equation. Schrödinger next made the bold assumption that the behavior of particles can also be described by the wave equation

$$\nabla^2 \Psi = \frac{1}{u^2} \frac{\partial^2 \Psi}{\partial t^2}.$$ (3.73)

He assumed a solution in the form

$$\Psi(x,y,z,t) = Ae^{i\phi} = A \exp\left(\frac{2\pi i S^*}{h}\right) \exp\left(-\frac{2\pi i Et}{h}\right)$$ (3.74)

or

$$\Psi(x,y,z,t) = \psi(x,y,z) \exp\left(-\frac{2\pi i Et}{h}\right),$$ (3.75)

where $\psi(x,y,z)$ is the amplitude function. When Equation 3.75 is substituted into Equation 3.73,

$$\nabla^2 \psi = -\frac{4\pi^2 E^2}{h^2 u^2}\,\psi.$$

But for matter waves the wave velocity is given by Equation 3.60 in the form

$$u = \frac{E}{\sqrt{2m(E - V)}},$$

hence

$$\nabla^2 \psi + \frac{8\pi^2 m}{h^2}(E - V)\psi = 0. \tag{3.76}$$

This is the famous *Schrödinger equation* for the amplitude of a matter wave. What has been presented here was not the exact procedure followed by Schrödinger in his original derivation of this equation. There are at least a dozen distinct ways of arriving at the Schrödinger equation, but all involve, at one point or another, an act of pure creation. By "act of pure creation" we mean a more or less remarkable assumption that can only be justified ultimately by the fact that the resultant equation, in this case the Schrödinger equation, predicts the outcome of experiments with great precision. Acts of pure creation, inspired guesses, are not new in science. It can be argued, for example, that Newton's second law of motion is just such an act. The doubtful reader is invited to try to derive $F = ma$ from something more basic that is not in itself an act of pure creation or that is not ultimately traceable to such an act.

Second Derivation of the Schrödinger Equation
We derived Schrödinger's equation by applying Einstein's equation $E = h\nu$ to define the frequency of a matter wave in terms of its total energy. It is equally easy to derive the Schrödinger equation in one dimension from the de Broglie relation $\lambda = h/p$. The classical wave equation in one dimension is

$$\frac{\partial^2 Y}{\partial x^2} = \frac{1}{u^2}\frac{\partial^2 Y}{\partial t^2}, \tag{3.77}$$

where $Y = Y(x,t)$. Assume a solution of the form

$$Y = \psi(x)e^{-i\omega t} \tag{3.78}$$

and substitute it into Equation 3.77 to obtain

$$\frac{d^2\psi}{dx^2} + \frac{\omega^2}{u^2}\psi = 0. \tag{3.79}$$

But $u = \lambda v$ and $\omega = 2\pi v$; so Equation 3.79 may be written

$$\frac{d^2\psi}{dx^2} + \frac{4\pi^2}{\lambda^2}\psi = 0. \tag{3.80}$$

Substituting the de Broglie relation $\lambda = h/p$ into Equation 3.80 results in

$$\frac{d^2\psi}{dx^2} + \frac{4\pi^2 p^2}{h^2}\psi = 0. \tag{3.81}$$

But $p^2 = (mv)^2 = 2mT = 2m(E - V)$; so Equation 3.81 becomes

$$\frac{d^2\psi}{dx^2} + \frac{8\pi^2 m}{h^2}(E - V)\psi = 0. \tag{3.82}$$

Equation 3.82 is Schrödinger's amplitude equation in one dimension. The extension to two or three dimensions is straightforward.

To avoid use of the Greek letter π, we define the quantity

$$\hbar \equiv \frac{h}{2\pi} \tag{3.83}$$

Equation 3.82, after rearrangement, becomes

$$\left(-\frac{\hbar^2}{2m}\frac{d^2}{dx^2} + V\right)\psi = E\psi. \tag{3.84}$$

If this equation is compared with the equation that expresses conservation of energy in Hamiltonian form, $H = E$ or

$$\frac{p^2}{2m} + V = E,$$

and if we now consider p^2 and V to be operating on a wave function ψ, we obtain

$$\left(\frac{p^2}{2m} + V\right)\psi = E\psi. \tag{3.85}$$

After comparing Equations 3.84 and 3.85, we conclude that the classical momentum p must be replaced in one-dimensional wave mechanics by the operator $-i\hbar(d/dx)$. In three dimensions, by similar argument, we obtain the operators

$$p_x \rightarrow -i\hbar \frac{\partial}{\partial x},$$

$$p_y \rightarrow -i\hbar \frac{\partial}{\partial y}, \tag{3.86}$$

$$p_z \rightarrow -i\hbar \frac{\partial}{\partial z}.$$

The minus sign is used in the preceding equation for reasons which will become apparent subsequently. Equation 3.76 is the time-independent Schrödinger equation in three dimensions. To obtain the time-dependent equation, we eliminate E between Equations 3.75 and 3.76. From Equation 3.75,

$$\frac{\partial \Psi}{\partial t} = -\frac{iE}{\hbar} \psi(x,y,z) \exp\left(-\frac{iEt}{\hbar}\right) \tag{3.87}$$

or

$$E\Psi = i\hbar \frac{\partial \Psi}{\partial t}. \tag{3.88}$$

Substituting Equations 3.75 and 3.88 into Equation 3.76, we obtain

$$\left(-\frac{\hbar^2}{2m}\nabla^2 + V\right)\Psi = i\hbar \frac{\partial}{\partial t}\Psi. \tag{3.89}$$

Equation 3.89 is the time-dependent Schrödinger equation in three dimensions, an equation with which we shall be spending a great deal of time throughout subsequent sections of this book.

Equation 3.88 suggests that the classical total energy E is to be replaced in wave mechanics by the differential operator $i\hbar(\partial/\partial t)$,

$$E \rightarrow i\hbar \frac{\partial}{\partial t}. \tag{3.90}$$

Formal Derivation of the Schrödinger Equation
Once the operators corresponding to the total energy, E, and the components of momenta, p_x, p_y, and p_z, have been identified, a formal derivation of the Schrödinger equation is almost trivial. Since $H = E$, it follows that

$$H\Psi = E\Psi$$

or

$$\left(\frac{p_x^2}{2m} + \frac{p_y^2}{2m} + \frac{p_z^2}{2m} + V(x,y,z) \right) \Psi = E\Psi. \tag{3.91}$$

Substituting the energy and momentum operators from Equations 3.90 and 3.86 into the above equation, we obtain the Schrödinger equation,

$$\left(-\frac{\hbar^2}{2m} \nabla^2 + V \right) \Psi = i\hbar \frac{\partial}{\partial t} \Psi. \tag{3.92}$$

It may prove interesting and helpful to note the formal similarity between the derivation of the Hamilton–Jacobi equation and the Schrödinger equation. The steps in the derivation of both equations are shown below:

Hamilton–Jacobi equation	*Schrödinger equation*
$[H(q_j, p_j) - E] = 0$	$[H(q_j, p_j) - E]\Psi = 0$
Replace p_j by $\dfrac{\partial S}{\partial q_j}$	Replace p_j by $\dfrac{\hbar}{i} \dfrac{\partial}{\partial q_j}$
Replace E by $-\dfrac{\partial S}{\partial t}$	Replace E by $-\dfrac{\hbar}{i} \dfrac{\partial}{\partial t}$

3.3 Postulates of Wave Mechanics

Over the course of the last three centuries, classical mechanics has provided answers to a wide range of physical problems from those in astronomy to those in gas dynamics. For atomic and subatomic phenomena, however, classical mechanics has proved to be inadequate. For these phenomena, quantum mechanics provides the answers. Since classical mechanics has demonstrated its validity when applied to large systems, we reason that, as quantum mechanics is applied to systems whose dimensions become larger and larger, the yielded behavioral predictions must correspond to classical predictions. This useful and quite reasonable statement is called the *correspondence principle*. In the postulates to follow, the correspondence principle is carried along as an unvoiced auxiliary postulate.

Convenient and instructive working rules of wave mechanics can be stated in the form of eight postulates. The choice of eight as opposed to any other small number of working postulates is quite arbitrary. This particular number was chosen to facilitate the ease with which the engineering student can begin to apply wave mechanics as a tool. These postulates are also intended to reveal the empirical restrictions on the line of reasoning used throughout the application of quantum mechanics.

Postulate 1: To each system there corresponds a function $\Psi(q_j,t)$, which is a function of the coordinates and, possibly, time. The function Ψ, which is called the *wave function*, the *Schrödinger function*, or simply the *Ψ-function* of the system, contains all the information that can be obtained about the system. The Ψ-function for a one-particle system moving in one dimension will be a function of one spatial coordinate; for a two-particle system moving in three dimensions, it will be a function of six spatial coordinates, three for each particle, etc.

Mindful that no high degree of rigor is presented or intended, we can proceed directly into the more or less heuristic postulations that provide rules for calculating Ψ for any particular system of interest.

Postulate 2: To every classical observable (dynamical variable), there corresponds an operator. The operators corresponding to coordinates, linear momenta, and total energy are as follows:

Observable	*Operator*
q_j	q_j
p_j	$-i\hbar\dfrac{\partial}{\partial q_j}$
E	$i\hbar\dfrac{\partial}{\partial t}$

The rule governing the momentum operators, namely,

$$p_j = -i\hbar\frac{\partial}{\partial q_j}, \tag{3.93}$$

applies only in Cartesian coordinates. The momentum operators in any other coordinate system can be obtained by transforming the Cartesian momentum operators. In plane polar coordinates, for example, we find $p_r = -i\hbar[(\partial/\partial r) + (1/r)]$ and $p_\theta = -i\hbar(\partial/\partial\theta)$.

Once a postulate has been laid down, the mathematical implications arising from it also present themselves as "rules of the game." As will be seen, the correspondence principle strongly supports the contention that quantum-mechanical operators must be taken to obey the *commutation rule*

$$[A,B] = i\hbar(a,b), \tag{3.94}$$

where $[A,B]$ is the quantum-mechanical *commutator* of the operators A and B, defined by

$$[A,B] \equiv AB - BA, \tag{3.95}$$

and where (a,b) is the Poisson bracket (see Section 3.1) of the corresponding classical observables a and b. If $[A,B] = 0$, operators A and B are said to commute.

Postulate 3: The wave function Ψ of a system is the solution to the Schrödinger equation for the system, $H\Psi = E\Psi$, where H, the Hamiltonian of the system, and E, the total energy of the system, have been converted by substituting operators for dynamical variables.

Postulate 4: To be physically acceptable, a wave function Ψ must have the following properties:

(a) The quantities $\Psi(q_j,t)$ and $\partial\Psi(q_j,t)/\partial q_j$ must be *continuous, finite*, and *single-valued* throughout configuration space, which may be of finite or infinite extent. The requirement that Ψ be single valued is actually too stringent for the complete development of quantum theory. Nevertheless, it will prove convenient to accept it for the time being and to modify it later as required. Eventually (in Chapter 7) this postulate will be replaced by the physically more reasonable and less stringent requirement that observables be single valued. There is one other exception to this postulate, namely, $\partial\Psi/\partial q$ is not necessarily continuous at an infinite potential step.

(b) The function Ψ, which may be complex, must be quadratically integrable over this configuration space; that is, $\int_{\text{all space}} \bar{\Psi}\Psi \, d\tau$ must be finite, where $d\tau$ is the element of volume in configuration space and $\bar{\Psi}$ is the complex conjugate of Ψ. As will be seen, the numerical value of the wave function Ψ will usually be adjusted so that the preceding integral has value unity.

Postulate 5: The average value of any dynamical variable ω, which corresponds to an operator Ω, is calculated from the Ψ-function of the system by the rule

$$\langle \omega \rangle = \int \bar{\Psi}\Omega\Psi \, d\tau, \tag{3.96}$$

where the integral extends over all configuration space. The symbol $\langle \ \rangle$ means the average value of the variable it encloses. This average value will also be referred to as the *expectation value* of the observable. In particular, the expectation value of the x coordinate of a particle would be given by

$$\langle x \rangle = \int_{-\infty}^{\infty} \bar{\Psi}x\Psi \, d\tau = \int_{-\infty}^{\infty} x\bar{\Psi}\Psi \, d\tau, \tag{3.97}$$

which indicates that the product $\bar{\Psi}\Psi$ is the probability density function for position. That is, $\bar{\Psi}\Psi \, d\tau$ is the probability of finding the particles in volume element $d\tau$ about position τ in configuration space. But, if $\bar{\Psi}\Psi$

is to be a probability density, its integral over all space must be unity; thus we arrive at the conclusion that wave functions must be normalized so that

$$\int \bar{\Psi}\Psi \, d\tau = 1. \tag{3.98}$$

This normalization requirement sometimes limits the possible extent of the representative configuration space; i.e., if the integral cannot be normalized over an infinite region, a finite region of integration must be imposed. This finite region can be chosen sufficiently large, however, that the resulting solutions are valid to within any desired accuracy at interior points far from the boundaries of the region.

Postulate 6: The only possible values which the measurement of an observable ω can yield are the *eigenvalues* ω_i of the equation

$$\Omega\Psi_i = \omega_i\Psi_i, \tag{3.99}$$

where Ω is the operator corresponding to ω and Ψ_i is a solution of Schrödinger's equation for the system under consideration.

A few words of explanation of the term "eigenvalue" are in order. The discussion will be restricted to one independent variable; its extension to more than one variable is quite straightforward. An equation of the form

$$(\text{Operator}) \, u(x) = (\text{constant}) \, u(x) \tag{3.100}$$

together with the boundary conditions which $u(x)$ must satisfy is called an *eigenvalue equation*. "Eigen" is a German adjective meaning "proper." For a given operator Ω, there may be an infinite number of functions that satisfy Equation 3.100. To each of these functions $u_i(x)$, there will correspond a constant ω_i such that

$$\Omega u_i(x) = \omega_i u_i(x). \tag{3.101}$$

The term $u_i(x)$ is called an *eigenfunction* of the operator Ω, and ω_i is called the eigenvalue of the operator Ω which belongs to the eigenfunction $u_i(x)$.

The problem with which one is usually faced is: Given an operator Ω and the conditions which its eigenfunctions must fulfill, find all possible eigenfunctions and their associated eigenvalues. As an example, given the differential operator d/dx and the conditions that all eigenfunctions must be real and must be bounded as $x \rightarrow \infty$, find the eigenfunctions and associated eigenvalues. The eigenvalue equation is

$$\frac{d}{dx} u(x) = \omega u(x). \tag{3.102}$$

The function $u_1(x) = c$, a real constant, is an obvious eigenfunction with eigenvalue zero, as can be verified by direct substitution into Equation

3.102. If $u_2(x) = Ae^{\alpha x}$ is tried in Equation 3.102, the result is

$$\frac{d}{dx}(Ae^{\alpha x}) = \alpha(Ae^{\alpha x}).$$

Thus $Ae^{\alpha x}$ seems to be an eigenfunction of the operator d/dx with eigenvalue α. This statement is not true if A is complex or if α is either positive or complex because the conditions of the problem state that $u(x)$ must be real and must be bounded as $x \rightarrow +\infty$. Thus α must be a negative real number. Are there any other eigenfunctions? If not, the complete *eigenfunction spectrum* and associated eigenvalue spectrum of the operator d/dx subject to the given conditions is:

	Eigenfunction	*Corresponding Eigenvalue*
$u_1 =$	any real constant c	0
$u_2 =$	$Ae^{\alpha x}$ where $\alpha < 0$	α

where A and α are real constants. In this case, we obtain a continuous spectrum of eigenfunctions and eigenvalues. There is, in fact, a doubly infinite set since for any value of α the coefficient A can have any arbitrary real value. If, however, we demanded that $u(x)$ be normalized in some fashion, A would be uniquely determined for each value of α.

Some of the wave-mechanical operators that will be encountered have continuous spectra; others have a finite number of eigenfunctions and associated eigenvalues. However, all wave-mechanical operators will be found to have real eigenvalues, as might be expected since eigenvalues represent real, physically measurable quantities.

The six postulates stated above are sufficient for the wave-mechanical analysis of any system. They are not necessary, however, in the sense of being the only possible postulates on which wave mechanics can be built. Just as there are numerous possible formulations of classical mechanics, there are numerous possible formulations of quantum mechanics.

Two other useful working rules of quantum mechanics will now be given. These rules are presented as postulates only in the interest of brevity. Their derivation can be found in most books on quantum mechanics.* A steady-state quantum-mechanical system (e.g., an atom or nucleus) is completely characterized by its eigenfunctions ψ_i. Rules governing the transition from one state, ψ_j, to another state, ψ_i, under the influence of a perturbing force are given by Postulates 7 and 8.

---

*See, for example, Leonard I. Schiff's *Quantum Mechanics*, McGraw-Hill Book Co., Inc., New York, 1949.

Postulate 7: The probability per unit time of a *direct transition* from state j to state i under the influence of a perturbing force is given by

$$\Lambda = \frac{2\pi}{\hbar} |H'_{ij}|^2 \frac{dn}{dE}, \tag{3.103}$$

where H'_{ij}, the so-called matrix element of the perturbation that produces the transition, is given by

$$H'_{ij} \equiv \int \bar{\Psi}_i H' \Psi_j \, d\tau. \tag{3.104}$$

In this expression H' is the Hamiltonian function of the perturbation, and Ψ_i and Ψ_j are eigenfunctions of the unperturbed Hamiltonian H_0; that is:

$$H_0 \Psi_i = E_i \Psi_i,$$

$$H_0 \Psi_j = E_j \Psi_j.$$

Equation 3.103 applies to a system in initial state Ψ_j that is perturbed by a force described by the Hamiltonian H', which is much smaller than H_0. As a result the system suffers a transition to state Ψ_i with accompanying emission or absorption of energy of amount $|E_j - E_i|$. The quantity dn/dE represents the energy density of final states, i.e., the number of final states per unit energy interval.

Postulate 8: The probability per unit time of an *indirect transition* from state j to state i by way of state k is given by

$$\Lambda = \frac{2\pi}{\hbar} \left| \frac{H'_{kj} H'_{ik}}{E_j - E_i} \right|^2 \frac{dn}{dE}, \tag{3.105}$$

where the matrix elements are defined as

$$H'_{kj} \equiv \int \bar{\Psi}_k H' \Psi_j \, d\tau$$

and

$$H'_{ik} \equiv \int \bar{\Psi}_i H' \Psi_k \, d\tau$$

with Ψ_k representing the unperturbed wave function of the intermediate state.

3.4 Heisenberg's Principle

In 1927 Werner Heisenberg* announced a rather startling conclusion from his early research in quantum mechanics. He pointed out that,

*W. Heisenberg, *Z. Physik.* **43**: 172 (1927).

contrary to all laws of classical mechanics, two canonically conjugate variables could not be measured simultaneously to any desired degree of accuracy.

Heisenberg's uncertainty principle in its most general form may be stated as follows: Given two dynamical variables a and b whose corresponding quantum-mechanical operators A and B do not commute, we cannot devise an experiment for the measurement of the variables a and b that would yield their simultaneously exact values. If a series of identical experiments are performed, the values of a and b found in the experiments will differ from experiment to experiment in such a way that the products of the standard deviations in a and b, namely, Δa and Δb, will always be greater than or equal to $\hbar/2$ no matter how many experiments are performed. If $AB \neq BA$, then

$$\Delta a\, \Delta b \geq \frac{\hbar}{2}. \tag{3.106}$$

If, on the other hand, the variables are not canonically conjugate, their operators do commute, and the variables can be measured simultaneously with absolute precision. The standard deviations in such a case may be expected to continue to decrease toward zero as the number of identical experiments performed increases toward infinity. That is, if $AB = BA$, then

$$\Delta a\, \Delta b \geq 0. \tag{3.107}$$

The inequality expressed in Equation 3.106 holds for all canonically conjugate variables. In particular,

$$\left.\begin{array}{l} \Delta x\, \Delta p_x \\ \Delta y\, \Delta p_y \\ \Delta z\, \Delta p_z \\ \Delta E\, \Delta t \\ \Delta \theta\, \Delta L \end{array}\right\} \geq \frac{\hbar}{2}, \tag{3.108}$$

where p_x, p_y, and p_z are the momenta conjugate to x, y, and z; E is the total energy; t is the time; and L is the angular momentum component conjugate to θ.

Heisenberg's inequalities can be derived mathematically from wave-mechanical considerations or by matrix methods. There are in existence certain simple gedanken experiments that help to illustrate Heisenberg's principle.*

Heisenberg's relationships are often written in the form

$$\Delta a \, \Delta b \gtrsim \hbar, \tag{3.109}$$

meaning that the product of the uncertainties in two canonically conjugate variables is greater than or equal to a number of the order $\hbar$. This sentence is vague only because the word "uncertainties" is vague. Once the exact statistical measure that will be used to define the "uncertainties" has been decided, a very definite inequality can be written. If, for example, the uncertainties are defined as standard deviations,† then $\Delta a \, \Delta b \geq \hbar/2$. If the uncertainties are defined by some other statistical measure, a different number will appear on the right-hand side, but certainly it will be a number of the order of $\hbar$, a number such as $\hbar/4$, $\hbar/\pi$, $2\hbar$, etc. There is no uncertainty in the number that appears on the right-hand side of Heisenberg's inequalities once the quantities on the left-hand side have been defined.

The uncertainty principle is a statistical statement; that is, it is a statement about the outcome of a very large number of experiments. It says nothing specific about the results of a single experiment. To illustrate: Suppose a simple experiment is set up to measure the position and momentum of an electron. The electron is accelerated through a fixed potential so that it travels with known total energy in a straight line, let us say along the z-axis. There is an apparatus in the path of the electron to simultaneously measure its position, z, and conjugate momentum, p_z. We could, for example, detect its interactions in a cloud chamber and so measure its position and momentum simultaneously. Until now, there seem to be no complications in the procedure. The measurements are recorded as z_1 and p_1, and the experiment is repeated. Another electron is accelerated through the same potential, and a different set of values, z_2 and p_2, are measured. The experiment is repeated many times, and each

---

*See W. Heisenberg, *The Physical Principles of the Quantum Theory* (The University of Chicago Press, Chicago, 1930), pp. 20–35.

† The standard, or root-mean-square, deviation of a statistical variable q is defined as

$$\Delta q \equiv \sqrt{\langle (q - \langle q \rangle)^2 \rangle}.$$

but, since

$$\langle (q - \langle q \rangle)^2 \rangle = \langle q^2 - 2\langle q \rangle q + \langle q \rangle^2 \rangle = \langle q^2 - \langle q \rangle^2 \rangle = \langle q^2 \rangle - \langle q \rangle^2,$$

the root-mean-square deviation is the square root of the average of the square of q minus the square of the average of q,

$$\Delta q = \sqrt{\langle q^2 \rangle - \langle q \rangle^2}.$$

time the results are tabulated and the standard deviations in z and p_z are calculated. It is found that the standard deviations do not approach zero but approach values such that Heisenberg's inequality ($\Delta z\, \Delta p_z \geq \hbar/2$) is satisfied.

Might not the standard deviation in z be reduced by making the cloud chamber extremely narrow? When the experiment is repeated many times with the narrower chamber, it is found that Δp_z has increased in inverse proportion to Δz so that once again $\Delta z\, \Delta p_z \geq \hbar/2$. These results suggest a drastic alternative procedure, namely, to fix z. A tiny paddle wheel is set up at a fixed position along the path of the electron, and the momentum of the electron is determined by the impulse it gives to the paddle. Now, the standard deviation in the position is zero. It is discovered that after a large number of measurements a uniform spread in momenta from zero to infinity is obtained so that $\Delta p_z \to \infty$. We can conclude that Heisenberg's principle is indeed an empirical truth.

The result of this little experiment suggests a question that would perhaps have seemed ridiculous had it been posed before 1927: Does an electron, or indeed any other particle, have a simultaneous position and conjugate momentum? The experiments outlined above seem to indicate that the answer cannot be a simple "yes" or "no," only a qualified "yes and no." The qualification is contingent on the temporal sense in which the question is posed. The answer is positive if the question is interpreted in an after-the-fact sense; that is, an experiment can indeed be set up which will determine what the position and momentum of the particle were. The answer is negative if the question is interpreted in a before-the-fact sense; there is no known method of predicting what the simultaneous position and momentum will be before the measurement is taken.

From a practical viewpoint, since we have no way of predicting the simultaneous exact values of conjugate variables describing a system, these exact values do not exist simultaneously. If they did exist simultaneously, they could be measured exactly. This viewpoint may be expressed succinctly by the fundamental postulate of the operational philosophy: Those quantities, properties, or attributes that cannot be measured may be considered, for all practical purposes, to be nonexistent.

Excellent discussions on the uncertainty principle and its philosophical consequences may be found in many books.*

---

*See, for example, Henry Margenau's *The Nature of Physical Reality* (Reference 3 at the end of this chapter); David Bohm's *Quantum Theory* (Prentice-Hall, Inc., Englewood Cliffs, N. J., 1951); or A. D'Abro's *The Rise of the New Physics* (Reference 1 at the end of this chapter).

3.5 One-Dimensional Wave Mechanics

In the remainder of this chapter some of the major consequences of wave mechanics will be considered. In particular, we will show that wave mechanics predicts two phenomena that are incomprehensible from the viewpoint of classical mechanics. These two classical impossibilities are the penetration of potential barriers and the existence of discrete energy states in potential wells. To simplify the mathematical details so attention may be more effectively concentrated on the physical content of the theory, we will restrict discussion to a single particle moving in one dimension. All the results from one dimension have analogies in two and three dimensions. Before we discuss wave mechanics proper, we will briefly review the nomenclature used to describe waves in one dimension.

Waves in One Dimension

If a complex variable y depends on the spatial coordinate x and the time t through an equation of the form

$$y(x,t) = A' \exp\left[2\pi i(\kappa x - vt)\right], \tag{3.110}$$

where κ is a real constant, v is a real positive constant, and A' is a constant which may be complex, then y is called a *monochromatic wave*. If A' is expressed in the form of a real, positive number times a complex phase factor,

$$A' = Ae^{i\phi_0}, \tag{3.111}$$

then y takes the form

$$y(x,t) = A \exp\left[2\pi i(\kappa x - vt + \phi_0/2\pi)\right] = Ae^{i\phi}, \tag{3.112}$$

where the last step in deriving Equation 3.112 follows simply by definition of the phase ϕ. In the nomenclature usually employed to describe waves, κ is called the propagation constant of the wave, v is the frequency, A is the amplitude, ϕ is the phase, and ϕ_0 is the initial phase. Other related constants sometimes employed are $|\kappa|$, which is called the wave number of the wave; $\lambda = 1/|\kappa|$, the wavelength; $T = 1/v$, the period; and $\omega = 2\pi v$, the angular velocity.

Consider now either the real or the imaginary components of $y(x,t)$ designated, respectively, by $u(x,t)$ and $w(x,t)$ with

$$u(x,t) = A \cos 2\pi(\kappa x - vt + \phi_0/2\pi),$$

$$w(x,t) = A \sin 2\pi(\kappa x - vt + \phi_0/2\pi).$$

Once the constant A is fixed, the numerical value of u or of w at a given point x and a given time t is determined by the value of the phase,

$$\phi = 2\pi(\kappa x - vt + \phi_0/2\pi). \tag{3.113}$$

Hence as t increases, any preassigned value of y drifts along the x-axis, as shown in Figure 3.1. The solid line is a graph of u at time t; the dashed line, a graph of u at time $t + dt$. The preassigned value of u has drifted from a to b, a distance dx, in time dt. To determine the velocity dx/dt of this drift, we note that the phases at a and b must be identical since u at a equals u at b. Therefore,

$$\phi(x + dx, t + dt) = \phi(x,t)$$

or, from Equation 3.113,

$$\kappa(x + dx) - v(t + dt) = \kappa x - vt,$$

which, when solved for dx/dt, yields

$$\frac{dx}{dt} = \frac{v}{\kappa}. \tag{3.114}$$

The velocity given by Equation 3.114 is called the phase velocity, v_ϕ, of the wave. Note that the algebraic sign of κ determines the sign of v_ϕ and, thus, the direction of motion of the wave along the x-axis.

Consider now the sum of two waves of slightly different frequency and slightly different propagation constants. For simplicity, assume that both their initial phases are zero and both their amplitudes are unity.

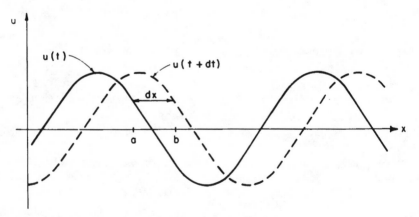

Figure 3.1—One-dimensional wave at two instants in time.

The sum of the two waves will be given by

$$y(x,t) = \exp\left[2\pi i(\kappa x - vt)\right] + \exp\left[2\pi i(\kappa'x - v't)\right]$$

$$= \left\{\exp\left[2\pi i\left(\frac{\kappa - \kappa'}{2}x - \frac{v - v'}{2}t\right)\right]\right.$$

$$+ \exp\left[-2\pi i\left(\frac{\kappa - \kappa'}{2}x - \frac{v - v'}{2}t\right)\right]\right\}$$

$$\times \left\{\exp\left[2\pi i\left(\frac{\kappa + \kappa'}{2}x - \frac{v + v'}{2}t\right)\right]\right\}$$

$$= 2\cos 2\pi\left(\frac{\kappa - \kappa'}{2}x - \frac{v - v'}{2}t\right)$$

$$\times \left\{\exp\left[2\pi i\left(\frac{\kappa + \kappa'}{2}x - \frac{v + v'}{2}t\right)\right]\right\}.$$

The composite wave, which is the sum of the two waves, has an amplitude $2\cos 2\pi\{[(\kappa - \kappa')/2]x - [(v - v')/2]t\}$, which varies very slowly with time since $v - v'$ is very small. The phase velocity of the composite wave is, using Equation 3.114, $(v + v')/(\kappa + \kappa')$, which is essentially the same as the phase velocity of either of the waves composing it since $v \simeq v'$ and $\kappa \simeq \kappa'$. Thus the new wave is essentially a monochromatic wave modulated by the amplitude. A graph of either the real or imaginary part of the composite wave at two instants in time is shown in Figure 3.2. Note that the composite wave consists at any instant in time of a series of groups of ripples and that as time goes on these groups maintain their shapes but move along the x-axis. The velocity with which a group moves is called the group velocity, v_g. The group velocity is the velocity of points of equal amplitude in the composite wave. The amplitudes will be equal when

$$\frac{\kappa - \kappa'}{2}x - \frac{v - v'}{2}t = \frac{\kappa - \kappa'}{2}(x + dx) - \frac{v - v'}{2}(t + dt)$$

or when

$$\left(\frac{dx}{dt}\right)_g = \frac{v - v'}{\kappa - \kappa'} = v_g.$$

In the limit as $v - v'$ and $\kappa - \kappa'$ both approach zero,

$$v_g = \frac{dv}{d\kappa}. \tag{3.115}$$

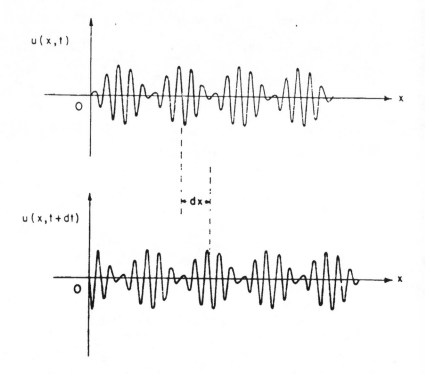

Figure 3.2—Composite wave at two instants in time.

The phase velocity and the group velocity have the following relation: Since $v_\phi = \nu\lambda$ and $\lambda = 1/\kappa$, we know that $\nu = \kappa v_\phi$. Now by Equation 3.115,

$$v_g = \frac{d\nu}{d\kappa} = \frac{d(\kappa v_\phi)}{d\kappa} = v_\phi + \kappa\frac{dv_\phi}{d\kappa} = v_\phi + \frac{1}{\lambda}\frac{dv_\phi}{d\lambda}\frac{d\lambda}{d\kappa},$$

and, therefore,

$$v_g = v_\phi - \lambda\frac{dv_\phi}{d\lambda}. \qquad (3.116)$$

The optical index of refraction of a transparent medium is defined in terms of the phase velocity of the medium by the equation $v_\phi = c/n$. Thus Equation 3.116 can be written

$$v_g = v_\phi - \lambda c\frac{d(1/n)}{d\lambda} = v_\phi + \frac{\lambda c}{n^2}\frac{dn}{d\lambda}.$$

In a vacuum or other nondispersive medium, the index of refraction is not a function of λ, and therefore $v_g = v_\phi$. In most transparent media, n decreases with increasing λ so that $dn/d\lambda < 0$ and $v_g < v_\phi$.

The Schrödinger Equation for One Particle in One Dimension
Let us assume a particle of mass m is moving along the x-axis under the influence of a force derivable from a potential, $V(x)$.

The first postulate states that this system can be described by a Schrödinger function Ψ that is a function of the coordinate x and the time t; that is, $\Psi = \Psi(x,t)$.

The second postulate states that the position, momentum, and total energy of this particle may be represented by the following operators:

$$x \to x,$$

$$p \to -i\hbar \frac{\partial}{\partial x},$$

$$E \to i\hbar \frac{\partial}{\partial t}.$$

Since the potential energy is a function only of the position x, it will represent itself, $V(x) \to V(x)$.

The third postulate states that Ψ is a solution of the operator equation

$$H\Psi = E\Psi$$

or

$$\left[\frac{p^2}{2m} + V(x) \right] \Psi = E\Psi.$$

This equation becomes, after substitution of the proper operators for p and E,

$$-\frac{\hbar^2}{2m} \frac{\partial^2 \Psi(x, t)}{\partial x^2} + V(x)\Psi(x, t) = i\hbar \frac{\partial \Psi(x, t)}{\partial t}. \tag{3.117}$$

This partial differential equation is the time-dependent Schrödinger equation for the system. The method of separation of variables is used to solve this equation. We assume that the solution may be written in the form of a product of a space-dependent function $\psi(x)$ and a time-dependent function $\alpha(t)$,

$$\Psi(x, t) = \psi(x)\alpha(t). \tag{3.118}$$

When this expression for Ψ is substituted into Equation 3.117,

$$\frac{1}{\psi(x)} \left[-\frac{\hbar^2}{2m} \frac{d^2\psi(x)}{dx^2} + V(x)\psi(x) \right] = i\hbar \frac{1}{\alpha(t)} \frac{d\alpha(t)}{dt}. \tag{3.119}$$

Since the left-hand side is a function of x alone and the right-hand side is a function of t alone, both sides must equal a constant. This constant obviously has the dimensions of energy; so it is called E. The meaning of this separation constant will be investigated shortly. The solution for $\alpha(t)$ is now straightforward. After equating the right-hand side of Equation 3.119 to E, we find that

$$\frac{d\alpha(t)}{dt} = -\frac{i}{\hbar} E\alpha(t)$$

or

$$\alpha(t) = \exp\left(-\frac{iEt}{\hbar}\right). \tag{3.120}$$

When the left-hand side of Equation 3.119 is equated to the constant E, we find

$$\left[-\frac{\hbar^2}{2m} \frac{d^2}{dx^2} + V(x) \right] \psi(x) = E\psi(x), \tag{3.121}$$

which is just the Schrödinger amplitude equation, whose solutions will depend on the specific form of the potential-energy function $V(x)$. Once the solution or solutions to Equation 3.121 have been found, the complete Schrödinger function of the system is given by the product of $\psi(x)$ and $\alpha(t)$,

$$\Psi(x, t) = \psi(x) \exp\left(-\frac{iEt}{\hbar}\right). \tag{3.122}$$

Note that the Schrödinger amplitude equation, Equation 3.121, has the form of an eigenvalue equation. We therefore expect that more than one eigenfunction $\psi_j(x)$ and associated eigenvalue E_j will satisfy this equation. Equations 3.118, 3.120, and 3.122 may be suitably modified to represent this spectrum of solutions; thus

$$\Psi_j(x,t) = \psi_j(x)\alpha_j(t),$$

$$\alpha_j(t) = \exp\left(-\frac{iE_j t}{\hbar}\right), \tag{3.123}$$

$$\left[-\frac{\hbar^2}{2m} \frac{d^2}{dx^2} + V(x) \right] \psi_j(x) = E_j\psi_j(x).$$

The constant E_j, which entered the analysis as a separation constant, will now be shown to be the total energy of the particle in a state whose wave function is Ψ_j. The expectation value of the total energy is given by the fifth postulate as

$$\langle E_j \rangle = \int \overline{\Psi}_j \left(i\hbar \frac{\partial}{\partial t} \right) \Psi_j \, dx, \tag{3.124}$$

where the total-energy operator is $i\hbar \, \partial/\partial t$. After substituting the wave function,

$$\Psi_j(x, t) = \psi_j(x) \exp \left(-\frac{iE_j t}{\hbar} \right),$$

and its complex conjugate,

$$\overline{\Psi}_j(x, t) = \overline{\psi}_j(x) \exp \left(\frac{iE_j t}{\hbar} \right),$$

into Equation 3.124 and carrying out the differentiation, we find that

$$\langle E_j \rangle = E_j \int \overline{\psi}\psi \, dx.$$

Since the integral of $\overline{\psi}\psi$ over all space is unity,

$$\langle E_j \rangle = E_j, \tag{3.125}$$

which says that the jth separation constant is indeed the expectation value of the energy for a particle whose wave function is the jth eigenfunction of the Schrödinger equation.

The separation constant E_j can also be shown to be the eigenvalue of the total energy of the particle which is in state Ψ_j. Since

$$E\Psi_j(x, t) = i\hbar \frac{\partial}{\partial t} \Psi_j(x, t) = i\hbar \frac{\partial}{\partial t} \left[\psi(x) \exp \left(-\frac{iE_j t}{\hbar} \right) \right]$$

$$= E_j \left[\psi(x) \exp \left(-\frac{iE_j t}{\hbar} \right) \right] = E_j \Psi_j(x, t),$$

it follows that

$$E\Psi_j(x, t) = E_j \Psi_j(x, t),$$

which implies, according to the sixth postulate, that E_j is the eigenvalue of the total energy of the particle whose state is described by the wave function $\Psi_j(x, t)$.

3.6 The Free Particle in One Dimension

The simplest possible dynamical system is a single particle on which no forces act, a free particle. To obtain the wave function for the free particle, we solve the Schrödinger amplitude equation with zero potential. The amplitude equation

$$\frac{d^2\psi(x)}{dx^2} + \frac{2m}{\hbar^2}E\psi(x) = 0$$

has the solution

$$\psi(x) = A \exp\left(\frac{i\sqrt{2mE}}{\hbar}x\right) + B \exp\left(-\frac{i\sqrt{2mE}}{\hbar}x\right).$$

The complete solution, including the time-dependent factor, is

$$\Psi(x, t) = A \exp\left[2\pi i\left(\frac{\sqrt{2mE}}{h}x - \frac{E}{h}t\right)\right]$$

$$+ B \exp\left[2\pi i\left(-\frac{\sqrt{2mE}}{h}x - \frac{E}{h}t\right)\right] \tag{3.126}$$

When Equation 3.126 is compared with the equation for a monochromatic wave traveling in the positive x direction (see Equation 3.110),

$$y(x, t) = A' \exp\left[2\pi i(\kappa x - \nu t)\right],$$

it is seen that the Schrödinger function for the free particle consists of two complex monochromatic waves, one traveling to the right and one to the left. For this reason, the Schrödinger function for a free particle is often called a monochromatic de Broglie wave. When a term-by-term comparison of Equations 3.126 and 3.110 is made, we discover that the wavelength of the de Broglie wave is $h/(2mE)^{1/2} = h/p$, as expected from de Broglie's hypothesis. The frequency of the de Broglie wave is E/h in conformity with Einstein's hypothesis that $E = h\nu$.

The Schrödinger amplitude equation for the free particle has a solution for any positive value of E, a result which is in agreement with classical results since a free particle can have any energy. The eigenfunction spectrum and associated eigenvalue spectrum for a free particle are both continuous.

Now suppose the system has been prepared so that the free particle is traveling with energy E in a definite direction along the x-axis toward increasing values of x. Its wave function is therefore of the form

$$\Psi(x, t) = A \exp\left(\frac{i\sqrt{2mE}}{\hbar} x\right) \exp\left(-\frac{iE}{\hbar} t\right).$$ (3.127)

The average value of the momentum of the particles in a number of identical systems is, by Postulate 5,

$$\langle p \rangle = \int_{-\infty}^{\infty} \bar{\Psi}\left(-i\hbar \frac{\partial}{\partial x}\right) \Psi \, dx = \sqrt{2mE}.$$

The last step follows by direct substitution of Ψ and $\bar{\Psi}$ from Equation 3.127, noting the normalization of Ψ. The only possible value a measurement of the particle's momentum can yield is, from Postulate 6, the eigenvalue in the operator equation

$$-i\hbar \frac{\partial}{\partial x} \Psi = p\Psi.$$

Substituting the expression for the wave function from Equation 3.127, we find this value to be

$$p = \sqrt{2mE},$$ (3.128)

which is the classical expression for the linear momentum. Having prepared the system, a single particle, to have both a definite energy, E, and direction of motion, we find that it has a definite linear momentum. Energy and linear momentum can be known simultaneously because their operators commute, as can be readily verified by substituting operator expressions for E and p into the commutator expression $(Ep - pE)\Psi$ and noting that the result is zero. Suppose we now attempt to determine the possible positions of the particle. From Postulate 6, the eigenvalue equation that applies is $x\Psi = x'\Psi$. But any eigenvalue x' will satisfy this equation. Thus if the momentum is known exactly, we can predict nothing about the position of the particle; it may lie anywhere along the x-axis. The probability that it lies in an interval dx about x is $\bar{\Psi}\Psi \, dx$, which becomes, using Equation 3.127, $\bar{A}A \, dx$. Since A is a constant, the probability density function is a constant. The particle has an equal probability of lying in any unit increment along the x-axis; so the standard deviation in position is infinite. Since the momentum is known exactly, its standard deviation is zero; but the standard deviation in the position is infinite—an example of the uncertainty principle at work.

Note that phase velocity of the monochromatic de Broglie wave (Equation 3.127) is

$$v_\phi = \frac{v}{\kappa} = \frac{E/h}{\sqrt{2mE}/h} = \frac{\frac{1}{2}mv^2}{mv} = \frac{1}{2}v$$

and the group velocity is

$$v_g = \frac{dv}{d\kappa} = \frac{d(E/h)}{d(\sqrt{2mE}/h)} = \frac{d(\frac{1}{2}mv^2)}{d(mv)} = \frac{1}{2}\frac{d(v^2)}{dv} = v.$$

Thus we see that the group velocity, not the phase velocity, represents the classical velocity of the particle.

3.7 One-Dimensional Potential Barriers

Attention is now directed to the first classical impossibility—the penetration of particles through potential barriers. Consider the experimental arrangement sketched in Figure 3.3. Electrons are boiled off a filament and accelerated through a potential difference of V_1 volts acquiring a kinetic energy $E = V_1$ electron volts. In the region between $x = A$ and $x = B$, they travel freely with kinetic energy E. At point B they enter a repulsive force field. If the potential of this field, V_2, is greater than V_1, then classically the electrons must loose kinetic energy, come to rest eventually at the point labelled C in Figure 3.3, and then be accelerated backward from C to B, from there to A, and from A to the origin. Thus, if

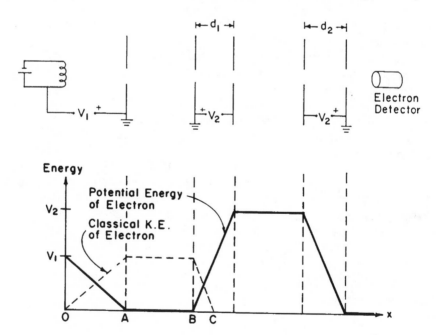

Figure 3.3—Experimental potential barrier for electrons.

$V_2 > E$ electrons cannot classically penetrate the potential barrier, they must be 100% reflected. We will presently demonstrate quantum mechanically that some electrons will indeed penetrate the barrier and be detected. To simplify the mathematical analysis, we replace the trapezoidal barrier of Figure 3.3 with a rectangular barrier (Figure 3.4). Such a barrier may be approximated experimentally by making the distances d_1 and d_2 (Figure 3.3) extremely small.

The following analysis is completely independent of the type of particle incident on the barrier. Electrons were introduced only to illustrate how to set up a one-dimensional potential barrier experimentally. The particle we will now consider may be any massive body, an electron, proton, neutron, billiard ball, or even an automobile. If we are given particles of mass m and kinetic energy E incident on a potential barrier of height $V_0 > E$ that extends from $x = 0$ to $x = b$, our problem will be to find the probability of transmission, i.e., the transparency, T, of the barrier. This quantity is defined as the ratio of the current of transmitted particles to the current of incident particles.

The wave function of the incident particle is the solution to the Schrödinger equation for a free particle (there are no forces in region 1 of Figure 3.4) of energy E traveling in the positive x direction. From Equation 3.127,

$$\Psi_{in} = A e^{i(kx - \omega t)}. \tag{3.130}$$

Note that k is defined here and throughout the remainder of the book as 2π times the propagation constant of the wave; the angular velocity ω is

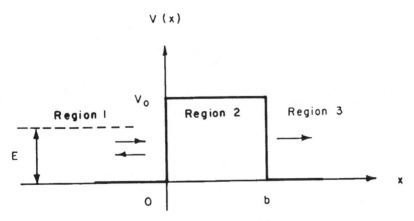

Figure 3.4—Rectangular potential barrier.

again defined as 2π times the frequency of the wave. As functions of energy, therefore,

$$k \equiv \frac{\sqrt{2mE}}{\hbar},$$

$$\omega \equiv \frac{E}{\hbar}. \tag{3.131}$$

The wave function for the transmitted particle is also the solution to the Schrödinger equation for a free particle of energy E, but with a different amplitude, traveling to the right,

$$\Psi_{tr} = Ce^{i(kx - \omega t)}. \tag{3.132}$$

The transparency of a barrier is defined in general as the ratio of the probability current densities (for definition, see Section 4.5) of the transmitted and incident particles, but, in this particular case, the velocities of the incident and transmitted particles are equal; so we may write

$$T = \frac{\Psi_{tr}(x,t)\Psi_{tr}(x,t)}{\Psi_{in}(x,t)\Psi_{in}(x,t)} = \frac{\bar{\psi}_{tr}(x)\psi_{tr}(x)}{\bar{\psi}_{in}(x)\psi_{in}(x)} = \frac{\bar{C}C}{\bar{A}A}. \tag{3.133}$$

Note that this is a steady-state problem both mathematically and physically. Mathematically, the time-dependent parts of the Schrödinger function cancel out of the expression for the transparency. Physically, we imagine a steady stream of incident and transmitted particles; so we are interested only in the ratio of the numbers of incident and transmitted particles. Since this is a steady-state problem, attention is directed to the solution of the Schrödinger amplitude equation in each of the three regions:

Schrödinger Amplitude Equation

Region 1: $\dfrac{d^2\psi_1}{dx^2} + \dfrac{2m}{\hbar^2}E\psi_1 = 0,$

Region 2: $\dfrac{d^2\psi_2}{dx^2} - \dfrac{2m}{\hbar^2}(V_0 - E)\psi_2 = 0.$ $\tag{3.134}$

Region 3: $\dfrac{d^2\psi_3}{dx^2} + \dfrac{2m}{\hbar^2}E\psi_3 = 0.$

The general solution in Region 2 is a sum of real exponentials; in Regions 1 and 3, the general solutions are sums of imaginary exponentials. Using the very natural physical requirement that particles to the right of the barriers be traveling only to the right, we can eliminate one of the mathematically possible solutions in Region 3, i.e., the solution representing a wave traveling in the negative x direction. We then have the solutions

Solution

Region 1: $\psi_1(x) = Ae^{ikx} + Be^{-ikx}$,

where $k \equiv \left(\dfrac{2mE}{\hbar^2}\right)^{1/2}$

Region 2: $\psi_2(x) = Ke^{gx} + Le^{-gx}$,   (3.135)

where $g \equiv \left[\dfrac{2m(V_0 - E)}{\hbar^2}\right]^{1/2}$

Region 3: $\psi_3(x) = Ce^{ikx}$,

where $k \equiv \left(\dfrac{2mE}{\hbar^2}\right)^{1/2}$

There are five arbitrary constants in these solutions, namely, A, B, C, K and L. The objective is to find the ratio of two of them; C/A. The others may be eliminated by applying the conditions that ψ and $d\psi/dx$ must be continuous throughout all space. Application of the continuity conditions at the interfaces $x = 0$ and $x = b$ gives

$\psi_1(0) = \psi_2(0)$ from which $A + B = K + L$,   (3.136a)

$\psi_1'(0) = \psi_2'(0)$ from which $ik(A - B) = g(K - L)$,   (3.136b)

$\psi_2(b) = \psi_3(b)$ from which $Ke^{gb} + Le^{-gb} = Ce^{ikb}$,   (3.136c)

$\psi_2'(b) = \psi_3'(b)$ from which $g(Ke^{gb} - Le^{-gb}) = ikCe^{ikb}$.   (3.136d)

We can eliminate B between Equations 3.136a and 3.136b,

$$A = \frac{K}{2}\left(1 + \frac{g}{ik}\right) + \frac{L}{2}\left(1 - \frac{g}{ik}\right),$$ (3.137)

and solve Equations 3.136c and 3.136d for K and L,

$$K = \frac{C}{2}e^{ikb}e^{-gb}\left(1 + \frac{ik}{g}\right),$$

$$L = \frac{C}{2}e^{ikb}e^{gb}\left(1 - \frac{ik}{g}\right).$$

These expressions for K and L can now be substituted into Equation 3.137 with the result that

$$\frac{A}{C} = \frac{1}{4} e^{ikb} \left[\left(1 + \frac{g}{ik}\right)\left(1 + \frac{ik}{g}\right)e^{-gb} + \left(1 - \frac{g}{ik}\right)\left(1 - \frac{ik}{g}\right)e^{gb} \right].$$

The absolute square of the inverse of this ratio is the desired expression for the transparency,

$$T = \left|\frac{C}{A}\right|^2 = \frac{16}{\left[12 - 2\left(\frac{g}{k}\right)^2 - 2\left(\frac{k}{g}\right)^2\right] + \left[2 + \left(\frac{g}{k}\right)^2 + \left(\frac{k}{g}\right)^2\right]\left[e^{-2gb} + e^{2gb}\right]}$$

$$(3.138)$$

Equation 3.138 is the general expression for the transparency of a one-dimensional rectangular potential barrier. The transparency is seen to be a function of two dimensionless parameters, $(g/k)^2$ and gb, where

$$(g/k)^2 = \frac{V_0 - E}{E} \tag{3.139}$$

and

$$gb = b\left[\frac{2m(V_0 - E)}{\hbar^2}\right]^{1/2}. \tag{3.140}$$

In the special case where $(g/k)^2$ is approximately unity and gb is much greater than unity, Equation 3.138 reduces to the approximate expression

$$T \simeq 4\exp(-2gb) = 4\exp\left(-\frac{2b}{\hbar}\sqrt{2m(V_0 - E)}\right). \tag{3.141}$$

As a numerical example, suppose electrons are accelerated to a kinetic energy $E = 1$ ev and then are allowed to impinge on a potential barrier of height $V_0 = 2$ volts and width $b = 1$ cm. Substituting the constants

$$m = 9.1 \times 10^{-28} \text{ g},$$

$$(V_0 - E) = 1 \text{ ev} = 1.6 \times 10^{-12} \text{ ergs},$$

$$\hbar = \frac{h}{2\pi} = 1.05 \times 10^{-27} \text{ erg-sec},$$

$$b = 1 \text{ cm},$$

into Equation 3.141, we find that the order of magnitude* of the transparency is

$$T \stackrel{\circ}{=} e^{-10^8} = 10^{-0.454 \times 10^8} = 10^{-45,400,000}.$$

*The symbol $\stackrel{\circ}{=}$ means "equal to order of magnitude."

This extremely small transparency helps to explain why quantum effects are so difficult to detect with macroscopic experiments.

To illustrate just how small this transparency is, suppose a 1-amp current of electrons were directed at this barrier. Let us calculate how much time, on the average, would elapse before a single electron penetrated it. One ampere is equal to 6×10^{18} electrons/sec, and the probability of penetration of each electron is T; therefore, the probability per second of penetration is

$$P = 6 \times 10^{18} T \doteq 10^{-45,399,982} \text{ sec}^{-1}.$$

The average time before a single electron penetrated would be $1/P$ or

$$\frac{1}{P} \doteq 10^{45,399,982} \text{ sec} \doteq 10^{45,399,975} \text{ years.}$$

No one would want to begin an experiment requiring this length of time for completion. On the other hand, a microscopic experiment would seem to be quite feasible. For example, if the width of the barrier were reduced to say 10^{-7} cm, then the transparency would be

$$T \doteq e^{-10} = 10^{-4.54},$$

and penetration of the barrier will be of the order of 1 electron in every 30,000. This transparency is more representative of those found in atomic and nuclear systems than the transparency based on a barrier width of 1 cm.

Barrier of Arbitrary Shape

In the general case in which the potential barrier has some complicated shape $V(x)$, it is not usually possible to solve the Schrödinger equation in closed form. If an exact solution for the transparency is required, we will be forced to a numerical solution of the differential equation. If, however, we are willing to settle for a rough estimate of the transparency, we can obtain it by converting the barrier to a rectangular one having the same "area" above E (Figure 3.5). Then the estimate is*

$$\text{Transparency} \doteq \exp\left\{ -\frac{2}{\hbar} \int_a^b \sqrt{2m[V(x) - E]}\, dx \right\}. \tag{3.142}$$

---

*Equation 3.142 can be derived by the WKB (Wentzel–Kramers–Brillouin) method. See, for example, David Bohm's *Quantum Theory*.

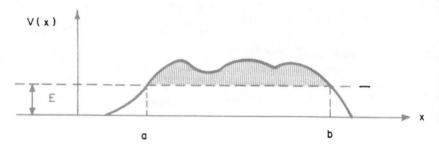

Figure 3.5—Potential barrier of arbitrary shape.

This equation was derived and used by G. Gamow in 1928 to explain the half-life for emission of alpha particles from radioactive nuclei. We will return to it when we discuss the emission of charged particles from nuclei in Chapter 10.

Potential Barrier with $V_0 < E$

If the height of the potential barrier is less than the kinetic energy of the incident particles, as in Figure 3.6, then classically all particles are transmitted. But, as will now be shown, quantum mechanics predicts that some will be reflected. Schrödinger's amplitude equation in Regions 1 and 3 has forms identical to those previously investigated. In Region 2, however, it now has the form

$$\frac{d^2\psi_2(x)}{dx^2} + \frac{2m}{\hbar^2}(E - V_0)\,\psi_2(x) = 0$$

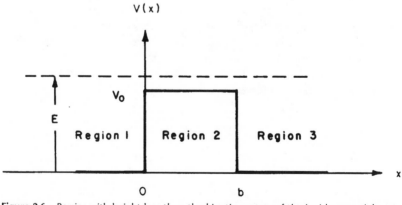

Figure 3.6—Barrier with height less than the kinetic energy of the incident particle.

or

$$\frac{d^2\psi}{dx^2} + q^2\psi_2 = 0,$$

where $q = [2m(E - V_0)/\hbar^2]^{1/2} > 0$. The general solution is

$$\psi_2(x) = Ke^{iqx} + Le^{-iqx}.$$

The solutions in Regions 1 and 3 are

$$\psi_1(x) = Ae^{ikx} + Be^{-ikx},$$

$$\psi_3(x) = Ce^{ikx}.$$

The interface continuity conditions on ψ and ψ' are now imposed, and the same algebraic procedure as before is carried out. The result is

$$T = \left|\frac{C}{A}\right|^2 = \frac{16}{\alpha^2 + \beta^2 + 2\alpha\beta \cos 2qb}, \tag{3.143}$$

where, by definition,

$$\alpha \equiv \left(1 + \frac{q}{k}\right)\left(1 + \frac{k}{q}\right),$$

$$\beta \equiv \left(1 - \frac{q}{k}\right)\left(1 - \frac{k}{q}\right). \tag{3.144}$$

The transparency of the barrier, given by Equation 3.143, is obviously not identically unity as would be predicted classically. However, in the event that $E \gg V_0$, then $q/k \simeq k/q \simeq 1$; so $\alpha \simeq 4$ and $\beta \simeq 0$ and the transparency is unity, as expected classically. One hundred percent transmission will also occur when the constants are such that $2qb$ is an integral multiple of 2π, as can be easily proved from Equation 3.143 and the definitions of α and β. Since $q = 2\pi/\lambda$, perfect transmission occurs when $b = \lambda/2, 2(\lambda/2)$, ..., that is, when the barrier contains an integral number of half-wavelengths of the de Broglie wave. This is an example of a resonance phenomenon common to wave motion but totally unexpected in the behavior of material particles before the advent of quantum mechanics.

Potential Barrier with $V < 0$

If $V < 0$, the potential barrier is called a potential well or, more specifically, a rectangular potential well (Figure 3.7). The potential is $V(x) = -V_0 < 0$, where $0 \leq x \leq b$; $V(x) = 0$ elsewhere.

The analysis of this problem is identical with that of the barrier just

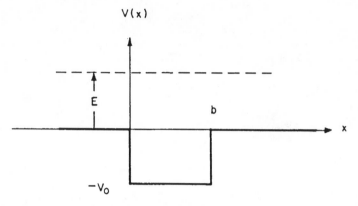

Figure 3.7—One-dimensional rectangular potential well.

considered except that

$$q = \left[\frac{2m(E + V_0)}{\hbar^2} \right]^{1/2}.$$ (3.145)

The transparency is given again by Equation 3.143. As in the previous example, we see that particles may be reflected or transmitted and that there are particular combinations of b and q for which perfect transmission occurs.

Potential Step

As the next example of the interaction of particles with potential barriers, a rectangular barrier that is semi-infinite in length is considered. A barrier of this sort, called a potential step, may be realized experimentally for electrons or other charged particles by an arrangement such as that in Figure 3.8, provided the separation of the last two plates is made vanishingly small.

Particles are incident with kinetic energy E less than the barrier height V_0. Classically, no particles can penetrate the barrier; all must be reflected at $x = 0$, the face of the barrier. To predict what really will happen, we must solve the Schrödinger equations that apply to the two regions. The equations and their general solutions have been investigated in the previous barrier problems. They are:

	Schrödinger Equation	*General Solution*
Region 1:	$\dfrac{d^2\psi_1}{dx^2} + \dfrac{2m}{\hbar^2} E\psi_1 = 0$	$\psi_1 = Ae^{ikx} + Be^{-ikx}$
Region 2:	$\dfrac{d^2\psi_2}{dx^2} - \dfrac{2m}{\hbar^2}(V_0 - E)\psi_2 = 0$	$\psi_2 = Ce^{-gx} + De^{gx}$

where, as before, $k = (2mE/\hbar^2)^{1/2}$ and $g = [2m(V_0 - E)/\hbar^2]^{1/2}$.

The constant D must be set equal to zero; otherwise $\psi_2(x)$ would approach infinity as x approaches infinity and thus would violate Postulate 4. The continuity conditions are applied at the interface

$$\psi_1(0) = \psi_2(0) \qquad \text{from which} \qquad A + B = C \qquad (3.146)$$

and

$$\frac{d\psi_1}{dx}\bigg|_0 = \frac{d\psi_2}{dx}\bigg|_0 \qquad \text{from which} \qquad ik(A - B) = -gC. \qquad (3.147)$$

When Equation 3.147 is solved for C in terms of A and B,

$$C = -\frac{ik}{g}(A - B) = -i\left(\frac{E}{V_0 - E}\right)^{1/2}(A - B). \qquad (3.148)$$

The limit of C as V_0 approaches infinity is zero; and, since ψ_2 is proportional to C, the wave function ψ_2 approaches 0 as V_0 approaches ∞, meaning that no particle can penetrate an infinite potential barrier. This is an example of an important rule: The wave function must vanish at any boundary at which the potential energy is infinite.

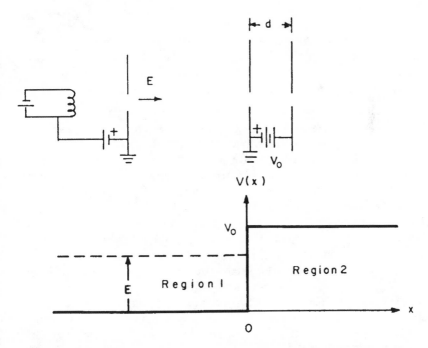

Figure 3.8—Experimental potential step for electrons as $d \to 0$.

Now let us calculate the probability of reflection, R. The incident and reflected wave functions are

$$\Psi_{in} = A e^{i(kx - \omega t)}$$

$$\Psi_r = B e^{i(-kx - \omega t)};$$

therefore

$$R = \frac{\bar{\Psi}_r \Psi_r}{\bar{\Psi}_{in} \Psi_{in}} = \frac{\bar{B}B}{\bar{A}A} = \left| \frac{B}{A} \right|^2. \qquad (3.149)$$

The ratio B/A can be obtained readily from Equations 3.146 and 3.147,

$$\frac{B}{A} = \frac{ik + g}{ik - g}.$$

From this,

$$R = \left| \frac{B}{A} \right|^2 = 1.$$

The probability of reflection is unity; all particles incident on the potential step are reflected. But C, which is given by Equation 3.148, is not zero since $B \neq A$. Therefore, although all particles are eventually reflected, some penetrate into the region beyond $x = 0$. The probability density of such particles is

$$\bar{\psi}_2 \psi_2 = \bar{C}C e^{-gx} e^{-gx} = |C|^2 e^{-2gx}. \qquad (3.150)$$

The probability density falls off exponentially to the right of the interface; hence the farther the detector is to the right of the interface, the fewer particles it will detect. In view of the result that all particles are eventually reflected, we must assume that the particles penetrate to various distances beyond $x = 0$ before being turned around. This is not understandable classically. There are no forces acting on the particles beyond $x = 0$; yet they all eventually turn around. Why? Do they remember the potential barrier they passed through? Furthermore, and even more shattering to classical notions, any particle in the region beyond $x = 0$ apparently has negative kinetic energy. What does this mean? Quantum mechanics offers no physical answers to these questions; instead it mathematically pontificates: This is the way it is because this is the way the mathematical formalism says it is, and the mathematical formalism has yielded remarkably accurate predictions in all instances in which it has been possible to compare it with experiments.

3.8 One-Dimensional Potential Wells

The first classical impossibility predicted by quantum mechanics is the penetration of potential barriers by particles. The reader's attention is now directed to the second classical impossibility—the existence of discrete energy levels for bound particles. Quantum mechanics predicts that particles bound within potential wells cannot have any arbitrary total energy, only certain discrete values. The number and specific values of the allowable energies will depend on the shape of the well, but discrete energy levels will be found to exist in wells of all shapes.

The Infinite Rectangular Box

A particle of mass m is assumed to be imprisoned in the region $-b < x < +b$ by perfectly rigid impenetrable walls at the points $x = \pm b$ (Figure 3.9). Classically, this particle may have any kinetic energy as it bounces back and forth between the walls. Since energy is conserved, it will have exactly the kinetic energy it is given when put into the well, and this, classically, may be any energy between zero and infinity. Quantum mechanically everything we can know about the motion of the particle is

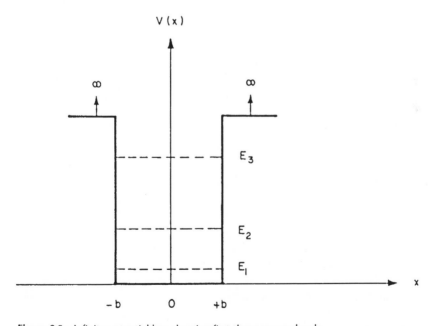

Figure 3.9—Infinite potential box showing first three energy levels.

contained in its wave function, which is obtained from the solution of Schrödinger's equation. The relevant equation in the region within the well is Schrödinger's amplitude equation with $V(x) = 0$.

$$\frac{d^2\psi}{dx^2} + \frac{2m}{\hbar^2} E\psi = 0.$$

As has been found previously, this equation has the general solution

$$\psi(x) = A'e^{+ikx} + B'e^{-ikx},$$

which, for present purposes, it will prove more convenient to express in the form of sines and cosines,

$$\psi(x) = A \sin kx + B \cos kx, \tag{3.151}$$

where, as usual, $k \equiv (2mE/\hbar^2)^{1/2}$. The constants A and B can be evaluated by applying the boundary conditions. Because there are infinite potential barriers at $x = \pm b$, the wave function must be zero at these points and everywhere outside the well:

$$\psi(b) = 0 = A \sin kb + B \cos kb,$$

$$\psi(-b) = 0 = -A \sin kb + B \cos kb.$$

By adding, then subtracting, these two equations, we find that they imply two conditions

$$A \sin kb = 0,$$

$$B \cos kb = 0.$$

Both A and B cannot be zero since this would make ψ identically zero, a result that would imply that the particle does not exist. Also, $\sin kb$ and $\cos kb$ cannot both be zero at the same value of kb. Thus we are forced to the conclusion that there are two sets of solutions,

First set: $A = 0$ and $\cos kb = 0$,

Second set: $B = 0$ and $\sin kb = 0$.

The solutions therefore are

First set: $\psi(x) = B \cos \dfrac{n\pi x}{2b},$ (3.152a)

where n is an odd integer, and

Second set: $\psi(x) = A \sin \dfrac{n\pi x}{2b},$ (3.152b)

where n is an even integer.

When these solutions are compared with the general solution (Equation 3.151), we find that

$$\frac{n\pi}{2b} = k = \left(\frac{2mE}{\hbar^2}\right)^{1/2},$$

where n is any integer. Solving for E yields

$$E = \frac{\pi^2\hbar^2}{8mb^2}n^2, \qquad (3.153)$$

where $n = 1, 2, 3, \ldots$.

The only kinetic energies a particle of mass m may have within an infinite box of width b are those given by Equation 3.153. This is a classically impossible result. It means that we cannot arbitrarily fix the energy of a particle in a one-dimensional box. Its energy must be one of the set of quantum mechanically permitted values

$$E = E_1, 4E_1, 9E_1, \ldots, n^2E_1, \qquad (3.154)$$

where E_1 is the lowest possible energy, the zero point energy, and is given by

$$E_1 = \frac{\pi^2\hbar^2}{8mb^2}. \qquad (3.155)$$

The first three levels of the energy-level diagram corresponding to the infinite rectangular box are shown in Figure 3.9. The levels are increasingly far apart as the energy increases. But the ratio of the energy gap between successive levels to the energy of the levels decreases with increasing energy; that is,

$$\frac{\Delta E}{E} = \frac{E_{n+1} - E_n}{E_n} = \frac{(n+1)^2 - n^2}{n^2} = \frac{2}{n} + \frac{1}{n^2} \xrightarrow{n \gg 1} \frac{2}{n},$$

which decreases with increasing n. Thus the levels actually crowd together at higher energies and thus become more and more like the continuity of classically possible energies. This can be recognized as an example of the correspondence principle which states that, in the limit of large quantum numbers (or large masses or dimensions), quantum-mechanical results must become identical with classical results.

The integers n in the solutions to the above equations are called quantum numbers. Quantum numbers are always found in the solutions of Schrödinger's equation for bound particles. These quantum numbers are simply indices that number the various eigenfunctions of the Schrödinger equation. To illustrate this, we can write the Schrödinger equation for the infinite rectangular box in the form of a typical eigenvalue equation,

$$\left(\frac{-\hbar^2}{2m}\frac{d^2}{dx^2}\right)\psi_n(x) = E_n\psi_n(x).$$

The eigenfunctions and associated eigenvalues of the operator $(-\hbar^2/2m)\,d^2/dx^2$, subject to the boundary conditions $\psi_n(\pm b) = 0$, are

n	Eigenfunction, $\psi_n(x)$	Eigenvalue, E_n
1	$B_1 \cos \dfrac{\pi x}{2b}$	$\dfrac{\pi^2\hbar^2}{8mb^2}$
2	$A_2 \sin \dfrac{2\pi x}{2b}$	$\dfrac{4\pi^2\hbar^2}{8mb^2}$
3	$B_3 \cos \dfrac{3\pi x}{2b}$	$\dfrac{9\pi^2\hbar^2}{8mb^2}$

and so on for all integer values of n.

Note that eigenvalues corresponding to two different eigenfunctions are different in this case. This is not necessarily true in other potentials. It sometimes happens that two or more linearly independent* eigenfunctions have the same eigenvalue. To discuss such cases, it is necessary to introduce the concept of degeneracy. If j linearly independent eigenfunctions of the wave equation have the same eigenvalue E_n, then the energy level E_n is said to have a degeneracy of order j.

We have seen that the energy levels in the infinite box are nondegenerate. As an example of degenerate levels, consider the free particle in one dimension: The eigensolutions to the Schrödinger amplitude equation are

$$\psi_1(x) = A \exp\left(+i\frac{\sqrt{2mE}}{\hbar}x\right)$$

and

$$\psi_2(x) = B \exp\left(-i\frac{\sqrt{2mE}}{\hbar}x\right),$$

where E can have any value greater than zero. The functions ψ_1 and ψ_2 are obviously linearly independent. Therefore, every eigenvalue E is doubly degenerate, corresponding to the two possible directions of motion of the particle. This degeneracy may be removed by specifying the direction of motion of the particle so that only one of the wave functions is a proper solution to the Schrödinger equation with auxiliary condition.

---

*A set of j functions $\psi_1(x), \psi_2(x), \ldots, \psi_j(x)$ is said to be linearly independent if it is impossible to find a set of constants other than zero such that, for all values of x, $C_1\psi_1(x) + C_2\psi_2(x) + \ldots + C_j\psi_j(x) = 0$.

The coefficients B_n and A_n of the eigenfunctions $\psi_n(x)$ for the infinite rectangular box are readily determined by application of the normalization condition

$$\int_{-b}^{+b} \overline{\psi}_n(x)\, \psi_n(x)\, dx = 1.$$

The coefficients are all equal to $e^{i\phi_n}/b^{1/2}$. Thus the normalized eigenfunctions, including the time-dependent part but not including the arbitrary phase factor $e^{i\phi_n}$, are

$$\Psi_n(x,t) = \frac{1}{\sqrt{b}} \cos \frac{n\pi x}{2b} \exp\left(-i\frac{E_n}{\hbar}t\right),$$

where $n = 1, 3, 5, \ldots$, and

$$\Psi_n(x,t) = \frac{1}{\sqrt{b}} \sin \frac{n\pi x}{2b} \exp\left(-i\frac{E_n}{\hbar}t\right),$$

where $n = 2, 4, 6, \ldots$.

It requires but a trivial bit of calculus to prove that the different eigenfunctions of this problem are all orthogonal to one another. Two functions are orthogonal by definition if the integral over all space of the product of one times the complex conjugate of the other is zero, that is, if

$$\int_{-\infty}^{\infty} \overline{\Psi}_j(x)\, \Psi_k(x)\, dx = 0,$$

where j is not equal to k. This is a special case of the general theorem which states that the eigenfunctions corresponding to two different energy eigenvalues of the Schrödinger equations are orthogonal. This very important theorem can be combined with the requirement of normalization in the form of a single equation,

$$\int_{\text{all space}} \overline{\Psi}_j \Psi_k \, d\tau = \delta_{jk}, \qquad (3.156)$$

where δ_{jk} is the *Kronecker delta.*

The Finite Rectangular Well

The finite rectangular potential well is defined by the conditions that

$$V(x) = \begin{cases} 0 & (-b < x < +b), \\ V_0 & (|x| \geq b), \end{cases}$$

where V_0 is a real, positive constant. The total energy of the particle, E, which is also its kinetic energy, is measured from the bottom of the well, as shown in Figure 3.10. The object of the analysis will be to find all the possible values of total energy which a particle of mass m may possess when bound in this potential well.

Regions 1 and 3 are classically forbidden regions of infinite extent. The wave functions must therefore have the form of decaying exponentials,

$$\psi_1(x) = Ae^{gx} \qquad (x \leq -b), \tag{3.157}$$

$$\psi_3(x) = De^{-gx} \qquad (x \geq +b), \tag{3.158}$$

where $g = [2m(V_0 - E)/\hbar]^{1/2}$.

The general solution of the Schrödinger equation in Region 2 is

$$\psi_2(x) = Be^{ikx} + Ce^{-ikx},$$

where $k = (2mE/\hbar^2)^{1/2}$, but, because of the symmetry of the potential, the probability density $\bar{\psi}\psi$ must be symmetric about $x = 0$. This limits the possible solutions in Region 2 to pure sines or pure cosines. As in the case of the infinite potential well, there are two sets of solutions:

First set: $\psi_2(x) = B \sin kx$

Second set: $\psi_2(x) = C \cos kx.$

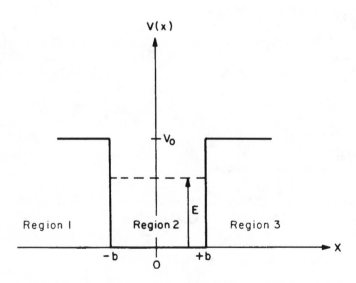

Figure 3.10—Finite rectangular one-dimensional potential well.

Upon applying the continuity conditions on the wave function and its derivative at the interface $x = b$, we find for the first set,

$$\psi_2(b) = \psi_3(b) \qquad B \sin kb = De^{-gb}, \qquad (3.159a)$$

$$\psi'_2(b) = \psi'_3(b) \qquad kB \cos kb = -gDe^{-gb}, \qquad (3.159b)$$

and for the second set,

$$\psi_2(b) = \psi_3(b) \qquad C \cos kb = De^{-gb}, \qquad (3.160a)$$

$$\psi'_2(b) = \psi'_3(b) \qquad -kC \sin kb = -gDe^{-gb}. \qquad (3.160b)$$

Dividing Equation 3.159a by Equation 3.159b and Equation 3.160a by Equation 3.160b eliminates the undetermined constants and results in

$$\tan kb = -k/g$$

and

$$\cot kb = k/g.$$

Upon replacing k and g by their definitions, there results,

$$\tan \sqrt{\frac{2mE}{\hbar^2}}\, b = -\sqrt{\frac{E}{V_0 - E}}, \qquad (3.161a)$$

$$\cot \sqrt{\frac{2mE}{\hbar^2}}\, b = \sqrt{\frac{E}{V_0 - E}}. \qquad (3.161b)$$

These equations for the energy eigenvalues cannot be solved in closed form; they must be solved numerically or graphically. The usual procedure is to plot both sides as a function of E, with E ranging from zero to V_0. The intersections of the two plots are the eigenvalues of E.

As can be seen by inspection of Equations 3.161a and 3.161b, the energy eigenvalues are functions of the mass of the particle m, the potential depth V_0, and the half-width b of the well. In particular, the parameters characterizing the well enter these equations in the form of the product $V_0 b^2$, which is called the well strength. The greater the well strength, the more bound states it can contain. As an example of this, the energy levels for particles of mass 10^{-24} g (approximate nucleon mass) in wells of half-width 10^{-12} cm (approximate nuclear radius) have been plotted in Figure 3.11. The well depths were chosen to be 1, 10, and 50 Mev to represent shallow, moderately deep, and deep nuclear wells. Parts a, b, and c of Figure 3.11, though drawn for rectangular wells, exhibit certain features common to energy-level diagrams for almost all potentials: (a) the lowest possible kinetic energy is not zero; (b) the levels tend to crowd closer

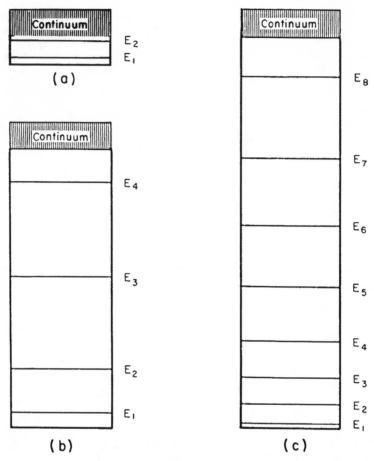

Figure 3.11—Energy levels in finite rectangular potential wells of half-width $b = 10^{-12}$ cm for a particle of mass $m = 10^{-24}$ g. (a) $V_0 = 1$ Mev, (b) $V_0 = 10$ Mev, (c) $V_0 = 50$ Mev.

together ($\Delta E/E$ decreases) toward the top of the well; and (c) beyond the top of the well, there is a continuum of free-particle levels.

The Linear Harmonic Oscillator

No introduction to quantum mechanics, regardless of its brevity, is permitted to omit mention of the linear harmonic oscillator. To avoid heresy, we present the problem and the solution here; for the detailed treatment, the reader may consult any quantum-mechanics textbook.

Consider a particle of mass m bound in a force field whose restoring force is proportional to the distance of the particle from its equilibrium point, i.e., $F = -Kx$. A macroscopic example of such a system is a mass on a spring with spring constant K. The potential corresponding to this force is $V(x) = Kx^2/2$, an infinite paraboloidal potential well. The Schrödinger amplitude equation within the well is

$$\left[-\frac{\hbar^2}{2m}\frac{d^2}{dx^2} + \tfrac{1}{2}Kx^2 \right]\psi(x) = E\psi(x). \tag{3.162}$$

The eigenfunctions of this equation, subject to the appropriate interface conditions, can be expressed in terms of Hermite polynomials

$$\psi_n(x) = N_n e^{-\beta^2/2} H_n(\beta), \tag{3.163}$$

where $n = 0, 1, 2, \ldots$, and where, by definition,

$$\beta \equiv \sqrt{\alpha}\, x,$$

$$\alpha \equiv \frac{2\pi m}{\hbar}\, v_0,$$

$$v_0 \equiv \frac{1}{2\pi}\sqrt{\frac{K}{m}}, \tag{3.164}$$

and the normalization constant is

$$N_n = \left(\sqrt{\frac{\alpha}{\pi}}\, \frac{1}{2^n n!} \right)^{1/2}.$$

The expression $H_n(\beta)$ is the nth Hermite polynomial; the first four are

$H_0 = 1,$

$H_1 = 2\beta,$

$H_2 = 4\beta^2 - 2,$

$H_3 = 8\beta^3 - 12\beta.$

The energy eigenvalues can be shown to be

$$E_n = (n + \tfrac{1}{2})hv_0, \tag{3.165}$$

where $n = 0, 1, 2, \ldots$ and v_0 is, from its definition in Equation 3.164, the classical frequency of the harmonic oscillator. Thus the energy levels form a discrete, nondegenerate set with zero-point energy $hv_0/2$. Note that, as usual, the levels crowd together at higher energies:

$$\frac{\Delta E}{E} = \frac{E_{n+1} - E_n}{E_n} = \frac{(n + \tfrac{3}{2}) - (n + \tfrac{1}{2})}{(n + \tfrac{1}{2})} \xrightarrow[n \gg 1]{} \frac{1}{n}.$$

3.9 Virtual Energy Levels

In Section 3.7 consideration was given to free particles impinging on potentials of various types—barriers, wells, and potential steps. The energy eigenvalues of the Schrödinger equations that described these particles were found to be continuous in that any positive energy was allowed. In Section 3.8 attention was directed to bound particles, particles that were enclosed in infinite potential boxes or in wells. It was found that in these cases the energy eigenvalues of the Schrödinger equation were discrete; i.e., only certain specific energies were allowed.

Now we shall consider a situation that has some of the aspects of both of the previous cases. Suppose we have a potential that consists of a well and a barrier (Figure 3.12). For simplicity assume that the potential has a rectangular shape and that it is infinite for all negative values of x. The results to be discussed are qualitatively independent of the shape of the well and barrier. We might guess, on the basis of previous results, that the energy eigenvalue spectrum for a potential of this sort consists of a discrete

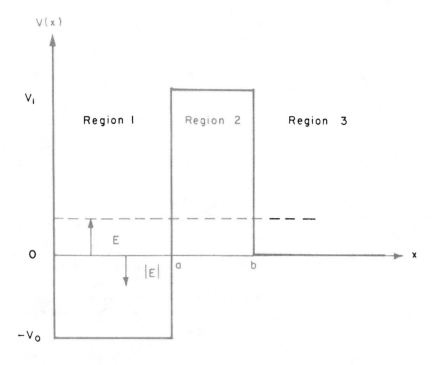

Figure 3.12—Potential consisting of well and barrier.

set of levels lying below zero and a continuous set of energy eigenvalues above zero. This is in fact the spectrum, as straightforward solution of the Schrödinger equation proves.

A detailed analysis of the states of positive energy will show that at certain values of E lying between 0 and V_1 the ψ-function inside the barrier will be much larger than the ψ-function outside the barrier. These values of energy are called virtual energy levels or scattering resonances. They correspond to states in which (a) a particle initially within the barrier will have small probability of getting out because the barrier transparency is so small in the outward direction (for this reason, a particle in a virtual energy level is said to be in a metastable state) and (b) a particle initially outside and moving toward the barrier will have high probability of getting in because the barrier transparency is almost unity in the inward direction.

The Schrödinger amplitude equations that apply to each of the three regions for the case $0 < E < V_1$ and their general solutions are:

Schrödinger Equation	*General Solution*

Region 1:

$$\frac{d^2\psi_1}{dx^2} + \frac{2m}{\hbar^2}(E + V_0)\psi_1 = 0 \qquad \psi_1 = Fe^{iqx} + Be^{-iqx}$$

$$\text{where} \quad q = \left[\frac{2m(E + V_0)}{\hbar^2}\right]^{1/2}$$

Region 2:

$$\frac{d^2\psi_2}{dx^2} - \frac{2m}{\hbar^2}(V_1 - E)\psi_2 = 0 \qquad \psi_2 = Ke^{gx} + Le^{-gx}$$

$$\text{where} \quad g = \left[\frac{2m(V_1 - E)}{\hbar^2}\right]^{1/2}$$

Region 3:

$$\frac{d^2\psi_1}{dx^2} + \frac{2m}{\hbar^2}E\psi_3 = 0 \qquad \psi_3 = Ce^{+ikx} + De^{-ikx}$$

$$\text{where} \quad k = \left[\frac{2mE}{\hbar^2}\right]^{1/2}$$

The complex numbers F, B, K, L, C, and D are undetermined. The wave function in Region 1 must vanish at $x = 0$ because of the infinite potential; therefore $F + B$ equals 0 or B equals $-F$ and

$$\psi_1(x) = F[e^{iqx} - e^{-iqx}] = 2iF \sin qx = A \sin qx.$$

The last step follows from replacing the undetermined constant $2iF$ with the undetermined constant A.

In Region 3 the amplitudes of the incoming and outgoing free waves must be equal since all incident particles are eventually reflected. Therefore $|C|$ equals $|D|$, and ψ_3 may be written, in arbitrarily normalized form, as

$$\psi_3 = \sin(kx + \delta),$$

where δ is an undetermined constant. The interface conditions on ψ and ψ' at $x = a$ and $x = b$ yield

$$\psi_1(a) = \psi_2(a) \qquad\qquad A \sin qa = Ke^{ga} + Le^{-ga}, \qquad (3.166a)$$

$$\psi_1'(a) = \psi_2'(a) \qquad\qquad qA \cos qa = g(Ke^{ga} - Le^{-ga}), \qquad (3.166b)$$

$$\psi_2(b) = \psi_3(b) \qquad\qquad Ke^{gb} + Le^{-gb} = \sin(kb + \delta), \qquad (3.166c)$$

$$\psi_2'(b) = \psi_3'(b) \qquad\qquad g(Ke^{gb} - Le^{-gb}) = k \cos(kb + \delta). \qquad (3.166d)$$

These four equations in four unknowns, A, K, L and δ, can be solved in general; however, the resulting expressions are algebraically complicated and not readily interpretable without numerical calculation. To simplify the problem, consider only the special case when the barrier width $(b - a)$ is sufficiently great that $e^{-g(b-a)}$ is negligible compared with $e^{g(b-a)}$. Solve Equations 3.166a and 3.166b for K and L in terms of A:

$$K = \frac{A}{2} e^{-ga} \left(\frac{q}{g} \cos qa + \sin qa \right), \qquad (3.167a)$$

$$L = -\frac{A}{2} e^{ga} \left(\frac{q}{g} \cos qa - \sin qa \right). \qquad (3.167b)$$

Next substitute these expressions into Equations 3.166c and 3.166d and drop terms multiplied by $e^{-g(b-a)}$. (Note that this procedure is valid except when the factor in parentheses in Equation 3.167a is near zero; this special case will be examined shortly.)

$$\sin(kb + \delta) = \frac{A}{2} e^{g(b-a)} \left(\frac{q}{g} \cos qa + \sin qa \right);$$

$$\cos(kb + \delta) = \frac{A}{2} \frac{g}{k} e^{g(b-a)} \left(\frac{q}{g} \cos qa + \sin qa \right). \qquad (3.168)$$

After squaring and adding, we obtain

$$A^2 = \frac{4e^{-2g(b-a)}}{[1 + (g/k)^2][(q/g) \cos qa + \sin qa]^2}. \qquad (3.169)$$

Thus we see that, if $g(b - a)$ is large compared to unity, then A is generally very small compared to unity; that is, the amplitude of the wave function is much smaller in Region 1 than in Region 3.

At certain exceptional values of the energy, the factor in parentheses in Equation 3.167a becomes zero. These are the energies at which

$$\tan qa = -\frac{q}{g}. \tag{3.170}$$

At these energies Equations 3.166 and 3.167 reduce to

$$K = 0,$$

$$L = Ae^{ga} \sin qa,$$

$$\sin (kb + \delta) = Ae^{-g(b-a)} \sin qa,$$

$$\cos (kb + \delta) = -A \left(\frac{g}{k}\right) e^{-g(b-a)} \sin qa.$$

Squaring and adding the last two equations,

$$A^2 = \frac{e^{2g(b-a)}}{[1 + (g/k)]^2 \sin^2 qa}. \tag{3.171}$$

Since the exponent in Equation 3.171 is much greater than unity, we can conclude that, at certain exceptional energies given implicitly by Equation 3.170, the amplitude of the inner wave function A^2 becomes much larger than the amplitude of the outer wave function. The series of energies determined by this condition are called the virtual energy levels. At these energies a particle initially inside the barrier has a minimum probability of escape, and a particle initially incident on the barrier from outside has a maximum probability of penetrating the barrier to the inside. "Minimum" and "maximum" in this context mean minimum and maximum with respect to nearby energies. If we plotted A^2 as a function of energy, we would find a series of sharp peaks centered at the virtual energy levels. The wave functions for systems close to and far from virtual levels would look roughly like those shown in Figure 3.13.

It is relatively easy to estimate the mean time T that the particle spends in Region 1. The probability of finding the particle in the range dx about x is $\bar\psi(x) \psi(x) dx$; therefore the relative probability that the particle lies in the range 0 to a is

$$\int_0^a A^2 \sin^2 qx \, dx = A^2 \left(\frac{a}{2} - \frac{\sin 2qa}{4q}\right),$$

and, if $qa \gg 1$, i.e., if there are many waves in Region 1, then this reduces to $A^2a/2$. The relative probability of finding the particle in a spatial interval of length a outside the barrier is

$$\int_x^{x+a} \sin^2(kx + \delta)\, dx,$$

which is approximately $a/2$ if we again assume many waves in the interval x to $x + a$. Far outside the barrier the particle will spend a time a/v in each region of width a, where v, the particle's velocity, is equal to $(2E/m)^{1/2}$. We can now set up the proportion

$$\frac{\text{Time inside}}{\text{Time outside}} = \frac{T}{a/\sqrt{2E/m}} = \frac{A^2a/2}{a/2}$$

and find that

$$T = A^2a \sqrt{\frac{m}{2E}}.$$

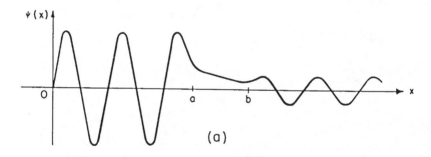

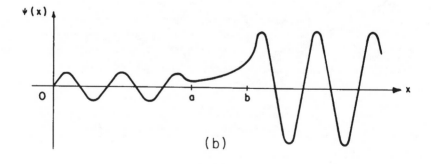

Figure 3.13—Approximate shape of wave function: (a) near a virtual level, (b) far from a virtual level.

Substituting for A^2 from Equations 3.169 and 3.171, we conclude that, for most values of energy, the particle will spend a mean time of the order of

$$T_{nv} \triangleq a \sqrt{\frac{m}{2E}} e^{-2g(b-a)} \qquad \text{(nonvirtual level)} \qquad (3.172)$$

within the well; at the virtual levels it will spend a time of the order of

$$T_v \triangleq a \sqrt{\frac{m}{2E}} e^{2g(b-a)} \qquad \text{(virtual level)} \qquad (3.173)$$

within the well. As a rough numerical example, consider a particle having a mass of the order of an alpha-particle mass, 4 amu, confined within a barrier approximately the size of a nucleus. Specifically, if we choose $a = 10^{-12}$ cm, $b = 2 \times 10^{-12}$ cm, $V_1 = 20$ Mev, and $E = 4$ Mev, we find that $2g(b - a) \simeq 17.5$ and $a(m/2E)^{1/2} \simeq 10^{-21}$ sec. Thus the mean life of the alpha particle in a nonvirtual level is less than 10^{-21} sec, but at a virtual level its mean life is some 10^{16} times as long. A theory much like this, but in three dimensions and with a more appropriate potential barrier shape, was employed in 1928 by Gamow and, independently, by Gurney and Condon to derive the experimentally observed relation between the mean life for alpha decay of radioactive nuclei and the energies of the emitted alpha particles.

Phase Shift of Reflected Wave

Consider now yet another aspect of the problem of particles incident on and reflected from a real potential. The potential is assumed to be infinite in the range $x < 0$ and to be zero beyond $x = b$ but otherwise arbitrary. It may have any shape $V(x)$ in the range $0 < x < b$. We may ask one of the most basic questions in interaction theory: What is the net effect of the potential $V(x)$ on the outgoing wave function in the region outside the potential? If $V(x)$ were identically zero, the outgoing wave function would be

$$(\psi_+)_0 = |C_0| e^{i\theta_0} e^{ikx}, \qquad (3.174)$$

where $|C_0| e^{i\theta_0}$ is a complex constant. With a potential, regardless of its form, the outgoing wave function beyond $x = b$ is of the form

$$(\psi_+)_1 = |C| e^{i\theta_1} e^{ikx}, \qquad (3.175)$$

where $|C| e^{i\theta_1}$ is a complex constant. But every particle incident on a real potential having an infinite step in the region $x < 0$ is eventually reflected;

therefore $|C| = |C_0|$, and, from Equations 3.174 and 3.175,

$$(\psi_+)_1 = (\psi_+)_0 \, e^{i(\theta_1 - \theta_0)} = (\psi_+)_0 \, e^{i\delta} \tag{3.176}$$

in the region $x > b$.

We can readily see that the net effect of the potential, whatever its shape, is to shift the phase of the outgoing wave relative to what it would have been without the potential.

The angle δ, defined by

$$\delta = (\theta_1 - \theta_0), \tag{3.177}$$

is called the phase shift of the wave function. A phase shift of π or an integral multiple of π changes, at most, the sign of the outgoing wave function; whereas a phase shift of $\pi/2$ or an odd-integral multiple of $\pi/2$ multiplies the outgoing wave by $\pm i$. In Chapter 4 we will show that cross sections can be determined uniquely from knowledge of the phase shifts and that the closer the phase shift is to an odd-integral multiple of $\pi/2$, the greater is the effect on the cross section.

3.10 Superposition of States

As has been seen, we can solve straightforwardly for the energy eigenvalues E_n and corresponding eigenfunctions Ψ_n for particles trapped in potential wells. Given a particle of energy E_n, we can state definitely that its wave function is the eigenfunction Ψ_n corresponding to the eigenvalue E_n. If the energy of the particle in the well is not known, what is its Ψ-function? In this case we would express its Ψ-function as a linear combination of its possible eigenfunctions,

$$\Psi = \sum_n c_n \Psi_n, \tag{3.178}$$

where $n = 1, 2, 3, \ldots, N$ and $|c_n|^2$ is the probability of finding the system in the nth eigenstate.

To show that this formula for the superposition of states is compatible with the postulates of wave mechanics and the ordinary rules of probability, we calculate the expectation value of the total energy,

$$\langle E \rangle = \int \bar{\Psi} \left(i\hbar \frac{\partial}{\partial t} \Psi \right) d\tau = \int \left(\sum_j \bar{c}_j \bar{\Psi}_j \right) \left(i\hbar \frac{\partial}{\partial t} \right) \left(\sum_n c_n \Psi_n \right) d\tau$$

$$= \int \left(\sum_j \bar{c}_j \bar{\Psi}_j \right) \left(\sum_n c_n E_n \Psi_n \right) d\tau = \sum_j \sum_n \bar{c}_j c_n E_n \int \bar{\Psi}_j \Psi_n \, d\tau.$$

Because of the orthonormality of the eigenfunctions, namely,

$$\int \bar{\Psi}_j \Psi_n \, d\tau = \delta_{jn},$$

the expectation value of the energy reduces to

$$\langle E \rangle = \sum_n E_n |c_n|^2,$$

which is just what one would expect if $|c_n|^2$ is interpreted as the probability of the system being in the nth energy eigenstate.

When a measurement of energy is made on a particular system, we must obtain one of the possible eigenvalues E_n of the system. However, if this measurement is repeated again and again on identical systems that have not been prepared to be in particular eigenstates, we will obtain a random selection of the eigenvalues such that the arithmetic mean energy is given by the previous equation. In M measurements the eigenvalue E_k will be found approximately $M|c_k|^2$ times, provided M is much greater than unity.

Equation 3.178 is a statement that contains all our knowledge about the system and all our ignorance about the system. It is a statistical statement. The same sort of description could be devised for any probabilistic system. For example, the possible eigenvalues of a pair of cubical dice are the numbers 2 to 12. Let the eigenstates corresponding to these eigenvalues be called $\psi(2)$, $\psi(3)$, ..., $\psi(12)$. The eigenstate $\psi(n)$ is any state of the pair of dice in which the sum of the numbers on their upturned sides is n. Most of the eigenstates are degenerate with respect to n; i.e., there is more than one way in which the sum can equal n. For example, the state with eigenvalue $n = 6$ has a fifth-order degeneracy since it can be obtained from the five distinct sums $5 + 1 = 1 + 5 = 4 + 2 = 2 + 4 = 3 + 3 = 6$.

Now suppose the dice are shaken in a box. What can be said about the state of the dice prior to looking at them? The only thing we can do is list the probabilities of various states. If the dice are perfectly balanced so that any face has equal probability of being on top, we can write formally that the state of the system is

$$\psi = \sum_{n=2}^{12} c_n \psi(n),$$

where c_n is the square root of the probability of the state $\psi(n)$ being obtained. Table 3.1 shows the probability and order of degeneracy for a pair of perfectly balanced dice.

Each time the dice are shaken, we will obtain one of the possible eigenvalues $n = 2, 3, \ldots, 12$. The mean value of n after M shakes, where $M \gg 1$, would be

$$\langle n \rangle = \sum_{n=2}^{12} n c_n^2 = \sum_{n=2}^{12} n P(n),$$

Table 3.1—Probabilities of States of Two Dice

Eigenvalue (n)	Order of Degeneracy $[D(n)]$	Probability of State $[P(n)]$	$c_n = \sqrt{P_n}$
2	1	1/36	$\sqrt{1}/6$
3	2	2/36	$\sqrt{2}/6$
4	3	3/36	$\sqrt{3}/6$
5	4	4/36	$\sqrt{4}/6$
6	5	5/36	$\sqrt{5}/6$
7	6	6/36	$\sqrt{6}/6$
8	5	5/36	$\sqrt{5}/6$
9	4	4/36	$\sqrt{4}/6$
10	3	3/36	$\sqrt{3}/6$
11	2	2/36	$\sqrt{2}/6$
12	1	1/36	$\sqrt{1}/6$
	36	1	

which is 7. For a pair of unbalanced dice the probabilities would differ from those given in Table 3.1, and the expectation value of n might be any number, not necessarily an integer, between 2 and 12.

Interference of Probabilities

From the discussion thus far, it may be concluded that there is a strong analogy between the superposition of quantum-mechanical eigenstates and the ordinary combinatorial analysis of probability theory. We will now examine a quantum effect for which there is no combinatorial analogy, the interference of probabilities. For simplicity, suppose that the system under consideration has but two nondegenerate eigenstates. We may, for example, assume that the system is a single particle trapped in a shallow one-dimensional potential well. The eigenfunctions corresponding to the two energy eigenvalues E_1 and E_2 are $\psi_1(x)$ and $\psi_2(x)$. If the particle were in eigenstate 1, its spatial probability density would be $P_1(x) = \bar{\psi}_1(x)\psi_1(x)$. If it were in eigenstate 2, its spatial probability density would be $P_2(x) = \bar{\psi}_2(x)\psi_2(x)$. Suppose we do not know which particular eigenstate it is in but do know that it has the same probability of being in either of the two eigenstates. According to the rules of combinatorial analysis, we would say immediately that its probability density is

$$P(x) = \tfrac{1}{2}P_1(x) + \tfrac{1}{2}P_2(x) \tag{3.179}$$

or

$$\bar{\psi}\psi = \tfrac{1}{2}\bar{\psi}_1\psi_1 + \tfrac{1}{2}\bar{\psi}_2\psi_2;$$

but this is wrong. The formula for the superposition of states (Equation 3.178) tells us that the state of the combined system is given by

$$\psi(x) = \sqrt{\tfrac{1}{2}}\,\psi_1(x) + \sqrt{\tfrac{1}{2}}\,\psi_2(x), \tag{3.180}$$

and Postulate 5 clearly states that the probability density is given by $\bar{\psi}(x)\psi(x)$; that is,

$$P(x) = \bar{\psi}\psi = [\sqrt{\tfrac{1}{2}}\,\bar{\psi}_1(x) + \sqrt{\tfrac{1}{2}}\,\bar{\psi}_2(x)][\sqrt{\tfrac{1}{2}}\,\psi_1(x) + \sqrt{\tfrac{1}{2}}\,\psi_2(x)]$$

$$= \tfrac{1}{2}\bar{\psi}_1\psi_1 + \tfrac{1}{2}\bar{\psi}_2\psi_2 + \tfrac{1}{2}[\bar{\psi}_1\psi_2 + \bar{\psi}_2\psi_1]$$

or

$$P(x) = \tfrac{1}{2}P_1(x) + \tfrac{1}{2}P_2(x) + \tfrac{1}{2}[\bar{\psi}_1\psi_2 + \bar{\psi}_2\psi_1]. \tag{3.181}$$

When Equation 3.181 is compared with Equation 3.179, we see that the correct probability density function contains additional terms which arise from the cross products of the eigenfunctions. These are the *interference terms*.

It is easy to prove that $P(x)$ as given by Equation 3.181 is still normalized to unity, as it must be to represent a probability density. If we integrate Equation 3.181 over the range of the x from minus infinity to plus infinity, the interference terms yield zero because of the orthogonality of the eigenfunctions, and the other two terms each integrate to one-half because of the normality of each of the eigenfunctions ψ_1 and ψ_2.

3.11 The Physical Meaning of the Wave Function

In this chapter and in future chapters, the use of the formalism of quantum mechanics to obtain the solutions to problems is stressed rather than its meaning. It is undoubtedly possible for a person to be quite skillful in the manipulation of quantum mechanics without deep knowledge of its philosophical basis, just as it is possible for a musician to play with great skill without knowledge of the physical theory of sound which forms the basis for the design of his instrument, and just as it is possible for most engineers and physicists to use Newton's second law without knowing that the basis of this law is still a matter of open philosophical dispute.

Fortunately, it is not absolutely necessary to know what Ψ means to solve quantum-mechanical problems because, as a matter of fact, there is still much debate regarding the meaning of the wave function. Alfred Landé lists the following seven opinions of what various authorities believe the Ψ-function describes*:

*Alfred Landé, *From Dualism to Unity in Quantum Physics*, Cambridge University Press, New York, 1960.

(1) The physical state of a continuous material medium in space and time, with particles as mere appearances (Schrödinger 1926);

(2) A continuous pilot wave which controls point events along its course (de Broglie);

(3) A fluid containing hypothetical quantum forces invented *ad hoc* so as to determine its own motion according to the laws of hydrodynamics (Bohm);

(4) One definite state or a sequence of many states of an object or of a statistical ensemble of objects (the textbooks vacillate among these opinions);

(5) A wave state in space (for example in the transmitted as well as in the reflected part of a 'wave packet' in case of partial reflection) contracting with superluminal velocity either when a point event takes place (in one of the two parts) or only when an observer gains knowledge thereof (Heisenberg, 'waves of expectation');

(6) A mathematical symbol incapable of pictorial representation, completely abstract and containing, so to speak, no physics at all, yet completely 'objective' since not referring to any observers' knowledge (Heisenberg, the Copenhagen language);

(7) A well-ordered list of betting odds based on past statistical experience, for the diverse outcomes of specific tests of a microscopic object with macroscopic instruments.

The last opinion in the preceding list is that of Landé himself. It makes a great deal of sense, as the reader can discover for himself by spending an evening with Landé's delightful little book.

We are accustomed to describing particles in classical mechanics using dynamical variables, such as the position r, the momentum p, the angular momentum L, and the total energy E. In quantum mechanics the dynamical variables are converted by a set of rules into mathematical operators that then operate on a function Ψ describing the system under consideration. We find that the mathematical form of Ψ is, in many instances, similar to the mathematical form used to describe a wave. We therefore define certain groups of constants as the wavelength, the frequency, etc., of the particle that Ψ describes. The particle has not become a wave by this association. It is true that particles sometimes exhibit characteristics, such as interference, that we usually associate with waves, but these characteristics follow directly from the postulates of quantum mechanics, and do not depend on our definitions of wavelength, frequency, and so on. Particles would exhibit these properties whether or not we ever bothered to draw the analogy between waves and the solution to the Schrödinger equation.

For the reader who cannot help but ask what neutrons, protons, and electrons really are—waves or particles—an excerpt from an article by Erwin Schrödinger* is presented. Although written some 27 years after

*Erwin Schrödinger, What is Matter?, *Sci. Amer.* **189**: 52 (September 1953).

his momentous creation of wave mechanics, it indicates that he too was still asking the same question.

According to Einstein a particle has the energy mc^2, m being the mass of the particle and c the velocity of light. In 1925 Louis de Broglie drew the inference, which rather suggests itself, that a particle might have associated with it a wave process of frequency mc^2 divided by h. The particle for which he postulated such a wave was the electron. Within two years the "electron waves" required by his theory were demonstrated by the famous electron diffraction experiment of C. J. Davisson and L. H. Germer. This was the starting point for the cognition that everything—anything at all— is simultaneously particle and wave field. Thus de Broglie's dissertation initiated our uncertainty about the nature of matter. Both the particle picture and the wave picture have truth value, and we cannot give up either one or the other. But we do not know how to combine them.

That the two pictures are connected is known in full generality with great precision and down to amazing details. But concerning the unification to a single, concrete, palpable picture opinions are so strongly divided that a great many deem it altogether impossible. I shall briefly sketch the connection. But do not expect that a uniform, concrete picture will emerge before you; and do not blame the lack of success either on my ineptness in exposition or your own denseness—nobody has yet succeeded.

One distinguishes two things in a wave. First of all, a wave has a front, and a succession of wave fronts forms a system of surfaces like the layers of an onion. You are familiar with the two-dimensional analogue of the beautiful wave circles that form on the smooth surface of a pond when a stone is thrown in. The second characteristic of a wave, less intuitive, is the path along which it travels—a system of imagined lines perpendicular to the wave fronts. These lines are known as the wave "normals" or "rays."

We can make the provisional assertion that these rays correspond to the trajectories of particles. Indeed, if you cut a small piece out of a wave, approximately 10 or 20 wavelengths along the direction of propagation and about as much across, such a "wave packet" would actually move along a ray with exactly the same velocity and change of velocity as we might expect from a particle of this particular kind at this particular place, taking into account any force fields acting on the particle.

Here I falter. For what I must say now, though correct, almost contradicts this provisional assertion. Although the behavior of the wave packet gives us a more or less intuitive picture of a particle, which can be worked out in detail (e.g., the momentum of a particle increases as the wavelength decreases; the two are inversely proportional), yet for many reasons we cannot take this intuitive picture quite seriously. For one thing, it is, after all, somewhat vague, the more so the greater the wavelength. For another, quite often we are dealing not with a small packet but with an extended wave. For still another, we must also deal with the important special case of very small "packelets" which form a kind of "standing wave" which can have no wave fronts or wave normals. . . .

If you finally ask me: "Well, what *are* these corpuscles, really?" I ought to confess honestly that I am almost as little prepared to answer that as to tell where Sancho Panza's second donkey came from. At the most, it may be permissible to say that one can think of particles as more or less temporary entities within the wave field whose form and general behavior are nevertheless so clearly and sharply determined by the laws of waves that many processes take place as if these temporary entities were substantial permanent beings.

SELECTED REFERENCES

1. A. D'Abro, *The Rise of the New Physics*, Vol. II, Dover Publications, New York, 1951. This book traces the historical development of quantum theory from Planck's introduction of h through Bohr's old quantum theory, de Broglie's wave mechanics, and Schrödinger's and Heisenberg's quantum theory, to Dirac's relativistic quantum theory. Mathematics is employed sparingly to make pleasant and rewarding reading for a rainy Sunday afternoon.

2. Vladimir Rojansky, *Introductory Quantum Mechanics*, Prentice-Hall, Inc., Englewood Cliffs, N. J., 1938. The elementary introduction to such topics as eigenvalue equations, linear operators and matrices, Schrödinger's wave mechanics, Heisenberg's matrix mechanics, and Dirac's relativistic quantum mechanics is excellent. Many illustrative examples are contained in this book which is well suited for self study.

3. Henry Margenau, *The Nature of Physical Reality*, McGraw-Hill Book Company, New York, 1950. This is one philosopher–physicist's attempt to explain the methodology of the exact physical sciences. Chapters 17 through 19, which deal with quantum mechanics, include some highly illuminating examples of the uncertainty principle.

4. P. A. M. Dirac, *The Principles of Quantum Mechanics*, 3rd ed., Oxford University Press, 1947. This mathematical development of the concepts and equations of quantum mechanics is beautiful and concise. After the reader has identified Dirac's bra and ket vectors with wave functions, he should find little difficulty in understanding the rest.

EXERCISES

1. Set up and solve the following problems by the Newtonian, Lagrangian, and Hamiltonian formulations of classical mechanics to find position as function of time.

(a) Simple harmonic oscillator: A particle of mass m on the end of a spring with force constant K, i.e., $F(x) = -Kx$, $V(x) = (K/2)x^2$.

(b) Double-bob pendulum: Two particles attached to a weightless string. The particles of masses m_1 and m_2 are distances L_1 and L_2, respectively, from the top of the string.

(c) Simple Atwood machine: A pulley of mass m, moment of inertia I, over which is a weightless string carrying two masses, m_1 and m_2, on its ends.

2. Prove the following Poisson bracket relations:

(a) $(q_j, q_k) = 0$,

(b) $(p_j, p_k) = 0$,

(c) $(q_j, p_k) = \delta_{jk}$,

(d) $(p_k, q_j) = -\delta_{jk}$,

(e) $(p_x, L_y) = p_z$.

3. Calculate the transparency of a one-dimensional real rectangular well of depth V_0 and width b for monoenergetic particles of mass m. Plot the transparency as a function of the well strength $V_0 b^2$ for fixed energy E. Also, plot the transparency as a function of b/λ for a fixed V_0, where λ is the wavelength within the region of the well. Comment on the significance of the graphs.

4. Arbitrarily normalize and plot either the real or the imaginary part of the time-independent wave function for a one-dimensional step potential in which $E < V_1$ and for one in which $E > V_1$. Comment on the differences between the wave functions in these two cases.

5. Plot either the real or imaginary part of the wave function for a one-dimensional rectangular potential barrier under each of the following conditions: (a) nearly perfect transparency and (b) very low transparency. Explain the differences between the wave functions.

6. As an example of a one-dimensional potential box with infinite walls, consider a mass m which is sliding back and forth within a frictionless closed cylinder of length $2b$. Show that, if m and b are of macroscopic size (say, $m \triangleq 10^2$ g and $b \triangleq 10$ cm), it is experimentally impossible to detect energy levels in this system. Show, in particular, that the velocity of the mass in the lowest energy level is experimentally indistinguishable from zero and that the energy levels near 1 ev are of the order of 10^{-23} ev apart and therefore are not experimentally separable.

7. A frictionless wagon of 10 kg mass rolls down a hill 20 m high under the influence of gravity and then encounters a sinusoidally shaped hill that is 25 m high and 30 m long at the base. If it fails to get over the hill, the wagon rolls back to its starting point, and the process repeats itself. (a) What is the order of magnitude of the time you would expect to elapse before the wagon makes it over the sinusoidal hill? (b) What would happen if the mass of the wagon and all distances were reduced by a factor of 10^{15}?

8. Plot the probability density functions $\bar{\psi}\psi$ for a particle in each of the lowest three energy levels and in the nth level, where $n \gg 1$, in a one-dimensional, infinite-wall potential box. Discuss the graphs.

9. Show that a particle in an infinite rectangular potential well of width $2b$ has the same energy eigenvalues (except for sign) as a particle in an infinite rectangular box of width $2b$.

10. Find the energy eigenvalues and corresponding eigenfunctions of a particle in a potential box of the form

$$
\begin{aligned}
V(x) = \quad & V_1 && (x < 0), \\
= & -V_0 && (0 \le x \le b), \\
= & \ 0 && (x > b),
\end{aligned}
$$

where V_0 and V_1 are real, positive constants.

11. Show that the expectation value of the momentum of a particle that is bound in the nth energy eigenstate of an infinite-height, finite-width potential box is zero but that the eigenvalues of the momentum operator are $\pm (2mE_n)^{1/2}$. Comment on the reasonableness of these results.

12. Consider a particle of mass m in a potential of the form

$$
\begin{aligned}
V(x) = \quad & 0 && (x < 0 \ \text{ and } \ x > b), \\
= & -g(x) && (0 \le x \le b),
\end{aligned}
$$

where $g(x)$ is a positive, real function of x. Develop the flow diagram for a digital computer program by which you could determine all the energy eigenvalues and corresponding eigenfunctions in this well, regardless of its shape.

13. Compare the energy-level densities for an electron confined in one-dimensional boxes of widths 10^{-8} cm and 1 cm and comment on the significance of the difference in density.

14. Show that the number of energy levels in a one-dimensional rectangular potential well increases with the depth of the well. Calculate the number of energy levels in wells of width 10^{-12} (typical diameter of a nucleus) and depths 2, 10, and 20 Mev for a particle of the approximate mass of a nucleon, 10^{-24} g.

15. Show that the virtual levels in a one-dimensional well–barrier potential become more sharply defined with increasing width of the barrier. Comment on why one should expect this behavior.

16. Prove the following commutation relations:

(a) $[x, p_x] = i\hbar$,   (b) $[p_x, x] = -i\hbar$,
(c) $[p_y, x] = 0 = [x, p_y]$,   (d) $[p_x, p_y] = 0$,
(e) $[E, t] = i\hbar$,   (f) $[E, p_x] = 0$.

17. Assume that a particle within a one-dimensional infinite-well potential box of width $2b$ has equal probability of being in one of the three lowest energy states. (a) What is the complete time-dependent Schrödinger function for this particle? (b) Calculate the expectation value of the energy using this wave function. (c) Calculate and plot the position probability density using this wave function. (d) What are the possible outcomes of measurement of position, momentum, and energy of the system? (e) Show by computation that $\Delta x \, \Delta p_x$ is indeed $\geq \hbar/2$. (Remember that momentum is a vector quantity.)

18. Calculate the minimum depth required to bind a neutron or a proton in a one-dimensional square well of width 10^{-12} cm. Repeat the calculation for an electron and discuss the relative possibility of binding nucleons and electrons in nuclei.

19. Calculate the lowest two energy eigenvalues of a ball bouncing elastically under the influence of the force of gravity. Suggestion: look up Airy functions. (Don't be surprised if you can't solve this one in closed form.)

20. As an elementary example of the behavior of phase shifts, consider the problem of a single particle of mass m and positive total energy E moving in one dimension under the influence of a square-well potential of the form

$$V(x) = \quad \infty \quad (x < 0),$$
$$= -V_0 \quad (0 < x < b \quad \text{where} \quad V_0 > 0),$$
$$= \quad 0 \quad (x > b).$$

(a) Show that the wave function in the region $0 < x < b$ can be written in the form $\psi_1(x) = A \sin qx$, where A is real.

(b) Show that the wave function in the region $x > b$, which is to represent an incoming and outgoing wave, can be written in the form $\psi_2(x) = B \sin(kx + \delta)$, where B and δ are real.

(c) Show that the amplitude of $\psi_1(x)$ reaches a maximum whenever $\sin qb = \pm 1$, which corresponds to $\delta = (2n + 1)\pi/2$ where $n = 0, 1, 2, \ldots$.

QUANTUM MECHANICS
OF COLLISIONS 4

Our principal objective in this chapter is to develop wave-mechanical expressions for the cross sections of two-body collisions. The quantum-mechanical two-body problem is first reduced to two tractable one-body problems, as was done classically. The Schrödinger amplitude equation in the R-system is then solved in spherical coordinates.

After a brief detour to examine certain peculiarities of angular momentum in quantum mechanics, we will demonstrate that linear momentum and total energy are constants of the motion for conservative quantum-mechanical systems just as they are for classical systems. Thus all the classical kinematic results previously derived are shown to apply equally to quantum-mechanical systems.

After an additional detour to derive a quantum-mechanical expression for the probability current density, we will plunge directly into the major topic of the chapter, the wave-mechanical description of scattering.

The remainder of the chapter consists of the development of two distinct and widely useful methods for calculating cross sections, namely, the *method of partial waves* and the *Born approximation*. Also included is a description of a numerical procedure for determining the radial wave function for any real potential. This procedure can be used in conjunction with the method of partial waves to determine the cross sections for scattering from any real central potential that is likely to be found in practice. Complex potentials, which are employed to describe absorptive interactions, may be treated by a straightforward extension of this procedure (Chapter 7).

4.1 The Two-Particle Problem in Three Dimensions

The Schrödinger equation for a system of two particles in three-dimensional space interacting through a potential that is a function only of their relative coordinates will now be shown to reduce to two tractable one-particle equations. One equation is the Schrödinger equation that describes the motion of the center of mass; the other is the Schrödinger equation that describes the motion of the reduced-mass particle.

The Schrödinger function of the system will be, according to Postulate 1 (Chapter 3), a function of all the coordinates and of time,

$$\Psi = \Psi(x_1, y_1, z_1, x_2, y_2, z_2, t),$$

where (x_1, y_1, z_1) and (x_2, y_2, z_2) are the coordinates of particles 1 and 2, respectively. The Schrödinger function of the system is, by Postulate 3, the solution of the Schrödinger equation $H\Psi = E\Psi$. This can be written, using Postulate 2, as

$$\left[-\frac{\hbar^2}{2m_1}\left(\frac{\partial^2}{\partial x_1^2} + \frac{\partial^2}{\partial y_1^2} + \frac{\partial^2}{\partial z_1^2}\right) - \frac{\hbar^2}{2m_2}\left(\frac{\partial^2}{\partial x_2^2} + \frac{\partial^2}{\partial y_2^2} + \frac{\partial^2}{\partial z_2^2}\right) + V \right]\Psi$$

$$= i\hbar\frac{\partial}{\partial t}\,\Psi, \quad (4.1)$$

where m_1 and m_2 are the masses of particles 1 and 2, respectively. Note that intrinsic angular momentum, *spin*, has been excluded from present considerations since any such contribution to the total energy would, if taken into account, necessarily appear in the Hamiltonian.

If the potential V depends only on the relative coordinates, i.e., if

$$V = V(x_1 - x_2, \, y_1 - y_2, \, z_1 - z_2), \quad (4.2)$$

then a great simplification can be made. The coordinates of the center of mass of the two particles are, from Equation 2.14,

$$X = \frac{m_1 x_1 + m_2 x_2}{M},$$

$$Y = \frac{m_1 y_1 + m_2 y_2}{M}, \quad (4.3)$$

$$Z = \frac{m_1 z_1 + m_2 z_2}{M},$$

where $M \equiv m_1 + m_2$. The position coordinates of particle 1 relative to

particle 2 are

$$x = x_1 - x_2,$$

$$y = y_1 - y_2, \tag{4.4}$$

$$z = z_1 - z_2.$$

Equation 4.1 can be transformed readily to the center-of-mass coordinates X, Y, and Z and the position coordinates x, y, and z with the result that

$$\left[-\frac{\hbar^2}{2M} \left(\frac{\partial^2}{\partial X^2} + \frac{\partial^2}{\partial Y^2} + \frac{\partial^2}{\partial Z^2} \right) \right.$$

$$\left. -\frac{\hbar^2}{2m} \left(\frac{\partial^2}{\partial x^2} + \frac{\partial^2}{\partial y^2} + \frac{\partial^2}{\partial z^2} \right) + V(x,y,z) \right] \Psi = i\hbar \frac{\partial}{\partial t} \Psi, \tag{4.5}$$

where m is the reduced mass of the system,

$$m = \frac{m_1 m_2}{m_1 + m_2}. \tag{4.6}$$

Since the potential $V(x,y,z)$ is not a function of the time, the wave function can be separated into a product of spatial and time-dependent functions. In addition, a separation can be made into products of functions of the relative coordinates and functions of the center-of-mass coordinates. Substitution of

$$\Psi(x,y,z,X,Y,Z,t) = \psi(x,y,z)\,\psi_c(X,Y,Z)e^{-iEt/\hbar}e^{-iE_ct/\hbar} \tag{4.7}$$

into Equation 4.5 results in two Schrödinger amplitude equations,

$$\left(-\frac{\hbar^2}{2M} \nabla_c^2 \right) \psi_c = E_c \psi_c \tag{4.8}$$

and

$$\left(-\frac{\hbar^2}{2m} \nabla^2 + V \right) \psi = E\psi. \tag{4.9}$$

Equation 4.8, the quantum statement of the well-known classical fact that the center of mass moves like a free particle of mass $m_1 + m_2$, is of little interest. Equation 4.9, the Schrödinger amplitude equation in the relative coordinate system, describes the relative motion of the two particles by replacing them with a single particle of reduced mass m moving in a potential that is a function of the relative coordinates. This equation is of great interest, and the next section is devoted to its solution.

The separation constants E_c and E can be identified readily as the kinetic energy associated with the movement of the center of mass of the two particles and the total available energy in the C-system, respectively.

That is,

$$E_c = \tfrac{1}{2}(m_1 + m_2)\bar{u}^2$$

and

$$E = \tfrac{1}{2}m_1 v_1^2 + \tfrac{1}{2}m_2 v_2^2,$$

where $\bar{u}$ is the laboratory speed of the center of mass and where v_1 and v_2 are the C-system speeds of particles 1 and 2 before or after the interaction, which is assumed to be an elastic-scattering interaction. The generalization of this formalism to include nonelastic processes is described in Chapter 6.

4.2 Solution of the Schrödinger Equation in Spherical Coordinates

The Schrödinger amplitude equation in the R-system, Equation 4.9, may be written as

$$\left[-\frac{\hbar^2}{2m} \nabla^2 + V(\mathbf{r}) \right] \psi(\mathbf{r}) = E\psi(\mathbf{r}), \tag{4.10}$$

where $\mathbf{r} = (x,y,z)$ is the vector position of particle 1 when particle 2 is the origin, m is the reduced mass, E is the total energy in the C-system, and ∇^2 is the Laplacian operator.

Subsequent developments will be restricted to spherically symmetric potentials, i.e., potentials that are a function only of the distance between the particles r and not, for example, their directions in space. This is equivalent to restricting the analysis to central forces. If V is a function of r only, the Schrödinger amplitude equation can be separated in spherical coordinates. The Laplacian in spherical coordinates is

$$\nabla^2 = \frac{1}{r^2} \frac{\partial}{\partial r} \left(r^2 \frac{\partial}{\partial r} \right) + \frac{1}{r^2 \sin \theta} \frac{\partial}{\partial \theta} \left(\sin \theta \frac{\partial}{\partial \theta} \right) + \frac{1}{r^2 \sin^2 \theta} \frac{\partial^2}{\partial \phi^2},$$

$$\tag{4.11}$$

where, as usual, θ is the center-of-mass, not the laboratory, polar angle. The Schrödinger amplitude equation in spherical coordinates follows from substitution of Equation 4.11 into Equation 4.10:

$$\frac{1}{r^2} \frac{\partial}{\partial r} \left(r^2 \frac{\partial \psi}{\partial r} \right) + \frac{1}{r^2 \sin \theta} \frac{\partial}{\partial \theta} \left(\sin \theta \frac{\partial \psi}{\partial \theta} \right)$$

$$+ \frac{1}{r^2 \sin^2 \theta} \frac{\partial^2 \psi}{\partial \phi^2} + \frac{2m}{\hbar^2} \left[E - V(r) \right] \psi = 0. \tag{4.12}$$

Equation 4.12 may be separated into radial and angular parts by substituting

$$\psi(r,\theta,\phi) = R(r)Y(\theta,\phi) \tag{4.13}$$

and dividing through by RY. The result is

$$\frac{1}{R}\frac{d}{dr}\left(r^2\frac{dR}{dr}\right) + \frac{2mr^2}{\hbar^2}[E - V(r)]$$

$$= -\frac{1}{Y}\left[\frac{1}{\sin\theta}\frac{\partial}{\partial\theta}\left(\sin\theta\frac{\partial Y}{\partial\theta}\right) + \frac{1}{\sin^2\theta}\frac{\partial^2 Y}{\partial\phi^2}\right] \quad (4.14)$$

The left-hand side of Equation 4.14 depends only on the variable r; the right-hand side depends only on the variables θ and ϕ. In order that the equation hold for all values of r, θ, and ϕ, both sides must equal a constant, β. Thus two equations are obtained:

Radial equation:

$$\frac{1}{r^2}\frac{d}{dr}\left(r^2\frac{dR}{dr}\right) + \left\{\frac{2m}{\hbar^2}[E - V(r)] - \frac{\beta}{r^2}\right\}R = 0, \quad (4.15)$$

Angular equation:

$$\frac{1}{\sin\theta}\frac{\partial}{\partial\theta}\left(\sin\theta\frac{\partial Y}{\partial\theta}\right) + \frac{1}{\sin^2\theta}\frac{\partial^2 Y}{\partial\phi^2} + \beta Y = 0.$$

The angular equation can be further separated by substituting

$$Y(\theta,\phi) = T(\theta)F(\phi)$$

and then dividing through by TF. The result is two ordinary differential equations,

θ-equation: $\qquad \dfrac{1}{\sin\theta}\dfrac{d}{d\theta}\left(\sin\theta\dfrac{dT}{d\theta}\right) + \left(\beta - \dfrac{\gamma^2}{\sin^2\theta}\right)T = 0, \quad (4.16)$

ϕ-equation: $\qquad \dfrac{d^2F}{d\phi^2} + \gamma^2 F = 0, \quad (4.17)$

where γ^2 is the separation constant, chosen as a square with an eye to the solution of the ϕ-equation, which will be given later.

The final solution for ψ will be of the form

$$\psi(r,\theta,\phi) = R(r)\,T(\theta)\,F(\phi), \quad (4.18)$$

where the functions R, T, and F are the solutions to Equations 4.15, 4.16, and 4.17, respectively. Before starting these solutions, note that only the radial equation depends on the potential; the θ- and ϕ-equations are independent of the potential. This means that the solutions $T(\theta)$ and $F(\phi)$ are the same for all central potentials but that $R(r)$ depends on the specific functional form of the potential $V(r)$.

In reading the solutions which follow, pay particular attention to the way in which quantum numbers, those magical integers that are the sign and nearly the substance of quantum mechanics, naturally emerge from the differential equations.

Solution of the φ-equation

Equation 4.17 for $F(\phi)$ can be recast as a typical eigenvalue equation,

$$\frac{d^2}{d\phi^2} F(\phi) = -\gamma^2 F(\phi). \tag{4.19}$$

The eigenfunctions of this equation are

$$F(\phi) = A e^{i\gamma\phi}, \tag{4.20}$$

where γ is, momentarily, any real constant and A is any number, real or complex. But a proper wave function should be single valued according to Postulate 4 (Chapter 3). This means, when applied to the azimuthal angle ϕ, that $\psi(r,\theta,\phi)$ should equal $\psi(r,\theta,\phi + n2\pi)$ where n is any integer. The only ϕ-dependent part of the wave function is $F(\phi)$; therefore this condition requires that

$$A e^{i\gamma\phi} = A e^{i\gamma(\phi + n2\pi)}. \tag{4.21}$$

Equation 4.21 is valid only if γ is an integer, i.e., only if $\gamma = m_l$, where

$$m_l = 0, \pm 1, \pm 2, \ldots. \tag{4.22}$$

The integers m_l associated with the coordinate ϕ are called the *azimuthal quantum numbers*.

The most straightforward way to ensure that ψ will be normalized is to normalize separately each of the functions R, T, and F. It is therefore required that

$$\int_0^{2\pi} \bar{F}(\phi) F(\phi) \, d\phi = 1. \tag{4.23}$$

If the constant A in Equation 4.20 is expressed in the form $A = A_0 e^{i\delta}$, where A_0 is the amplitude of A and δ is the constant phase of A, then Equation 4.23 becomes

$$\int_0^{2\pi} (A_0 e^{-i\delta} e^{-im_l\phi})(A_0 e^{i\delta} e^{im_l\phi}) \, d\phi = 1,$$

which immediately yields $A_0 = 1/\sqrt{2\pi}$. Thus the normalized eigenfunctions of the ϕ-equation are

$$F_{m_l}(\phi) = \frac{1}{\sqrt{2\pi}} e^{im_l\phi}, \tag{4.24}$$

where $m_l = 0, \pm 1, \pm 2, \ldots$. The constant phase factor $e^{i\delta}$ has been dropped from the expression for $F_{m_l}(\phi)$; it will eventually be incorporated in the arbitrary phase of the complete wave function $\psi(r,\theta,\phi)$. Note that the eigenfunctions are now being subscribed with integers m_l to call attention to the fact that the solution to the ϕ eigenvalue equation is a family of eigenfunctions $F_0, F_{-1}, F_{+1}, F_{-2}, F_{+2}$, etc.

Solution of the θ-equation

Equation 4.16 for the function $T(\theta)$ becomes somewhat less cumbersome if a familiar change of variables is introduced, namely,

$$\mu \equiv \cos \theta. \tag{4.25}$$

The change of variables is carried out to give

$$\frac{d}{d\theta} = \frac{d\mu}{d\theta}\frac{d}{d\mu} = -\sin\theta\frac{d}{d\mu}$$

so that the first term in Equation 4.16 becomes

$$\frac{1}{\sin\theta}\frac{d}{d\theta}\left(\sin\theta\frac{dT}{d\theta}\right) = (-1)\frac{d}{d\mu}\left(-\sin^2\theta\frac{dT}{d\mu}\right)$$

$$= \frac{d}{d\mu}\left[(1-\mu^2)\frac{dT}{d\mu}\right]$$

$$= (1-\mu^2)\frac{d^2T}{d\mu^2} - 2\mu\frac{dT}{d\mu},$$

and the entire equation is in typical eigenvalue form,

$$\left[-(1-\mu^2)\frac{d^2}{d\mu^2} + 2\mu\frac{d}{d\mu} + \frac{m_l^2}{1-\mu^2}\right]T(\mu) = \beta T(\mu). \tag{4.26}$$

To solve Equation 4.26, we must find both the eigenfunctions $T(\mu)$ and the eigenvalues β. Following the procedure of Rojansky,* $T(\mu)$ is written in the form

$$T(\mu) = f(\mu)(1-\mu^2)^{|m_l|/2}, \tag{4.27}$$

where $f(\mu)$ is an undetermined function of μ. Substituting this expression for $T(\mu)$ into Equation 4.26, we obtain

$$(1-\mu^2)f'' - 2(|m_l|+1)\mu f' + (\beta - |m_l| - m_l^2)f = 0, \tag{4.28}$$

where $f' \equiv df/d\mu$ and $f'' \equiv d^2f/d\mu^2$. Assume that $f(\mu)$ can be expressed

---

*V. Rojansky, *Introducing Quantum Mechanics*, Prentice-Hall, Inc., Englewood Cliffs, N. J., 1938.

as a power series in μ; that is,

$$f(\mu) = \sum_j a_j \mu^j, \qquad (4.29)$$

where $j = \tau, \tau + 1, \tau + 2, \ldots$ and $a_\tau \neq 0$. The values of τ and the coefficients a_j are constants still to be determined. After substituting $f(\mu)$ from Equation 4.29 into Equation 4.28, we find that

$$\sum_j \{a_j(j)(j-1)\mu^{j-2} + a_j[\beta - (j + |m_l|)(j + |m_l| + 1)]\mu^j\} = 0. \qquad (4.30)$$

If Equation 4.30 is to hold for all values of μ, the coefficients of each μ^κ must vanish; that is,

$$a_{\kappa+2}(\kappa + 2)(\kappa + 1) + a_\kappa[\beta - (\kappa + |m_l|)(\kappa + |m_l| + 1)] = 0$$

or, equivalently,

$$\frac{a_{\kappa+2}}{a_\kappa} = \frac{(\kappa + |m_l|)(\kappa + |m_l| + 1) - \beta}{(\kappa + 2)(\kappa + 1)}, \qquad (4.31)$$

where $\kappa = 0, 1, 2, \ldots$.

We will return to Equation 4.31 shortly. First, to determine τ, the lowest power of μ in our series solution for $f(\mu)$, note that the coefficient of the lowest power of μ in Equation 4.30 is $a_\tau(\tau)(\tau - 1)$. Since $a_\tau \neq 0$, the value of τ must be 0 or 1. Let the two series obtained from $\tau = 0$ and $\tau = 1$ be denoted by $f_0(\mu)$ and $f_1(\mu)$, respectively:

$$f_0(\mu) = a_0 + a_2\mu^2 + a_4\mu^4 + \ldots,$$
$$f_1(\mu) = a_1\mu + a_3\mu^3 + a_5\mu^5 + \ldots. \qquad (4.32)$$

The general solution for $f(\mu)$ then has the form

$$f(\mu) = Af_0(\mu) + Bf_1(\mu). \qquad (4.33)$$

If either f_0 or f_1 were an infinite series, $f(\mu)$, $T(\mu)$, and ψ would be infinite at $\mu = +1$. Since this would violate Postulate 4, it is necessary that f_0 and f_1 be finite polynomials, and we are forced to the conclusion that the eigenvalues are those values of β that produce finite series. From Equation 4.31, we see that the condition

$$\beta = (\kappa + |m_l|)(\kappa + |m_l| + 1) \qquad (4.34)$$

causes the vanishing of $a_{\kappa+2}, a_{\kappa+4}$, and so on. Thus, to obtain well-behaved solutions of the θ-equation, we must (1) choose a β given by Equation 4.34 where κ is an integer greater than or equal to zero causing f_0 to terminate if κ is even or f_1 to terminate if κ is odd and (2) set the arbitrary multiplier (A or B) of the nonterminating series equal to zero.

An investigation of the form of the eigenvalues β is fruitful and easily accomplished. Since $|m_l|$ and κ both take on the values 0, 1, 2, ..., the quantity $(\kappa + |m_l|)$ takes on the values 0, 1, 2, ... also. Therefore we naturally define

$$l \equiv (\kappa + |m_l|), \tag{4.35}$$

where $l = 0, 1, 2, \ldots$. Equation 4.34 then becomes

$$\beta = l(l + 1), \tag{4.36}$$

where $l = 0, 1, 2, \ldots$, and we obtain the important result that the eigenvalues of the θ-equation are the integers $l(l + 1)$, where l takes on all integer values greater than or equal to zero. The integers l associated with the coordinate θ are called the polar quantum numbers.

It follows from Equation 4.35 that, for a given value of the polar quantum number l, the azimuthal quantum number m_l can only take on the $2l + 1$ integer values between $-l$ and $+l$. For example, if $l = 2$, κ can take on the values 0, 1, or 2 and m_l can take on the values ± 2, ± 1, or 0.

The eigenfunctions of the θ-equation remain to be found. The well-behaved solutions of the θ-equation are those for which $\beta = l(l + 1)$, therefore Equation 4.26 may be written

$$(1 - \mu^2)\frac{d^2 T}{d\mu^2} - 2\mu \frac{dT}{d\mu} + \left[l(l + 1) - \frac{m_l^2}{1 - \mu^2} \right] T = 0. \tag{4.37}$$

Equation 4.37 is Legendre's differential equation of degree l and order m_l. The solutions are the well-known, tabulated, associated Legendre functions of the first kind, which are given* by

$$P_l^{m_l}(\mu) = \frac{1}{2^l l!} (1 - \mu^2)^{|m_l|/2} \frac{d^{l + |m_l|}}{d\mu^{l + |m_l|}} (\mu^2 - 1)^l. \tag{4.38}$$

It will be left as an exercise for the student to show that these are indeed the functions of μ that would have been obtained had the solution by means of finite power series been pursued to completion.

An associated Legendre function of the first kind that is of order zero $(m_l = 0)$ is called a Legendre polynomial. It is customarily written without the superscript zero,

$$P_l(\mu) = \frac{1}{2^l l!} \frac{d^l}{d\mu^l} (\mu^2 - 1)^l. \tag{4.39}$$

---

*For the differential equation, its solutions, and their tabulated values see either the *Handbook of Mathematical Functions* (AMS-55, National Bureau of Standards, 1964) or E. Jahnke and F. Emde's, *Tables of Functions* (Dover Publications, Inc., New York, 1945).

This particular expression for the Legendre polynomials is known as Rodriques' formula. The first four Legendre polynomials are

$$P_0(\mu) = 1 \qquad\qquad P_1(\mu) = \mu,$$

$$P_2(\mu) = (3\mu^2 - 1)/2 \qquad P_3(\mu) = (5\mu^3 - 3\mu)/2.$$

It can be readily verified that the Legendre polynomials satisfy the following orthogonality relation:

$$\int_{-1}^{+1} P_l(\mu)P_{l'}(\mu)\,d\mu = \frac{2}{2l+1}\,\delta_{ll'}. \tag{4.40}$$

The eigenfunctions of Equation 4.37 are of the form $T_l^{m_l} = CP_l^{m_l}$, where C remains to be determined by the normalizing condition

$$\int_{-1}^{+1} \overline{T}_l^{m_l} T_l^{m_l}\,d\mu = 1.$$

The normalization* results in

$$T_l^{m_l}(\mu) = (-1)^{(m_l + |m_l|)/2}\left[\frac{(2l+1)(l - |m_l|)!}{2(l + |m_l|)!}\right]^{1/2} P_l^{m_l}(\mu). \tag{4.41}$$

These are the eigenfunctions of the θ-equation. We can readily verify that they are orthonormal, i.e., that they satisfy the relation

$$\int_{-1}^{+1} \overline{T}_l^{m_l} T_{l'}^{m_l}\,d\mu = \delta_{ll'}. \tag{4.42}$$

The explicit forms of $T_l^{m_l}(\mu)$ for $l = 0$ to 3 are shown in Table 4.1.

Table 4.1—Normalized Associated Legendre Functions of the first kind $T_l^{m_l}(\mu)$ for $l = 0,1,2,$ and 3

m_l	$l = 0$	$l = 1$	$l = 2$	$l = 3$
0	$\sqrt{\tfrac{1}{2}}$	$\sqrt{\tfrac{3}{2}}\,\mu$	$\sqrt{\tfrac{5}{8}}(3\mu^2 - 1)$	$\sqrt{\tfrac{7}{8}}(5\mu^3 - 3\mu)$
± 1		$\mp\sqrt{\tfrac{3}{4}}\sqrt{1 - \mu^2}$	$\mp\sqrt{\tfrac{15}{4}}\,\mu\sqrt{1 - \mu^2}$	$\mp\sqrt{\tfrac{21}{32}}(5\mu^2 - 1)\sqrt{1 - \mu^2}$
± 2			$\sqrt{\tfrac{15}{16}}(1 - \mu^2)$	$\sqrt{\tfrac{105}{16}}\,\mu(1 - \mu^2)$
± 3				$\mp\sqrt{\tfrac{35}{32}}(1 - \mu^2)\sqrt{1 - \mu^2}$

*See E. U. Condon and G. H. Shortley, *The Theory of Atomic Spectra* (Cambridge University Press, New York, 1935), p. 52.

The r-equation

The next step in the development of the complete solution for $\psi(r,\theta,\phi)$ is the substitution of the expression $\beta = l(l + 1)$ into the radial equation, Equation 4.15, to obtain the eigenvalue equation for the radial wave function,

$$\left\{ \frac{-\hbar^2}{2m} \frac{1}{r^2} \frac{d}{dr} \left(r^2 \frac{d}{dr} \right) + \left[V(r) + \frac{\hbar^2}{2m} \frac{l(l + 1)}{r^2} \right] \right\} R_l(r) = E_l R_l(r).$$

(4.43)

The subscript l is attached to R and E in Equation 4.43 to call attention to the fact that the eigenvalues and eigenfunctions of the radial differential equation depend on the value of l.

The first thing to notice is that the radial eigenfunctions are determined by an effective potential that is the sum of the regular potential between the two particles $V(r)$ and a repulsive centrifugal potential $V_c(r)$ where

$$V_c(r) = \frac{l(l + 1)\hbar^2}{2mr^2}.$$

(4.44)

When we compare this with the classical centrifugal potential given by

$$V_c(r) = \frac{L^2}{2mr^2},$$

we find that the total orbital angular momentum L, which classically can have any value, can have quantum mechanically only the eigenvalues

$$L = \hbar\sqrt{l(l + 1)},$$

(4.45)

where $l = 0, 1, 2, \ldots$. The total orbital angular momentum is quantized; i.e., it has one of the values $L = 0, \sqrt{2}\,\hbar, \sqrt{6}\,\hbar, \ldots, \sqrt{l(l + 1)}\,\hbar$. The allowed values of the various components of the vector $\mathbf{L}$ will be examined in the next section.

Our present goal is the determination of the eigenvalues and associated eigenfunctions of Equation 4.43 for a given potential $V(r)$. However, until the potential is specified, no further progress is possible. There are a limited number of potentials for which Equation 4.43 may be solved in closed form. In this section, only the two most widely useful of these potentials are considered, the spherical square-well potential and the inverse-r potential. In Section 4.8, a numerical method of solution is developed that has sufficient generality to yield the eigenvalues and eigenfunctions for any real potential of practical interest.

The Spherical Square Well

The spherical square-well potential is defined by the conditions that

$$V(r) = \begin{cases} -V_0 & \text{(for } 0 < r < b), \\ 0 & \text{(for } r > b), \end{cases}$$

where V_0 is a real positive constant. The effective potential function for the spherical square well, including the centrifugal potential, is

$$V_e(r) = V(r) + \frac{l(l+1)\hbar^2}{2mr^2}.$$

If the total orbital angular-momentum quantum number l is zero, the effective potential is simply $V(r)$, a spherical square well. But for all $l > 0$, the effective potential includes the centrifugal potential term. The larger the value of l, the greater the centrifugal potential. A typical effective potential for a nonzero l is sketched in Figure 4.1.

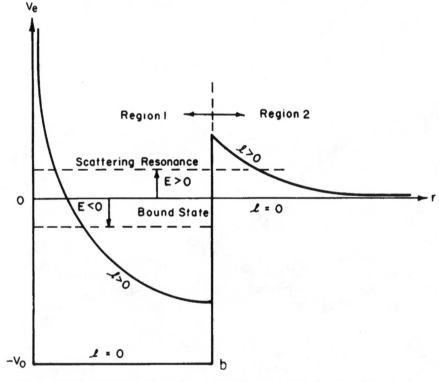

Figure 4.1—The effective spherical square-well potentials for $l = 0$ and $l > 0$. Also shown are a bound level ($E < 0$) and a virtual level ($E > 0$).

We will presently show that the eigenvalues and eigenfunctions of the spherical square-well potential are qualitatively similar to the eigenvalues and eigenfunctions of the analogous one-dimensional potentials. Two cases must be distinguished, namely, $l = 0$ and $l > 0$. When $l = 0$, the eigenvalues and eigenfunctions of the Schrödinger equation for the spherical square well are qualitatively similar to eigenvalues and eigenfunctions of the one-dimensional square well. If the total energy E is negative, the particle is actually trapped in the well, the eigenvalues are discrete, and the eigenfunctions fall exponentially to zero in Region 2, the classically inaccessible region outside the well. All positive values of E are eigenvalues, but there may be certain positive values of E for which the inside wave function becomes large with respect to its value at nearby energies. These correspond to broad scattering resonances, energies at which the particle tends to spend more time within the potential than outside it.

When $l > 0$, the eigenvalues and eigenfunctions of the Schrödinger equation for the spherical square well are qualitatively similar to the eigenvalues and eigenfunctions of the one-dimensional square well plus barrier of Figure 3.12. There are bound states of negative total energy unless the repulsive centrifugal potential is sufficiently great to decrease the effective well depth to such an extent that a bound state cannot exist. All positive values of E are eigenvalues; however, there may be scattering resonances (virtual levels) in this energy region.

Bound States in the Spherical Square Well

First consider the case of a particle whose total energy is negative. We wish to find the energy eigenvalues and corresponding eigenfunctions of the bound states. When the independent variable is changed from r to the dimensionless variable ρ, defined by

$$\rho \equiv qr \equiv \sqrt{\frac{2m(V_0 - |E|)}{\hbar^2}}\, r, \tag{4.46}$$

the radial Schrödinger equation (Equation 4.43) takes the form

$$\rho^2 \frac{d^2 R_l}{d\rho^2} + 2\rho \frac{dR_l}{d\rho} + [\rho^2 - l(l + 1)]R_l = 0. \tag{4.47}$$

Equation 4.47 is called the spherical Bessel equation.* Before solving the problem at hand, we must briefly digress into the properties of the solutions of this equation. Particular solutions of the spherical Bessel equation are of three types:

---

*See the National Bureau of Standards *Handbook of Mathematical Functions* (AMS-55, 1964), for definitions, properties, graphs, and tables of spherical Bessel functions.

1. Spherical Bessel functions of the first type (also called simply *spherical Bessel functions*),

$$j_l(\rho);$$

2. Spherical Bessel functions of the second type (also called *spherical Neumann functions*),

$$n_l(\rho);$$

3. Spherical Bessel functions of the third type (also called *spherical Hankel functions*),

$$h_l^{(+)}(\rho) = j_l + in_l$$

$$h_l^{(-)}(\rho) = j_l - in_l,$$

where $i = \sqrt{-1}$.

Expressions for spherical Bessel functions in terms of sines and cosines may be found by use of Rayleigh's formulas,

$$j_l(\rho) = \rho^l \left(-\frac{1}{\rho} \frac{d}{d\rho} \right)^l \frac{\sin \rho}{\rho}$$

and

$$n_l(\rho) = -\rho^l \left(-\frac{1}{\rho} \frac{d}{d\rho} \right)^l \frac{\cos \rho}{\rho}. \tag{4.48}$$

Explicit expressions for the first three j's and n's are

$$j_0(\rho) = \frac{\sin \rho}{\rho} \qquad\qquad n_0(\rho) = -\frac{\cos \rho}{\rho}$$

$$j_1(\rho) = \frac{\sin \rho}{\rho^2} - \frac{\cos \rho}{\rho} \qquad\qquad n_1(\rho) = -\frac{\cos \rho}{\rho^2} - \frac{\sin \rho}{\rho} \tag{4.49}$$

$$j_2(\rho) = \left(\frac{3}{\rho^3} - \frac{1}{\rho} \right) \sin \rho \qquad\qquad n_2(\rho) = -\left(\frac{3}{\rho^3} - \frac{1}{\rho} \right) \cos \rho$$

$$\qquad\quad - \frac{3}{\rho^2} \cos \rho \qquad\qquad\qquad\quad - \frac{3}{\rho^2} \sin \rho$$

For small ρ the leading terms are

$$j_l(\rho) \xrightarrow[\rho \to 0]{} \frac{\rho^l}{1 \cdot 3 \cdot 5 \cdots (2l+1)},$$

$$n_l(\rho) \xrightarrow[\rho \to 0]{} \frac{-1 \cdot 1 \cdot 3 \cdot 5 \cdots (2l-1)}{\rho^{l+1}}. \tag{4.50}$$

For large ρ the asymptotic forms are

$$j_l(\rho) \xrightarrow[\rho \to \infty]{} \frac{1}{\rho} \cos\left[\rho - \frac{(l+1)\pi}{2}\right] = \frac{1}{\rho} \sin\left(\rho - \frac{l\pi}{2}\right),$$

$$n_l(\rho) \xrightarrow[\rho \to \infty]{} \frac{1}{\rho} \sin\left[\rho - \frac{(l+1)\pi}{2}\right] = -\frac{1}{\rho} \cos\left(\rho - \frac{l\pi}{2}\right). \tag{4.51}$$

Graphs of the first few spherical Bessel and Neumann functions are shown in Figure 4.2. The reader is forewarned that these and the related spherical

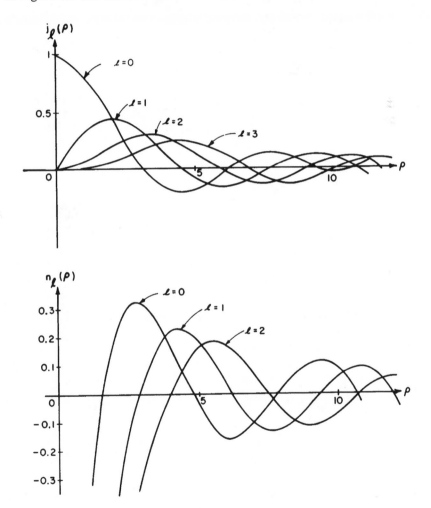

Figure 4.2—The spherical Bessel functions $j_l(\rho)$ and spherical Neumann functions $n_l(\rho)$.

Hankel functions will be encountered over and over again in subsequent sections. Much time will be saved by simply memorizing the properties of these functions as given by Equations 4.48 to 4.51.

We now return to the calculation of the bound states in the spherical square well. Since $R_l(r)$ must remain finite as r approaches zero, the wave function within the well must not contain spherical Bessel functions of the second or third types because these functions approach minus infinity at the origin. Thus the inside wave function must be of the form

$$R_l^{in}(r) = A_l j_l(qr) \tag{4.52}$$

The wave equation in Region 2 can be put into the form of a spherical Bessel equation (Equation 4.47) by redefining ρ to be

$$\rho \equiv igr \equiv i \sqrt{\frac{2m|E|}{\hbar^2}} \, r. \tag{4.53}$$

Since the domain of ρ outside the well does not extend to zero, spherical Bessel functions of the second or third types may appear in the solution. The function $h_l^{\{+\}}(\rho)$ is chosen because it decreases exponentially for large values of ρ. This is readily seen by writing the asymptotic form of $h_l^{\{+\}}(\rho)$,

$$h_l^{\{+\}}(\rho) = j_l(\rho) + i n_l(\rho) \xrightarrow[\rho \to \infty]{} \frac{1}{\rho} \exp\left\{ i \left[\rho - \frac{(l+1)\pi}{2} \right] \right\},$$

and, since $\rho = igr$, the term $h_l^{\{+\}}(r)$ approaches e^{-gr}/gr times a constant phase factor as r approaches infinity. Thus the outside wave function for a bound state with total orbital angular-momentum quantum number l is

$$R_l^{out}(r) = B_l[j_l(igr) + i n_l(igr)]. \tag{4.54}$$

The energy levels for any fixed value of l are now obtained by applying the usual continuity conditions on R_l and the derivative of R_l at the interface. It will be left as an exercise to show that, when $l = 0$, the implicit expression for the energy eigenvalues is

$$\xi = -\alpha \cot \alpha, \tag{4.55}$$

where

$$\alpha = b \sqrt{\frac{2m(V_0 - |E|)}{\hbar^2}}$$

and

$$\xi = b \sqrt{\frac{2m|E|}{\hbar^2}}, \tag{4.56}$$

so that

$$\alpha^2 + \xi^2 = \frac{2m}{\hbar^2} V_0 b^2. \tag{4.57}$$

The eigenvalues are obtained from the intersections on an α vs. ξ plot of the circles defined by Equation 4.57 with the curves defined by Equation 4.55. We can easily show that there is no intersection, and therefore no bound state, if the well strength $V_0 b^2$ is less than $(\hbar^2/2m)(\pi/2)^2$.

Comparing Equations 4.55 and 3.161, we conclude that the $l = 0$ energy eigenvalues for the spherical square well are identical to one-half of those of the one-dimensional square well. The one-dimensional square well has additional solutions because it extends from $-b$ to $+b$ and therefore may contain both even and odd eigenfunctions.

When $l = 1$, the implicit expression for the energy eigenvalues can readily be shown by imposing the usual interface conditions to be

$$\frac{1}{\xi} + \frac{1}{\xi^2} = \frac{\cot \alpha}{\alpha} - \frac{1}{\alpha^2}. \tag{4.58}$$

The intersections of the curves defined by Equations 4.57 and 4.58 on a plot of α vs. ξ determine the $l = 1$ energy eigenvalues. It is easy to show that there is no intersection, hence no bound state, if the well strength $V_0 b^2$ is less than $(\hbar^2/2m)\pi^2$. The corresponding minimum well strength for $l = 0$ particles was only one-fourth this amount. We therefore conclude that a deeper well is required to bind particles of higher orbital angular momentum. This result might have been expected since the greater the value of l, the shallower the effective potential well. In fact, the repulsive centrifugal force both raises the effective well and narrows it, as can be seen in Figure 4.1. Both effects raise the height of the lowest bound level.

Similar, though algebraically more complicated, implicit expressions can be obtained for the bound-energy levels corresponding to higher orbital angular-momentum quantum numbers.

Scattering Resonances in the Spherical Square Well
Now consider the case of a particle whose total energy E is positive. The objective of our analysis is to find the ratio of the inside to the outside probability density as a function of energy so that we can determine the energies at which scattering resonances (virtual levels) occur. The radial Schrödinger equation (Equation 4.43) is reduced to the form of the spherical Bessel equation (Equation 4.47) by defining new independent variables:

$$\rho \equiv qr \equiv \sqrt{\frac{2m(V_0 + E)}{\hbar^2}}\, r \tag{4.59a}$$

inside the well ($r < b$) and

$$\rho \equiv kr \equiv \sqrt{\frac{2mE}{\hbar^2}}\, r \tag{4.59b}$$

outside the well ($r > b$). Inside the well the appropriate solution for a given l value is the well-behaved spherical Bessel function of the first type of order l

$$R_l^{in} = A_l j_l(qr). \tag{4.60}$$

Outside the well the appropriate solution is the linear combination of particular solutions that asymptotically represents incoming and outgoing free particles of angular-momentum quantum number l. These solutions are the spherical Hankel functions $h_l^{(-)}$ and $h_l^{(+)}$. These functions exhibit the proper asymptotic behavior, namely,

$$h_l^{(-)}(kr) \equiv j_l(kr) - in_l(kr) \xrightarrow[r \to \infty]{} \frac{1}{kr}\exp\left\{-i\left[kr - \frac{(l+1)\pi}{2}\right]\right\}$$

and

$$h_l^{(+)}(kr) \equiv j_l(kr) + in_l(kr) \xrightarrow[r \to \infty]{} \frac{1}{kr}\exp\left\{+i\left[kr - \frac{(l+1)\pi}{2}\right]\right\}.$$

The complete solution outside the well has the form

$$R_l^{out}(kr) = B_l h_l^{(-)}(kr) + C_l h_l^{(+)}(kr),$$

where B_l and C_l are undetermined complex constants. Since no attempt is being made at present to describe absorptive processes, every particle incident on the well must be reflected. Hence the magnitudes of the incoming and outgoing waves must be equal, i.e., $|B_l| = |C_l|$; and C_l can differ from B_l at most in phase. To account for this possible phase difference, we may write

$$C_l = B_l e^{2i\delta_l},$$

where δ_l is a phase angle whose principal part has a value between 0 and π. When the expression for C_l is substituted into the equation above, the wave function outside the well takes the form

$$
\begin{aligned}
R_l^{out}(kr) &= B_l[h_l^{(-)}(kr) + e^{2i\delta_l}\, h_l^{(+)}(kr)] \\
&= B_l e^{i\delta_l}[e^{-i\delta_l}\, h_l^{(-)}(kr) + e^{i\delta_l}\, h_l^{(+)}(kr)] \\
&= B_l e^{i\delta_l}[(\cos\delta_l - i\sin\delta_l)(j_l - in_l) \\
&\qquad\qquad + (\cos\delta_l + i\sin\delta_l)(j_l + in_l)] \\
&= 2B_l e^{i\delta_l}[(\cos\delta_l)j_l - (\sin\delta_l)n_l].
\end{aligned}
$$

Note that the imaginary components of the wave function have vanished. This might have been expected since the interior wave function $A_l j_l$ is real (except for the possibly complex multiplicative constant) and since all parameters in the problem, i.e., k, q, V_0, and l, are real.

The factor $2e^{i\delta_l}$ now will be absorbed in the unknown constant B_l, and the outside wave function will be expressed as

$$R_l^{\text{out}}(kr) = B_l[(\cos \delta_l)j_l(kr) - (\sin \delta_l)n_l(kr)]. \tag{4.61}$$

There are now three unknowns in the solution: A_l, B_l, and δ_l. The desired ratio of A_l to B_l can be obtained by the usual procedure of imposing the continuity conditions on the wave function and its first derivative at the interface between the inner and outer regions.

$$A_l j_l(qr)|_{r=b} = B_l[(\cos \delta_l)j_l(kr) - (\sin \delta_l)n_l(kr)]|_{r=b} \tag{4.62a}$$

$$A_l \frac{d}{dr} j_l(qr)\bigg|_{r=b} = B_l\left[(\cos \delta_l)\frac{d}{dr}j_l(kr) - (\sin \delta_l)\frac{d}{dr}n_l(kr)\right]\bigg|_{r=b} \tag{4.62b}$$

Dividing one of these equations by the other, we eliminate both A_l and B_l; so the resulting equation can be solved for $\tan \delta_l$:

$$\tan \delta_l = \frac{j'_l(kb)j_l(qb) - j'_l(qb)j_l(kb)}{n'_l(kb)j_l(qb) - j'_l(qb)n_l(kb)}, \tag{4.63}$$

where the primes indicate differentiation with respect to r and the arguments indicate substitution after the differentiation. For example,

$$j'(kb) \equiv \frac{d}{dr} j(kr)\bigg|_{r=b}.$$

Since δ_l is a phase angle whose principal part lies in the range 0 to π, $\tan \delta_l$ uniquely determines both $\sin \delta_l$ and $\cos \delta_l$. The sine must be positive, and the cosine must have the same algebraic sign as the tangent. After solving Equation 4.63 for $\sin \delta_l$ and $\cos \delta_l$ and substituting the solutions into Equation 4.62a, we find that

$$\left|\frac{A_l}{B_l}\right|^2 = \frac{[j'_l(kb)n_l(kb) - n'_l(kb)j_l(kb)]^2}{[j'_l(kb)j_l(qb) - j'_l(qb)j_l(kb)]^2 + [n'_l(kb)j_l(qb) - j'_l(qb)n_l(kb)]^2}. \tag{4.64}$$

To determine the behavior of the ratio of wave-function amplitudes as a function of E for a given l value, we substitute the appropriate spherical Bessel functions into Equation 4.64 and carry out the elementary but tedious algebra. As an example, for $l = 0$ we substitute $j_0(\rho) = (\sin \rho)/\rho$ and $n_0(\rho) = (-\cos \rho)/\rho$. From this

$$\left|\frac{A_0}{B_0}\right|^2 = \frac{(q/k)^2}{\sin^2 qb + (q/k)^2 \cos^2 qb}$$

where $(q/k)^2 = (V_0 + E)/E$. This equation has maxima whenever $\sin^2 qb = 1$; that is, when qb is an odd multiple of $\pi/2$. Thus, from the definition of q, the equation has maxima when

$$\frac{\sqrt{2m(V_0 + E)}}{\hbar} b = n\frac{\pi}{2},$$

where n is an odd integer whose least value is not necessarily unity. It is instructive to write this equation in terms of the de Broglie wavelength of the particle within the well. Since $\lambda = h/\sqrt{2m(V_0 + E)}$, we write

$$\frac{b}{\lambda} = \frac{n}{4},$$

where n is an odd integer.

Hence, a maximum in the internal wave function occurs whenever the energy is such that the well contains an odd integral multiple of quarter-wavelengths. It is interesting to note that this condition for virtual levels is the same as the condition for acoustical resonances in a pipe having one open end. Such a pipe will resonate when it contains any odd number of quarter-wavelengths.

The values of the $l = 0$ resonance energies can be found by solving the equation above for E:

$$E_n = \frac{1}{32m}\left(\frac{nh}{b}\right)^2 - V_0,$$

where n is any odd integer such that $E > 0$. Do not jump to the conclusion that these scattering resonances are responsible for the sharp, closely spaced peaks that have been observed in neutron–nucleus cross sections. To refute this conclusion, we need only insert reasonable nuclear parameters into the preceding equation. For example, if we insert $m = 1.67 \times 10^{-24}$ g, $b \doteq 10^{-12}$ cm, and $V_0 \doteq 10$ Mev, we quickly find that these scattering resonances have separations of the order of 1 to 10 Mev and so cannot account for the much more closely spaced neutron–nucleus resonances. To treat the theory of the fine structure of these cross sections, we must consider the internal structure of the nucleus, particularly the energy levels, in considerably more detail (see Chapter 6). Actually, the broad scattering resonances are associated with certain aspects of the gross structure of neutron–nucleus cross sections, as explained in Chapter 7.

It will be left as an exercise for the student to calculate the resonance energies for higher l values. The results will be found to be similar to those for $l = 0$ both qualitatively and in order of magnitude.

The Inverse-r Potential

The *inverse-r* or *Coulomb potential* has the form

$$V(r) = \frac{a}{r},$$ (4.65)

where $0 < r < \infty$. We naturally expect to find bound states only if the force corresponding to the Coulomb potential is attractive, that is, only if the constant a is negative.

The Schrödinger radial equation (Equation 4.43) may be simplified by substituting a new dependent variable,

$$W_l(r) = rR_l(r),$$ (4.66)

so that the equation becomes

$$\frac{d^2W_l(r)}{dr^2} + \frac{2m}{\hbar^2}\left[E - \frac{a}{r} - \frac{\hbar^2}{2mr^2}l(l+1)\right]W_l(r) = 0.$$ (4.67)

Now if we define a new independent variable ρ by writing

$$\rho \equiv kr \equiv \sqrt{\frac{2mE}{\hbar^2}}\, r,$$ (4.68)

then Equation 4.67 becomes, using Equations 4.65 and 4.68,

$$\frac{d^2W_l(\rho)}{d\rho^2} + \left[1 - \frac{(a/E)}{\rho} - \frac{l(l+1)}{\rho^2}\right]W_l(\rho) = 0.$$ (4.69)

Equation 4.69 is called the *Coulomb wave equation*. The general solution to this equation has the form

$$W_l(\eta,\rho) = AF_l(\eta,\rho) + BG_l(\eta,\rho),$$ (4.70)

where $\eta = ka/2E$. The function $F_l(\eta,\rho)$ is called the *regular Coulomb wave function*, and $G_l(\eta,\rho)$ is called the *irregular* or *logarithmic Coulomb wave function*.* The terms A and B are constants to be determined by the conditions of the physical problem.

The solution for the energy eigenvalues in attractive Coulomb wells follows from the usual procedure of matching appropriate eigensolutions across the interface between the inside and the outside of the well. We shall not carry out the detailed algebra here but merely describe the results. The energy eigenvalues are found to be given by

$$E = -\frac{ma^2}{2\hbar^2N^2},$$ (4.71)

---

*See National Bureau of Standards *Handbook of Mathematical Functions* for definitions, properties, graphs, and tables of the Coulomb wave functions.

where N, the *principal quantum number*, equals 1, 2, 3, ... These energy levels are identical to those derived by Bohr on the basis of his semiclassical theory of the hydrogen atom. (In the case of the hydrogen atom, $a = -e^2$, where e is the charge on the electron.)

The principal quantum number is given by

$$N = n + l,$$

where n is the number of nodes in the radial wave function, including the node at the origin, and l is the total angular-momentum quantum number. For a fixed N, the polar quantum number may range from $l = 0$ to $l = N - 1$ by integers. Furthermore, the azimuthal quantum number m_l may range from $-l$ to l by integers. The order of the degeneracy of an energy level with principal quantum number N is N^2,

$$\text{Degeneracy} = \sum_{l=0}^{N-1} (2l + 1) = N^2.$$

The lowest energy level ($N = 1$) is nondegenerate. The next highest ($N = 2$) has a fourth-order degeneracy, and so on. Azimuthal degeneracy is common to all central fields. Polar degeneracy is rare; it occurs only for the Coulomb potential and the three-dimensional harmonic oscillator potential.

As can be seen by inspection of Equation 4.71, as n approaches infinity, the energy levels crowd closer and closer together. There are an infinite number of bound levels near the top of the potential well. This is in sharp distinction to the finite number of levels found in the spherical square well. It can be proved, in general, that if $V(r)$ approaches zero as r^{-p}, where $p \leq 2$, there are an infinite number of energy eigenvalues in the well whereas, if $p > 2$, there are only a finite number.

4.3 Orbital Angular Momentum in Quantum Mechanics

In this section we will show that the polar quantum number l and the azimuthal quantum number m_l, which were introduced in the previous section, are related, respectively, to the expectation value of the total angular momentum and the expectation value of one of the three components of the total angular momentum.

In the R-system, $\mathbf{r}$ is the radius vector from the center of coordinates (particle 2) to the position of the particle of reduced mass m (particle 1), which has linear momentum $\mathbf{p} = m\mathbf{v}$ (see Figure 4.3). Thus

$$\mathbf{r} = (x,y,z)$$

and

$$\mathbf{p} = (p_x, p_y, p_z).$$

The classical definition of the angular momentum vector of the reduced mass particle is

$$\mathbf{L} = \mathbf{r} \times \mathbf{p}, \tag{4.72}$$

where $\times$ denotes the *cross product*. The term $\mathbf{L}$ is the vector angular momentum of the particle of reduced mass m about the origin. From the general expression for $\mathbf{L}$ above, we can obtain immediately the classical expressions for the components of $\mathbf{L}$. (Recall the vector rules: $\mathbf{i} \times \mathbf{i} = 0$, $\mathbf{i} \times \mathbf{j} = \mathbf{k}, \mathbf{j} \times \mathbf{i} = -\mathbf{k}$, etc.)

$$L_x = yp_z - zp_y,$$

$$L_y = zp_x - xp_z, \tag{4.73}$$

$$L_z = xp_y - yp_x.$$

These classical expressions are converted to quantum-mechanical operators, observing postulate 2, by substituting the appropriate operator expressions for the coordinates and momenta $[x \to x, p_x \to -i\hbar(\partial/\partial x), \text{etc.}]$, being careful to maintain the proper order since operators do not necessarily commute. The resulting expressions are:

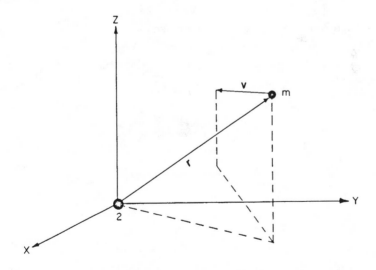

Figure 4.3—Particle of reduced mass m has linear momentum $\mathbf{p} = m\mathbf{v}$ in the R-system in which Particle 2 is at rest. The angular momentum of the system is $\mathbf{L} = \mathbf{r} \times \mathbf{p}$.

Classical Variable	*Quantum-Mechanical Operator*	
L_x	$-i\hbar\left(y\dfrac{\partial}{\partial z} - z\dfrac{\partial}{\partial y}\right)$	
L_y	$-i\hbar\left(z\dfrac{\partial}{\partial x} - x\dfrac{\partial}{\partial z}\right)$	(4.74)
L_z	$-i\hbar\left(x\dfrac{\partial}{\partial y} - y\dfrac{\partial}{\partial x}\right)$	

The operator corresponding to the square of the angular-momentum vector,

$$L^2 = \mathbf{L}\cdot\mathbf{L} = L_x L_x + L_y L_y + L_z L_z,$$

is readily obtained with the help of Equations 4.74. For example, the x component is

$$L_x L_x = -\hbar^2\left(y\frac{\partial}{\partial z} - z\frac{\partial}{\partial y}\right)\left(y\frac{\partial}{\partial z} - z\frac{\partial}{\partial y}\right)$$

$$= -\hbar^2\left(y^2\frac{\partial^2}{\partial z^2} + z^2\frac{\partial^2}{\partial y^2} - yz\frac{\partial^2}{\partial z\partial y} - zy\frac{\partial^2}{\partial y\partial z} - y\frac{\partial}{\partial y} - z\frac{\partial}{\partial z}\right).$$

Similar expressions can be obtained for the y and z components of L^2.

Commutation Rules Governing Angular Momenta

The commutation relations among the components of orbital angular momentum may be obtained by straightforward calculation using Equations 4.74. For example,

$$[L_x, L_y] \equiv L_x L_y - L_y L_x$$

$$= -\hbar^2\left[\left(y\frac{\partial}{\partial z} - z\frac{\partial}{\partial y}\right)\left(z\frac{\partial}{\partial x} - x\frac{\partial}{\partial z}\right)\right.$$

$$\left. - \left(z\frac{\partial}{\partial x} - x\frac{\partial}{\partial z}\right)\left(y\frac{\partial}{\partial z} - z\frac{\partial}{\partial y}\right)\right]$$

$$= -\hbar^2\left(y\frac{\partial}{\partial x} - x\frac{\partial}{\partial y}\right)$$

$$= i\hbar L_z.$$

Noting that the variables x, y, and z appear in cyclic order in the expressions for L_x, L_y, and L_z, we obtain without further calculation

$$[L_x,L_y] = i\hbar L_z,$$
$$[L_y,L_z] = i\hbar L_x, \qquad (4.75)$$
$$[L_z,L_x] = i\hbar L_y.$$

The components of orbital angular momentum do not commute among themselves; however, it can readily be shown that each of the components commutes with the square of the total angular momentum. As an example, consider the commutator of L_x and L^2:

$$[L_x,L^2] = L_x L^2 - L^2 L_x$$
$$= L_x(L_y^2 + L_z^2) - (L_y^2 + L_z^2)L_x.$$

After adding and subtracting identical terms, we obtain

$$[L_x,L^2] = (L_x L_y - L_y L_x)L_y + L_y(L_x L_y - L_y L_x)$$
$$+ (L_x L_z - L_z L_x)L_z + L_z(L_x L_z - L_z L_x).$$

Using Equations 4.75, we get

$$[L_x,L^2] = (i\hbar L_z L_y) + (i\hbar L_y L_z) - (i\hbar L_y L_z) - (i\hbar L_z L_y)$$
$$= 0,$$

and we can conclude that the x component of orbital angular momentum commutes with the total orbital angular momentum. Once again noting the cyclic symmetry among the operator expressions for L_x, L_y, and L_z, we safely conclude that each component of L commutes with L^2. Thus

$$[L_x,L^2] = [L_y,L^2] = [L_z,L^2] = 0. \qquad (4.76)$$

Equations 4.75 and 4.76, when coupled with Heisenberg's uncertainty principle, imply the following important rules:

1. No two components of the orbital angular momentum can be measured simultaneously with absolute precision.

2. Any single component of the orbital angular momentum can be measured simultaneously with the magnitude of the total orbital angular momentum.

Angular Momenta in Spherical Polar Coordinates

Expressions for the angular-momentum operators in spherical polar coordinates are readily obtained by changing variables in Equations 4.74

from x, y, and z to r, θ, and ϕ, where

$$r^2 = x^2 + y^2 + z^2,$$

$$x = r \sin \theta \cos \phi,$$

$$y = r \sin \theta \sin \phi,$$

$$z = r \cos \theta.$$

When this change is made, the following expressions result:

Classical Variable	Quantum-Mechanical Operator
L_x	$-i\hbar \left(-\sin \phi \dfrac{\partial}{\partial \theta} - \cot \theta \cos \phi \dfrac{\partial}{\partial \phi} \right)$
L_y	$-i\hbar \left(\cos \phi \dfrac{\partial}{\partial \theta} - \cot \theta \sin \phi \dfrac{\partial}{\partial \phi} \right)$
L_z	$-i\hbar \dfrac{\partial}{\partial \phi}$
L^2	$-\hbar^2 \left[\dfrac{1}{\sin \theta} \dfrac{\partial}{\partial \theta} \left(\sin \theta \dfrac{\partial}{\partial \theta} \right) + \dfrac{1}{\sin^2 \theta} \dfrac{\partial^2}{\partial \phi^2} \right]$

$$(4.77)$$

Henceforth the discussion will be restricted to central potentials. When $V(\mathbf{r}) = V(r)$, it is found that the wave function separates, in spherical coordinates, into eigenfunctions of r, θ, and ϕ, and, of course, t. The eigenfunctions are given by

$$\Psi(r,\theta,\phi,t) = R(r)\, T(\theta)\, F_{m_l}(\phi)\, e^{-iEt/\hbar}. \tag{4.78}$$

Expectation Value of L_z

The expectation value of the z component of angular momentum, L_z, is, by Postulate 5,

$$\langle L_z \rangle = \iiint \bar{\Psi} L_z \Psi \, d\tau.$$

This integral becomes, after substituting the operator representing L_z from Equation 4.77 and the wave function from Equation 4.78,

$$\langle L_z \rangle = \iint \int \bar{R}\,\bar{T}\,\bar{F}_{m_l} \left(-i\hbar \frac{\partial}{\partial \phi} \right) R T F_{m_l} \, r \sin \theta \, dr \, d\theta \, d\phi, \tag{4.79}$$

where the volume element in spherical coordinates is $d\tau = r^2 \sin \theta \, dr \, d\theta \, d\phi$. But the operator $-i\hbar(\partial/\partial \phi)$ operates only on $F_{m_l}(\phi)$ since both R and T are functions that are independent of ϕ. By Equation 4.24,

$$-i\hbar \frac{\partial}{\partial \phi} (F_{m_l}) = -i\hbar \frac{\partial}{\partial \phi} \left(\frac{1}{\sqrt{2\pi}} e^{im_l \phi} \right) = m_l \hbar F_{m_l}. \tag{4.80}$$

When Equation 4.80 is substituted into Equation 4.79,

$$\langle L_z \rangle = m_l \hbar \iint \bar{\Psi}\Psi \, d\tau = m_l \hbar,$$

where the last step follows from the normalization of the eigenfunctions. Thus the first of several important results regarding angular-momentum quantum numbers is shown to be

$$\langle L_z \rangle = m_l \hbar, \tag{4.81}$$

where $m_l = 0, \pm 1, \pm 2, \ldots, \pm l$. Stated verbally, the expectation value of the z component of the angular momentum is $m_l \hbar$ provided the state of the system is described by an eigenfunction $F_{m_l}(\phi)$. In the general case, the state of the system may be described by a superposition of eigenfunctions in ϕ. In this case, measurement of L_z will yield a number of different values, hence an average value of L_z that is not necessarily an integral multiple of $\hbar$.

Expectation Value of L^2

The expectation value of the square of the total orbital angular momentum is, by Postulate 5 and Equation 4.78,

$$\langle L^2 \rangle = \iiint \bar{R}\,\bar{T}_l^{m_l}\,\bar{F}_{m_l}$$

$$\times \left\{ -\hbar^2 \left[\frac{1}{\sin\theta} \frac{\partial}{\partial\theta}\left(\sin\theta \frac{\partial}{\partial\theta}\right) + \frac{1}{\sin^2\theta} \frac{\partial^2}{\partial\phi^2} \right] \right\} R T_l^{m_l} F_{m_l} \, d\tau.$$

The operation $(\partial^2/\partial\phi^2)F_{m_l}$ produces $-m_l^2 F_{m_l}$; so the operator within the integral above may be replaced by

$$-\hbar^2 \left[\frac{1}{\sin\theta} \frac{\partial}{\partial\theta}\left(\sin\theta \frac{\partial}{\partial\theta}\right) - \frac{m_l^2}{\sin^2\theta} \right]$$

But this operator operating on $T_l^{m_l}$, the only part of the wave function that contains the variable θ, produces the eigenvalues $l(l+1)\hbar^2$ times $T_l^{m_l}$, as can be seen by inspection of Equation 4.16. Thus

$$\langle L^2 \rangle = \iiint \bar{R}\,\bar{T}\,\bar{F}\,\{\hbar^2 l(l+1)\}\,R\,T\,F\,d\tau$$

$$= \hbar^2 l(l+1) \iiint \bar{\Psi}\Psi \, d\tau$$

$$= l(l+1)\hbar^2$$

or

$$\langle L^2 \rangle = l(l + 1)\hbar^2. \tag{4.82}$$

Again stating this important result verbally, the expectation value of the square of the total orbital angular momentum is $l(l + 1)\hbar^2$ provided the state of the system is described by an eigenfunction of the total angular momentum $T_l^{m_l}$. If, however, the ψ-function of the system is a superposition of angular-momentum eigenfunctions, a measurement of L^2 will yield one of the set of possible values $0, 2\hbar^2, 6\hbar^2, 12\hbar^2$, etc., and the average value $\langle L^2 \rangle$ will not necessarily be an integral multiple of $\hbar^2$.

Vector Model of Angular Momenta

Suppose the system is in an eigenstate of definite total orbital angular momentum. The quantum number l is a fixed integer, and the total orbital angular momentum is given by

$$L = \sqrt{l(l + 1)}\, \hbar.$$

The component of L in the z direction may have any value up to $\pm l\hbar$; that is,

$$L_z = m_l\hbar = (0, \pm 1, \ldots, \pm l)\hbar.$$

The possible values of L_z can be represented as the projection of a vector of length L on the z-axis, as shown in Figure 4.4, where, to be definite, we assume that $l = 3$.

In a state in which both L and L_z are known, there is an irreducible uncertainty in the other two components; L_x and L_y can have any values, subject only to the condition that

$$L_x^2 + L_y^2 = L^2 - L_z^2. \tag{4.83}$$

A pictorial representation of this situation may be obtained by imagining that the vector representing L in Figure 4.4 is rotating about the z-axis so that L_x and L_y are continuously changing but L and L_z are fixed. It should be obvious that the choice of the direction of the polar axis is completely arbitrary. If the azimuthal angle ϕ were measured around the x-axis rather than the z-axis, we would conclude that L_x and L were simultaneously measurable and L_y and L_z uncertain. The quantum numbers l and m_l are frequently called *good quantum numbers* in contrast to the nonexistence of quantum numbers for the other two components of the total angular momentum.

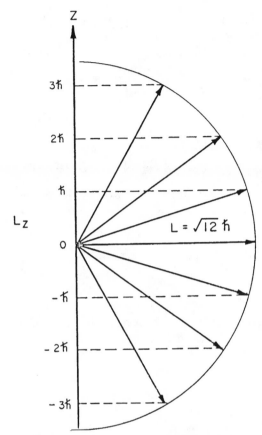

Figure 4.4—Possible orientations of the vector **L** relative to the z-axis for the case $l = 3$. The vector **L** may be assumed to be rotating about the z-axis.

4.4 Constants of Motion

In any classical central-force collision between two isolated point particles, there are 10 physical quantities that are unchanged regardless of the type of interaction. These 10 constants of motion are (1) the total energy, (2) the 3 components of the velocity of the center of mass in the L-system, (3) the 3 components of the total linear momentum in the L-system, and (4) the 3 components of the total angular momentum in the L-system.

In the classical C-system, six of the above quantities vanish identically and therefore can be dropped from further consideration. The center of mass, being the origin of the coordinate system, is taken to be at rest; therefore the three components of velocity of the center of mass vanish. By definition the total linear momentum through the origin of the

C-system is zero; this condition forces one component of the linear momentum to zero. The angular momentum corresponding to this component of linear momentum must therefore also vanish. Since central-force motion always takes place in a plane, one further component of angular momentum is taken to be zero. Thus only four constants of motion remain, the total energy, two components of linear momentum, and one component of angular momentum.

In the classical R-system only energy and angular momentum are conserved. Linear momentum is not conserved during the interaction, although it has the same magnitude after the interaction as before the interaction.

In quantum mechanics, if $(d/dt)\langle \Omega \rangle = 0$, where $\langle \Omega \rangle$ is the expectation value of the operator Ω, then the dynamical variable corresponding to the operator Ω is called a *constant of motion* of the system.

Classically, the time rate of change of a dynamical variable Ω that does not depend explicitly on the time is given by the Poisson bracket relation (see Equation 3.34)

$$\dot{\Omega} = (\Omega, H),$$

where H is the Hamiltonian of the system. In quantum mechanics the Poisson bracket is replaced by the commutator, according to Postulate 2, giving

$$(\Omega, H) = \frac{i}{\hbar} [H, \Omega], \tag{4.84}$$

so that

$$\dot{\Omega} = \frac{i}{\hbar} [H, \Omega]. \tag{4.85}$$

If Ω commutes with H, then $\dot{\Omega} = 0$ and Ω is a constant of motion. Thus it is seen that any operator which does not depend explicitly on the time and which commutes with the Hamiltonian operator is a constant of motion of the system. It follows directly from the three equations above that every constant of motion of a classical system is also a constant of motion of the system when described by quantum mechanics.

If a system is conservative, H does not depend explicitly on the time, $\dot{H} = 0$, and the total energy is a constant of motion, just as it is classically.

If the potential V between two particles is a function of r alone and not a function of θ and ϕ, then L^2 and all components of $\mathbf{L}$ will commute with the Hamiltonian operator. This can be proved readily by noting that the Hamiltonian operator in the R-system in spherical coordinates is

$$H = -\frac{\hbar^2}{2m} \left(\frac{\partial^2}{\partial r^2} + \frac{2}{r} \frac{\partial}{\partial r} \right) + \frac{L^2}{2mr^2} + V(r), \tag{4.86}$$

where L^2 is the operator given by Equation 4.77. Since H contains θ and ϕ only in the form of the operator L^2 and since L^2 and all components of L commute with L^2, the term L^2 and all components of L commute with H. Although all components of the angular momentum are constants of the motion in a central force field, the quantum-mechanical state of the system can only be specified by the magnitude of the orbital angular momentum, L^2, and one component of the orbital angular momentum, say, L_z.

Thus, in practice, we can specify as quantum-mechanical constants of motion in the R-system only (1) the eigenvalue of the total energy H, (2) the eigenvalue of the orbital angular momentum L, and (3) the eigenvalue of L_z, one component of the orbital angular momentum.

Linear momentum behaves much the same way in quantum mechanics as in classical mechanics. In the case of a free particle, the momentum operator in Cartesian coordinates, $\mathbf{p} = -i\hbar\nabla$, commutes with the Hamiltonian operator $H = -(\hbar^2/2m)\nabla^2$. Hence the momentum of a free particle is a constant of motion. On the other hand, the momentum of a particle moving in a field of force described by a potential $V(r)$ is not a constant of the motion since the operators $\mathbf{p}$ and $V(r)$ do not in general commute. This is the quantum-mechanical analogue of the classical fact that linear momentum is not conserved in the R-system during the interaction. Before and after the interaction, the distance r between the two particles is sufficiently great that $V(r)$ can be taken to be zero. Therefore the linear-momentum operator does commute with the Hamiltonian operator. But during the interaction $V(r)$ has appreciable magnitude compared with the other operators in the Hamiltonian; so the linear-momentum operator does not commute with the Hamiltonian operator.

The most significant lesson to be learned from this discussion is that classical and quantum kinematics, based solely on the conservation laws of energy and linear momentum, are identical. The conservation laws that are the basis of kinematics have survived the drastic reorganization of mechanics effected during the present century. All the purely kinematic relations derived in Chapter 2 apply with equal validity to all collisions, whether the colliding bodies be large or small.

4.5 Probability Current Density

Conservation equations of the general form

$$\frac{\partial P(\mathbf{r},t)}{\partial t}\, d\tau = -\nabla \cdot \mathbf{S}(\mathbf{r},t)\, d\tau \tag{4.87}$$

appear quite extensively throughout the physical analysis of the motion of material particles or fluids. Such equations are interpreted to mean that the rate at which the density P changes in volume element $d\tau$ is equal to the net inflow into the volume element $d\tau$. For example, the flow of an incompressible fluid of density $\rho(\mathbf{r},t)$ and average current density $\mathbf{S}(\mathbf{r},t)$ through a region in which there are no sources or sinks is governed by the conservation law

$$\frac{\partial\rho}{\partial t} + \mathbf{\nabla}\cdot\mathbf{S} = 0.$$

The diffusion of one-speed particles (e.g., neutrons) in a region having no sources or sinks is described by the familiar

$$\frac{\partial n}{\partial t} + \mathbf{\nabla}\cdot\mathbf{J} = 0,$$

where $n(\mathbf{r},t)$ and $\mathbf{J}(\mathbf{r},t)$ are the number density and current density of particles, respectively.

A similar conservation relation for the flow of probability density of a single particle will now be derived. According to Postulate 5, $\overline{\Psi}\Psi \, d\tau$ is the probability of finding a particle in volume element $d\tau$ at time t. The partial rate of change of $\overline{\Psi}\Psi$ with respect to time is

$$\frac{\partial}{\partial t}(\overline{\Psi}\Psi) = \overline{\Psi}\frac{\partial}{\partial t}\Psi + \Psi\frac{\partial}{\partial t}\overline{\Psi}. \tag{4.88}$$

We can eliminate the time derivatives from the right-hand side of Equation 4.88 using Schrödinger's equation

$$i\hbar\frac{\partial}{\partial t}\Psi = \left(\frac{-\hbar^2}{2m}\nabla^2 + V\right)\Psi \tag{4.89}$$

and the complex conjugate equation

$$-i\hbar\frac{\partial}{\partial t}\overline{\Psi} = \left(\frac{-\hbar^2}{2m}\nabla^2 + \overline{V}\right)\overline{\Psi}. \tag{4.90}$$

Substituting Equations 4.89 and 4.90 into Equation 4.88, we obtain

$$\frac{\partial}{\partial t}(\overline{\Psi}\Psi) = \overline{\Psi}\left[\frac{i}{\hbar}\left(\frac{\hbar^2}{2m}\nabla^2 - V\right)\right]\Psi - \Psi\left[\frac{i}{\hbar}\left(\frac{\hbar^2}{2m}\nabla^2 - \overline{V}\right)\right]\overline{\Psi}. \tag{4.91}$$

Real Potentials

If the potential is real, then $\overline{V}$ equals V and Equation 4.91 reduces to

$$\frac{\partial}{\partial t}(\overline{\Psi}\Psi) = -\frac{\hbar}{2mi}(\overline{\Psi}\,\nabla^2\Psi - \Psi\,\nabla^2\overline{\Psi})$$

$$= -\mathbf{\nabla}\cdot\left[\frac{\hbar}{2mi}(\overline{\Psi}\,\nabla\Psi - \Psi\,\nabla\overline{\Psi})\right]. \tag{4.92}$$

The probability current density can now be identified as that vector $\mathbf{S}$ which satisfies the conservation equation for probability density,

$$\frac{\partial}{\partial t}(\bar{\Psi}\Psi) + \nabla \cdot \mathbf{S} = 0. \tag{4.93}$$

Comparing Equations 4.92 and 4.93, we see that the probability current density is given by

$$\mathbf{S}(\mathbf{r},t) = \frac{\hbar}{2mi}(\bar{\Psi}\,\nabla\Psi - \Psi\,\nabla\bar{\Psi}). \tag{4.94}$$

If the Schrödinger function of the particle is an eigenfunction of the Hamiltonian operator, then we can write

$$\Psi(\mathbf{r},t) = \psi(\mathbf{r})e^{-iEt/\hbar},$$

and Equation 4.94 reduces to

$$\mathbf{S}(\mathbf{r}) = \frac{\hbar}{2mi}(\bar{\psi}\,\nabla\psi - \psi\,\nabla\bar{\psi}). \tag{4.95}$$

As an elementary application of Equation 4.95, let us calculate the probability current density in the one-dimensional rectangular potential barrier problem (see Figure 3.4). It has been found that the wave function is given by

Wave function   Where by definition

Region 1: $\psi_1(x) = Ae^{ikx} + Be^{-ikx}$ $k = \left[\dfrac{2mE}{\hbar^2}\right]^{1/2}$

Region 2: $\psi_2(x) = Ke^{gx} + Le^{-gx}$ $g = \left[\dfrac{2m(V-E)}{\hbar^2}\right]^{1/2}$

Region 3: $\psi_3(x) = Ce^{ikx}$

The terms K and L in the functions above were related to C by the equations

$$K = \frac{C}{2}e^{ikb}e^{-gb}\left(1 + \frac{ik}{g}\right)$$

and

$$L = \frac{C}{2}e^{ikb}e^{gb}\left(1 - \frac{ik}{g}\right).$$

The current in each of the three regions is obtained by substitution of the appropriate wave function into the one-dimensional form of Equation 4.95, i.e., into

$$S(x) = \frac{\hbar}{2mi} \left[\bar{\psi} \frac{d}{dx} \psi - \psi \frac{d}{dx} \bar{\psi} \right]. \tag{4.96}$$

The results are

Region 1: $S_1 = \frac{k\hbar}{m} (\bar{A}A - \bar{B}B)$

Region 2: $S_2 = \frac{g\hbar}{mi} (\bar{L}K - \bar{K}L) = \frac{k\hbar}{m} (\bar{C}C)$

Region 3: $S_3 = \frac{k\hbar}{m} (\bar{C}C).$

But $k\hbar/m = \sqrt{2E/m} = v$, the velocity of the incident particles, and $\bar{A}A$ can be assumed to represent the number density of incident particles, N. After multiplying and dividing each of the above equations by $\bar{A}A$ and noting that $\bar{B}B/\bar{A}A = R$, the probability of reflection, and that $\bar{C}C/\bar{A}A = T$, the probability of transmission, we obtain

$S_1 = Nv(1 - R) = \phi_{inc} - \phi_{refl},$

$S_2 = S_3 = NvT = \phi_{trans}$

where ϕ_{inc}, ϕ_{refl}, and ϕ_{trans} are the incident, reflected, and transmitted fluxes. The probability current density in Region 1 is therefore seen to be just the net current, the algebraic sum of the incident and reflected fluxes. In Regions 2 and 3, the probability current density is equal to the transmitted flux, which is the net flux in Region 2 and the only flux in Region 3.

Complex Potentials

In deriving Equation 4.94 for the probability current density, we assumed that the potential was real. It is instructive to note that absorption of particles from a beam may be represented mathematically by assuming the potential has a negative imaginary part. That is, we assume the potential can be written

$$V(r) = V_0(r) - i V_1(r), \tag{4.97}$$

where V_0 and V_1 are real functions of r. When this potential is substituted into Equation 4.91, there results

$$\frac{\partial}{\partial t} (\bar{\Psi}\Psi) = \frac{\hbar}{2mi} (\bar{\Psi} \nabla^2 \Psi - \Psi \nabla^2 \bar{\Psi}) - \frac{2V_1}{\hbar} \bar{\Psi}\Psi,$$

and, using Equations 4.92 and 4.94 for the first term on the right-hand side, we obtain

$$\frac{\partial}{\partial t}(\bar{\Psi}\Psi) = -\nabla\cdot\mathbf{S} - \frac{2V_1}{\hbar}\bar{\Psi}\Psi. \tag{4.98}$$

Equation 4.98 is a generalization of Equation 4.93, the conservation equation for probability. Since the left-hand side is the time rate of change of probability density and the first term on the right-hand side is the net rate of inflow of probability density, the final term may be interpreted as a rate of absorption of probability density. Thus the imaginary part of a potential, if negative, is seen to give rise to a *sink* for probability density. This fact will be used in subsequent chapters to describe neutron interactions which include absorption.

4.6 Wave-Mechanical Description of Scattering

In Chapter 2, we showed that we could calculate the angle of scattering of a particle in a known potential classically from knowledge of two quantities, the energy E of the incident particle and the impact parameter b. Given E and b, we are able to calculate $\theta = \theta(b)$ and finally the differential cross section $\sigma(\theta)$.

Quantum mechanically, it is impossible to prepare projectile particles such that they have a definite impact parameter b and simultaneously have zero momentum parallel to b. If the uncertainty in b were zero, the uncertainty in the momentum would be infinite. Thus, in the quantum-mechanical description of scattering, the concept of a definite impact parameter must necessarily be abandoned.

Consider an experimental arrangement in which particles are collimated into an almost monodirectional beam that strikes and scatters from a center of force. In a typical experiment the beam will have a diameter of the order of millimeters. Thus the uncertainty of the transverse position of any particle Δy is of the order of 10^{-1} cm. The uncertainty in the transverse momentum can then be made as small as

$$\Delta p_y \cong \frac{\hbar}{\Delta y} \cong \frac{10^{-27} \text{ erg-sec}}{10^{-1} \text{ cm}}$$

or

$$m\,\Delta v_y \cong 10^{-26} \text{ g-cm/sec.}$$

If the projectile particles are nucleons so that $m \cong 10^{-24}$ g, then

$$\Delta v_y \cong 10^{-2} \text{ cm/sec.}$$

Now if the beam were collimated to within, say, 10^{-2} radian and if the speed of the particles is greater than 10^5 cm/sec, the spread in transverse speeds is at least of order 10^3 cm/sec, this spread being the purely classical effect of imperfect collimation. Comparing this with Δv_y calculated above, it is seen that the quantum-mechanical uncertainty in transverse velocity is very small compared to the classical uncertainty. Thus one can safely use a classical picture for the formation of the beam, its composition and angular spread being determined by classical mechanics. However, the language and formalism of quantum mechanics must be used to describe the scattering.

Consider a steady stream of particles incident, scattered, and being detected. Notice that this is a time-independent situation; only the spatial part of the Schrödinger function need be considered. The reader who is dubious is invited to replace ψ by $\psi \exp(-iEt/\hbar)$ in any of the following equations and verify that the time-dependent function cancels. As usual, the direction of the incident particles is taken as the z-direction. All the calculations are carried out in the R-system; m is the reduced mass of the projectile–target particle system, θ and ϕ are the polar and azimuthal angles of scattering in the R-system, and E is the available energy.

The scattering process is described in the language of quantum mechanics as follows: A plane wave moves along the z-axis until it encounters a potential. A radial wave (the scattered particles) travels outward from the scattering center. A plane wave slightly diminished in amplitude (the unscattered particles) continues along the z-axis beyond the region containing the potential. A detector records the flux of particles in the scattered wave at angles θ and ϕ. (See Figure 4.5.)

It is most important to recognize the vast inequalities of dimensions in the typical scattering experiment. The range of effective potential in atomic and nuclear processes is less than 10^{-8} cm. The transverse dimension of the collimated beam is of the order of 10^{-1} cm. The distance between the scattering center and the detector is of the order of 10^2 cm. Thus the diameter of the incident beam is effectively infinite relative to the dimension of the potential and is effectively zero relative to the distance to the counter. A clear conception of these magnitudes will greatly facilitate the understanding of what is to follow.

The Incident Wave

The incident wave is the solution to the Schrödinger amplitude equation for a free particle moving parallel to the z-axis with energy E,

$$\psi_{\text{inc}} = e^{ikz}, \tag{4.99}$$

where $k = \sqrt{2mE}/\hbar$. Equation 4.99 describes a wave of infinite transverse dimension. This is a very good approximation because the incident wave is effectively infinite relative to the dimension of the potential.

The probability current density, which is the flux of incident particles, is, using Equation 4.96,

$$S_{inc} = \frac{\hbar}{2mi}\left(\overline{\psi}_{inc}\frac{d}{dz}\psi_{inc} - \psi_{inc}\frac{d}{dz}\overline{\psi}_{inc}\right)$$

$$= \frac{\hbar}{2mi}(ik + ik) = \frac{\hbar}{m}k = \frac{\hbar}{m}\frac{\sqrt{2mE}}{\hbar} = v.$$

Thus the incident flux is equal to the speed of one of the particles in the incident beam. The incident number density is therefore normalized to 1 particle/unit volume.

The Scattered Wave

The scattered wave at distances far from the region where the potential exists must be chosen to represent a free particle moving radially outward with (1) the same energy as the incident particle (because the analysis is made in the R-system), (2) amplitude dependent, in general, on θ and ϕ,

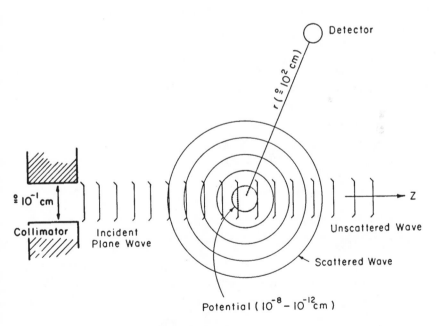

Figure 4.5—Quantum-mechanical picture of the scattering of a collimated beam of particles by a center of force. Note that the scale of dimensions is highly distorted.

and (3) magnitude inversely proportional to r (since to conserve particles the radial flux, which is proportional to ψ squared, must fall off as the inverse square of the distance from the scattering center). Thus the scattered wave must have the form

$$\psi_{sc} = f(\theta,\phi)\,\frac{e^{ikr}}{r}, \tag{4.100}$$

where $f(\theta,\phi)$ is the *amplitude* of the scattered wave. The scattered flux, which is directed radially outward and is therefore the scattered current in the r direction, is given by

$$S_{sc} = \frac{\hbar}{2mi}\left(\bar{\psi}_{sc}\frac{\partial}{\partial r}\psi_{sc} - \psi_{sc}\frac{\partial}{\partial r}\bar{\psi}_{sc}\right)$$

$$= \frac{\hbar k}{m}\frac{|f(\theta,\phi)|^2}{r^2}$$

$$= v\frac{|f(\theta,\phi)|^2}{r^2}.$$

The differential cross section at angles θ and ϕ is, by definition,

$$\sigma(\theta,\phi) = \frac{r^2 \times \text{scattered flux at point } (r,\theta,\phi)}{\text{incident flux}}$$

$$= r^2\frac{S_{sc}}{S_{inc}} = |f(\theta,\phi)|^2$$

or, in terms of the complex conjugate of the amplitude function,

$$\sigma(\theta,\phi) = \overline{f(\theta,\phi)}f(\theta,\phi). \tag{4.101}$$

It will be assumed in subsequent sections that the scattering amplitude is independent of the azimuthal angle ϕ; so Equation 4.101 reduces to

$$\sigma(\theta) = \overline{f(\theta)}f(\theta). \tag{4.102}$$

Thus we can see that the problem of finding a differential elastic-scattering cross section via quantum mechanics reduces to the problem of determining the amplitude, $f(\theta)$, of the scattered wave. In the remaining sections of this chapter, two procedures by which the amplitude may be found will be investigated. Before moving to the first of these procedures, note that the complete wave function far from the scattering center must represent both an incident wave and a scattered wave. That is, for large r

$$\psi \sim e^{ikz} + \frac{f(\theta)e^{ikr}}{r}. \tag{4.103}$$

The symbol $\sim$ is to be read "is asymptotically equal to." The word asymptotic is used here in the sense of "far from the scattering center." Expressing z as a function of r and θ by the relation $z = r \cos \theta$, we obtain for the asymptotic function in spherical coordinates

$$\psi(r,\theta) \sim e^{ikr \cos \theta} + \frac{f(\theta)e^{ikr}}{r}. \tag{4.104}$$

4.7 The Method of Partial Waves

An expression for $f(\theta)$ will now be developed by the procedure known, for reasons that will soon be obvious, as the method of partial waves. The wave function for the incident particles can be expanded in a series of Legendre polynomials,

$$\psi_{inc}(r,\theta) = e^{ikr \cos \theta} = \sum_{l=0}^{\infty} a_l B_l(r) P_l(\cos \theta). \tag{4.105}$$

To determine the unknown coefficients a_l and the unknown functions $B_l(r)$, we multiply Equation 4.105 by $P_{l'}(\cos \theta) \sin \theta \, d\theta = P_{l'}(\mu)(-d\mu)$ and integrate over the range of θ from 0 to π or μ from $+1$ to -1. Because of the orthogonality of the Legendre polynomials,

$$\int_{-1}^{+1} P_l(\mu)P_{l'}(\mu) \, d\mu = \frac{2}{2l + 1} \delta_{ll'}, \tag{4.106}$$

only the lth term survives in the sum. Since $l' = l$, we find that

$$a_l B_l(r) \frac{2}{2l + 1} = \int_{-1}^{+1} e^{ikr\mu} P_l(\mu) \, d\mu. \tag{4.107}$$

But the right-hand side of Equation 4.107 is related to the integral representation of the spherical Bessel function encountered in Section 4.2.* In fact, the right-hand side is just $2i^l j_l(kr)$; therefore

$$a_l B_l(r) \frac{2}{2l + 1} = 2i^l j_l(kr), \tag{4.108}$$

where $j_l(kr)$ is the spherical Bessel function of order l and argument kr. Using Equation 4.108, we may write Equation 4.105 as

$$\psi_{inc}(r,\theta) = \sum_{l=0}^{\infty} (2l + 1)i^l j_l(kr) P_l(\cos \theta). \tag{4.109}$$

---

*See E. Jahnke and F. Emde, *Tables of Functions* (Dover Publications, Inc., New York, 1945).

What has been accomplished? We readily recognize that the incident wave function has been expanded in a series of partial waves, an $l = 0$ wave plus an $l = 1$ wave plus an $l = 2$ wave, etc. This expansion is not only quite useful but also intuitively appealing since the incident, practically infinite beam of particles contains particles with all quantum mechanically permitted values of angular momentum relative to the scattering center. From a classical viewpoint this expansion is equivalent to breaking down the incident beam into partial beams, each of which is characterized by a given impact parameter. The factor $(2l + 1)$ in Equation 4.109 accounts for the fact that a partial wave that is characterized by a total angular-momentum quantum number l consists of $2l + 1$ subwaves, each of which is characterized by a distinct value of the azimuthal quantum number m_l. In other words, the lth partial wave has a $(2l + 1)$-fold degeneracy.

Suppose the projectile particles are slow neutrons ($v \doteq 10^6$ cm/sec) that are incident on a nucleus with a radius of 5×10^{-13} cm. Classically, one would say that neutrons cannot strike the nucleus if they have an impact parameter greater than 5×10^{-13} cm. This corresponds to an angular momentum of

$$L = mbv \doteq (1.67 \times 10^{-24} \text{ g})(5 \times 10^{-13} \text{ cm})(10^6 \text{ cm/sec})$$

$$\doteq 10^{-30} \text{ erg-sec.}$$

Thus it would be argued classically that a slow neutron must have an angular momentum less than 10^{-30} erg-sec to be scattered. Quantum mechanically, a neutron with $l = 1$ has an angular momentum $\sqrt{l(l + 1)}\hbar = \sqrt{2}\hbar \doteq 10^{-27}$ erg-sec; so the guess would be that only slow neutrons with $l = 0$ would be scattered. The same line of semiclassical-semiquantum reasoning leads us to the conclusion that fast neutrons (say, $v \doteq 10^9$ cm/sec) with both $l = 0$ and $l = 1$ will contribute to the scattering cross section. This argument, though not rigorous, happens to have a general conclusion which is true. It will be found that, in general, the greater the velocity of the incident particles, the higher the number of partial waves that contribute appreciably to the scattering cross section.

Scattering of $l = 0$ particles is called *s-wave scattering*; scattering of $l = 1$ particles is called *p-wave scattering*. The letters corresponding to the various l values are borrowed from atomic spectroscopic notation. The first four letters are $s(l = 0)$, $p(l = 1)$, $d(l = 2)$, and $f(l = 3)$. Letters corresponding to l values beyond 3 are in alphabetical order beginning with g.

The complete wave function far from the scattering center is given by the sum of the incident and scattered waves,

$$\psi(r,\theta) \sim \sum_{l=0}^{\infty} (2l + 1)i^l \frac{1}{kr} \sin\left(kr - \frac{l\pi}{2}\right) P_l(\cos\theta) + \frac{f(\theta)e^{ikr}}{r}, \quad (4.110)$$

where $j_l(kr)$ in Equation 4.109 has been replaced by its asymptotic form, $(1/kr)\sin[kr - (l\pi/2)]$. Note that this wave function was derived on the basis of physical arguments as to the form the asymptotic wave function must have. This is not the procedure used to find wave functions in previous examples. Quantum mechanics is cookbook mechanics, like the Lagrangian and Hamiltonian formulations of classical mechanics. Problems in quantum mechanics are solved more or less automatically; one simply substitutes the potential function of a given problem into the Schrödinger amplitude equation and solves for the eigenfunctions, the ψ's of the differential equation, from which all physically meaningful answers can be determined. Little thinking is required, and physics is thus reduced to the solution of differential equations. To obtain the wave function which describes the system, merely turn the crank and grind it out of the Schrödinger equation. The $\psi(r,\theta)$ found this way, when equated to the $\psi(r,\theta)$ given by Equation 4.110, produces an expression for $f(\theta)$ and the differential scattering cross section.

We want the solution to the Schrödinger equation in the R-system far from the scattering center. The solution must be independent of the azimuthal angle ϕ since azimuthal symmetry is assumed. But this is just a special case of the general solution obtained in Section 4.2. The appropriate Schrödinger equation is, by comparison with Equation 4.12,

$$\left\{\frac{1}{r^2}\frac{\partial}{\partial r}\left(r^2 \frac{\partial}{\partial r}\right) + \frac{1}{r^2 \sin\theta}\frac{\partial}{\partial\theta}\left(\sin\theta \frac{\partial}{\partial\theta}\right) + \frac{2m}{\hbar^2}[E - V(r)]\right\}\psi(r,\theta) = 0.$$

$$(4.111)$$

The solution of Equation 4.111 is identical to the solution obtained in Section 4.2, except that the azimuthal dependence is missing or, equivalently, that the azimuthal quantum number m_l is zero. Therefore, for any fixed value of the polar quantum number l, the solution has the form

$$\psi_l(r,\theta) = R_l(r)\, P_l(\cos\theta),$$

where the Legendre polynomial P_l appears, rather than the associated Legendre function $P_l^{m_l}$, because $m_l = 0$.

Since the particles under consideration are taken to have all values of l, the most general solution is

$$\psi(r,\theta) = \sum_{l=0}^{\infty} C_l R_l(r)\, P_l(\cos\theta), \quad (4.112)$$

where the coefficients C_l are arbitrary constants and the functions $R_l(r)$ are solutions to the r-dependent part of the Schrödinger equation (Equation 4.43)

$$\left\{ \frac{1}{r^2} \frac{d}{dr}\left(r^2 \frac{d}{dr} \right) - \frac{l(l+1)}{r^2} + \frac{2m}{\hbar^2}[E - V(r)] \right\} R_l(r) = 0.$$

Henceforth, attention will be limited to potentials which fall to zero faster than $1/r$ for larger values of r and which have no pole at the origin that is of higher order than $1/r$. All potentials that will be considered obey these two conditions except the Coulomb potential, which must be treated separately.

The solution to the radial Schrödinger equation in the region outside the well was found, in Section 4.2, to be given by a sum of spherical Hankel functions of the form

$$R_l(r) = B_l[h_l^{(-)}(kr) + e^{2i\delta_l} h_l^{(+)}(kr)], \tag{4.113}$$

where, as usual, $k = \sqrt{2mE}/\hbar$. In the region where $kr \gg 1$, the spherical Hankel functions take on their asymptotic forms:

$$h_l^{(-)}(kr) \sim \frac{1}{kr} \exp\left\{ -i\left[kr - \frac{(l+1)\pi}{2} \right] \right\}, \tag{4.114a}$$

$$h_l^{(+)}(kr) \sim \frac{1}{kr} \exp\left\{ i\left[kr - \frac{(l+1)\pi}{2} \right] \right\}. \tag{4.114b}$$

When these expressions are substituted into Equation 4.113, there results

$$R_l(r) \sim A_l \frac{1}{kr} \sin\left(kr - \frac{l\pi}{2} + \delta_l \right), \tag{4.115}$$

where A_l is an undetermined constant. Equation 4.115 is now substituted into Equation 4.112 with the result

$$\psi(r,\theta) \sim \sum_{l=0}^{\infty} A_l \frac{1}{kr} \sin\left(kr - \frac{l\pi}{2} + \delta_l \right) P_l(\cos \theta), \tag{4.116}$$

where the product of undetermined constants, $A_l C_l$, has been replaced by a single undetermined constant, A_l.

Now, as previously proposed, expressions given in Equations 4.110 and 4.116 for $\psi(r,\theta)$ are equated:

$$\sum_{l=0}^{\infty} (2l+1)i^l \frac{1}{kr} \sin\left(kr - \frac{l\pi}{2} \right) P_l(\cos \theta) + \frac{1}{r} f(\theta) e^{ikr}$$

$$= \sum_{l=0}^{\infty} A_l \frac{1}{kr} \sin\left(kr - \frac{l\pi}{2} + \delta_l \right) P_l(\cos \theta). \tag{4.117}$$

In this equation the A_l and δ_l are unknown constants and $f(\theta)$ is an unknown function. An expression for $f(\theta)$ that involves only the unknown phase shifts δ_l can be found by eliminating the A_l. To accomplish this, we express the sine functions in complex exponential form, i.e., in the form $\sin \alpha = (e^{i\alpha} - e^{-i\alpha})/2i$, and equate coefficients of e^{-ikr} and e^{ikr} on each side of the equation. The resulting two equations are

$$\sum_{l=0}^{\infty} (2l + 1)i^l e^{i(l\pi/2)} P_l(\cos \theta) = \sum_{l=0}^{\infty} A_l e^{-i(\delta_l - l\pi/2)} P_l(\cos \theta) \tag{4.118}$$

and

$$2ik f(\theta) + \sum_{l=0}^{\infty} (2l + 1)i^l e^{-i(l\pi/2)} P_l(\cos \theta) = \sum_{l=0}^{\infty} A_l e^{i(\delta_l - l\pi/2)} P_l(\cos \theta). \tag{4.119}$$

Multiplying each side of Equation 4.118 by $P_l(\cos \theta) \sin \theta \, d\theta$ and integrating over θ from 0 to π, we obtain, because of the orthogonality of the Legendre polynomials, $A_l = (2l + 1)i^l e^{i\delta_l}$. When this expression is substituted into Equation 4.119, it produces

$$f(\theta) = \frac{1}{2ki} \sum_{l=0}^{\infty} (2l + 1)(e^{2i\delta_l} - 1) P_l(\cos \theta).$$

Then the differential elastic-scattering cross section is given by

$$\sigma(\theta) = |f(\theta)|^2 = \frac{1}{4k^2} \left| \sum_{l=0}^{\infty} (2l + 1)e^{i\delta_l}(e^{i\delta_l} - e^{-i\delta_l}) P_l(\cos \theta) \right|^2$$

or

$$\sigma(\theta) = \frac{1}{k^2} \left| \sum_{l=0}^{\infty} (2l + 1)e^{i\delta_l}(\sin \delta_l) P_l(\cos \theta) \right|^2 \tag{4.120}$$

The total elastic-scattering cross section can now be computed by integration of Equation 4.120 over all solid angles. To facilitate this computation, note that the squared absolute value in Equation 4.120 may be expressed as a double sum,

$$\sigma(\theta) = \frac{1}{k^2} \sum_{l=0}^{\infty} \sum_{l'=0}^{\infty} (2l + 1)$$

$$\times (2l' + 1)e^{i(\delta_l - \delta_{l'})} \sin \delta_l \sin \delta_{l'} P_l(\cos \theta) P_{l'}(\cos \theta).$$

The total elastic-scattering cross-section is given by

$$\sigma_s = \int_0^{\pi} \sigma(\theta) 2\pi \sin \theta \, d\theta,$$

but only the Legendre polynomials depend on θ, and, because of their orthogonality, the integral becomes

$$\sigma_s = \frac{2\pi}{k^2} \sum_{l=0}^{\infty} (2l + 1)^2 \left(\frac{2}{2l + 1}\right) \sin^2 \delta_l$$

or

$$\sigma_s = \frac{4\pi}{k^2} \sum_{l=0}^{\infty} (2l + 1) \sin^2 \delta_l. \tag{4.121}$$

For subsequent use, note that the total asymptotic wave function (Equation 4.116) can be written, using $A_l = (2l + 1)i^l e^{i\delta_l}$ and expressing the sine function in terms of exponentials, in the form

$$\psi(r,\theta) \sim \sum_{l=0}^{\infty} \frac{(2l + 1)}{2kr} i^{l+1} [e^{-i(kr - l\pi/2)} - e^{2i\delta_l} e^{i(kr - l\pi/2)}] P_l(\cos \theta).$$

In this form it is clear that the outgoing wave is identical to the incoming wave except for the phase-shift term $e^{2i\delta_l}$, which has modulus unity since δ_l is a real number. Thus the intensity of the outgoing wave is identical to the intensity of the incoming wave. In Chapter 6, it will be shown that absorptive processes can be described by allowing δ_l to be an imaginary number so that the modulus of $e^{2i\delta_l}$, and hence the amplitude of the outgoing wave, is less than unity.

Equations 4.120 and 4.121 represent the solution for the differential and total elastic-scattering cross sections for any central potential that falls off faster than $1/r$. The solutions are expressed in terms of phase shifts with δ_0 the phase shift for the s-wave, δ_1 the phase shift for the p-wave, and so on. The net effect of the potential is to shift the phases of the outgoing partial waves relative to what they would have been were the potential absent. If the phase of a particular partial wave is not shifted or is shifted by an integral multiple of π, this partial wave contributes nothing to either the differential or the total cross section. Conversely, the more closely a phase shift approaches $\pi/2$ or an odd integral multiple of $\pi/2$, the greater the contribution to the cross sections.

Equations 4.120 and 4.121 may be written in terms of the de Broglie wavelength of the particle rather than its propagation constant. Since $\lambdabar \equiv \lambda/2\pi \equiv 1/k$,

$$\sigma(\theta) = \lambdabar^2 \left| \sum_{l=0}^{\infty} (2l + 1)e^{i\delta_l} \sin \delta_l P_l(\cos \theta) \right|^2 \tag{4.120}$$

and

$$\sigma_s = 4\pi\lambdabar^2 \sum_{l=0}^{\infty} (2l + 1) \sin^2 \delta_l. \tag{4.121}$$

Examining the equations written in this form, we can see that the smaller the de Broglie wavelength of a particle, the smaller its scattering cross section, all else being equal. The de Broglie wavelength of a particle is thus very roughly equivalent to its classical radius. Note that the total scattering cross section can be expressed as a sum of partial cross sections ($\sigma_s = \sum_l \sigma_{s,l}$, where $\sigma_{s,l}$ is the contribution to σ_s from particles with angular momentum l). The differential cross section $\sigma(\theta)$ cannot be so expressed because of the interference between terms of different l values.

The potential $V(r)$ affects the cross sections through the phase shifts δ_l. If the potential were zero, all the phase shifts would be zero. To complete the solution, we must seek an expression for the phase shifts that depends on the details of the potential.

Calculation of Phase Shifts

It will be assumed that the potential $V(r)$ is either zero or is so small relative to E that it may be neglected beyond some distance $r = b$. The lth phase shift may be calculated by applying the usual continuity conditions on the lth wave function and on its radial derivative at $r = b$. The outside radial wave function, that is, the radial wave function at values of $r \geq b$, is the solution of the Schrödinger radial equation (Equation 4.43) with $V(r) = 0$ which is given by

$$\left[\frac{-\hbar^2}{2m} \frac{1}{r^2} \frac{d}{dr}\left(r^2 \frac{d}{dr} \right) + \frac{\hbar^2}{2m} \frac{l(l+1)}{r^2} \right] R_l^{out} = E R_l^{out}.$$

When the independent variable is changed from r to ρ, where $\rho \equiv kr \equiv r\sqrt{2mE}/\hbar$, then the preceding equation is reduced to the form of the spherical Bessel equation,

$$\rho^2 \frac{d^2 R_l^{out}}{d\rho^2} + 2\rho \frac{d R_l^{out}}{d\rho} + [\rho^2 - l(l+1)] R_l^{out} = 0.$$

The most general solution of this equation is a sum of spherical Bessel functions and spherical Neumann functions, i.e., $R_l^{out}(kr) = a_l j_l(kr) + b_l n_l(kr)$, where a_l and b_l are arbitrary constants. Now it must be demanded that the asymptotic form of R_l^{out} represent an outgoing wave shifted in phase. That is, far from the origin it must have the form $(1/kr) \sin [kr - (l\pi/2) + \delta_l] = (1/kr) \cos [kr - (l+1)\pi/2 + \delta_l]$. Since the asymptotic forms of j_l and n_l are, from Equation 4.51,

$$j_l(kr) \xrightarrow[r \to \infty]{} \frac{1}{kr} \cos\left[kr - \frac{(l+1)\pi}{2} \right]$$

$$n_l(kr) \xrightarrow[r \to \infty]{} \frac{1}{kr} \sin\left[kr - \frac{(l+1)\pi}{2} \right],$$

the quantities a_l and b_l must be given by $A_l \cos \delta_l$ and $-A_l \sin \delta_l$, respectively, where A_l is an arbitrary constant. Finally then,

$$R_l^{out}(kr) = A_l[(\cos \delta_l)j_l(kr) - (\sin \delta_l)n_l(kr)].$$

The inside radial wave function depends on the specific form of $V(r)$. It will be given, in general, by some function $R_l^{in}(r)$. Equating the inside and outside wave functions and their derivatives at $r = b$, we obtain

$$A_l[(\cos \delta_l)j_l(kb) - (\sin \delta_l)n_l(kb)] = R_l^{in}(b) \tag{4.122}$$

and

$$A_l[(\cos \delta_l)j_l'(kb) - (\sin \delta_l)n_l'(kb)] = \left. \frac{dR_l^{in}}{dr} \right|_b, \tag{4.123}$$

where the primes indicate differentiation with respect to r. To get rid of the arbitrary constant A_l, divide Equation 4.123 by Equation 4.122:

$$\frac{(\cos \delta_l)j_l'(kb) - (\sin \delta_l)n_l'(kb)}{(\cos \delta_l)j_l(kb) - (\sin \delta_l)n_l(kb)} = \left[\frac{1}{R_l^{in}} \frac{dR_l^{in}}{dr} \right]_{r=b}. \tag{4.124}$$

Let the symbol D_l be assigned to the logarithmic derivative of the inner radial wave function at $r = b$; i.e.,

$$D_l \equiv \left[\frac{1}{R_l^{in}} \frac{dR_l^{in}}{dr} \right]_{r=b}. \tag{4.125}$$

When Equation 4.125 is substituted into Equation 4.124 and the numerator and denominator are divided through by $\cos \delta_l$, the result is

$$\tan \delta_l = \frac{j_l'(kb) - D_l j_l(kb)}{n_l'(kb) - D_l n_l(kb)}. \tag{4.126}$$

It is clear that we need only determine $\tan \delta_l$ to solve Equation 4.120 for $\sigma(\theta)$ since $e^{i\delta_l} \sin \delta_l$ is identical to $e^{i(\delta_l + \pi)} \sin(\delta_l + \pi)$. Given $\tan \delta_l$, we know only that δ_l itself is either δ_l or $\delta_l + \pi$, but either one yields the same cross sections.

Everything on the right-hand side of Equation 4.126 except D_l is known. It seems that all we have found is the expression of the unknown phase shifts δ_l in terms of the equally unknown logarithmic derivatives D_l. But the logarithmic derivatives can always be determined since the radial wave function $R_l(r)$ can always be found by solution of Equation 4.43. For any but the most simple potentials, Equation 4.43 cannot be solved in closed form, but it can always be solved numerically or graphically by transforming it to a finite difference equation, as is shown in Section 4.8.

Alternative Procedure for Finding Phase Shifts

Far from the origin and from the potential, in the region where $kr \gg 1$ and $r \gg b$, the lth radial wave function in the absence of a potential will be given by the asymptotic form

$$R_l^{no} \sim \frac{1}{kr} \sin\left(kr - \frac{l\pi}{2}\right).$$

Assuming the lth radial wave function has been calculated, numerically or otherwise, and assuming the effect of the potential has been included in this calculation, the wave function is certain to have the asymptotic form

$$R_l^{yes} \sim \frac{1}{kr} \sin\left(kr - \frac{l\pi}{2} + \delta_l\right),$$

where δ_l remains to be determined. Plots of rR_l^{no} and rR_l^{yes} will be identical except for the shift in phase (see Figure 4.6). The phase shift can be determined by noting that at the origins of equal (zero) phase of R_l^{no} and R_l^{yes}, designated r_1 and r_2 in Figure 4.6, it is necessary that

$$\sin\left(kr_1 - \frac{l\pi}{2}\right) = 0$$

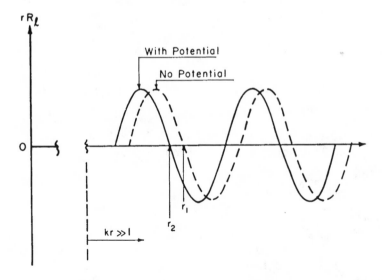

Figure 4.6—The asymptotic radial wave functions with and without a potential. Adjoining points of zero phase are at r_1 and r_2.

and

$$\sin\left(kr_2 - \frac{l\pi}{2} + \delta_l\right) = 0;$$

hence

$$kr_1 - \frac{l\pi}{2} = kr_2 - \frac{l\pi}{2} + \delta_l$$

or, finally,

$$\delta_l = k(r_1 - r_2).$$

Thus, to find the phase shifts, we need only find the adjacent asymptotic zeros of the radial wave functions with and without the potential. In fact, the zeros need not be adjacent since nonadjacent zeros will simply add ineffectual factors of π to δ_l.

Cross Section of an Impenetrable Sphere

As an elementary example of the use of the method of partial waves, let us calculate the scattering cross section of particles of mass m and energy E which are interacting with a perfectly rigid elastic sphere of radius b. By "perfectly rigid" we mean that the potential is infinite in the region $r < b$ and zero elsewhere:

$$V(r) = \infty \qquad (r < b)$$

$$V(r) = 0 \qquad (r > b).$$

The radial wave function outside the potential is always given by

$$R_l^{out}(kr) = A_l[(\cos \delta_l)j_l(kr) - (\sin \delta_l)n_l(kr)],$$

but, in this case, because the potential is infinite in the region $r < b$, the wave function must vanish at $r = b$. Therefore

$$\tan \delta_l = \frac{j_l(kb)}{n_l(kb)}, \tag{4.127}$$

where $k \equiv \sqrt{2mE}/\hbar$. This equation determines the phase shifts of all the partial waves; hence Equations 4.120 and 4.121 may be employed to determine the differential and total scattering cross sections, and the problem is solved in general. It is particularly interesting to examine two extreme special cases of scattering from a perfectly rigid sphere: low-energy scattering and high-energy scattering, which are defined, respectively, by the conditions $kb \ll 1$ and $kb \gg 1$.

At low energies, that is, when $kb \ll 1$, the leading terms in the spherical Bessel and Neumann functions are given by Equation 4.50. When these approximate expressions are substituted into the preceding general expressions for $\tan \delta_l$, there results

$$\tan \delta_l = \frac{\dfrac{(kb)^l}{1 \cdot 3 \cdot 5 \cdots (2l + 1)}}{\dfrac{-1 \cdot 3 \cdot 5 \cdots (2l - 1)}{(kb)^{l+1}}} = \frac{-(kb)^{2l+1}}{(2l + 1)[1 \cdot 3 \cdot 5 \cdots (2l - 1)]^2} \; .$$

Since kb is much less than unity, the phase shifts for the p-wave, and higher order waves are much less than the phase shift for the s-wave. Thus one need only consider the s-wave phase shift, δ_0, which is given by $\tan \delta_0 = -kb$. Furthermore, since $kb \ll 1$, $\tan \delta_0$ is approximately equal to the angle in radians, and $\delta_0 \simeq -kb$. The differential elastic-scattering cross section is given by Equation 4.120 with the summation containing only the $l = 0$ term.

$$\sigma(\theta) = \frac{1}{k^2} \left| e^{-ikb} \sin (-kb) P_0(\cos \theta) \right|^2.$$

But $P_0(\cos \theta) = 1$, and, since $kb \ll 1$, $\sin (-kb) \simeq -kb$; so

$$\sigma(\theta) = \frac{(-kb)^2}{k^2} \left[e^{-ikb} e^{+ikb} \right] = b^2.$$

There is no angular dependence. Low-energy scattering from a rigid sphere is isotropic in the C-system. Pure s-wave scattering is always isotropic in the C-system because P_0 is not a function of θ. The total scattering cross section, given by Equation 4.121, is

$$\sigma_s = \frac{4\pi}{k^2} \sum_l (2l + 1) \sin^2 \delta_l = \frac{4\pi}{k^2} (-kb)^2 = 4\pi b^2.$$

The total low-energy cross section of the rigid sphere is seen to be four times the classical value. It is also energy independent as long as $kb \ll 1$.

At high energies, i.e., when $kb \gg 1$, many partial waves will contribute to the differential scattering cross section, which can then be shown by direct but laborious computation to be highly peaked in the forward direction. The total scattering cross section can be obtained without great difficulty as follows: Since $\tan \delta_l = j_l(kb)/n_l(kb)$.

$$\sin^2 \delta_l = \frac{j_l^2(kb)}{j_l^2(kb) + n_l^2(kb)},$$

which becomes, after substituting the asymptotic expressions (Equation

4.51) for the spherical Bessel and Neumann functions,

$$\sin^2 \delta_l = \cos^2 \left[kb - \frac{(l + 1)\pi}{2} \right].$$

Thus the total scattering cross section is

$$\sigma_s = \frac{4\pi}{k^2} \sum_{l=0}^{\infty} (2l + 1) \cos^2 \left[kb - \frac{(l + 1)\pi}{2} \right].$$

Now it is reasonable to replace the upper limit in this sum by the largest value of l likely to result in appreciable scattering. One argues as follows: Particles with classical impact parameters greater than b would not be (classically) scattered. These are particles with classical angular momentum pb; but $p = k\hbar$, and particles with angular momenta greater than $kb\hbar$ should contribute little to the cross section. Thus quantum-mechanical particles with angular momenta greater than $[l(l + 1)]^{1/2}\hbar = kb\hbar$ should contribute little. Or, since $[l(l + 1)]^{1/2} \simeq l$ for $l \gg 1$, we reach the conclusion that the sum in the expressions for σ_s can be cut off at $l = kb$ with little error. Now, since $kb \gg 1$, the argument of the cosine is a rapidly varying function of k, and it is therefore reasonable to compute the sum by replacing the cosine-square terms by their average value of one-half. Hence

$$\sigma_s \simeq \frac{2\pi}{k^2} \sum_{l=0}^{kb} (2l + 1),$$

which, for $kb \gg 1$, is approximately

$$\sigma_s \simeq \frac{2\pi}{k^2} \int_0^{kb} 2l \, dl = 2\pi b^2.$$

Thus it is seen that the high-energy total elastic-scattering cross section of a rigid sphere is twice the classical value, that is, twice the geometric cross section of the sphere. This is an important result but one which must be used with caution. If one were to plot the differential scattering cross section for this case, it would be found to be extremely peaked in the forward direction, as in Figure 4.7. In fact, one-half the total area under the curve would be found to be contributed by extremely small angle scattering. This small-angle scattering is analogous to the diffraction of light around a small opaque sphere which produces the well-known Arago bright spot. If it were possible to set up a detector that could distinguish between unscattered particles and particles scattered through extremely small angles, the measured total cross section would indeed be $2\pi b^2$. From a practical viewpoint, however, particles that scatter forward have not scattered at all since they retain their original energy and direction. Thus for practical purposes, e.g., reactor calculations, the high-energy scattering

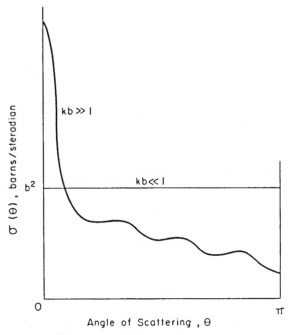

Figure 4.7—Typical differential elastic-scattering cross sections for impenetrable spheres at low energies ($kb \ll 1$) and high energies ($kb \gg 1$).

cross section of a rigid sphere is πb^2, not $2\pi b^2$. The reader is invited to carry out the elementary but laborious calculation of the high-energy differential cross section of the rigid sphere to verify the truth of the above arguments.

4.8 Numerical Solution of the Radial Schrödinger Equation

In this section we exhibit one of the simplest and most direct of the numerous possible numerical procedures for finding the radial Schrödinger function. It is assumed that the potential $V(r)$ has the form of a real well or a real barrier that is effectively zero beyond some radius $r = b$. It is further assumed that the potential has no pole of higher order than $1/r$ at the origin. Except for these restrictions, the well or barrier may have any shape.

The radial Schrödinger equation for a single particle of reduced mass m with total angular-momentum quantum number l and total available energy E is

$$\left\{ \frac{1}{r^2} \frac{d}{dr} \left(r^2 \frac{d}{dr} \right) + \frac{2m}{\hbar^2} \left[E - V(r) \right] - \frac{l(l+1)}{r^2} \right\} R_l(r) = 0, \qquad (4.43')$$

which is simplified by making the substitution $W_l(r) \equiv rR_l(r)$,

$$\frac{d^2 W_l(r)}{dr^2} + \frac{2m}{\hbar^2}\left[E - V(r) - \frac{\hbar^2}{2m}\frac{l(l + 1)}{r^2} \right] W_l(r) = 0. \qquad (4.67')$$

This can be further simplified by defining $\rho \equiv r\sqrt{2mE}/\hbar$ in order to measure lengths in units of the reduced de Broglie wavelength (λ) of the particle outside the potential and by defining $U(\rho) \equiv V(\rho)/E$ in order to measure the potential in units of the total energy of the particle. When these definitions are substituted in Equation 4.67', the dimensionless radial Schrödinger equation results:

$$\frac{d^2 W_l(\rho)}{d\rho^2} + \left[1 - U(\rho) - \frac{l(l + 1)}{\rho^2} \right] W_l(\rho) = 0. \qquad (4.128)$$

Since all terms within the brackets are real, the differential equations for the real and imaginary parts of $W_l(\rho)$ are identical, and one need only solve for the real part.

The second derivative in Equation 4.128 will now be converted to a finite difference form using the central difference approximation.* The ρ-axis is divided into a mesh of points, each separated by an increment Δ from the previous point. The first point, designated $n = 0$, is at the origin. Succeeding points, designated by the integers $n = 1, 2, 3, \ldots$ are at $\rho = n\Delta$, where Δ is an increment in ρ chosen to be sufficiently small that the percentage change in $W_l(\rho)$ between mesh points is very small. The value of $W_l(\rho)$ at mesh point n will be called $W_l^{(n)}$. Finite difference expressions can now be derived for the first and second derivatives of W_l at point n. Referring to Figure 4.8, we see that the first derivative at point n may be approximated by

$$\frac{dW_l}{d\rho}\bigg|_n \simeq \frac{W_l^{(n+1)} - W_l^{(n-1)}}{2\Delta}$$

The second derivative is given by $1/\Delta$ times the difference of derivatives at the points marked X in Figure 4.8,

$$\frac{d^2 W_l}{d\rho^2}\bigg|_n \simeq \frac{1}{\Delta}\left(\frac{W_l^{(n+1)} - W_l^{(n)}}{\Delta} - \frac{W_l^{(n)} - W_l^{(n-1)}}{\Delta} \right)$$

$$= \frac{W_l^{(n+1)} + W_l^{(n-1)} - 2W_l^{(n)}}{\Delta^2}.$$

---

*Other more exact but more complicated methods of numerical solutions, such as those of Runge and Kutta, may be found in most textbooks on numerical methods. See, for example, H. Levy and E. A. Baggott's *Numerical Solutions of Differential Equations* (Dover Publications, Inc., New York, 1950).

Thus Equation 4.128 is converted to a finite difference equation by the following substitutions:

$$\rho = n\Delta,$$

where $n = 0, 1, 2, \ldots,$

$$U(\rho) = U(n\Delta) \equiv U^{(n)},$$

$$W_l(\rho) = W_l(n\Delta) \equiv W_l^{(n)},$$

and

$$\frac{d^2 W_l}{d\rho^2} = \frac{W_l^{(n+1)} + W_l^{(n-1)} - 2W_l^{(n)}}{\Delta^2}.$$

The result of these substitutions is

$$W_l^{(n+1)} = A_l^{(n)} W_l^{(n)} - W_l^{(n-1)}, \tag{4.129}$$

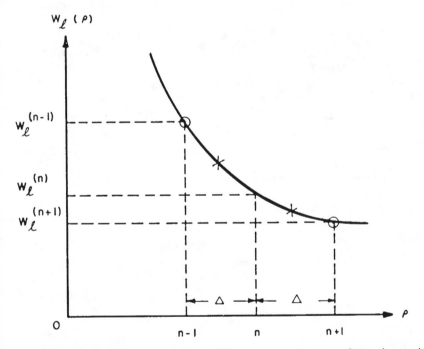

Figure 4.8—Construction to obtain central difference approximation to first and second derivatives of the function $W_l(\rho)$ at point n. The 0 points were used to evaluate the first derivative, and the X points to evaluate the second derivative.

where

$$A_l^{(n)} = \left\{ 2 + \frac{l(l+1)}{n^2} \quad \Delta^2 [1 \quad U^{(n)}] \right\}.$$
(4.130)

Equation 4.129 is a three-point recursion formula. Given the values of W_l at two points, $n-1$ and n, we can find its value at point $n+1$ since $A_l^{(n)}$ is known at all values of n. At the origin, $W_l(\rho)$ must be zero because $W_l(\rho) = rR_l(\rho)$ and $R_l(\rho)$, to be a valid wave function, must not be infinite at any point (postulate 4). Thus $W_l^{(0)} = 0$, and the recursion formula then says that

$$W_l^{(2)} = A_l^{(1)} W_l^{(1)}.$$

It seems at first glance that we need a value of $W_l^{(1)}$ in order to proceed. Indeed we would need this value if we wished to calculate the normalized wave function, but there is no need to calculate the normalized function because the logarithmic derivatives $(1/R_l)(dR_l/dr)$, which must be found in order to calculate the phase shifts, are independent of the absolute magnitude of the wave function. Furthermore, if for any reason the normalized wave function were desired, we could always obtain it from the nonnormalized wave function by a straightforward numerical integration. Thus the value of $W_l^{(1)}$ can be arbitrarily fixed. We could, for example, simply set $W_l^{(1)} = 1.0$ and proceed to find all other $W_l^{(n)}$ using the recursion relation (Equation 4.129). If this were done, however, the nonnormalized wave functions for various l values would be wildly different in magnitude. A somewhat better prescription for $W_l^{(1)}$ may be obtained by noting that as r approaches zero the dominant term in the bracket in Equation 4.128 is the centrifugal potential

$$\frac{d^2 W_l(\rho)}{d\rho^2} - \frac{l(l+1)}{\rho^2} W_l(\rho) = 0$$

if $l > 0$ and if $r \to 0$. We can verify by direct substitution that this equation has the solution $W_l(\rho) = C_l \rho^{l+1}$, where C_l is an arbitrary constant. Thus it would seem appropriate that we set $C_l = 1.0$ and $W_l^{(1)} = \Delta^{l+1}$. The larger the value of l, the stronger the repulsive centrifugal potential near the origin and therefore the smaller the wave function. The behavior of $W_l(\rho)$ for $l = 0$ will depend on the form of the potential $U(\rho)$ near the origin. If $U(\rho)$ is a constant near the origin, $W_0(\rho)$ will behave like $\rho^1 = \rho^{0+1} = \rho^{l+1}$ also. Since, as previously pointed out, $W_l^{(1)}$ may be arbitrarily fixed, we impose the condition that

$$W_l^{(1)} = \Delta^{l+1}$$

for all values of l. The resulting functions $W_l(\rho)$ will not be normalized but will all be of the same order of magnitude.

The procedure described above for numerical integration of the radial Schrödinger equation may be used directly to obtain the phase shifts for a positive energy state (scattering problem). A slight extension of the same procedure may be used to determine the energy levels, both bound and virtual, in a potential well.

Determination of Phase Shifts

In scattering problems, interest centers on the calculation of the logarithmic derivative D_l of each of the partial waves at the edge of the potential. Having determined the values of D_l for $l = 0, 1, \ldots$, we may calculate the phase shifts using Equation 4.126. The differential and total elastic-scattering cross sections may then be calculated using Equations 4.120 and 4.121.

The procedure for finding each of the logarithmic derivatives is identical. For a fixed value of l:

1. Set up a mesh by choosing a small value of Δ, say $\Delta_1 = 10^{-2}$, in such a way that one of the mesh points falls on the point $\rho = b/\lambda$. That is, set $\Delta = b/N\lambda$ where N is an integer so that the Nth mesh point falls on $\rho = b/\lambda$.

    2. Set $W_l^{(0)} = 0$ and $W_l^{(1)} = \Delta_1^{l+1}$.

    3. Solve the recursion relation (Equation 4.129) for $W_l^{(n)}$ at $n = 2, 3, \ldots, N + 1$.

    4. Calculate the wave function $R_l(\rho)$ at the final three points, $n = N - 1, N,$ and $N + 1$ using $R_l^{(n)} = W_l^{(n)}/n\Delta_1$.

    5. Calculate the logarithmic derivative of the wave function using the central difference approximation

$$D_l^{(1)} = \frac{1}{R_l^{(N)}} \frac{R_l^{(N+1)} - R_l^{(N-1)}}{2\Delta_1}.$$

    6. Repeat the entire calculation with a smaller value of Δ, say Δ_2. Calculate D_l again, calling it $D_l^{(2)}$. Continue this process of inner iteration until two successive values of $D_l^{(J)}$ agree within the limits of error you allow, say ε. Thus, when

$$\left| \frac{D_l^{(J)} - D_l^{(J-1)}}{\frac{1}{2}[D_l^{(J)} + D_l^{(J-1)}]} \right| \le \varepsilon,$$

then $D_l^{(J)} = D_l$, the converged logarithmic derivative.

    7. Calculate the phase shift δ_l using Equation 4.126.

The entire procedure is carried out for $l = 0, 1, 2, \ldots$ until successive values of δ_l begin to become so small as to be ignored. We may proceed with impunity, being certain that eventually the δ_l will begin to decrease toward zero since any differential cross section can be expressed by the sum of a finite number of Legendre polynomials.

Determination of Energy Levels

Given a potential well $V(r)$ that extends from $r = 0$ to $r = b$, we wish to find all the energy eigenvalues within the well, i.e., all eigenvalues $E < 0$. To see clearly how this is done, we rewrite the recursion relation to contain E explicitly. That is, in Equation 4.67' we set

$$r = n\Delta \text{ for } n = 0, 1, 2, \ldots$$

$$V(r) = V^{(n)}$$

$$W_l(r) = W_l^{(n)}$$

$$\frac{d^2 W_l(r)}{dr^2} = \frac{W_l^{(n+1)} + W_l^{(n-1)} - 2W_l^{(n)}}{\Delta^2}$$

and find that

$$W_l^{(n+1)} = B_l^{(n)} W_l^{(n)} - W_l^{(n-1)}, \tag{4.131}$$

where

$$B_l^{(n)} = \left\{ 2 - \frac{2m}{\hbar^2} \Delta^2 \left[E - V^{(n)} - \frac{\hbar^2}{2m} \frac{l(l+1)}{n^2 \Delta^2} \right] \right\}. \tag{4.132}$$

What we are now seeking are those particular negative values of E lying between 0 and the maximum depth of the well for which the wave function is well behaved. From the general discussion in Section 4.2, we concluded that the outside wave function must behave at large values of r as

$$W_l(r) \sim C_l e^{-gr},$$

where C_l is an arbitrary constant and $g = \sqrt{2m|E|}/\hbar$. Thus the general procedure is to guess at values of E; determine by the usual step-by-step procedure using Equations 4.131 and 4.132 the values of $W_l^{(n)}$ inside and to some distance outside the well. If we guess E wrong, the outside wave function will not decrease exponentially. We then make a second guess and proceed to close in on the correct value of E by iteration. For each outer iteration on E, we must either have chosen a Δ that is known to be sufficiently small or we must perform inner iteration on Δ to assure the accuracy of the numerical integration procedure.

For each l, we must survey the entire well, from 0 to its maximum depth, to find all the negative energy eigenvalues associated with this l, i.e., all the E_{nl}. This survey must be repeated for $l = 0, 1, 2, \ldots$. The number of bound states will decrease with increasing l because of the centrifugal potential. Eventually there will be an l beyond which no bound states can be found.

A somewhat faster procedure, one that avoids the necessity of carrying the integration far beyond the edge of the well, is the following: It is known that the outside wave function for a bound state must go as the Hankel function $h_l^{(+)}(igr)$, thus

$$R_l^{out} = C_l h_l^{(+)}(igr)$$

in the region $r \geq b$ where C_l is an arbitrary constant. Therefore, we can determine the exact value of the logarithmic derivative of R_l^{out} at $r = P$, some point beyond the edge of the well. The numerical integration is then stopped at the point P and a check made to see if the logarithmic derivative of the wave function agrees with the value it must have. If not, another value of E is chosen and another attempt made. If it does agree within some predetermined epsilon, we would continue the calculation of the wave function to a greater distance to ensure that it is indeed falling off exponentially.

4.9 The Born Approximation

The method of partial waves discussed in Section 4.7 has very general applicability; there are but few potentials met in practical problems to which it cannot be applied. Unfortunately, it is a laborious procedure, particularly if a large number of phase shifts must be calculated. The Born approximation, in contrast, has only a limited range of applicability, but it is the simplest method to apply in those situations in which it is valid. The primary condition for its validity is that the incident-wave function be much larger than the scattered-wave function within the potential region. This will be true, for example, if the potential energy $V(r)$ is much smaller than the total energy E of the scattered particle.

The derivation of the Born approximation begins with the Schrö-dinger amplitude equation for a particle of reduced mass m in the R-system where the target particle is at rest,

$$\left[\frac{-\hbar^2}{2m} \nabla^2 + V(\mathbf{r}) \right] \psi(\mathbf{r}) = E\psi(\mathbf{r}), \qquad (4.133)$$

where E is the total available energy of the particle in the C-system and $V(\mathbf{r})$ is the potential energy at relative position $\mathbf{r}$. After rearranging terms in Equation 4.133 and defining $k = (2mE)^{1/2}/\hbar$, we have

$$(\nabla^2 + k^2)\, \psi(\mathbf{r}) = \frac{2m}{\hbar^2}\, V(\mathbf{r})\, \psi(\mathbf{r}) = F(\mathbf{r}), \qquad (4.134)$$

where $F(\mathbf{r})$ has been defined to be

$$F(\mathbf{r}) \equiv \frac{2m}{\hbar^2} V(\mathbf{r}) \, \psi(\mathbf{r}). \tag{4.135}$$

The wave function expressed as a sum of incident and scattered waves, $\psi(\mathbf{r}) = \exp(ikz) + \psi_{sc}(\mathbf{r})$, is substituted into the left-hand side of Equation 4.134. Since $(\nabla^2 + k^2) \exp(ikz) = 0$, the result is

$$(\nabla^2 + k^2) \, \psi_{sc}(\mathbf{r}) = F(\mathbf{r}). \tag{4.136}$$

If its right-hand side were zero, Equation 4.136 would have the form of Laplace's equation for a field without sources or sinks. Thus $F(\mathbf{r})$ may be regarded as the source of the scattered wave function.

Now suppose there were a unit isotropic point source of scattered particles emanating from point $\mathbf{r}'$. What would be the form of the wave function at $\mathbf{r}$? The wave function would consist of an undulating term $e^{ik|\mathbf{r}-\mathbf{r}'|}$, where $|\mathbf{r} - \mathbf{r}'|$ is the distance between the source and the point r, and an "inverse-square" term $1/4\pi|\mathbf{r} - \mathbf{r}'|$. Thus

$$\psi_0(\mathbf{r},\mathbf{r}') = \frac{e^{ik|\mathbf{r}-\mathbf{r}'|}}{4\pi|\mathbf{r} - \mathbf{r}'|}. \tag{4.137}$$

The function $\psi_0(\mathbf{r}.\mathbf{r}')$, which is the scattered-wave function at $\mathbf{r}$ due to a unit isotropic point source of scattered particles at $\mathbf{r}'$, is called the *Green's function* of the differential equation (Equation 4.136).*

The total scattered intensity at $\mathbf{r}$ is the integral of $F(\mathbf{r}')$ times $\psi_0(\mathbf{r},\mathbf{r}')$ over the volume in which the potential is nonzero. That is,

$$\psi_{sc}(\mathbf{r}) = \int \frac{e^{ik|\mathbf{r}-\mathbf{r}'|}}{4\pi|\mathbf{r} - \mathbf{r}'|} F(\mathbf{r}') \, d\tau'$$

or

$$\psi_{sc}(\mathbf{r}) = \frac{2m}{\hbar^2} \int \frac{e^{ik|\mathbf{r}-\mathbf{r}'|}}{4\pi|\mathbf{r} - \mathbf{r}'|} V(\mathbf{r}') \, \psi(\mathbf{r}') \, d\tau'. \tag{4.138}$$

Thus far the result is exact but of little value since the wave function $\psi(\mathbf{r}')$, which appears under the integral sign, is not known.

We can introduce a very useful and accurate approximation based on the facts that (1) we are interested exclusively in potentials $V(\mathbf{r}')$ that have an appreciable value only when r' is of the order of 10^{-8} cm or less and (2) in all practical scattering experiments, the distance r to the detector is at least of the order of centimeters.

---

*For a rigorous derivation of this Green's function see Mott and Massey's *The Theory of Atomic Collisions* (Reference 2 at the end of this chapter), pp. 114–117.

Since $r \gg r'$, it is clear that the term $|\mathbf{r} - \mathbf{r}'|$ in the denominator of Equation 4.138 can be replaced with negligible error by r. However, the numerator must be treated with more respect because a slight error in $|\mathbf{r} - \mathbf{r}'|$ may produce an appreciable difference in phase.

Define $\mathbf{u_0}$ as the unit vector along the initial direction of motion of the incident particles and $\mathbf{u}$ as the unit vector along the direction $\mathbf{r}$. The relation between the various vectors is depicted in Figure 4.9. The length $|\mathbf{r} - \mathbf{r}'|$ can then be written, using the law of cosines, as

$$|\mathbf{r} - \mathbf{r}'| = (r^2 + r'^2 - 2\mathbf{r} \cdot \mathbf{r}')^{1/2}$$

$$= r\left[1 + \left(\frac{r'}{r}\right)^2 - 2\frac{\mathbf{r} \cdot \mathbf{r}'}{r^2}\right]^{1/2}.$$

However, the second term in the brackets is orders of magnitude smaller than the third term; and, since $\mathbf{r}/r \equiv \mathbf{u}$, to a high degree of accuracy

$$|\mathbf{r} - \mathbf{r}'| \simeq r\left(1 - 2\frac{\mathbf{r}' \cdot \mathbf{u}}{r}\right)^{1/2} \simeq r - \mathbf{r}' \cdot \mathbf{u}, \tag{4.139}$$

where the final approximation results from neglecting terms of order $(r'/r)^2$ and higher in the expansion of the square root. These terms are of the order 10^{-16} or smaller compared with unity in typical problems, making this a very good approximation. It maintains an accuracy of about 1 in 10^8 parts provided the potential is effectively zero beyond 10^{-8} cm.

When Equation 4.139 is substituted into the numerator of Equation 4.138 and the denominator is replaced with r, we obtain

$$\psi_{sc}(\mathbf{r}) = \left[\frac{m}{2\pi\hbar^2}\int e^{-ik(\mathbf{r}' \cdot \mathbf{u})} V(\mathbf{r}')\,\psi(\mathbf{r}')\,d\tau'\right]\frac{e^{ikr}}{r}. \tag{4.140}$$

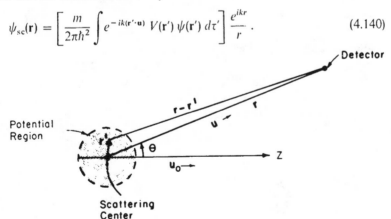

Figure 4.9—The various vectors which enter into the derivation of the Born approximation of the differential elastic-scattering cross section.

Comparing Equation 4.140 with the general expression for ψ_{sc},

$$\psi_{sc} = f(\theta)\frac{e^{ikr}}{r},$$

we see that the bracketed term in Equation 4.140 is $f(\theta)$; and, since $\sigma(\theta) = |f(\theta)|^2$,

$$\sigma(\theta) \quad \frac{m}{2\pi\hbar^2}\left|\int e^{-ik(\mathbf{r}'\cdot\mathbf{u})}V(\mathbf{r}')\psi(\mathbf{r}')\,d\tau'\right|^2 \tag{4.141}$$

The First Born Approximation

Equation 4.141 is an almost exact expression for $\sigma(\theta)$. The term in the exponential is in error by less than 1 in 10^8 parts. But this equation is of little value as it stands because the wave function $\psi(\mathbf{r}')$ is unknown. The first Born approximation is based on the assumption that the incident particle is scattered very weakly so that the complete wave function within the potential region is approximately equal to the incident wave function,

$$\psi(\mathbf{r}') \simeq \psi_{inc}(\mathbf{r}') = e^{ikz'} = e^{ikr'\cdot\mathbf{u_0}}.$$

Since the wave function provides a probabilistic description, we can equivalently state that the first Born approximation applies where only a small fraction of the incident ensemble of particles is scattered. When the above approximation for $\psi(\mathbf{r}')$ is substituted into Equation 4.141, there results

$$\sigma(\theta) = \left|\frac{m}{2\pi\hbar^2}\int e^{ikr'\cdot(\mathbf{u_0}-\mathbf{u})}V(\mathbf{r}')\,d\tau'\right|^2 \tag{4.142}$$

Equation 4.142 is the general expression of the first Born approximation. From this point on, the development is restricted to the single most important special case—that resulting when the potential is central, i.e., when $V(\mathbf{r}') = V(r')$. Then, in spherical coordinates (r',α,β) with the polar axis along the direction $(\mathbf{u_0} - \mathbf{u})$, we have

$$\mathbf{r}'\cdot(\mathbf{u_0} - \mathbf{u}) = r'|\mathbf{u_0} - \mathbf{u}|\cos\alpha$$

and

$$d\tau' = (r')^2\sin\alpha\,dr'\,d\alpha\,d\beta,$$

so that

$$f(\theta) = \frac{m}{2\pi\hbar^2}\int_0^\infty\int_0^\pi\int_0^{2\pi}\exp(ikr'|\mathbf{u_0} - \mathbf{u}|\cos\alpha)V(r')(r')^2\sin\alpha\,dr'\,d\alpha\,d\beta.$$

After the elementary integrations over β and α,

$$f(\theta) = \frac{2m}{\hbar^2} \int_0^\infty \frac{1}{2ikr'|\mathbf{u}_0 - \mathbf{u}|} \left[\exp\left(ikr'|\mathbf{u}_0 - \mathbf{u}|\right) \right.$$

$$\left. - \exp\left(-ikr'|\mathbf{u}_0 - \mathbf{u}|\right) \right] V(r')(r')^2 \, dr'$$

$$= -\frac{2m}{\hbar^2} \int_0^\infty \frac{\sin\left[kr'|\mathbf{u}_0 - \mathbf{u}|\right]}{kr'|\mathbf{u}_0 - \mathbf{u}|} V(r')(r')^2 \, dr'. \tag{4.143}$$

The unit vectors $\mathbf{u}_0$ and $\mathbf{u}$ are separated by the scattering angle θ, as seen in Figure 4.9. Thus

$$|\mathbf{u}_0 - \mathbf{u}| = [(\mathbf{u}_0 - \mathbf{u}) \cdot (\mathbf{u}_0 - \mathbf{u})]^{1/2} = [2 - 2(\mathbf{u}_0 \cdot \mathbf{u})]^{1/2}$$

$$= [2 - 2\cos\theta]^{1/2} = [4\sin^2(\theta/2)]^{1/2}$$

$$= 2\sin(\theta/2).$$

When this expression for $|\mathbf{u}_0 - \mathbf{u}|$ is substituted into Equation 4.143 and the absolute square taken, the result is the very useful first Born approximation for the differential scattering cross section,

$$\sigma(\theta) = \left| \frac{2m}{\hbar^2} \int_0^\infty \frac{\sin\left[2kr'\sin(\theta/2)\right]}{2kr'\sin(\theta/2)} V(r')(r')^2 \, dr' \right|^2. \tag{4.144}$$

In this equation θ is the scattering angle in the C-system, $\sigma(\theta)$ is the differential cross section in units of area per unit solid angle, m is the reduced mass of the particle, and $k = (2mE)^{1/2}/\hbar$.

The major approximation that was made in deriving the first Born approximation was the replacement (within the potential region) of the complete-wave function $\psi(\mathbf{r}')$ by the incident-wave function $\psi_{inc}(\mathbf{r}')$. This is a reasonable approximation provided the potential energy is very small relative to the kinetic energy so that within the potential the amplitude of the scattered wave is small relative to the amplitude of the incident wave. To see more clearly how the condition that $V(r') \ll E$ enters into the expression for the differential cross section, we rewrite Equation 4.144 in terms of the dimensionless variable $\rho = kr'$; that is, we measure lengths in units of $\lambdabar$, the reduced de Broglie wavelength of the free particles. It can readily be verified that Equation 4.144 becomes

$$\sigma(\theta) = \left| \frac{1}{k} \int_0^\infty \frac{\sin \xi\rho}{\xi\infty} \frac{V(\rho)}{E} \rho \, d\rho \right|^2, \tag{4.145}$$

where $\xi \equiv 2\sin\theta/2$ and $V(\rho) = V(\rho/k) = V(r')$. In this form it can be clearly seen that the integral will be infinite unless $V(\rho)$ decreases at least as rapidly as $1/\rho$ when ρ approaches infinity. The condition that $V(\rho)$

decrease at least as rapidly as $1/\rho$ for large ρ is a necessary condition for convergence of the integral. On the other hand, the condition that $V(\rho)/E$ be everywhere small is only a sufficient condition for the validity of the first Born approximation. It will be found, for example, that correct cross sections can be obtained for potentials that go to infinity at the origin in such a way that over a small volume of space $V(\rho)/E \gg 1$. Physically this makes sense since the contribution to the scattered wave function from a small volume may be expected to be negligible compared to the incident-wave function.

Second and higher order Born approximations may be obtained by iterating on $\psi(\mathbf{r}')$ in Equation 4.140. The first Born approximation is used to find a first approximation to the scattered-wave function $\psi_{sc}^{(1)}$. This is added to the incident-wave function, and the sum $[\psi_{inc}(\mathbf{r}') + \psi_{sc}^{(1)}(\mathbf{r}')]$ is inserted into the integral in Equation 4.140 to produce a second approximation to the scattered wave function $\psi_{sc}^{(2)}$, and so on. This process generally becomes so cumbersome beyond the first approximation that it is rarely employed. In fact, the words "Born approximation" without further qualification have become almost synonymous with "first Born approximation."

As an elementary but important example of the use of the Born approximation, consider the inverse-r potential $V(r) = a/r$, where $0 \leq r < \infty$. When the potential $V(\rho) = ak/\rho$ is substituted into Equation 4.145, the result is

$$\sigma(\theta) = \left| \frac{1}{k} \int_0^\infty \frac{\sin \xi\rho}{\xi} \left(\frac{1}{E} \frac{ak}{\rho} \right) \rho \, d\rho \right|^2 = \left| \frac{a}{E\xi} \int_0^\infty \sin \xi\rho \, d\rho \right|^2.$$

The integral can be evaluated using an integrating factor $e^{-c\rho}$ as follows:

$$\int_0^\infty \sin \xi\rho \, d\rho = \lim_{c \to 0} \int_0^\infty e^{-c\rho} \sin \xi\rho \, d\rho = \lim_{c \to 0} \left(\frac{\xi}{\xi^2 + c^2} \right) = \frac{1}{\xi}.$$

Hence

$$\sigma(\theta) = \left| \frac{a}{E\xi^2} \right|^2 = \frac{a^2}{16E^2 \sin^4 (\theta/2)}.$$

For a Coulomb potential the constant a is equal to zZe^2, so

$$\sigma(\theta) = \frac{(zZe^2)^2}{16E^2 \sin^4 (\theta/2)}.$$

This quantum-mechanical expression for the Rutherford cross section is identical to the expressions derived classically in Section 2.4. Furthermore, it is exactly the same expression that is obtained from a rigorous solution of the Schrödinger equation for the inverse-r potential.* In retrospect it is perhaps not too surprising that the Born approximation yields the correct cross section for Rutherford scattering. Rutherford scattering is extremely forward directed; it goes as $1/(\sin^4 \theta/2)$; hence the total wave function within the potential region is indeed very much like the incident-wave function as it must be for the Born approximation to be accurate.

We should be neither surprised nor discouraged to learn that there are relatively few potentials which allow the integral in the Born approximation expression for $\sigma(\theta)$ to be evaluated in closed form. But whether or not this integral can be evaluated in closed form is somewhat beside the point because it can always be evaluated numerically. We can always use Equation 4.145 to obtain a plot of $\sigma(\theta)$ vs. θ, regardless of the shape of the potential but provided, of course, that the potential meets the criteria for the applicability of the first Born approximation. Hence the Born approximation can, in fact, be used to find the differential scattering cross sections for a large class of scattering problems. Those problems that do not meet the criteria for the applicability of the Born approximation can be solved by the very general method of partial waves. The method of partial waves and the Born approximation are to some extent complementary. If the incoming particle has low energy, only a few partial waves are needed to calculate the cross section. At energies sufficiently high that the Born approximation is applicable, the method of partial waves usually becomes prohibitively laborious. At intermediate energies careful examination will usually distinguish the more appropriate method.

SELECTED REFERENCES

1. David Bohm, *Quantum Theory*, Prentice-Hall, Inc., Englewood Cliffs, N. J., 1951. Chapters 14 and 15 on the three-dimensional wave equation and Chapter 21 on the theory of scattering are particularly good references for the material in this chapter. Chapters 7 and 8, which summarize the most widely accepted version of the quantum nature of matter, provide profitable background for the reader interested in the philosophical foundations of quantum mechanics or, as Bohm suggested, "quantum nonmechanics."

2. N. F. Mott and H. S. W. Massey, *The Theory of Atomic Collisions*, 3rd ed., Oxford University Press, Fairlawn, N. J., 1965. Since its publication, Mott and Massey's book has been *the* standard reference for the theory of scattering. The first three chapters discuss, among other things, scattering by potential barriers, impenetrable spheres, and Coulomb potentials. The Born approximation is considered in Chapter V, and nuclear collisions are considered in Chapter XX.

* For the exact solution see Mott and Massey, *The Theory of Atomic Collisions*, Chapter III.

EXERCISES

1. Prove that the polynomial solutions to the θ-equation, given by Equation 4.27 and subsequent equations, agree with the tabulated solutions given by Equation 4.41.

2. Transform the Schrödinger equation from x,y,z coordinates into r,θ,ϕ coordinates. A reference on intermediate calculus might be consulted for an explanation of the use of Jacobians.

3. Demonstrate that, if a potential is a function of r alone and not a function of θ and ϕ, L^2 is a constant of the motion of the spherically symmetric quantum system whose Hamiltonian is

$$H = -\frac{\hbar^2}{2m}\left(\frac{\partial^2}{\partial r^2} + \frac{2}{r}\frac{\partial}{\partial r}\right) + \frac{L^2}{2mr^2} + V(r).$$

4. The angular momentum of a classical particle has been defined as

$$\mathbf{L} = \mathbf{r} \times \mathbf{p}.$$

Using the commutation rule presented in this chapter show that the quantum-mechanical angular momentum is represented by a pseudovector which satisfies

$$i\hbar\mathbf{L} = \mathbf{L} \times \mathbf{L},$$

where the components of $\mathbf{L}$ are the operators (L_x, L_y, L_z). Show how some of the known properties of the angular-momentum operators fit this expression.

5. Consider a quantum state of fixed energy described by the wave function

$$\Psi(r,t) = \psi(r,0)\exp\left(\frac{-iEt}{\hbar}\right).$$

Here $\psi(r,0)$ is a solution to the time-independent Schrödinger equation. Verify that $\nabla\cdot\mathbf{S} = 0$, where $\mathbf{S}$ is the probability current density, and give a physical interpretation of this result.

6. Show that

$$(\nabla^2 + k^2)\frac{e^{ikr}}{r} = 0.$$

What restriction on the value of r must accompany this equation?

7. Using the Born approximation, (a) calculate the differential cross section for a shallow $[(V_0/E) \ll 1]$ spherical square well

$$V(r) = -V_0 \quad (r < R),$$
$$V(r) = 0 \quad (r > R),$$

and (b) compare this differential cross section with the one obtained classically for the same potential.

8. Calculate and sketch the differential elastic-scattering cross section of an impenetrable sphere of radius R for $kR = 0.01$, 1, and 10.

9. Assume that the neutron–proton potential can be represented by a spherical square well. From the measured elastic cross section in the region 10 ev to 20 Mev, obtain estimates of the well radius and depth. Use the method of partial waves.

10. Show that the low-energy ($kR \ll 1$) scattering cross section of a particle incident on a spherical potential barrier of radius R and height V_0 is given by

$$\sigma_s = 4\pi R^2 \left[\frac{\tanh gR}{gR} - 1 \right]^2,$$

where $g = \sqrt{2mV_0}/\hbar$.

11. Use the Born approximation to calculate $\sigma(\theta)$ for the screened Coulomb potential $V(r) = (zZe^2/r)e^{-r/d}$. Let d approach infinity and show that $\sigma(\theta)$ approaches the normal Coulomb differential scattering cross section.

12. Show that the low-energy cross section of a particle incident on a spherical potential well of radius R exhibits infinite discontinuities at certain values of the potential V_0. Show further that each of these discontinuities corresponds to an allowed energy level in the potential well.

13. Derive the equation for determining the energies of the $l = 1$ virtual levels of a spherical square well. Calculate the energies of the lowest three levels for a particle having a reduced mass of 10^{-24} g in potentials having width 10^{-12} cm and depth 2 Mev and 20 Mev.

14. Sketch the differential cross section as a function of μ for (a) pure s-wave scattering, (b) pure p-wave scattering, and (c) a 50–50 admixture of s- and p-wave scattering.

5 SURVEY OF NEUTRON AND NUCLEUS PROPERTIES AND INTERACTIONS

The remainder of this book is devoted to the interactions of neutrons with nuclei. To understand these interactions, we must first become acquainted with some of the intrinsic properties of the interacting bodies. The first two sections of this chapter are devoted to a brief outline of those properties of neutrons and nuclei which are most pertinent to the understanding of their interactions. The third section is devoted to a general discussion of what is presently known of the specifically nuclear force.

After examining the shell model of atomic structure, we will develop the somewhat analogous shell model of nuclear structure. A brief historical sketch of the development of theories of neutron–nucleus interactions follows, and the nomenclature denoting the cross sections for such interactions is displayed. The last three sections are devoted to the very general conservation and reciprocity laws that are obeyed in all neutron–nucleus interactions.

The material in this chapter is presented in extremely abbreviated form. The serious student is urged to supplement this outline with selected reading from the references listed at the end of this chapter, particularly from Evans' *The Atomic Nucleus*.

5.1 Properties of the Neutron

The discovery of the neutron was announced in a note sent to the British periodical *Nature* by J. Chadwick* in February 1932. Chadwick's experiments clearly indicated that the then unknown penetrating radiation

---

* J. Chadwick, *Nature* 129: 312 (1932); *Proc. Roy. Soc. (London), Ser. A,* 136: 692 (1932).

emitted when beryllium was bombarded with alpha particles was capable of imparting kinetic energies greater than 1 Mev to recoil atoms of small mass, such as lithium, carbon, and nitrogen. He proved by simple kinematic analysis that these recoil energies were much greater than those which could be imparted by the photons of 50 Mev energy that previously had been postulated by Curie and Joliot* as the mysterious radiation. All difficulties in explanation of the observed effects vanished with the assumption that the beryllium nucleus when bombarded by alpha particles emitted a neutral particle with mass approximately equal to that of the proton in the reaction ${}^9_4\text{Be} + {}^4_2\text{He} \rightarrow {}^{12}_6\text{C} + {}^1_0 n$.

Chadwick's experiments rather conclusively proved the existence of a new particle, the neutron. Since over half the mass of the earth consists of neutrons, it might seem remarkable that the neutron stayed hidden so long. But practically all these neutrons are bound so tightly within nuclei that they could not be released by man until he discovered million-electron-volt probes, first in the form of alpha particles from radioactive decay and later in the form of accelerated charged particles. The neutron is now known to be a particle with the following attributes:

Mass. The mass of the neutron is 1.008665 *atomic mass units* (amu). One atomic mass unit is defined as one-twelfth the mass of the ${}^{12}\text{C}$ atom† or 1.66043×10^{-24} g.

Charge. The net charge of the neutron, if there is a charge, is known experimentally to be less than 10^{-18} times the charge of an electron. The net charge is therefore assumed to be zero.

Intrinsic angular momentum. The intrinsic angular momentum, commonly called the *spin*, of the neutron has magnitude $\hbar\sqrt{s(s+1)}$ where s, the spin quantum number, equals 1/2. The observable z component of the spin is either $+\hbar/2$ or $-\hbar/2$.

Magnetic moment. The magnetic moment of the neutron is -1.91 nuclear magnetons, where the nuclear magneton is 5.05×10^{-24} erg/gauss. The negative sign indicates that the neutron's magnetic moment is opposite to its intrinsic angular momentum. Presumably the neutron has equal positive and negative charges that are so distributed in space that upon rotation they give rise to its observed magnetic moment. The exact nature of this distribution of charge is presently unknown. According to the most widely accepted theory, the neutron exists in nature in a dynamic state. It is believed that the charge distribution within a neutron is in a

---

* I. Curie and F. Joliot, *Compt. Rend.* **194**: 708 (1932).

† Most atomic mass tables published before 1965 were based on 1 amu ≡ 1/16 of the mass of the ${}^{16}\text{O}$ atom. To convert these old values to the ${}^{12}\text{C}$ scale, use the formula $M({}^{12}\text{C scale}) = [M({}^{16}\text{O scale})]/1.000318$, where M represents mass.

constant state of motion. While the time average of the total charge is always equal to zero, the internal charge arrangement provides for a negative magnetic moment. The negative charge is associated with the virtual π^- meson which can be visualized as spending most of its time orbiting the periphery of the neutron boundary. The π^- meson orbiting the proton core presumably has a total angular momentum greater than the proton's angular momentum and antiparallel to it, which gives rise to the negative magnetic moment of the neutron.

Stability. The neutron is radioactive. It decays with a half-life of 12 min into a proton, an electron of 0.782 Mev maximum kinetic energy, and an antineutrino:

$$n \to p + e^- + \tilde{v} + 0.782 \text{ Mev.}$$

5.2 Properties of Nuclei

Every atom consists of a small, extremely dense, positively charged nucleus surrounded by enough electrons to make the atom as a whole electrically neutral. Since the discovery of the neutron in 1932, it has been recognized that atomic nuclei consist of two basic particle types, protons and neutrons. Because they are the building blocks of the nucleus, protons and neutrons are called *nucleons.*

Each proton, a stable particle, carries a positive charge equal in magnitude, but opposite in sign, to the charge on the electron. The mass of the proton is 1.00777 amu; its spin quantum number is the same as that of the neutron, namely, 1/2; and its magnetic moment is $+2.79$ nuclear magnetons.

The number of protons in the atomic nucleus is called the *atomic number* of the nucleus and is indicated by the symbol Z. The total number of nucleons in the nucleus is called the *nucleon number,* or the *mass number,* of the nucleus and is indicated by the symbol A. A specific nucleus is indicated by its chemical symbol with superscript A and subscript Z. For example, the nucleus of oxygen-17 would be designated $^{17}_{8}O$. The subscript is redundant and may be omitted since there is a one-to-one correspondence between the chemical symbol and the number of protons in the nucleus. In the above example, the number 8 may be omitted because oxygen is by definition the element with $Z = 8$.

A *nuclide* is a nuclear species with a specific number of neutrons and protons. More than 700 distinct nuclides, ranging in mass from ^1_1H to $^{257}_{103}Lr$, are known. About 270 of these nuclides are stable, 30 are naturally radioactive, and the rest are artificial radioactive species.

Isotopes are atoms having the same number of protons, Z, and hence the same numbers of orbital electrons and the same chemical properties.

For example, the atoms $^8_4$Be, $^9_4$Be, and $^{10}_4$Be, all of which have 4 protons, are isotopes of the element beryllium and follow the chemistry of beryllium.

Isotones are nuclides having the same number of neutrons, $N = A - Z$. Examples are $^9_3$Li, $^{10}_4$Be, $^{11}_5$B, all of which have 6 neutrons.

Isobars are nuclides having the same total number of nucleons, A. Examples are $^{16}_6$C, $^{16}_7$N, $^{16}_8$O, all of which contain 16 nucleons.

Isomers are nuclides having the same Z and A but different degrees of internal excitation. Examples are ^{117}Sn and ^{117m}Sn. Tin-117m is a metastable excited state of ^{117}Sn which decays with a half-life of 14 days to stable ^{117}Sn.

It is relatively easy to remember the distinction between the various iso's since the words isotoPe, isotoNe, isobAr, and isoMer contain a mnemonic device in the form of the key letters P, N, A, and M, which can be remembered to mean, respectively, equal-Proton, equal-Neutron, equal-A, and Metastable states.

The chemical properties of an atom are determined primarily by the number of its orbital electrons, hence by its atomic number and not by its atomic weight. Isotopes of a given element have almost identical chemical properties. Nuclear properties, in contrast, are determined primarily by the total number of nucleons in a nucleus, hence the mass number and not the atomic number. Isobars are to nuclear physics what isotopes are to atomic chemistry. The primary reason is that nuclear forces are at least approximately charge independent. That is, the specifically nuclear force between any two nucleons, whether two protons, two neutrons, or a neutron and a proton, is essentially the same. We shall return to the subject of nuclear forces in Section 5.3. The remainder of this section comprises a brief synopsis of what is presently known of the properties of nuclei.

A *nucleus* is a group of nucleons held together by attractive nuclear forces. Since each of the nucleons is bound in a potential well, the nucleus can exist in a series of quantum-mechanical energy eigenstates, the lowest of which is called the *ground state*. Many of the properties of a nucleus depend on its state of internal excitation. Unless specifically excepted, the properties we discuss are the properties of the nucleus in its ground state.

Nuclear Charge

The electrical charge of a nuclide is the sum of the elemental charges on the protons it contains. Thus a nuclide with atomic number Z has an electrical charge of $+Ze$, where e is 4.8×10^{-10} esu. The charge is one property that is the same whether the nuclide is in the ground state or in an excited state.

Nuclear Mass

It has been found experimentally that the mass of a nucleus is less than the masses of its constituents by as much as 1%. A nucleus containing Z protons and $N = A - Z$ neutrons has a mass M' that can be written

$$M' = Zm_p + Nm_n - M'_d, \tag{5.1}$$

where m_p and m_n are the rest masses of the free proton and free neutron, respectively, and M'_d is the *mass defect*. An energy is quantitatively associated with the mass defect by Einstein's mass–energy relation, $E = mc^2$. Application of this relation to the mass defect M'_d yields

$$B' = M'_d c^2, \tag{5.2}$$

which gives the *binding energy*, B', of the nucleus. The quantity B' is the total amount of energy that must be added to the nucleus to separate all its nucleons an infinite distance from one another. It also represents the energy that would be released (in the form of photons) if the constituent nucleons were brought together to form the nucleus. For example, when a neutron and a proton are brought together, they coalesce to form a nucleus of ^{2}H with the release of a 2.2-Mev gamma ray. The binding energy of the nuclide ^{2}H is thus 2.2 Mev. To separate the neutron from the proton, we must supply an energy of at least 2.2 Mev to the ^{2}H nucleus.

Mass spectrometers measure the mass of the atom, not the nucleus, and, since essentially all calculations can be carried out using atomic masses, nuclear masses are rarely tabulated. The relation between the atomic mass M and the nuclear mass M' is

$$M = M' + Zm_0 - B_Z, \tag{5.3}$$

where m_0 is the rest mass of an electron and B_Z is the total binding energy of all electrons in the atom. The total binding energy, B_Z, which is given approximately by the empirical relation

$$B_Z = 15.73 \, Z^{7/3} \text{ ev}, \tag{5.4}$$

is a very small correction except for the highest Z elements.

In terms of atomic masses, the nuclear mass defect is given by

$$M_d = Zm_H + Nm_n - M - B_Z, \tag{5.5}$$

where m_H is the mass of the neutral hydrogen atom.

A quantity more interesting than the total mass defect of a nucleus is the **binding fraction, f, defined as the mass defect per nucleon expressed in energy units,**

$$f = (M_d/A)c^2. \tag{5.6}$$

Figure 5.1 is a graph of the binding fraction versus the nucleon number A. It shows that all except a few light nuclei are bound together with an energy of approximately 9 Mev per nucleon. Since the mass of a neutron or proton is approximately 931 Mev in energy units, the mass defect of most nuclides is only of the order of 9/931, i.e., about 1%.

The binding fraction f is the average energy that must be given to each nucleon to completely separate all the nucleons from each other. The energy necessary to remove the first nucleon, called the *separation energy*, may be more or less than the binding fraction.

The neutron separation energy, S_n, is given by

$$S_n = B(Z,N) - B(Z,N - 1), \tag{5.7}$$

where $B(Z,N)$ is the total binding energy of the nucleus with Z protons and N neutrons and $B(Z,N - 1)$ is the total binding energy of the residual nucleus.

The proton separation energy, given by

$$S_p = B(Z,N) - B(Z - 1,N), \tag{5.8}$$

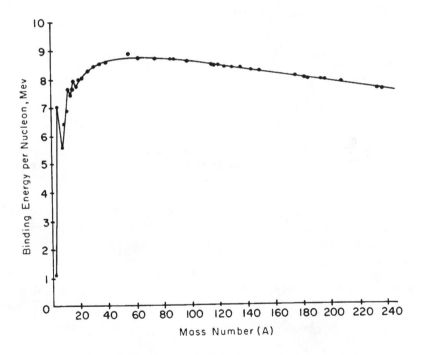

Figure 5.1—Binding fractions of stable nuclei as a function of mass number.

may be less than, equal to, or greater than the neutron separation energy, depending on the relative binding energies of the residual nuclei.*

For calculations it is convenient to use the previous expressions for the binding energies in rewriting the neutron and proton separation energies:

$$S_n \text{ (Mev)} = 931[M(Z,N - 1) + m_n - M(Z,N)], \tag{5.9}$$

$$S_p \text{ (Mev)} = 931[M(Z - 1,N) + m_H - M(Z,N)], \tag{5.10}$$

where $M(Z,N)$ is the mass, expressed in atomic mass units, of the atom that has Z protons and N neutrons in its nucleus. The coefficient 931 approximately converts atomic mass units to million electron volts (1 amu $\times$ c^2 = 931.478 Mev).

To illustrate the distinction between the binding fraction f and the separation energies S_n and S_p, we consider the nuclide $^{12}_6C$. From a table of atomic masses, we obtain the following data

$$M(^{12}_6C) = 12.000\,0000 \text{ amu}$$

$$M(^{11}_6C) = 11.011\,4374 \text{ amu}$$

$$M(^{11}_5B) = 11.009\,3051 \text{ amu}$$

$$m_n = 1.008\,6654 \text{ amu}$$

$$m_H = 1.007\,8252 \text{ amu}$$

Using Equations 5.5 and 5.6 with this data, we obtain for the binding fraction of $^{12}_6C$

$$f = 7.68 \text{ Mev/nucleon.}$$

The energy required to separate the first neutron from ^{12}C in the process $^{12}_6C \rightarrow {}^{11}_6C + {}^1_0n$ is given by Equation 5.9,

$$S_n = 931[M(^{11}_6C) + m_n - M(^{12}_6C)] = 18.71 \text{ Mev.}$$

The energy required to separate the first proton from ^{12}C in the process $^{12}_6C \rightarrow {}^{11}_5B + {}^1_1H$ is given by Equation 5.10,

$$S_p = 931[M(^{11}_5B) + m_H - M(^{12}_6C)] = 15.92 \text{ Mev.}$$

Note that less energy is needed to remove the first proton from ^{12}C than to remove the first neutron. Furthermore the average binding energy is much less than either of the nucleon separation energies in this particular nuclide.

---

* For a fuller discussion see Blatt and Weisskopf's *Theoretical Nuclear Physics* (Reference 2 at the end of this chapter), pp. 16–21.

Nuclear Dimensions

There is a substantial amount of experimental evidence from a variety of different experiments indicating that nuclei are approximately spherical and that they have volumes approximately proportional to the number of nucleons they contain. The nuclear radius is therefore expected to be approximately proportional to the cube root of the nucleon number A,

$$R \simeq R_0 A^{1/3}. \tag{5.11}$$

The constant of proportionality R_0 has been obtained in a variety of ways:

Alpha-particle decay:	$R_{0,\alpha} \simeq 1.45$ fm*
Mirror nuclei:	$R_{0,m} \simeq 1.28$ fm
High-energy-electron scattering:	$R_{0,e} = 1.25$ fm
High-energy-neutron scattering:	$R_{0,n} \simeq 1.5$ fm

The variation of about 20% among these values of R_0 reflects the fact that there is no precise way of defining a nuclear surface. Since the wave function describing the nucleons is not sharply confined in space, the nuclear radius may be defined, somewhat vaguely, as that radius beyond which there is a "negligible probability" of finding a nucleon.

In the calculation of neutron–nucleus cross sections, R_0 is often treated as a parameter that is varied in the range 1.25 to 1.6 fm to obtain the best theoretical fit to experimental data. When not treated as an adjustable parameter, R_0 is most often given a fixed value between 1.4 and 1.6 fm. In this book, unless otherwise stated, it will be assumed that $R_0 = 1.5$ fm.

The specific volume of nuclear matter is approximately

$$\frac{V}{A} = \frac{4}{3}\pi R_0^3 = 1.4 \times 10^{-38} \text{ cm}^3/\text{nucleon},$$

and the density of nuclear matter is approximately

$$\rho = \frac{m}{V/A} = \frac{1.67 \times 10^{-24} \text{ g/nucleon}}{1.4 \times 10^{-38} \text{ cm}^3/\text{nucleon}}$$

$$= 1.2 \times 10^{14} \text{ g/cm}^3 = 1.2 \times 10^8 \text{ metric tons/cm}^3.$$

Because of the extreme density of nuclear matter, it may seem rather remarkable that experiments indicate that a nucleon incident on a nucleus

---

*One femtometer or fermi, abbreviated fm, is equal to 10^{-13} cm.

may in certain instances pass through it with little or no interaction. It seems that intuition, gained as it is from everyday experience with macroscopic objects, is not the most trustworthy guide in the microscopic world.

The density of nucleons is apparently uniform throughout most of the nuclear volume. The most direct evidence for this assertion comes from the nuclear scattering of very high energy, approximately 900-Mev, electrons. These experiments actually measure the distribution of charge within the nucleus. However, it is commonly expected because of the charge independence of nuclear forces that the distribution of neutrons will not be appreciably different from the distribution of protons; so the mass density may be taken to be the same as the charge density. Figure 5.2 shows a typical curve of charge density as a function of nuclear radius.

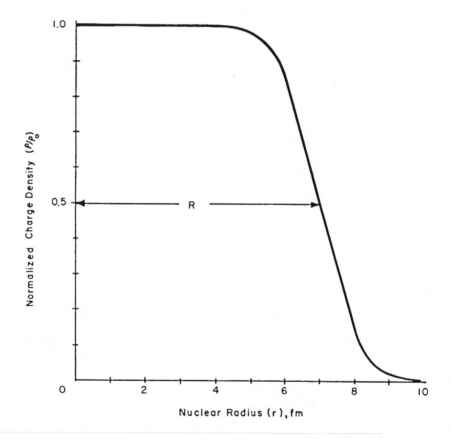

Figure 5.2—Typical nuclear charge density as a function of nuclear radius.

The density is seen to be approximately constant except in the outer 2 to 3 fm, the *skin* of the nucleus. This charge-density curve can be described by the empirical relation

$$\rho = \rho_0 \frac{1}{1 + \exp\left[(r - R)/a\right]},\tag{5.12}$$

where R is the nuclear radius at which the charge density is one-half its maximum value and a is an adjustable parameter called the *diffuseness*, chosen to best fit the experimental curve. Best fits to the charge-density distributions are obtained by assuming $R = R_0 A^{1/3}$ with R_0 approximately 1.3 fm for light nuclei and 1.2 fm for heavy nuclei. The diffuseness is found to be approximately 0.5 fm.

Nuclear Intrinsic Angular Momentum

Each nucleus has intrinsic angular momentum resulting from the vector addition of the individual intrinsic and orbital angular momenta of its nucleons. The absolute magnitude of the total nuclear angular momentum has the value $[I(I + 1)]^{1/2}\hbar$, where I, which is either an integer or a half-integer, is the total nuclear angular-momentum quantum number, often called the *nuclear spin*. A nucleus with nuclear spin I is commonly said to have an intrinsic angular momentum $I\hbar$, although, rigorously, $I\hbar$ is not the magnitude but rather the largest observable component of the intrinsic angular momentum. This component can take on the $2I + 1$ values of $m_I\hbar$ where m_I ranges from $+I$ to $-I$ by integers, i.e., where

$$m_I = I, (I - 1), \ldots, -(I - 1), -I.\tag{5.13}$$

The quantum-mechanical rules governing addition of angular momenta of a group of nucleons (each with spin 1/2) require that the resultant nuclear spin I be an odd multiple of 1/2 if the number of nucleons is odd and an integer if the number of nucleons is even. In fact, it has been observed that all nuclei with both even-Z and even-N have zero spin. Nuclei with an odd number of nucleons, either odd-Z or odd-N, have been found to have spins ranging from 1/2 to 9/2, always half-integer. Nuclei with both Z and N odd are relatively rare; however, those examined have been found to have integer spins ranging from 1 to 6.

It is convenient to divide all nuclei into four categories: *even-even*, *even-odd*, *odd-even*, and *odd-odd*, where the first word refers to the character of the number of protons, Z, and the second to the character of the number of neutrons, N, in the nucleus. With the aid of this nomenclature, Table 5.1 summarizes the experimentally measured spins of nuclei in their ground states. Also tabulated are the numbers of stable nuclides of each type.

Note that about 60% of all stable nuclides are even-even and that most of the others are either even-odd or odd-even; only about 2% are odd-odd. Nature seems to favor the building of nuclei out of pairs of protons and pairs of neutrons. This is one of the empirical facts that any successful nuclear theory must explain.

Table 5.1—Measured Nuclear-Spin Quantum Numbers

Nuclide type	Nuclear spin (I)	Number of stable nuclides
Even-even	0	159
Even-odd or odd-even	1/2, 3/2, 5/2, 7/2, 9/2	54 + 50
Odd-odd	1, 2, 3, 4, 6	4
	Total	267

The largest observable component of the total intrinsic angular momentum of a nucleus in an excited state will either be the same as its ground-state value or will differ by an integral multiple of $\hbar$. This rule is a consequence of the algebraic properties of the quantum-mechanical angular-momentum operators, which are discussed in Chapter 6. The rule is well supported by experimental observations. If the ground state has zero or integer spin, all excited states have zero or integer spins. If the ground state has half-integer spin, all excited states have half-integer spins.

Nuclear Magnetic Dipole Moment

Recall the following basic laws and definitions of classical magnetostatics:

1. Magnetic poles appear in nature in pairs and are called the *positive* or *north-seeking pole* and the *negative* or *south-seeking pole.* ·

2. Like poles repel; unlike poles attract.

3. The force between two poles is directed along the line between them and is directly proportional to the product of their pole strengths and inversely proportional to the square of the distance between them. This empirical law can be expressed in the form

$$F = \frac{p_1 p_2}{r^2}, \tag{5.14}$$

where p_1 and p_2 are the pole strengths of the two magnetic poles and r is distance between them *in vacuo.*

4. In the cgs system the unit of pole strength is defined by using $F = 1$ dyne and $r = 1$ cm in Equation 5.14. Then $p_1 = p_2 = 1$ unit pole. The unit pole is then seen to be the pole that repels an equal like pole at a distance of 1 cm *in vacuo* with a force of 1 dyne.

5. The force on a pole of strength p placed at a point where the magnetic field strength is **H** is given by

$$\mathbf{F} = p\mathbf{H}. \tag{5.15}$$

This equation can be considered the defining equation for the magnetic field strength **H**. The units of the vector **H** are dynes per unit pole, or gauss. If p is positive, i.e., a north-seeking pole, the force has the direction of **H**; otherwise it has the opposite direction.

6. A magnetic dipole that has poles of strength p and $-p$ separated by distance X has a magnetic dipole moment, μ, defined as

$$\mu \equiv pX. \tag{5.16}$$

7. Using Equations 5.15 and 5.16, we will now show that the dipole moment of a magnetic dipole is equal to the torque exerted on the dipole when it is placed at right angles to a uniform magnetic field of unit strength. In general, when a magnetic dipole of length X is placed at angle α to a uniform field of strength H, the forces on the two poles are identical in magnitude, namely, pH, but opposite in direction (Figure 5.3). The torque is given by the sum of the moments of the forces about the central pivot

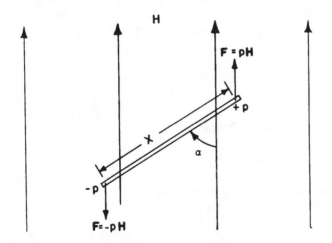

Figure 5.3—The forces acting on a magnetic dipole with moment $\mu = pX$ in a uniform magnetic field of strength H.

point,

$$\text{Torque} = 2(pH)(\tfrac{1}{2}X \sin \alpha) = \mu H \sin \alpha. \tag{5.17}$$

In the special case when $H = 1$ gauss and $\alpha = \pi/2$, according to Equation 5.17 the magnetic moment of the dipole does indeed equal the torque exerted on it by the field, as was to be proved.

Since torque has the dimension of energy (dyne-centimeter $\equiv$ erg) and H has the dimension gauss, the dimension of magnetic moment is erg/gauss. The common unit of nuclear magnetic moment is the *nuclear magneton*, defined as follows:

$$1 \text{ nuclear magneton} \equiv \frac{\hbar e}{2m_p c} = 5.05 \times 10^{-24} \text{ erg/gauss} \tag{5.18}$$

where e is the charge on the proton, m_p is the mass of the proton, and c is the velocity of light *in vacuo*. If neutrons and protons obeyed Dirac's relativistic wave equation, the neutron, being uncharged, would have no magnetic moment, and the proton would have a magnetic moment of 1 nuclear magneton. Actually the magnetic moments of the neutron and proton are -1.91280 and $+2.79255$ nuclear magnetons, respectively. As mentioned before, these anomalous magnetic moments are expected to be explained eventually on the basis of the internal mesonic structures of the neutron and proton; present meson theories do not entirely account for these anomalies in a quantitative manner.

All odd-A nuclei have nonzero magnetic moments. Measured moments range from about -2 to $+6$ nuclear magnetons. It was pointed out by T. Schmidt[*] and H. Schüler[†] that a remarkable correlation exists when the magnetic moments of these odd-A nuclei are plotted against their spins. If we make two such Schmidt diagrams, one for odd-Z nuclides and one for odd-N nuclides, we discover that:

1. For odd-Z nuclides the magnetic moment μ increases with increasing spin I. Furthermore, almost all odd-Z nuclides have positive magnetic moments in the range from 0 to $+6$ nuclear magnetons.

2. For odd-N nuclides μ shows no tendency to increase or decrease with I. Moreover μ itself has both negative and positive values ranging from about -2 to $+1$ nuclear magnetons.

Schmidt suggested that these observations could be explained by assuming that both the angular momentum and the magnetic moment of a nucleus with odd-Z or odd-N are essentially due to the odd particle. It is assumed that each pair of neutrons and each pair of protons separately combine in antiparallel states to effectively cancel both their spins

[*] T. Schmidt, *Z. Physik*, **106**: 358 (1937).
[†] H. Schüler, *Z. Physik*, **107**: 12 (1937).

and magnetic moments. Then both the total angular momentum and the magnetic moment of the nucleus are assumed to be completely determined by the unpaired particle.

Schmidt's model for odd-even and even-odd nuclides is an example of a single-particle model in that we visualize the nucleus as an even-even core orbited by a single proton or neutron. The core is assumed to contribute no angular momentum and no magnetic dipole moment to the nucleus. This is an attractive suggestion since it is in obvious agreement with the fact that even-even nuclei have zero spin and zero magnetic dipole moment.

The angular momentum of the odd particle is the vector sum of its orbital angular momentum with maximum component $l\hbar$ and its spin angular momentum with maximum component $\hbar/2$. It is assumed that the total angular momentum of the nucleus is given by the total angular momentum of the odd particle. Expressed in terms of quantum numbers, the assumption is that

$$I = l \pm \tfrac{1}{2}, \tag{5.19}$$

where the plus sign obtains when the orbital and spin momenta are parallel and the minus sign when they are antiparallel. The striking differences between the Schmidt diagrams for odd-Z and odd-N nuclides are then explained as follows:

1. If the odd particle is a proton, the magnetic moment of the nucleus will consist of the vector sum of the intrinsic magnetic moment of the proton and a contribution from the orbital momentum. This contribution results because the orbiting proton is, in effect, a current ring and therefore has an associated magnetic moment. Hence, the larger the l value of the proton, the larger the I of the nucleus, and the nuclear magnetic moment. Furthermore, under this assumption the magnetic moment will be positive, as observed, since a circulating positive charge has a magnetic moment in the same direction as its orbital angular momentum.

2. If the odd particle is a neutron, its orbital motion will have no electric current, hence no magnetic moment, associated with it; therefore μ will be independent of I, as is observed for odd-N nuclides. The relative orientations of the orbital motion and spin of the neutron will, however, determine the sign of the magnetic moment of the nucleus. If they are parallel, the magnetic moment will have the same negative sign as that of the free neutron. If they are antiparallel, μ will be positive. This explains why some odd-N nuclides have negative magnetic moments and others have positive magnetic moments.

Unfortunately, the model proposed by Schmidt, though beautiful in its simplicity and containing obvious elements of truth, has proved to be inadequate to explain all the observed facts. If the Schmidt model were rigorously accurate, we could immediately calculate the two possible magnetic moments of each odd-Z and odd-N nuclide. From the quantum rules for addition of angular-momentum vectors,* one obtains the values in Table 5.2. Actually, the hundred or more measured values of nuclear magnetic moments have been found to fall between the two extreme Schmidt limits with few exceptions. Furthermore, the measured μ rarely corresponds exactly to either of the values predicted in Table 5.2.

At present it is believed that the single-particle model is basically valid but that at least two modifications must be made to obtain quantitative agreement between observed and calculated nuclear magnetic moments. It is believed that the intrinsic magnetic moment of the odd neutron or proton may be reduced below its free value through meson effects when the nucleon is bound to the nuclear core. Furthermore, it seems rather certain that we must abandon the assumption that the core has spherical symmetry. The observed electric quadrupole moments of nuclei suggest that odd-A nuclear cores have, in fact, some slight asymmetry and therefore might be expected to contribute to the observed magnetic moments of these nuclides.

Table 5.2—Magnetic Moments of Odd-Z and Odd-N Nuclides Based on the Schmidt Model*

Nuclide type	Parallel state $(I = l + 1/2)$	Antiparallel state $(I = l - 1/2)$
Odd-proton	$\mu = I + \mu_p$	$\mu = I - \mu_p \dfrac{I}{I + 1}$
Odd-neutron	$\mu = \mu_n$	$\mu = -\mu_n \dfrac{I}{I + 1}$

*μ_p = 2.79 nuclear magnetons; μ_n = -1.91 nuclear magnetons.

In the Schmidt model an odd-odd nucleus would be visualized as a neutron and a proton orbiting an even-even core. Most odd-odd nuclides are radioactive, the exceptions being $^2_1\mathrm{H}$, $^6_3\mathrm{Li}$, $^{10}_5\mathrm{B}$, and $^{14}_7\mathrm{N}$. These four stable nuclides, as well as those radioactive odd-odd nuclides which have been investigated, have been found to have nonzero spin and nonzero magnetic dipole moment. This is evidence that a neutron and a proton do not tend to line up to cancel their angular momenta as two protons or two neutrons do.

*See R. D. Evans', *The Atomic Nucleus* (Reference 1 at the end of this chapter).

Magnetic moments of nuclides other than the proton are determined indirectly by use of the relation

$$\mu = g \frac{eh}{2m_p c} I. \tag{5.20}$$

The measured quantities are the maximum observable component of the nuclear spin, I, and the nuclear gyromagnetic ratio, g, defined as the ratio between the actual magnetic moment of the nucleus and the magnetic moment that would be classically calculated if the entire nuclear angular momentum were due to the orbital motion of a single proton.

Nuclear Electric Quadrupole Moment

A nonspherical nucleus may be expected to have a nonzero electric quadrupole moment. Furthermore the magnitude of the quadrupole moment of a nucleus is a measure of its departure from sphericity. To clearly understand just why this is so, we should perhaps recall some elementary classical electrostatic theory.

Suppose there exists in a region of space τ a charge density ρ that is a function of the three spatial coordinates but not of time. The Coulomb potential V at an external point P due to this charge distribution is found by integrating the differential potential dV, which is due to the charge in each differential volume element $d\tau$ over the entire volume τ,

$$V = \int_\tau dV. \tag{5.21}$$

The potential dV at an external point on the z-axis, $P(0,0,a)$, due to the charge within the differential volume element $d\tau$ is given by Coulomb's law,

$$dV = \frac{\rho \, d\tau}{b}, \tag{5.22}$$

where, referring to Figure 5.4, $\rho = \rho(r,\theta,\phi)$ is the charge per unit volume at point (r,θ,ϕ) and b is the distance between the volume element $d\tau$ and the point P. When the variable distance b is expressed in terms of the constant distance a, the above equation becomes

$$dV = \frac{\rho \, d\tau}{a} \left(1 - 2\frac{r}{a}\cos\theta + \frac{r^2}{a^2}\right)^{-1/2}. \tag{5.23}$$

Expanding the square root and collecting terms in powers of a, we get an infinite series,

$$dV = \frac{(\rho \, d\tau)}{a} + \frac{(\rho \, d\tau)r}{a^2}[\cos\theta] + \frac{(\rho \, d\tau)r^2}{a^3}\left[\frac{3}{2}\cos^2\theta - \frac{1}{2}\right] + \cdots. \tag{5.24}$$

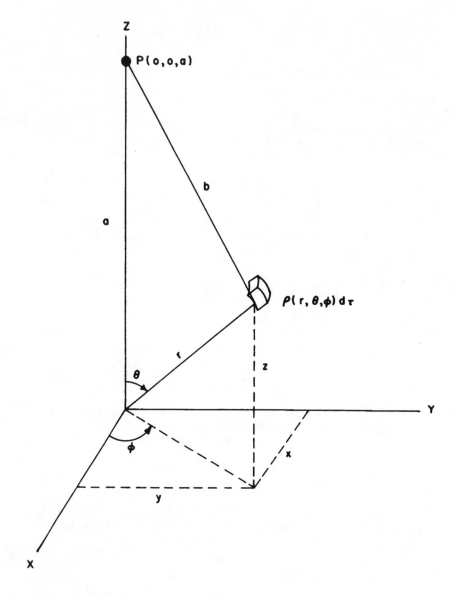

Figure 5.4—Construction used to find electrostatic potential at point *P* due to a differential charge in volume element $d\tau$ about point (r,θ,ϕ).

The functions in brackets are the Legendre polynomials $P_n(\cos\theta)$. Hence, the above equation can be written

$$dV = \sum_{n=0}^{\infty} \frac{(\rho \, d\tau)r^n}{a^{n+1}} P_n(\cos\theta). \qquad (5.25)$$

The electrostatic potential due to any distributed charge is given by the integral of Equation 5.25 over the volume that contains the charge,

$$V = \sum_{n=0}^{\infty} \frac{1}{a^{n+1}} \int_\tau \rho r^n P_n(\cos\theta) \, d\tau. \qquad (5.26)$$

The coefficient of $1/a$ in Equation 5.26 is called the *monopole strength* of the distributed charge, the coefficient of $1/a^2$ is called the *z component of the electrostatic dipole moment*, the coefficient of $1/a^3$ is called the *z component of the electrostatic quadrupole moment*, and so on. In general, the coefficient of $1/a^{n+1}$ is called the *z component of the electrostatic 2^n-pole moment*.

Let us now consider some examples of the classical multipole strengths of simple charge distributions.

Example 5.1. The Point Charge
For a single point charge, we need only replace $\rho \, d\tau$ by the charge q in Equation 5.24 to see by inspection that a point charge located at the origin has only a monopole strength. All other moments vanish because $r = 0$. It is also obvious that a charge not located at the origin will exhibit all dipole, quadrupole, and higher moments, except if it happens to be at one of those angles at which a particular $P_n(\cos\theta)$ vanishes. In this case the 2^n-pole moment vanishes.

Example 5.2. The Homogeneous Spherical Charge
A homogeneous spherical charge distribution has zero electric dipole and higher order moments. This is readily proved by substituting $\rho = \rho_c =$ a constant and $d\tau = r^2 \sin\theta \, dr \, d\theta \, d\phi$ into Equation 5.26 so that the potential expression is

$$V = \rho_c \sum_{n=0}^{\infty} \frac{1}{a^{n+1}} \int_0^R \int_0^\pi \int_0^{2\pi} P_n(\cos\theta)r^{n+2} \sin\theta \, dr \, d\theta \, d\phi.$$

The integration over ϕ yields 2π. The integration over θ yields zero because of the orthogonality of the Legendre polynomials except for the term in the summation with $n = 0$, in which case it yields 2. The integration of r yields $R^3/3$. Finally then,

$$V = \frac{(\frac{4}{3}\pi R^3 \rho_c)}{a} \equiv \frac{q}{a}, \qquad (5.27)$$

where q is defined as the total charge contained in the sphere of radius R. We see that only the term involving $1/a$ survives the integration. Hence, a homogeneous spherical charge distribution possesses only a monopole strength, q. Its dipole, quadrupole, and higher moments are all zero.

Example 5.3. Ring Charges Outside Spherical Cores
Consider now the charge distributions that we might naïvely picture for the time average of a proton orbiting a uniform spherical nuclear core. The two extreme cases considered are shown in Figure 5.5. In part a of 5.5, the time-averaged charge distribution is pictured as a ring lying parallel to the z-axis; in part b of 5.5, this distribution is

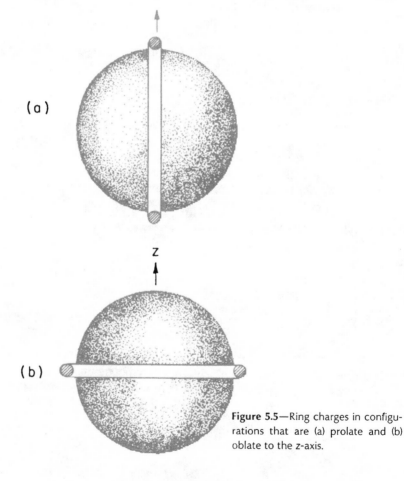

Figure 5.5—Ring charges in configurations that are (a) prolate and (b) oblate to the z-axis.

shown as a ring lying perpendicular to the z-axis. These will be called the *prolate* and *oblate configurations*, respectively. Both rings are assumed to circle a homogeneous spherical charge distribution.

The quadrupole moment of the entire charge distribution is due entirely to the ring charge since the homogeneous sphere has zero quadrupole moment. The classical quadrupole moment, Q_c, is given by the coefficient of $1/a^3$ in Equation 5.26,

$$Q_c = \int \rho(r,\theta,\phi) r^2 (\tfrac{3}{2} \cos^2 \theta - \tfrac{1}{2}) \, d\tau. \tag{5.28}$$

The cross-sectional diameter of the ring charge will be assumed so small relative to the distance a that the toroidal ring may be considered a circular line. Then ρ will be interpreted as the charge per unit length along the line and $d\tau$ as the element of length along the line:

$\rho = \lambda_c =$ constant charge per unit length,

$d\tau = dL =$ element of length along the line.

When these quantities are substituted into Equation 5.28 along with the relation $r = R =$ a constant along the ring, there results

$$Q_c = \lambda_c R^2 \oint (\tfrac{3}{2} \cos^2 \theta - \tfrac{1}{2}) \, dL, \tag{5.29}$$

where the line integral is taken around the ring.

In the prolate configuration, $dL = R \, d\theta$; hence

$$Q_c = 2\lambda_c R^3 \int_0^\pi (\tfrac{3}{2} \cos^2 \theta - \tfrac{1}{2}) \, d\theta = \tfrac{1}{2}\pi R^3 \lambda_c$$

$$= \tfrac{1}{4} q R^2, \tag{5.30}$$

where q is the total charge in the ring ($q \equiv 2\pi R \lambda_c$). In the oblate configuration, $\cos \theta = 0$; hence Equation 5.29 becomes simply

$$Q_c = \lambda_c R^2 \oint -\tfrac{1}{2} \, dL = -\tfrac{1}{2}\lambda_c R^2 (2\pi R)$$

$$= -\tfrac{1}{2} q R^2. \tag{5.31}$$

Thus it has been proved that a prolate ring charge gives rise to a positive quadrupole moment and an oblate ring charge to a negative quadrupole moment. Moreover, the absolute value of the quadrupole moment in both cases is of order qR^2, where q is the total charge contained in the ring and R is the radius of the ring.

Example 5.4. The Homogeneous Ellipsoidal Charge

The final example we shall consider is the one that probably comes closest to representing the time-averaged charge distribution in non-spherical nuclei. Picture a homogeneous ellipsoidal charge with semiaxis α lying along the z-axis and semiaxis β lying perpendicular to the z-axis. If $\alpha > \beta$, as in part a of Figure 5.6, the ellipsoid is called *prolate;* if $\alpha < \beta$, as in part b of Figure 5.6, the ellipsoid is called *oblate*. A prolate ellipsoid is shaped like a football, an oblate ellipsoid like a partially flattened sphere. The charge density within the ellipsoids is assumed to have the constant value ρ_c.

It will be left as an exercise for the reader to show that the integration of Equation 5.28 over the ellipsoidal volume yields as the expression for the z component of the quadrupole moment

$$Q_c = \frac{q}{5}(\alpha^2 - \beta^2), \tag{5.32}$$

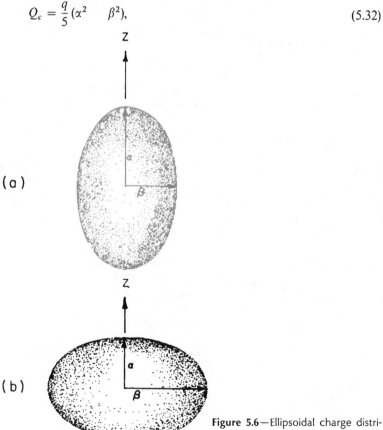

Figure 5.6—Ellipsoidal charge distributions that are (a) prolate and (b) oblate with respect to the z-axis.

where $q = \frac{4}{3}\pi\alpha\beta^2\rho_c$ is the total electric charge contained in the ellipsoid. Measured quadrupole moments for nuclei indicate that α and β differ at most by about 20%. Hence it is convenient to define the effective radius of the ellipsoid as the mean value of α and β,

$$R \equiv \frac{\alpha + \beta}{2}. \tag{5.33}$$

The ellipticity, η, of the ellipsoid is defined as

$$\eta \equiv \frac{\alpha - \beta}{R}. \tag{5.34}$$

This quantity is positive for prolate ellipsoids and negative for oblate ellipsoids. In terms of R and η, Equation 5.32 can be written

$$Q_c = \tfrac{2}{5}\eta q R^2. \tag{5.35}$$

This is the desired expression for the z component of the classical quadrupole moment of a homogeneous ellipsoidal charge distribution. It is seen that η, and therefore the quadrupole moment, is positive for prolate ellipsoids, negative for oblate ellipsoids, and, of course, zero for spheres. Note the similarity between the quadrupole moments of ellipsoids as given by Equation 5.35 and of ring charges as given by Equations 5.30 and 5.31. All are proportional to qR^2 and a numerical coefficient of order one-half. The major difference is the ellipticity η in the expression for Q_c of ellipsoids. It will shortly be seen how measured nuclear quadrupole moments can be used to infer the ellipticity of nuclei.

Thus far the discussion of electrostatic quadrupole moments has been purely classical. But nuclear quadrupole moments must be calculated quantum mechanically, and the quantum-mechanical definition of Q differs in several respects from the classical definition. In quantum mechanics (a) the charge distribution $\rho(r,\theta,\phi)$ is directly proportional to the spatial density of the protons in the nucleus $|\psi(\mathbf{r}_1, \mathbf{r}_2, \ldots, \mathbf{r}_z)|^2$, where $\mathbf{r}_1, \mathbf{r}_2, \ldots, \mathbf{r}_z$ are the coordinates of the Z protons in the nucleus; (b) the nuclear quadrupole moment is taken about the z-axis, which is defined not as the direction of nuclear spin vector but rather as the direction of the largest component I of the spin vector (see Figure 5.7); and (c) the nuclear electric quadrupole moment is commonly defined as $2/e$ times the classical expression for the quadrupole moment, where e is the charge on the proton.

The nuclear quadrupole moment is thus given by

$$Q_n = \left\langle \frac{1}{e} \int \rho r^2 (3\cos^2\theta - 1)\, d\tau \right\rangle,$$
(5.36)

where the triangular brackets indicate the quantum-mechanical expectation value. Another equivalent and common form for Q_n is obtained by substituting $\cos\theta = z/r$ into the above expression,

$$Q_n = \left\langle \frac{1}{e} \int \rho (3z^2 - r^2)\, d\tau \right\rangle.$$
(5.37)

Equation 5.37 is, in fact, just the quantum-mechanical statement that the observable Q_n, whose operator is $(1/e)(3z^2 - r^2)$, is given by the usual integral of $\bar{\psi} Q_n \psi$ over the volume in question. This is readily seen by substituting $\bar{\psi}\psi$ for ρ and noting that the operator representing Q_n, which contains only coordinates, commutes with the wave function so that

$$Q_n \equiv \langle Q_n \rangle = \int_{\text{nucleus}} \bar{\psi}(\mathbf{r}_1, \mathbf{r}_2, \ldots, \mathbf{r}_z) \left[\frac{1}{e}(3z^2 - r^2) \right] \psi(\mathbf{r}_1, \mathbf{r}_2, \ldots, \mathbf{r}_z)\, d\tau.$$

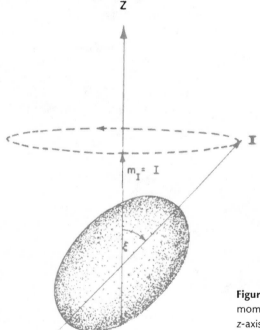

Figure 5.7—The nuclear quadrupole moment is calculated along the z-axis, defined as the axis along which $m_I = I$. The time average is taken as $\mathbf{I}$ precesses around the z-axis.

The nuclear quadrupole moment Q_n is seen to be a function of the state ψ of the nucleus. In practice, interest centers on the quadrupole moment in the ground state, the *static quadrupole moment*. Even in the ground state, however, there is not one unique wave function; there are $2I + 1$ eigenfunctions corresponding to the $2I + 1$ possible orientations of the angular-momentum vector of the nucleus relative to the z-axis. There are thus $2I + 1$ static quadrupole moments for each nucleus. The term "quadrupole moment of the nucleus" without further qualification is uniformly interpreted to mean the z component of the quadrupole moment in the ground state for which $m_I = I$.

It will now be shown that the quadrupole moment of an ellipsoid that is oriented at angle ξ relative to the z-axis is related to the quadrupole moment of the ellipsoid whose semiaxis lies along the z-axis. Referring to Figure 5.7, we see that ξ is the angle between the z-axis and the body-axis of the ellipsoidal charge distribution.

The quadrupole moment about the z-axis is given by Equation 5.36,

$$Q_n(z) = \frac{1}{e} \int \rho(r,\theta,\phi) \, r^2 (3 \cos^2 \theta - 1) \, d\tau. \tag{5.38}$$

But $\cos \theta$ may be expressed by the trigonometric identity

$$\cos \theta = \cos \theta' \cos \xi + \sin \theta' \sin \xi \cos (\phi - \gamma), \tag{5.39}$$

where r, θ', and ϕ' are the polar coordinates relative to the body axis of the ellipsoid and where $\theta = \xi$ and $\phi = \gamma$ are the directions of the body axis relative to the z-axis. Substitution of Equation 5.39 into Equation 5.38 gives

$$Q_n(z) = \tfrac{1}{2}(3 \cos^2 \xi - 1) Q_n(z'), \tag{5.40}$$

where $Q_n(z')$, the quadrupole moment of the charge distribution taken along the body axis, is

$$Q_n(z') = \frac{1}{e} \int \rho(r,\theta',\phi') \, r^2 (3 \cos^2 \theta' - 1) \, d\tau'. \tag{5.41}$$

As we pointed out, the nuclear quadrupole moment is taken in the state $m_I = I$; hence $\cos \xi = I/[I(I + 1)]^{1/2}$ and

$$Q_n(z) = \left[\frac{3}{2} \frac{I^2}{I(I + 1)} - \frac{1}{2} \right] Q_n(z'). \tag{5.42}$$

If the nuclear charge is assumed to be uniform in density and ellipsoidal in shape, then, combining Equations 5.42 and 5.35 and remembering that $q = Ze$ and $Q_n = (2/e)Q_c$, we obtain

$$Q_n(z) = \left[\frac{2}{5} \frac{3I^2}{I(I + 1)} - 1 \right] Z\eta R^2. \qquad (5.43)$$

For a nucleus of known Z, R, and I, a measurement of the quadrupole moment $Q_n(z)$ can be used with the above equation to determine the nuclear ellipticity η; thus the quadrupole moment of a nucleus contains information about its shape.

Nuclear quadrupole moments are determined experimentally by measurement of the effect of these moments on the hyperfine structure of atomic and molecular optical spectra. Most measured values of Q_n have been found to be within a factor of 10 of the value 10^{-25} cm^2, which is somewhat smaller than the square of the nuclear radius. The values of η determined from Equation 5.43 are mostly of the order of 0.05 or smaller, indicating that the charge distribution in most nuclei is almost perfectly spherical. The largest positive value of quadrupole moment found thus far is that of $^{176}_{71}$Lu whose $Q_n = 7 \times 10^{-24}$ cm^2 implies a prolate charge distribution with an ellipticity of $+0.174$. The largest negative Q_n is that of $^{123}_{51}$Sb whose $Q_n = -1.20 \times 10^{-24}$ cm^2 implies an oblate charge distribution with an ellipticity of -0.053. Even in these extreme cases, the nuclei deviate from spherical shape by less than 18 % and 6 %, respectively.

The neutron and the proton, each of which has $I = 1/2$, have zero quadrupole moments. The quadrupole moment of the deuteron, for which $I = 1$, is 0.273×10^{-26} cm^2, implying an ellipticity of about 10%. The deuteron, which we naïvely tend to picture as a dumbbell, is actually almost spherical in a time-averaged sense.

It can be shown* quite generally from quantum-mechanical calculations that all static electric multipoles of odd order vanish identically. This includes the dipole ($n = 1$), octupole ($n = 3$), 2^5-pole ($n = 5$), etc. It can also be shown that all static electric multipoles of even order are zero unless $I \geq n/2$. This means, for example, that a nucleus must have an I greater than or equal to 1 to exhibit a quadrupole moment. All nuclei with $I = 0$ or 1/2 should have, and experimentally do have, zero quadrupole moments.

Nuclear Isospin

The neutron and the proton have so many shared characteristics (such as identical spin and almost identical masses and nuclear forces) that it has become common to consider them as the two *charge states* of a single

*For this proof see Blatt and Weisskopf's *Theoretical Nuclear Physics* (Reference 2 at the end of this chapter), pp. 26–30.

fundamental particle, the nucleon. To formalize this concept, we assign to the nucleon a *total isobaric spin* (*isospin*) quantum number $T = 1/2$. To distinguish between the two charge states of the nucleon, we conventionally assign to the proton the isospin projection $T_\zeta = -1/2$ and to the neutron the isospin projection $T_\zeta = +1/2$.* This entire convention is arbitrary but quite convenient since it puts electrical charge into much the same quantum-mechanical framework as intrinsic spin. Just as the spin of a nucleon ($s = 1/2$) may have either of two projections on an axis ($s_z = \pm 1/2$), the isospin of a nucleon ($T = 1/2$) may have either of two projections ($T_\zeta = \pm 1/2$) on an axis that is, in this case, not a real physical axis but a hypothetical axis in a hypothetical "isospin-space." Thus the nucleon is considered to be a particle with isospin $T = 1/2$ and with either the projection $T_\zeta = +1/2$ (neutron) or $T_\zeta = -1/2$ (proton).

The isospin projection of any nucleus is obtained by adding the isospin projections of all its constituent nucleons,

$$T_\zeta = \tfrac{1}{2}(N - Z), \tag{5.44}$$

where N and Z are the numbers of neutrons and protons in the nucleus. Thus T_ζ is simply 1/2 times the neutron excess ($N - Z$) in the nucleus. The neutron excess is, of course, independent of the degree of excitation of the nucleus; hence, the ground state and all excited states, bound or unbound, of a nucleus have the same T_ζ.

The value of the total isospin T may vary from level to level in a nucleus. The only constraint on T is that one of its quantum mechanically allowed projections ($T_\zeta = -T, -T + 1, \ldots, T - 1, T$) must equal ($N - Z$)/2, the fixed T_ζ of all levels.

Isobaric spin has proved to be a valuable concept because of the discovery that there are definite selection rules regarding the possible changes in isobaric spin in β and γ decay. Furthermore, there are no known cases of nuclear reactions in which the total isospin T is not conserved. Hence the conservation of total isospin has become another of the constraints that any nuclear reaction must obey.

Nuclear Parity

The parity of a system of elementary particles, such as an atom or a nucleus, is said to be even if the algebraic sign of the wave function that describes the system does not change when all the space coordinates in the wave function are reflected about the origin. Parity is said to be odd if the wave function does change sign. In other words, if $\psi(-\mathbf{x}) = +\psi(\mathbf{x})$,

---

*The opposite sign convention for T_ζ is sometimes used, particularly in high-energy nuclear physics.

the parity of ψ is even and, if $\psi(-x) = -\psi(x)$, the parity of ψ is odd. In these equations the symbol x is an abbreviation for the coordinates of all the particles in the system under consideration; that is, $x = x_1, y_1, z_1, x_2, y_2, z_2, \ldots$.

In general, a system of particles does not have to possess any definite parity. But, when the potential of a system does not change when x is replaced by $(-x)$, then this system must have a definite parity. This may be proved as follows: If $V(-x) = V(x)$, the Hamiltonian operator does not change when we replace x by $-x$. Therefore, if $\psi(x)$ is an eigenfunction of the Hamiltonian with eigenvalue E,

$$H\psi(x) = E\psi(x).$$

This equation is also true if we replace x by $-x$,

$$H\psi(-x) = E\psi(-x).$$

Provided the energy level is nondegenerate, there is only one linearly independent eigenfunction; so these two solutions can differ only by a multiplicative constant

$$\psi(-x) = A\,\psi(x).$$

To evaluate the constant A, we replace x by $-x$ again and find

$$\psi(x) = A\,\psi(-x).$$

From these two equations, it follows immediately that

$$A^2 = 1 \quad \text{or} \quad A = \pm 1.$$

Thus all nondegenerate eigenfunctions of a symmetric potential have even or odd parity.

If the eigenfunctions are degenerate, then they need not have a definite parity. Nevertheless, it can be shown[*] that the eigenfunctions of a degenerate eigenvalue may be written as linear combinations of two functions, each of which has a definite parity.

The infinite one-dimensional potential well is an example of a potential in which $V(x) = V(-x)$. In this case it was found that the energy levels were nondegenerate, and the eigenfunctions were given by

$$\psi(x) = B \cos \frac{n\pi x}{2b} \qquad (n = \text{odd integer}),$$

$$\psi(x) = A \sin \frac{n\pi x}{2b} \qquad (n = \text{even integer}).$$

[*] Albert Messiah, *Quantum Mechanics*, Vol. 1, North Holland Publishing Company, Amsterdam, 1964, p. 113.

These functions have even parity if n is odd and odd parity if n is even. In other words, each energy level has a definite parity, the lowest being even, the next odd, and so on.

As a second example, consider the one-particle problem with a central potential, $V(r)$. Since $r = (x^2 + y^2 + z^2)^{1/2}$, it is clear that $V(-\mathbf{x}) = V(\mathbf{x})$. Thus we expect that each eigenstate has a definite parity. The eigenfunctions for the case $m_l = 0$ have the form $R_{nl}(r)\, P_l(\cos \theta)$. The radial wave function $R_{nl}(r)$ does not change when $\mathbf{x}$ is replaced by $-\mathbf{x}$ since it depends only on r. The Legendre polynomials, however, may or may not change sign. The even polynomials are functions of even powers of $\cos \theta$; the odd polynomials are functions of odd powers of $\cos \theta$. Since $\cos \theta = z/r$ does change sign when z is replaced by $-z$, the Legendre polynomial will have odd parity when l is odd and even parity when l is even. Thus the wave function of a single particle in a central potential has even or odd parity depending on whether l is even or odd.

The parity of a system of particles is even if the sum of the individual angular-momentum quantum numbers $\sum_i l_i$ is even; the parity is odd if $\sum_i l_i$ is odd. The parity of each energy level in a bound system depends on the wave function for that specific level. Thus an excited state of a nucleus may or may not have the same parity as the ground state.

The intrinsic parity of the electron is defined as even. The intrinsic parities of other fundamental particles have been determined experimentally from the properties of simple systems, such as the hydrogen atom, the deuteron, and the alpha particle. The intrinsic parities of the neutron, proton, and neutrino are all even.

Parity, like isospin, has proved to be a valuable concept because of the discovery that it is always conserved in a nuclear reaction. For example, a neutron can be emitted from the nucleus with an angular momentum $l = 1$ relative to the nucleus only if the nuclear parity changes. Since the intrinsic parity of the neutron is even, the total parity of the neutron is odd in the $l = 1$ state; hence the nuclear parity must change from even to odd, or vice versa.

Parity is not conserved in beta decay. In the next section nuclear forces will be considered. It will be pointed out that beta decay proceeds via the weak interaction force which is distinct from the strong nuclear force.

5.3 Nuclear Forces

Before the advent of nuclear physics, the known types of forces that two bodies could exert on one another at a distance were limited to the well-known gravitational and electromagnetic forces, the latter including the electrostatic, magnetostatic, and electrodynamic forces. It may prove helpful to recall some of the properties of these better-known forces before delving into the complexities of nuclear forces.

Ordinary Forces

The gravitational force exerted by mass m_1 on mass m_2 is given by Newton's universal law of gravitation

$$\mathbf{F}_g = \frac{Gm_1m_2}{r_{12}^2}\left(\frac{\mathbf{r}_{12}}{r_{12}}\right), \tag{5.45}$$

where $\mathbf{r}_{ij}$ is the vector originating at particle j and terminating at particle i and G is the universal gravitational constant 6.670×10^{-8} dyne-cm^2/g^2.

The electrostatic force exerted by charge q_1 on charge q_2 is given by Coulomb's law of electrostatics

$$\mathbf{F}_e = \frac{q_1q_2}{r_{12}^2}\left(\frac{\mathbf{r}_{21}}{r_{21}}\right). \tag{5.46}$$

The gravitational and the electrostatic forces are central forces; they always point along the direction between the masses or the charges. Moreover, they both vary as the inverse of the square of the distance between the bodies.

The magnetostatic force between two isolated magnetic poles is also central and inverse square. But the force between two magnets, each of which consists of two poles, has an entirely different character. The vector sum of the four inverse-square forces exerted by magnet 2 on magnet 1 has the form

$$\mathbf{F}_m = \sum_{i=1}^{2}\sum_{j=1}^{2}\frac{p_{1i}p_{2j}}{r_{ij}^2}\left(\frac{\mathbf{r}_{ji}}{r_{ji}}\right), \tag{5.47}$$

where p_{12}, for example, is the pole strength of the second pole of magnet 1. Now it is clear that the magnitude and direction of the net force between the magnets depends not only on the distance between their centers but also on the relative orientation of their north-south axes. Forces, like this one, that depend not only on the distance between bodies but also on the relative orientation of the bodies are called *tensor forces* to distinguish them from central forces, such as the gravitational and electrostatic forces.

The electrodynamic force exerted on charge q moving with velocity $\mathbf{v}$ at a point where there exists a magnetic field of strength $\mathbf{H}$ is given by

$$\mathbf{F}_{ed} = q(\mathbf{v} \times \mathbf{H}), \tag{5.48}$$

where the magnetic field **H** may have been created, for example, by a stationary magnet or by a second moving charge. The electrodynamic force is seen to be both noncentral and velocity dependent.

We cannot explain the known properties of the nucleus on the basis of the ordinary forces described above; even the most simple and apparent property is inexplicable on this basis. Every nucleus contains protons which repel each other electrostatically. This repulsive force must be counterbalanced by an attractive force; otherwise nuclei could not exist as stable entities. The gravitational force might seem to be an obvious candidate for this distinction, but the gravitational force between nucleons is some 36 orders of magnitude weaker than the electrostatic force,

$$\frac{F_g}{F_e} = \frac{Gm_1m_2}{q_1q_2}$$

$$= \frac{(6.67 \times 10^{-8} \text{ dyne-cm}^2/\text{g}^2)(1.67 \times 10^{-24} \text{ g})^2}{(4.8 \times 10^{-10} \text{ esu})^2} \cong 10^{-36}.$$

Therefore the gravitational force cannot explain the binding of nucleons in nuclei. Similar order-of-magnitude arguments rule out the other ordinary forces. All ordinary forces are too weak to explain both the stability of nuclei and the patterns of scattering of nucleons by nuclei.

Four basically different forces are now recognized to exist in nature: (1) the gravitational force, (2) the weak interaction force, which was postulated by Enrico Fermi in the 1930s in his development of a quantitative theory of beta decay, (3) the electromagnetic force, and (4) the nuclear force. The ratio of the strengths of these four forces between nucleons at distances of the order of 1 fm is approximately 1 (gravitational) to 10^{25} (weak interaction) to 10^{36} (electromagnetic) to 10^{38} (nuclear).

The most direct evidence for the existence and properties of the nuclear force comes from nucleon–nucleon scattering experiments. Experiments in which beams of protons or neutrons are scattered off targets of hydrogen or deuterium have enabled physicists to study the forces that obtain between all combinations of nucleons: *proton–proton, neutron–neutron,* and *neutron–proton.* Discussion of the ingenious experimental arrangements and theoretical analyses that are required to obtain information about the nuclear potential from the scattering patterns is beyond the scope of this book. Only the results will be summarized.

The Proton–Proton Nuclear Force
When the scattering effects due to strictly electromagnetic forces have been subtracted, the specifically nuclear force between two protons is found to have about as much complexity as nature will allow. To explain

all the experimental facts, we must assume that there is not one proton–proton force but, actually, that there are many such forces. The magnitudes and directions of these forces depend not only on the distance between the protons but also on the relative orientations of the spins of the two protons and on the magnitude and orientation of the total spin relative to the orbital angular momentum.

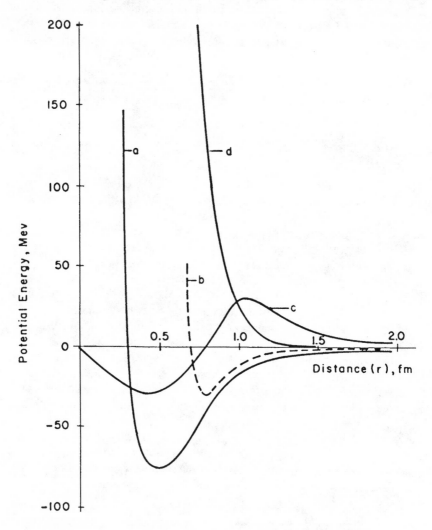

Figure 5.8—The nuclear force: The central potential between two nucleons in (a) anti-parallel spin state and (b) parallel spin state. Examples of (c) tensor potential and (d) spin-orbit potential.

Protons with antiparallel spins interact only through a central force. The potential describing this force is shown in curve a of Figure 5.8. Note that the force is short range and attractive from 1.5 to 0.5 fm but strongly repulsive at closer distances. Protons with parallel spins are less weakly attracted and more strongly repulsed at small distances than those with antiparallel spins (curve b of Figure 5.8). In addition, there is evidence that two types of noncentral forces are operative. These tensor and spin-orbit forces depend on the magnitude and relative orientations of the proton spins and the total orbital angular momentum. The curves for the tensor potential (curve c, Figure 5.8) and the spin-orbit potential (curve d, Figure 5.8) are drawn for one particular orientation of s relative to L. Since L may have any of the quantum mechanically allowed values ($l = 1, 3, 5, \ldots$ in the case of parallel spin nucleons), there are actually infinite possible spin-orbit and tensor forces. Some are attractive and some repulsive, depending on the relative orientation of s and L.

The Neutron–Neutron Nuclear Force

So far as has been determined, the nuclear forces between two neutrons are identical with the nuclear forces between two protons. This charge independence is evidence for the strong belief that the neutron and the proton are merely two different charge states of the same fundamental particle, the nucleon.

The Neutron–Proton Nuclear Force

The neutron–proton nuclear force also seems to verify the charge independence of nuclear forces; that is, the neutron–proton force seems to have components identical to those of the proton–proton and neutron–neutron forces. But, in the neutron–proton force, there are additional components that are ruled out by the Pauli principle in the case of identical nucleons. The neutron and the proton can scatter in all combinations of spin state (parallel or antiparallel) and orbital angular momentum ($l = 0, 1, 2, \ldots$), identical nucleons with parallel spins can scatter only in odd orbital angular-momenta states ($l = 1, 3, 5, \ldots$), and identical nucleons with antiparallel spin can scatter only in even states ($l = 0, 2, 4, \ldots$).

Saturation of Nuclear Forces

Forces may be classified as unsaturated or saturated according to whether the force on a given particle continues to increase as the number of other particles increases or reaches a maximum and remains unchanged regardless of the number of added particles. The gravitational force, for example, is unsaturated; it continues to increase in direct proportion to the attracting mass. The electrostatic force is likewise unsaturated; it continues to increase in direct proportion to the attracting or repulsing charge. In contradistinction, the force that holds an atom in its site within a crystal is saturated. No matter how large the crystal, provided it contains at least a score or so of atoms, the force is the same. The explanation for this is well known: The atom can bind itself by means of its valance electrons to only a few of the surrounding atoms. Once it has saturated all its possible bonds by means of the exchange forces, its binding is independent of all those other atoms in the solid with which it does not interact.

The force on a nucleon within all but the lightest of nuclei is saturated. The most direct evidence for this assertion is the fact that the binding energy per nucleon is quite insensitive to the number of nucleons in the nucleus, all the way from $A = 10$ to $A = 120$ (as shown in Figure 5.1).

The property of saturation has been explained on the basis of nuclear exchange forces, first suggested by W. Heisenberg in 1932. Basically the concept is much like that which explains the saturation of atomic forces within a solid except that, instead of electrons being exchanged between atoms, π mesons are exchanged between nucleons. Pi mesons are particles postulated by Yukawa in 1935 as those responsible for the strong nuclear force and actually discovered some years later. Pi mesons, which have a rest mass approximately 285 times as large as that of the electron, are now commonly generated in the laboratory by high-energy (of the order of 10^2 Mev) interactions between nucleons and nuclei. They exist in three distinct charge states, π^+, π^0, and π^-, the charges being respectively $+1$, 0, and -1 times the charge on the proton. The π^0 meson has a rest mass less than that of the π^+ and π^- mesons by some 9 electron masses. In a free state all π mesons are unstable and decay according to

$$\pi^+ \rightarrow \mu^+ + \tilde{\nu},$$

$$\pi^- \rightarrow \mu^- + \nu,$$

$$\pi^0 \rightarrow \gamma + \gamma,$$

where $\mu^\pm$ are the plus and minus μ mesons, ν and $\tilde{\nu}$ are the neutrino and antineutrino, and γ is a photon. These remarks and equations are introduced only to emphasize the unsettled nature and the seemingly diverging complexity encountered in the investigation of elementary forces and the particles associated with these forces. In the present rudimentary state of neutron interaction theory, the actual structure of the neutron and the underlying causes for the nuclear forces are not explicitly taken into account. For all practical purposes the neutron and the proton are treated as though they were structureless entities, "elementary particles."

5.4 The Pauli Exclusion Principle and Atomic Shell Theory

During the past few decades, it has become quite clear that there is a close analogy between the accepted views of atomic and nuclear structure. Specifically, the neutrons and protons within a nucleus seem to be grouped in series of concentric shells much as the electrons are grouped in an atom. Before examining the shell theory of nuclear structure, we will summarize the electron shell theory of atomic structure beginning with the simplest of all atoms and proceeding through the periodic table.

Energy Levels of the Hydrogen Atom

The single electron in the hydrogen atom is in a central Coulomb potential $V(r) = -e^2/r$. If electron spin is neglected for the moment, the possible energy eigenvalues of the hydrogen atom are given by the well-known equation

$$E_{n,l} = \frac{-me^4}{2\hbar^2(n + l)^2} = \frac{-me^4}{2\hbar^2 N^2}, \tag{5.49}$$

where n is the number of nodes in the radial wave function, including the node at the origin ($n \geq 1$); l is the total orbital angular-momentum quantum number ($l = 0, 1, \ldots$); and N is the *principal quantum number*, defined as the sum of n and l.

In addition to the quantum numbers n and l, there is the azimuthal quantum number associated with the allowable components of the angular momentum along any given axis. The azimuthal quantum number, m_l, can take on any of the $2l + 1$ integral values between $-l$ and $+l$.

Because there are n combinations of the integers n and l that will add to any given value of the principal quantum number N, an energy level with a given N has an nth-order degeneracy in l. This accidental degeneracy is peculiar to the Coulomb potential and will be removed if any small non-$1/r$ term is added to the Coulomb potential. The splitting of the accidentally degenerate energy levels is shown schematically in parts a and b of Figure 5.9.

A more fundamental type of degeneracy in energy occurs when the energy states are independent of the spatial orientation of the angular momentum of the system. Because m_l does not appear in Equation 5.49, the expression for the energy eigenvalues, it is obvious that an energy level with a given l has a degeneracy of order $2l + 1$. In other words, there are $2l + 1$ states, each represented by a linearly independent eigenfunction, for each level of fixed l. This fundamental degeneracy in m_l follows immediately from the fact that the Schrödinger eigenvalue equation for the energy of a particle in a central potential does not contain m_l. This particular degeneracy, which is common to all central potentials, is removed

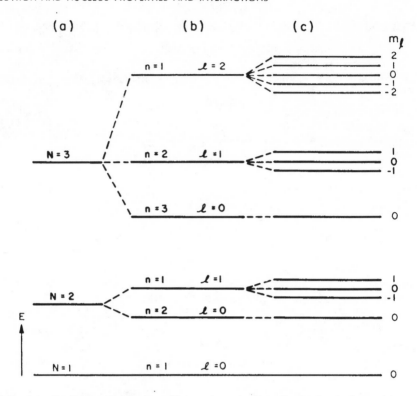

Figure 5.9—Qualitative distribution of lowest bound energy levels of a single particle in a (a) Coulomb potential, (b) non-Coulomb central potential, and (c) noncentral potential. All levels have negative total energy, the lowest being the ground-state level $N = 1$.

when a noncentral force field is applied. The force could be, for example, an externally applied electric or magnetic field.* An external field provides a preferred direction in space and removes the fundamental degeneracy. Parts b and c of Figure 5.9 qualitatively shows the splitting of levels of fixed l into $2l + 1$ sublevels in a noncentral potential.

A slight additional complexity arises when we take into account the spin of the electron. The spin quantum number of the electron is $s = 1/2$; its intrinsic angular momentum is $\hbar[s(s + 1)]^{1/2}$. The component of **s** along any space axis is restricted to have one of the values $m_s\hbar$ where $m_s = \pm 1/2$. This means that each of the energy levels specified by given n, l, and m_l is doubly degenerate because m_s can take on either of two

---

*The splitting of energy levels in electric and magnetic fields gives rise to the *Stark* and *Zeeman effects*, respectively.

values. Thus the degeneracy of a level of given n and l is $2(2l + 1)$ when the particle in question has spin 1/2.

The total angular momentum of the electron $\mathbf{j} =. \mathbf{L} + \mathbf{s}$ is restricted to the discrete values $\hbar[j(j + 1)]^{1/2}$, where $j = l + 1/2$ or $j = l - 1/2$, except that in the $l = 0$ state only $j = +1/2$ is possible. The component of $\mathbf{j}$ along any space axis is restricted to one of the values $m_j\hbar$, where m_j can take on any of the odd half-integer values from $-j$ to $+j$. Thus, the degeneracy of a level of given n, l, and j is $2j + 1$. This result will be further clarified in Chapter 6, where the angular momentum eigenfunctions will be discussed.

In the hydrogen atom and other heavier atoms, the intrinsic magnetic moment of the electron μ interacts with the magnetic field $\mathbf{H}$ that is associated with the motion of the electron through the static electric field and gives rise to an additional potential energy of the form

$$V_{so} = -\mu \cdot \mathbf{H}.$$

An electron moving with velocity $\mathbf{v}$ in an electric field $\mathbf{E}$ will experience a magnetic field $\mathbf{H} = -(1/2c^2)\mathbf{v} \times \mathbf{E}$. The electric field $\mathbf{E} = -(dV/dr)(\mathbf{r}/r)$, where $V = V(r)$ is the electrostatic potential. Furthermore $\mathbf{v} = \mathbf{p}/m$, $\mathbf{L} = \mathbf{r} \times \mathbf{p}$, and the magnetic moment of the electron, $\mu = (e/m)\mathbf{s}$, where $\mathbf{s}$ is the intrinsic angular momentum. Hence, the potential takes the form

$$V_{so}(r) = \frac{e}{2m^2c^2} \frac{dV}{dr} \mathbf{L} \cdot \mathbf{s}, \tag{5.50}$$

where c is the speed of light and m and e are the electron mass and charge, respectively.

We call V_{so} the spin-orbit potential because it depends on the relative orientations of the spin and orbital angular-momentum vectors of the electron. When these vectors are replaced by their quantum-mechanical operators and the resulting expression is added to the Coulomb potential in the Hamiltonian operator, it is found that the spin-orbit potential results in a splitting of all levels (except those with $l = 0$) into two distinct energy levels. If m_l is positive, the level with $m_s = -1/2$ has the lower energy because the spin-orbit force is attractive in the "antiparallel" state and repulsive in the "parallel" state. Thus, each of the $l > 0$ levels shown on the right-hand side of Figure 5.9 should be further split into two levels differing very slightly in energy. This splitting is the fine structure of electronic energy levels. The difference in energy of the spin doublet resulting from this spin-orbit splitting is given by

$$E_{(l+1/2)} - E_{(l-1/2)} = (2l + 1)\left(\frac{\hbar}{2mc}\right)^2 \left\langle \frac{1}{r} \frac{dV}{dr} \right\rangle \tag{5.51}$$

where the symbol $\langle \; \rangle$ stands for the quantum-mechanical expectation value of the quantity it contains.

The Designation of Energy Levels

To designate an atomic energy level completely, we must state four quantum numbers. Numerous combinations of these numbers may be used, for example, (n,l,m_l,m_s) or (n,l,m_j,m_s) and so on. The combination commonly chosen in nuclear physics is (n,l,j,m_j). In atomic physics the foursome usually chosen is (N,l,j,m_j), where N is the principal quantum number.

In a central potential the states with various m_j values are degenerate; so only three quantum numbers are needed to designate completely the energy of a level. Those numbers chosen in nuclear physics are n, l, and j, written in the form nl_j. Numerical values are written for n and j, but l is expressed by the usual spectroscopic notation: s for $l = 0$, p for $l = 1$, d for $l = 2$, etc. Thus, for example, the level with $n = 2$, $l = 1$, and $j = 1/2$ would be called $2p_{1/2}$. The level with $n = 3$, $l = 2$, and $j = 5/2$ would be called $3d_{5/2}$, and so on. Each level specified by these three quantum numbers has a degeneracy of order $2j + 1$. If a noncentral potential were added, each level would split into $2j + 1$ levels that would be separated in energy by an amount proportional to the strength of the perturbing potential. The multiplicity of a level of given j is thus $2j + 1$.

The Pauli Exclusion Principle

To understand the structure of bodies, such as atoms and nuclei, which contain many identical particles, we must invoke the Pauli exclusion principle. In its most general form, Pauli's principle may be stated as: *In one and the same body, no two identical particles can have identical sets of quantum numbers.* This principle applies to electrons, protons, and neutrons, all of which have spin 1/2 and are known collectively as *fermions*. It does not apply to particles with integer spin, *bosons*.

The Pauli principle is another example of an inspired guess. No complete explanation for it has ever been developed; but, perhaps more significantly, no exception to it has ever been found. It is one of the great unifying principles of modern science; without it, modern chemistry and nuclear theory would be impossible.

Wolfgang Pauli discovered this rule in the days of the Bohr atom, before the advent of quantum mechanics. When he first announced the principle, each electron in the Bohr atom was assumed to have only three quantum numbers, and he had to assign arbitrarily a fourth quantum number, representing the spin, to use his principle in explaining the periodic table. The subsequent discovery of electron spin justified his boldness.

The Many-Electron Atom

Each electron in a many-electron atom can be pictured as being in a potential field created by the combined potentials of the nucleus and all other electrons. This combined potential can be approximated by a central potential that goes to zero more rapidly than the nuclear Coulomb potential because of the screening effect of the other electrons. Since the potential is not a Coulomb potential, the degeneracy in l is removed, and the energy levels are qualitatively like those in part b of Figure 5.9.

The Pauli exclusion principle for the electrons in an atom may be stated in a number of equivalent forms:

1. An energy eigenstate characterized by the four quantum numbers N, l, j, and m_j can be occupied by not more than one electron.
2. An energy eigenstate characterized by the three quantum numbers N, l, and j can be occupied by not more than $2j + 1$ electrons.
3. An energy eigenstate characterized by the two quantum numbers N and l can be occupied by not more than $2(2l + 1)$ electrons.
4. An energy eigenstate characterized by the principal quantum number N can be occupied by not more than $2N^2$ electrons.

In all but the last statement of the Pauli principle, the principal quantum number N could be replaced by the radial quantum number n.

The structure of the periodic table of elements can be explained on the basis of the Pauli principle and one further observation, namely: Every electron in an atom seeks to be in the lowest unfilled energy state. Thus, atomic states are filled from the lowest energy state upward.

The single electron in the hydrogen atom goes into the lowest state. It has quantum numbers N, l, j, and m_j that equal 1, 0, 1/2, and $-1/2$, respectively.

Helium has two electrons. The first electron occupies the lowest energy state with the quantum numbers already assigned to hydrogen. The second electron takes the next lowest energy state allowed by the Pauli principle, $(N, l, j, m_j) = (1, 0, 1/2, 1/2)$.

The first two electrons in lithium occupy the same quantum states as the two electrons in helium. The third electron goes into the next lowest state, with $(N, l, j, m_j) = (2, 0, 1/2, -1/2)$.

It is possible to proceed in this manner through the periodic table adding one electron at a time and building up successively larger atoms. There is, of course, no way to determine which of two levels is lower than the other simply by examination of the quantum numbers associated with the levels. It is necessary to solve the Schrödinger equation with the proper potential to determine exactly how the sequence of energy eigenvalues depends on the quantum numbers. The results of such a detailed analysis of the first 37 elements are shown in Table 5.3. Note that the simple sequence of quantum numbers finally breaks down at potassium, whose $4s$ state lies lower than its $1d$ state.

As previously explained, in a pure Coulomb potential the energy of an atomic state is determined by the principal quantum number N. In atomic physics a shell is defined as a group of electrons having the same value of N. The shell with $N = 1$ is called the K shell; the shells with $N = 2, 3, 4, \ldots$ are called the $L, M, N, \ldots$ shells, respectively. According to the Pauli principle, each shell can contain a maximum of $2N^2$ electrons. Thus, filled K, L, M, and N shells contain 2, 8, 18, and 32 electrons, respectively.

A subshell is a group of electrons having the same n and l (or N and l) quantum numbers. Each subshell can contain a maximum of $2(2l + 1)$ electrons. Thus, filled s, p, d, and f subshells contain 2, 6, 10, and 14 electrons, respectively.

The successive filling of subshells and shells of the lighter atoms in the periodic table is shown in Table 5.3. Also tabulated are the ionization potentials of the atoms. The ionization potential is the amount of energy that must be supplied to separate the least tightly bound electron from the atom. Note that the large ionization potentials of the inert gases He, Ne, Ar, and Kr are associated with the completion of either a shell or a subshell of electrons. The ionization potentials of the atoms with one more electron than the inert gases are extremely low, implying that there is a large energy gap between the energy level at which a shell or a subshell is filled and the next higher energy level.

The occupancy number of an atomic level is defined as the total number of electrons in the atom that have energies less than or equal to the energy of the level in question. Thus, for example, the occupancy numbers of the $1s, 1p, 2p$, and $3p$ levels are 2, 10, 18, and 36, respectively.

The chemical behavior of atoms is directly related to their shell structure. For example, chlorine, which has one electron too few to complete a subshell, tends to form compounds in which it can partially supply its electron affinity by sharing an electron belonging to another atom. In solution, it takes on an extra electron and becomes a negative ion. Sodium, on the other hand, has one electron in excess of a closed shell. It readily shares this electron with an electron-hungry atom like chlorine or loses it entirely in a solution and becomes a positive ion. Atoms like phosphorus, which have half-filled shells, possess ambiguous valence. They are equally likely to lose or gain electrons to close their subshells. The inert gases, with closed shells or subshells, exhibit no tendency to either accept or donate an electron.

Table 5.3—Electron Configurations and Ionization Potentials of Atoms

		(N,l,m_l,m_s) of least-bound electron	Number of electrons in shells and subshells										Ionization potential, ev
			K	L		M			N				
Z	Element		1s	2s	1p	3s	2p	1d	4s	3p	2d	1f	
1	H	$(1,0,\ 0,-1/2)$	1										13.59
2	He	$(1,0,\ 0,+1/2)$	2										24.58
3	Li	$(2,0,\ 0,-1/2)$	2	1									5.39
4	Be	$(2,0,\ 0,+1/2)$	2	2									9.32
5	B	$(2,1,-1,-1/2)$	2	2	1								8.30
6	C	$(2,1,-1,+1/2)$	2	2	2								11.26
7	N	$(2,1,\ 0,-1/2)$	2	2	3								14.54
8	O	$(2,1,\ 0,+1/2)$	2	2	4								13.61
9	F	$(2,1,\ 1,-1/2)$	2	2	5								17.42
10	Ne	$(2,1,\ 1,+1/2)$	2	2	6								21.56
11	Na	$(3,0,\ 0,-1/2)$	2	2	6	1							5.14
12	Mg	$(3,0,\ 0,+1/2)$	2	2	6	2							7.64
13	Al	$(3,1,-1,-1/2)$	2	2	6	2	1						5.98
14	Si	$(3,1,-1,+1/2)$	2	2	6	2	2						8.15
15	P	$(3,1,\ 0,-1/2)$	2	2	6	2	3						10.55
16	S	$(3,1,\ 0,+1/2)$	2	2	6	2	4						10.36
17	Cl	$(3,1,\ 1,-1/2)$	2	2	6	2	5						13.01
18	Ar	$(3,1,\ 1,+1/2)$	2	2	6	2	6						15.76
19	K	$(4,0,\ 0,-1/2)$	2	2	6	2	6		1				4.34
20	Ca	$(4,0,\ 0,+1/2)$	2	2	6	2	6		2				6.11
21	Sc	$(3,2,-2,-1/2)$	2	2	6	2	6	1	2				6.56
22	Ti	$(3,2,-2,+1/2)$	2	2	6	2	6	2	2				6.83
23	V	$(3,2,-1,-1/2)$	2	2	6	2	6	3	2				6.74
24	Cr	$(3,2,-1,+1/2)$	2	2	6	2	6	5	1				6.76
25	Mn	$(4,0,\ 0,+1/2)$	2	2	6	2	6	5	2				7.43
26	Fe	$(3,2,\ 0,-1/2)$	2	2	6	2	6	6	2				7.90
27	Co	$(3,2,\ 0,+1/2)$	2	2	6	2	6	7	2				7.86
28	Ni	$(3,2,+1,-1/2)$	2	2	6	2	6	8	2				7.63
29	Cu	$(3,2,+1,+1/2)$	2	2	6	2	6	10	1				7.72
30	Zn	$(4,0,\ 0,+1/2)$	2	2	6	2	6	10	2				9.39
31	Ga	$(4,1,-1,-1/2)$	2	2	6	2	6	10	2	1			6.00
32	Ge	$(4,1,-1,+1/2)$	2	2	6	2	6	10	2	2			7.88
33	As	$(4,1,\ 0,-1/2)$	2	2	6	2	6	10	2	3			9.81
34	Se	$(4,1,\ 0,+1/2)$	2	2	6	2	6	10	2	4			9.75
35	Br	$(4,1,+1,-1/2)$	2	2	6	2	6	10	2	5			11.84
36	Kr	$(4,1,+1,+1/2)$	2	2	6	2	6	10	2	6			14.00
37	Rb	$(5,1,-1,-1/2)$	2	2	6	2	6	10	2	6	(4p)		4.18

5.5 The Nuclear Shell Model

A number of theoretical nuclear models have been constructed in attempting to understand the properties of nuclei. A nuclear model is a conceptual picture of the nucleus from which it is possible, in a systematic way and without impossibly lengthy calculations, to predict one or more of the observable properties of the nucleus. Nuclear models are required because the forces acting within the nucleus are not fully known. Even if the forces were known, nuclear models would be needed because the extreme complexity of a many-nucleon dynamical system would prohibit the calculation of the total wave function of any but the lightest nuclei.

Various nuclear models utilize different and often contradictory sets of simplifying assumptions. Each emphasizes certain aspects of nuclear structure to the exclusion of others. Thus, each model is capable of explaining only a part of the wealth of known experimental properties of nuclei, usually with rather limited accuracy.

Four nuclear models are of particular interest to nuclear engineers: the liquid-drop model, the nuclear shell model, the collective model, and the statistical model. The liquid-drop model is of interest primarily because it explains the kinematics of nuclear fission, the nuclear shell model because it explains the properties of nuclear ground states and low-lying energy level structure, and the collective model because it explains certain features of low-lying nuclear energy level structure not explained by the more simple shell model. All three also interest us because they are closely allied with various interaction models.

These three models represent an interesting example of thesis, antithesis, and synthesis. The liquid-drop model, proposed around 1930, assumes that the nucleus is like a drop of liquid in which the nucleons move randomly, much as do the molecules in a drop of water. The nuclear shell model, developed in the 1950s, takes a diametrically opposite view: The nucleons are visualized as occupying shells in the nucleus much as the orbital electrons occupy shells in the atom in the Bohr theory. The collective model combines features of the two older models. The nucleus is pictured as having a liquid-drop core of oscillating nucleon shells orbited in some nuclei by valence nucleons.

The shell model is an example of an *independent-particle model*. An independent-particle model attempts to calculate an empirical property of the nucleus on the basis of the quantized properties of a single nucleon moving in a potential determined by all the other nucleons. The Schmidt model for the magnetic moments of odd-Z and odd-N nuclides and the optical model are other examples of independent-particle models. The liquid-drop model is the extreme opposite of an independent-particle model. The collective model is a semi-independent-particle model.

Although the shell model and the more advanced collective model have been reasonably successful in explaining the distribution and properties of energy levels lying just above the ground state of the nucleus, they fail completely to predict the large numbers of closely spaced levels in the region 8 Mev or more above the ground state. These virtual levels, lying above the neutron separation energy, are responsible for the resonances in neutron cross sections. At the higher energies, independent-particle models fail completely to predict the details of the level spacings. It is believed that these levels are due to the quantization of the energy of the nucleus as a whole, not the quantization of a single neutron or proton. Because there are large numbers of such levels, the methods of statistical mechanics and thermodynamics are employed to obtain semiquantitative predictions of the distribution and properties of the levels. The resulting model is called the statistical model of the nucleus.

In addition to nuclear models, which attempt to predict specifically nuclear properties, there are also what might be called *interaction models*, which attempt to predict cross sections. Examples of the latter are the compound-nucleus model and the optical model. In this section we restrict consideration to nuclear models; interaction models are discussed in Section 5.6. Note, however, that nuclear models and interaction models are not completely independent. For example, the compound-nucleus model of neutron–nucleus interactions has much the same basis as the liquid-drop model of nuclear structure.

We cannot begin to understand neutron–nucleus interactions without taking into account the distribution and properties of low-lying energy levels in the nucleus. These are given, at least semiquantitatively, by the nuclear shell model and the collective model, to which the remainder of this section is devoted. The statistical model is discussed in the next chapter, and the liquid-drop model is discussed in Chapter 11, Section 11.5.

Magic Numbers

The nuclear shell model, proposed by T. H. Bartlett in 1933 and developed in great detail by M. Mayer and J. H. D. Jensen in the years following 1948, had its origin in certain similarities between atomic and nuclear properties, the most striking of which is the existence in both of *magic numbers*. Magic numbers are integers, associated with the atom or the nucleus, at which distinctive changes in properties occur.

Atoms of 2, 10, 18, 36, 54, or 80 electrons are remarkably stable compared with neighboring atoms in the periodic table. These are the inert-gas atoms; they show no tendency to accept or donate electrons like their neighbors. Similarly, nuclei of 2, 8, 20, 28, 50, 82, or 126 neutrons or protons are also remarkably stable. There are many experimental data which indicate that at these nuclear magic numbers something occurs which is closely analogous to the closing of shells in the inert-gas atoms. A few of the more striking of these empirical evidences for the existence of nuclear shell structure are:

1. Nuclei with magic Z (or magic N) have more stable isotopes (or isotones) than nearby nuclei. For example, calcium ($Z = 20$) has six stable isotopes, whereas potassium ($Z = 19$) has only two and scandium ($Z = 21$) has only one. There are five stable isotones with $N = 20$ but none with $N = 19$ or 21.

2. The neutron separation energy from magic-N nuclides and the proton separation energy from magic-Z nuclides are more than those from nearly nonmagic nuclides. In contrast, the separation energy of a magic-plus-one nucleon tends to be very low. The nuclide formed by adding a nucleon, particularly a proton, to a magic nucleus is often unstable. A striking example of this behavior may be found in the doubly magic nuclide ${}^{4}_{2}\text{He}$. The separation energies of a proton and a neutron from ${}^{4}_{2}\text{He}$ are 19.8 Mev and 20.6 Mev, respectively, both much higher than those of nearby nuclides. The magic-plus-one nuclei ${}^{5}_{2}\text{He}$ and ${}^{5}_{3}\text{Li}$ are extremely unstable; they decay with half-lives of the order of 10^{-21} sec into alpha particles and the odd nucleon.

3. Neutron-capture cross sections in the million-electron-volt region are a factor of about ten lower for nuclides with $N = 50, 82,$ and 126 than for nearby nonmagic nuclides. This again shows the tendency of magic-N nuclides to resist the addition of another neutron.

4. At least two of the delayed-neutron groups emitted following the fission of uranium and plutonium can be traced to a magic-number effect. The fission product ${}^{87}_{35}\text{Br}$ sometimes decays by β^- emission with a 55.6 sec half-life to an isomer of krypton, ${}^{87m}_{36}\text{Kr}$. This nuclide has 51 neutrons, one more than the magic number 50. It decays by emitting the extra neutron to become stable and neutron-magic ${}^{86}_{36}\text{Kr}$. Similarly, the fission product ${}^{137}_{54}\text{Xe}$, which has 83 neutrons, one more than the magic number, decays by emission of its extra neutron into stable ${}^{136}_{54}\text{Xe}$, a magic-N nuclide.

Many of the early attempts to explain the nuclear magic numbers were based on much the same procedure used so successfully to explain the atomic magic numbers. Each nucleon was assumed to move in a potential created by all the other nucleons. A reasonable guess was made as to the form of this potential, generally short range and attractive, and then the energy levels were calculated. Next, the Pauli exclusion principle was invoked to fill the calculated levels with either neutrons or protons. Since the neutron and the proton are not identical particles, they are treated separately when the exclusion principle is applied. We therefore imagine two sets of noninteracting shells, one for neutrons and one for protons.

In early studies it was naturally hoped that large energy gaps in the level structure would appear between those levels whose occupancy numbers were magic and the next higher levels. Large energy gaps were indeed found but generally not at the magic numbers. In a spherical square-well potential, for example, the energy gaps occur at occupancy numbers 2, 8, 20, 40, 70, 112, and 168, only the lowest three of which correspond to nuclear magic numbers. Many potentials were tried, but none predicted the entire magic-number sequence. Interest in such attempts waned, and, in the decade between about 1938 and 1948, little attention was paid to magic numbers and to their ultimate explanation.

In 1949 and 1950, the mystery of the magic numbers was finally resolved by Maria G. Mayer* and, independently, by Haxel, Jensen, and Suess.† They added one new element to the procedure used in the early attempts to derive the magic numbers, namely, a spin-orbit force term.

The early attempts had been based on the assumption that the nuclear potential was independent of the orientation of the spins and angular-momentum vectors of the nucleon moving within the potential. This was a natural assumption in view of the atomic analogy. The spin-orbit force on the electron is so weak relative to the Coulomb force that it produces only a very fine doublet splitting of each level with $l > 0$. This splitting in no way affects the atomic magic-number sequence. There was no theoretical reason to assume that the spin-orbit force on the nucleon would be large enough relative to the rest of the nuclear force to cause the spin-orbit splitting to seriously affect the energy level gap structure of the nucleus. However, it turns out that the nuclear spin-orbit force is larger than expected and does produce sufficient level splitting to furnish an explanation for the observed shell structure. Before investigating how this happens, let us examine the direct experimental evidence of the existence of the nuclear spin-orbit force.

Direct Evidence of Nuclear Spin-Orbit Forces

The most direct evidence of the existence of nuclear spin-orbit forces is obtained through double-scattering experiments. A schematic diagram of a typical experiment is shown in Figure 5.10. In such experiments we observe that (a) the incoming beam of high-energy neutrons or protons is scattered in all directions by target A; (b) detectors B and C, arranged

*Maria G. Mayer, *Phys. Rev.*, **78**: 16 (1950).

†O. Haxel, J. H. D. Jensen, and H. E. Suess, *Phys. Rev.*, **75**: 1766L (1949); *Z. Physik* **128**: 295 (1950).

at equal scattering angles θ but different azimuthal angles ϕ, record equal numbers of scattered nucleons in beams B and C; (c) when beam B is directed against a second target, detectors D and E, again placed at equal scattering angles but different azimuthal angles with reference to the second target, record different intensities of nucleons; and (d) the azimuthal asymmetry in the second scattering is qualitatively independent of the composition of targets A and B. They may contain identical or different nuclei with any spin, including zero.

How do we explain the fact that a single scattering produces an azimuthally independent scattered flux but a second scattering introduces an azimuthal asymmetry? The explanation cannot be based on some special property of the target nuclei, such as spin or magnetic moment, since all second scatterings produce asymmetry, regardless of the target materials. There must be some intrinsic difference between beams A and B. It cannot be the difference of the kinetic energies of the nucleons in the two beams since the effect is found to be qualitatively independent of these energies. The only variable that remains is the spin vector of the

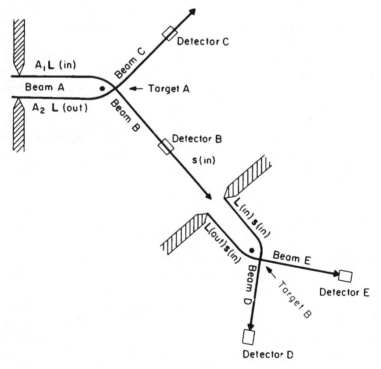

Figure 5.10—Schematic drawing of a double-scattering experiment.

nucleons. We are, by this line of reasoning, led to the conclusion that nucleons in beams A and B differ in their spin directions. Beam A, which scatters equally in all azimuthal directions, is presumably unpolarized; it contains equal numbers of spin-in and spin-out nucleons (where in and out mean into and out of the scattering plane). Beam B, which exhibits azimuthal asymmetry, must be at least partially polarized.

Now the major question is: How does the scattering of an unpolarized beam (beam A) produce a partially polarized beam (beam B)? This question has been straightforwardly answered by invoking the assumption that there is a force between the target nucleus and the nucleon that depends on the relative orientation of the spin and angular-momentum vectors of the nucleon. This spin-orbit force, analogous to the atomic electron spin-orbit force except for sign, is assumed to be central and to be proportional to $-\mathbf{s} \cdot \mathbf{L}$. If the spin and angular-momentum vectors are parallel, i.e., if they differ in direction by less than $\pi/2$, the spin-orbit potential is negative and adds to the attractive nuclear force. On the other hand, if $\mathbf{s}$ and $\mathbf{L}$ are antiparallel, the spin-orbit force is positive and decreases the net force between the nucleus and the nucleon.

The incoming beam of nucleons (beam A) is unpolarized. It contains equal numbers of nucleons with spins parallel and antiparallel to their individual angular-momentum vectors $\mathbf{L}$. With reference to a single nucleus in target A, the nucleons in beam A have orbital angular-momentum vectors $\mathbf{L} = \mathbf{r} \times \mathbf{p}$ that point both in and out of the plane of scattering. Those nucleons in the part of the beam labelled A_1 (Figure 5.10) have $\mathbf{L}$ in; those in A_2 have $\mathbf{L}$ out. Those nucleons in beam A_1 that have $\mathbf{L}$ and $\mathbf{s}$ parallel will be deflected through a greater angle than those with $\mathbf{L}$ and $\mathbf{s}$ antiparallel. Thus the selection of that part of the scattered flux which bends downward (beam B) means a selection of nucleons whose spins are preferentially directed into the scattering plane. The second scattering of this partially polarized beam will exhibit an azimuthal asymmetry because the spin-orbit force between the nucleons in beam B and the nuclei in target B will be stronger on the side where the orbital angular momentum with respect to the nucleus of target B has the same direction as the spin. Since more spins are in than out, more nucleons will be scattered downward into detector D than upward into detector E.

It should be noted that the degree of polarization produced in scattering depends on the magnitude of $\mathbf{L}$. In low-energy events, where the elastic scattering is entirely due to s-wave interactions, there will be no polarization simply because $l = 0$ and therefore the spin-orbit force which is proportional to l is zero. At higher energies, where p and higher order waves are responsible for part of the scattering, polarization can and does occur.

Information about the angular momenta of virtual energy levels can be obtained from double-scattering experiments. In a classic experiment of this type, Heusinkveld and Freier* scattered protons off targets of ^{4}He. The scattering showed a maximum at a 2.4-Mev laboratory proton energy. The angular distribution of the first scattered protons indicated that those with $l = 1$ were scattered preferentially. Analysis of the polarization data indicated that protons with spin parallel to their angular momentum ($j = l + 1/2$) were responsible. This data proved the existence of an unbound energy level in ^{5}Li (^{4}He $+ p$) of the type $p_{3/2}$. Subsequent experiments at energies several million electron volts higher have shown the existence of a second resonance of the type $p_{1/2}$. These two levels in ^{5}Li constitute a spin doublet of the type discussed in Section 5.4 in reference to atomic energy levels. The observed differences between the separations of spin doublets in the nucleus and the atom, a million electron volts compared with fractions of an electron volt, reflects the difference in magnitudes of the nuclear and atomic forces. Another interesting difference is in the order of the doublet. In atomic-level structure the $j = 1/2$ level occurs at a lower energy than the $j = 3/2$ level; in nuclear-level structure the opposite ordering of spin doublets occurs. This effect can be traced to the difference in sign of the atomic and nuclear spin-orbit forces, which are repulsive in the atomic case and attractive in the nuclear case for **L** and **s** parallel.

It is easy to explain the existence of spin-orbit forces for charged particles on the basis of purely classical physics. The forces are simply due to the interaction between the intrinsic magnetic moment of the charged particle and the magnetic field created by its orbital motion. This mechanism is sufficient to explain qualitatively and quantitatively the spin-orbit force on the electron. It explains qualitatively, but not quantitatively, the spin-orbit force on the proton and fails completely to explain the spin-orbit force on the uncharged neutron. Some deeper theory of nuclear forces, probably some form of meson theory, is expected to solve this mystery eventually. But presently we can only state that the empirical evidence for the existence of nuclear spin-orbit forces is so strong as to be almost certain; no one has yet, however, succeeded in deriving a quantitative expression for the nuclear spin-orbit force.

The Energy Level Sequence with Spin-Orbit Coupling

The first problem that faced Mayer and Jensen and his coworkers was that of finding a reasonable nuclear potential, $V(r)$. Two basic empirical facts strongly limit the choice of this potential: (1) Nucleons are bound

* M. Heusinkveld and G. Freier, *Phys. Rev.*, 85: 80 (1952).

within nuclei; therefore the potential must have the form of a well. (2) Scattering experiments indicate that the nuclear force is short-range; therefore the well must have a finite radius.

The model being set up is an independent-particle model. The potential $V(r)$ is to represent the action of all but one of the nucleons on that single nucleon. Since the nucleons are assumed to be more or less uniformly distributed in the nucleus, there should be no net force on a nucleon at the center of the nucleus. This further restricts $V(r)$ to a function with zero slope at $r = 0$.

There are infinitely many potentials satisfying these three conditions. With such a wide choice, the intelligent procedure is to choose the one most easily handled mathematically. If the simplest potential does not work, a more complicated one can be tried. The objective is to find a potential which, when perturbed by a spin-orbit term, produces a level sequence with wide energy gaps at those occupation numbers corresponding to the nuclear magic numbers. Actually, any relatively simple potential will produce the magic numbers provided it is accompanied by a judicious choice of parameters such as the spin-orbit interaction strength.

From a mathematical viewpoint the ideal potential to choose is the three-dimensional harmonic-oscillator potential

$$V(r) = -V_0 \left[1 - \left(\frac{r}{R} \right)^2 \right] \tag{5.52}$$

for $0 \le r \le \infty$ because the energy levels in this potential are given by the simple closed-form expression

$$E_{n,l} = [2(n - 1) + l]h\nu_0, \tag{5.53}$$

where n is the radial quantum number ($n = 1, 2, 3, \ldots$), l is the polar quantum number ($l = 0, 1, 2, \ldots$), and ν_0 is the frequency of the corresponding classical harmonic oscillator $2\pi\nu_0 = (2V_0/mR^2)^{1/2}$. In Equation 5.53, the energy $E_{n,l}$ is measured above the ground state. The actual total energy of the ground state is $-V_0 + (3h\nu_0/2)$.

The harmonic-oscillator potential does not vanish beyond $r = R$ as a reasonable nuclear potential should. However, if V_0 is large enough, the relative positions of the low-lying levels will not be seriously affected if the potential is clipped at R, i.e., is forced to zero at all points beyond $r = R$. The major effect on the low-lying levels will be to split those particular levels that have an accidental degeneracy in n and l into separate levels. For example, the $1d$ and $2s$ states, which (Equation 5.53) have the same energy in a pure harmonic-oscillator potential, will separate into two distinct levels in a clipped harmonic-oscillator potential.

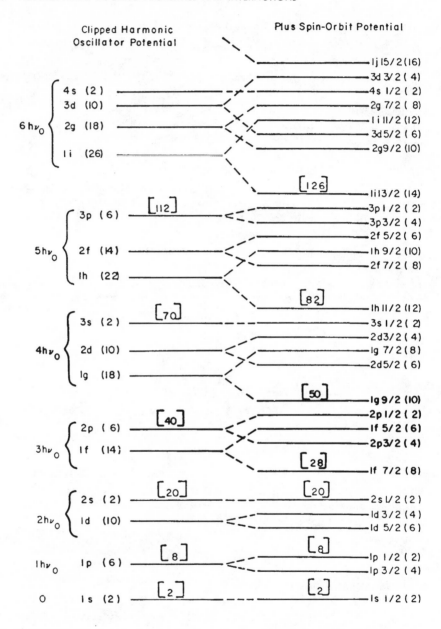

Figure 5.11—Qualitative diagram of effect of spin-orbit splitting on low-lying levels in clipped harmonic-oscillator potential. Numbers in parentheses are the number of nucleons in a level; numbers in brackets are the number of nucleons in states below a pronounced energy gap.

The level structure in a clipped harmonic-oscillator potential is shown schematically in the left side of Figure 5.11. Those groups of states which would have been degenerate in a pure oscillator potential but are separated in the clipped oscillator potential are $(1d, 2s)$, $(1f, 2p)$, $(1g, 2d, 3s)$, $(1h, 2f, 3p)$, and $(1i, 2g, 3d, 4s)$. The separation between levels within these groups is small relative to the distances between the groups; so large energy gaps occur above the $1s$, $1p$, $2s$, $2p$, $3s$, and $3p$ levels. The clipped harmonic-oscillator levels must now be filled with nucleons by use of the Pauli exclusion principle.

The Pauli exclusion principle for the nucleons in a nucleus may be stated in a number of completely equivalent forms:

1. An energy eigenstate characterized by the five quantum numbers n, l, j, m_j, and T_ζ can be occupied by not more than one nucleon.

2. A neutron (proton) energy eigenstate characterized by the four quantum numbers n, l, j, and m_j can be occupied by not more than one neutron (proton).

3. A neutron (proton) energy eigenstate characterized by the three quantum numbers n, l, and j can be occupied by no more than $2j + 1$ neutrons (protons).

4. A neutron (proton) energy eigenstate characterized by the two quantum numbers n and l can be occupied by not more than $2(2l + 1)$ neutrons (protons).

The Pauli principle in form 4 is used to determine the number of nucleons (meaning neutrons or protons) which fill each of the (n, l) oscillator levels. These numbers are shown in parentheses after the level designations on the left side of Figure 5.11. The important thing to notice is that large energy gaps occur at *occupancy numbers* 2, 8, 20, 40, 70, and 112, which are not the observed nuclear magic numbers.

When a spin-orbit potential proportional to $-\mathbf{s} \cdot \mathbf{L}$ is added to the clipped oscillator potential, each level with $l > 0$ splits into two levels. The state with $j = l + 1/2$ splits downward because the spin-orbit term increases the negativeness of the potential in a state in which $\mathbf{s}$ and $\mathbf{L}$ are parallel. Conversely, the level with $j = l - 1/2$ splits upward. The larger the value of l, the greater the splitting. Thus, for example, an f-state splits more than a d-state, which, in turn, splits more than a p-state. An s-state does not split at all. The Pauli principle is now invoked in form 3 to place $2j + 1$ nucleons into each of the sublevels of given j. The number of nucleons in each filled sublevel is enclosed in parentheses after the level designation on the right side of Figure 5.11.

As we see in Figure 5.11, the energy gaps with and without spin-orbit coupling occur at different occupancy numbers. The gaps with the lower occupancy numbers 2, 8, and 20 are not affected by the splitting. But the $1f_{7/2}$ level is lowered to such an extent that it is effectively isolated from its neighboring levels, giving rise to the magic number 28. The $1g_{9/2}$, $1h_{11/2}$, and $1i_{13/2}$ levels all split downward to such an extent that they intermingle with the next lower oscillator levels and give rise to the magic numbers 50, 82, and 126. Thus the magic numbers can be explained immediately on the basis of the spin-orbit interaction without any additional arbitrary assumptions.

The actual order of the levels within each shell will most likely differ somewhat from that shown schematically in Figure 5.11. Often, adjacent levels lie very close to one another; so the order that actually exists within any given nucleus depends on the exact form of the nuclear potential, usually unknown, but may be assumed to differ from nucleus to nucleus. Furthermore, the level structure for protons will differ somewhat from that for neutrons because of the additional Coulomb force on the protons. Despite these quantitative and detailed complexities, the gross shell structure exhibited in Figure 5.11 is unaffected. Large energy gaps between shells do occur at the observed nuclear magic numbers.

Triumphs and Shortcomings of the Shell Model

The shell model has been at least partially successful in systematizing and in explaining a vast variety of experimental nuclear data. Among the data it treats with varying degrees of success are (a) the existence of magic numbers and the ground-state properties of magic and near-magic nuclei, (b) the relative parity of nuclear levels as evidenced in β and γ decay, (c) the nuclear spins I of ground levels, (d) the relation between the magnetic dipole moments μ and the nuclear spins I of odd-N and odd-Z nuclides, summarized in Schmidt diagrams, and (e) the existence of electric quadrupole moments Q and their variation from nuclide to nuclide. A thorough discussion of these and other evidences of the basic validity of the shell model, which is beyond the scope of this book, may be found in the book by Mayer and Jensen listed in Selected References at the end of this chapter.

The nuclear shell model fails in the area that is perhaps of greatest interest to the nuclear engineer. It does not predict the correct number and distribution of excited nuclear energy levels, particularly at energies several million electron volts in excess of the ground level.

There are four fundamentally different types of energy levels in a nucleus: filled levels, partially filled levels, empty bound levels, and virtual levels. Filled levels are, as the name implies, energy levels filled with the maximum number of neutrons (or protons) allowed by the exclusion principle. A partially filled level contains less than the number of neutrons (or protons) allowed by the exclusion principle. Empty bound levels lie above the highest level that contains a neutron (or proton) but below the least excitation energy that allows separation of the neutron (or proton) from the nucleus. Virtual levels are those whose excitation energy exceeds the separation energy of a neutron (or proton) from the nucleus.

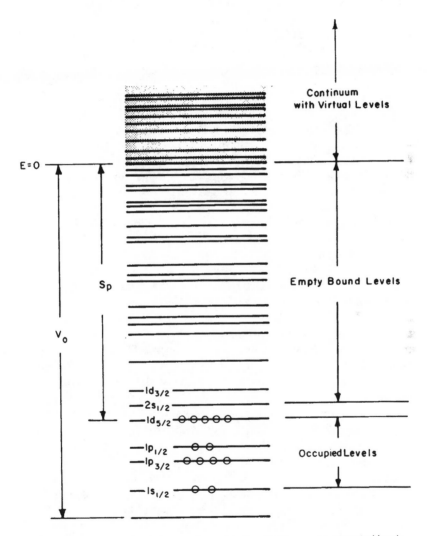

Figure 5.12—Proton energy levels in a nucleus with $Z = 13$. Shown are occupied levels, *empty bound levels, and the continuum with virtual levels.* (This figure is highly schematic; the actual number of levels is much greater.)

This distinction is exemplified in Figure 5.12, which is a schematic representation of the shell-model picture of the proton energy levels in an odd-even nucleus containing 13 protons, for example, $^{27}_{13}$Al. The neutron shells in $^{27}_{13}$Al have the same sequence but slightly different energies. All 14 neutrons pair and so contribute nothing to the angular momentum of the nucleus. The protons, indicated by open circles in Figure 5.12, fill the lowest three levels. The fourth level contains 5 protons, 4 of which pair to cancel their angular momenta. The fifth proton has a total angular-momentum quantum number $j = 5/2$.

The *ground state* of a nucleus is defined as the highest filled or partially filled energy level. Thus the ground state of the nucleus in Figure 5.12 is a $d_{5/2}$ state. Since $l = 2$, the parity of the ground state is even. Experimentally, it is found* that the ground state of the $^{27}_{13}$Al nucleus has spin and parity $5/2^+$ in agreement with the shell model.

The shell model has been quite successful in predicting the properties, which depend only on the configuration of the filled and partially filled levels. It has had partial success in predicting the number and properties of the low-lying bound excited states, those immediately above the ground level, but has had very little success in predicting the number and properties of the bound and virtual levels at high excitation energies.

The Collective Model

The number of bound levels detected experimentally is generally much greater than that predicted by the shell model. The additional bound levels and their properties are explained by the collective model of the nucleus.

The collective model, or unified model as it is sometimes called, is a synthesis of the liquid-drop and the shell models. The nucleus is pictured as a shell structure capable of vibrating and rotating as a whole. The single nucleon whose energy levels are to be determined is pictured as moving in a field that has only spheroidal symmetry, not spherical symmetry. The parameters describing the deformation of the nuclear field from a spherical field are quantized along with the single-particle variables. The result is that each of the single-particle (shell-model) energy levels is split into a series of vibration and rotation levels. This splitting of levels can be looked on as being due to surface waves and time-dependent distortions of the generally ellipsoidal surfaces of the various shells. A further discussion of the collective model may be found in P. E. Nemirovskii's *Contemporary Models of the Atomic Nucleus* (Reference 5 at the end of this chapter).

* C. M. Lederer, J. M. Hollander, and I. Perlman, *Table of Isotopes*, 6th ed., John Wiley & Sons, Inc., New York, 1967.

5.6 Models for Neutron–Nucleus Interactions

It is clear that the rigorous analysis of the interaction of a neutron with a system as complex as any but the lightest nucleus poses formidable difficulties. Even if the forces between individual nucleons were known exactly, as they presently are not, the theoretician would still be faced with a multibody problem whose solution could not be worked out in closed form classically, let alone quantum mechanically. Thus the theoretician has been forced to develop interaction models, simplified pictures of the neutron–nucleus interaction containing relatively few parameters. The ultimate validity of a model is measured by the extent to which experiment confirms its predictions. Ideally, a model should contain no freely adjustable parameters, no quantities that are not fixed a priori. Such ideal models are rare in the present state of nuclear physics; indeed, none of the models that we shall discuss is completely free from parameters that are adjusted to bring predictions into line with experimental data.

In 1935 Bethe, Fermi, and others began to use the then new toy quantum mechanics in attempts to explain the early crude observations on the interactions of neutrons with nuclei. Naturally they chose the simplest possible model to describe the interaction. They assumed that the potential between the neutron and the nucleus had the form of a spherical square well, i.e., that

$$V(r) = \begin{cases} -V_0 & (r < R), \\ 0 & (r > R), \end{cases}$$

where R is the radius of the nucleus.

This real potential well model, which contained two adjustable parameters, V_0 and R, predicted only two types of interactions, elastic scattering and radiative capture. The scattering has a much greater cross section because the neutron spends too little time within the nucleus to acquire an appreciable probability for radiative transition to a bound state. Resonances occur in theory, but they are widely spaced, broad, and dominated by scattering, not absorption. The real potential well model was soon proved inadequate. Experiments by Szilard, Fermi, and others in 1935 proved conclusively that neutron resonances at low energies are mainly radiative-capture resonances and are extremely narrow and closely spaced in direct disagreement with real potential well predictions.

In 1936 Niels Bohr proposed the compound-nucleus model to explain the observed resonances. Since measured capture resonances had widths only of the order of electron volts, Bohr pointed out that the uncertainty principle indicated the half-life of the nucleus after absorbing the neutron must be of the order of 10^{-16} sec. That is, since $\Delta E\, \Delta t \simeq \hbar \triangleq 10^{-15}$ ev-sec and the measured half-width ΔE is of the order of perhaps 10 ev, then Δt is of the order of 10^{-16} sec. But the velocity of, say, a 100-ev neutron is of the order of 10^7 cm/sec. It would cross the nucleus in about $(10^{-12}$ cm)/$(10^7$ cm/sec$) = 10^{-19}$ sec if it were not captured. Thus the compound nucleus exists for a time approximately 1000 times as long as a typical crossing time. This crude argument leads us to expect that the compound nucleus has "forgotten" how it was formed.

Bohr suggested that the interaction be considered to occur in two distinct steps, the formation of the compound nucleus and its subsequent decay. In the formation step, we assume that the total energy of the incoming neutron (kinetic energy plus separation energy) is immediately shared with all the other nucleons so that the identity of the incoming neutron is lost. Decay of the excited compound nucleus subsequently occurs through either photon emission, neutron emission, or other massive-particle emission, if kinematically allowed.

Bohr further simplified the compound-nucleus model through his famous independence hypothesis: *The decay of the compound nucleus is independent of its mode of formation.* Under this hypothesis the cross section for any interaction that proceeds through a compound nuclear state can be expressed in the form of the product

$$\sigma(E,b) = \sigma_c(E)\left(\frac{\Gamma_b}{\Gamma}\right)_c, \tag{5.54}$$

where $\sigma_c(E)$ is the cross section for formation of the compound nucleus by an incident neutron of energy E and $(\Gamma_b/\Gamma)_c$ is the probability that the compound nucleus will decay by emission of a particle of type b. The quantity $(\Gamma_b/\Gamma)_c$ is the branching ratio for the emission of particle b.

The Bohr independence hypothesis is based on the premise that the incoming neutron interacts so strongly with the nucleons it encounters that its mean free path Λ within the nucleus is much smaller than the nuclear radius. If we assume that the neutron–proton cross section is the same within the nucleus as outside, then we can readily estimate that $\Lambda \doteq 0.4 \times 10^{-13}$ cm, which is indeed much smaller than the nuclear radius. However, the assumption on which this estimate is based is doubtful. The nucleons within the nucleus are not free but can take up energy only by transitions to unoccupied energy levels. Hence the mean free path may be considerably larger than estimated above, and, in fact, incident neutrons may have an appreciable probability of passing through the nucleus without interaction.

The black-nucleus model was developed in an effort to estimate the cross section for compound-nucleus formation envisioned in the Bohr theory. Assume that the nucleus is a sphere of radius R and that all neutrons which are not reflected from the nuclear surface are absorbed to form a compound nucleus. Thus we estimate that

$$\sigma_c(E) = \pi(R + \lambda)^2 T, \tag{5.55}$$

where T is the transmission coefficient of the surface. That the position of the neutron is undefined within a wavelength λ is taken approximately into account by replacing the classical cross section πR^2 by $\pi(R + \lambda)^2$.

The transmission coefficient, or penetrability, T may be estimated to order of magnitude by considering a simple one-dimensional approximation to the three-dimensional problem. A neutron of kinetic energy E moves in a potential $V(x)$ given by

$$V(x) = \begin{cases} 0 & (x < 0), \\ -V_0 & (x > 0). \end{cases}$$

The neutron is partially reflected at $x = 0$. The wave functions are therefore given by

$$\psi_1(x) = Ae^{ikx} + Be^{-ikx}$$

for $x < 0$ and

$$\psi_2(x) = Ce^{iKx}$$

for $x > 0$, where the wave number outside the nucleus is

$$k = \frac{\sqrt{2mE}}{\hbar}$$

and inside the nucleus is

$$K = \frac{\sqrt{2m(E + V_0)}}{\hbar}.$$

The transmission coefficient, given by

$$T = \frac{|A|^2 - |B|^2}{|A|^2},$$

is obtained, as usual, by enforcing continuity of $\psi(x)$ and its derivative at $x = 0$. The result is

$$T = \frac{4kK}{(K + k)^2}; \tag{5.56}$$

so, finally,

$$\sigma_c(E) \doteq \pi(R + \lambda)^2 \frac{4kK}{(k + K)^2}. \tag{5.57}$$

The wave number inside the nucleus K depends on the choice of the nuclear potential depth V_0, one of the adjustable parameters in the theory. It has generally been found that a V_0 of the order of 40 Mev is optimum. Thus K is of the order of 10^{13} cm^{-1}.

Equation 5.57 was roughly derived and so can be expected to be little more than qualitatively correct. The cross section for compound-nucleus formation does approach πR^2 at high neutron energies as predicted by this equation. Furthermore, $\sigma_c(E)$ is indeed proportional to $1/E^{1/2}$ at very low energies, as predicted. This is the well-known empirical $1/v$ law of neutron absorption. Despite these minor successes, it must be emphasized that the black-nucleus model is only a very rough approximation. One major defect is that it predicts a monotonically varying $\sigma_c(E)$. Because it does not reproduce resonances, this model is sometimes referred to as continuum theory.

Since neither the crystal-clear nucleus, as envisioned in the real potential-well model, nor the black-nucleus model was entirely successful in predicting even the gross behavior of neutron cross sections, although each seemed to have some validity, it was perhaps only natural that an intermediate model, one which treated the nucleus as a gray sphere, be tried. The optical model was introduced by Bethe in 1940 but was not fully exploited until Feshbach, Porter, and Weisskopf applied it to neutron interactions in 1954. The optical model replaces the nucleus by a potential well that partially transmits and partially absorbs neutrons, hence its descriptive nickname, the cloudy crystal-ball model. Partial absorption and transmission is accomplished mathematically by describing the effect of the nucleus on the incident neutron by a potential well having real and imaginary parts: $V(r) = V_0(r) - iV_1(r)$.

The optical model has been remarkably successful in describing the behavior of certain cross sections, in particular, elastic-scattering differential cross sections, and in describing the gross behavior (averaged over individual resonances) of the total cross sections. It does not reproduce the shapes of individual resonances, but it does reproduce the wavelike average variation of the cross sections as functions of energy. In this respect, it is far superior to the old black-nucleus model. The optical model is described in greater detail in Chapter 7.

5.7 Classification of Neutron Cross Sections

The total neutron cross section of a nucleus, which is the sum of the individual cross sections for all interactions, can be separated into an *elastic-scattering cross section*, σ_n, and a *reaction cross section*, σ_r:

$$\sigma_T = \sigma_n + \sigma_r. \tag{5.58}$$

The elastic-scattering cross section is defined as the cross section for an interaction in which the total kinetic energy of the neutron–nucleus system is conserved. It is sometimes defined as the cross section for a scattering interaction that does not change the internal state of the nucleus. These two definitions are not rigorously the same since a neutron can exchange angular momentum with the target nucleus and still emerge from the compound nucleus with its original energy. We shall adopt the first definition because it corresponds to the elastic-scattering cross section that is measured experimentally and is, moreover, the elastic-scattering cross section of greatest interest to the nuclear engineer. Thus elastic scattering without further modification shall mean an interaction without energy transfer in the C-system.

The elastic cross section has been divided by theory into shape elastic and compound elastic cross sections,

$$\sigma_n = \sigma_{se} + \sigma_{ce}. \tag{5.59}$$

The shape elastic cross section is a measure of that part of the elastic scattering that occurs without formation of a compound nucleus; the compound elastic cross section is a measure of that part of the elastic scattering that takes place through the formation of a compound nucleus and the subsequent emission of a neutron by decay of the compound nucleus into the ground state of the original target nucleus. Shape elastic scattering is sometimes called *potential scattering;* compound elastic scattering is sometimes called *capture elastic scattering.*

The reaction cross section, σ_r, is the sum of the cross sections for all processes in which the product nucleus is either different from the target nucleus or is left in an internal energy state different from that of the target nucleus; this includes (n,n'), (n,γ), (n,p), (n,α), $(n,2n)$, and (n,f) reactions. The reaction cross section is sometimes called the *nonelastic cross section.* Most reaction processes, in the neutron energy range below 20 Mev at least, are assumed to proceed through formation and decay of a compound nucleus.

The cross section for compound-nucleus formation is the sum of the compound elastic-scattering cross section and the reaction cross section, both of which are assumed to proceed through the compound-nucleus intermediate state

$$\sigma_c = \sigma_{ce} + \sigma_r. \tag{5.60}$$

The total cross section can also be expressed as the sum of the shape elastic and compound-nucleus-formation cross sections,

$$\sigma_T = \sigma_{se} + \sigma_c. \tag{5.61}$$

Figure 5.13 should help to clarify the relations among the various cross sections.

There is no universally accepted notational scheme to distinguish between neutron cross sections for various interactions. The scheme used in this book is as common as any and perhaps more logical than most. The cross section for any interaction from which a particle emerges has as subscript the symbol for the particle. By particle we mean either a massive particle (n,p, or α) or a photon. Thus the following correspondence between interaction and cross-section symbol will be employed:

Interaction	*Cross-section symbol*
(n,n)	σ_n
(n,γ)	σ_γ
(n,n')	$\sigma_{n'}$
$(n,2n)$	σ_{2n}
(n,p)	σ_p
(n,d)	σ_d
(n,α)	σ_α
$(n,\text{fission})$	σ_f

In addition, special symbols are used for certain combined and partial cross sections.

Cross-section type	*Symbol*
Total	σ_T
Reaction	σ_r
Shape elastic	σ_{se}
Compound elastic	σ_{ce}
Compound-nucleus formation	σ_c

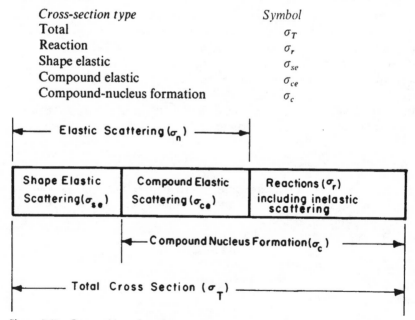

Figure 5.13—Composition of total neutron cross section.

Cross sections will sometimes be expressed as sums over partial-wave cross sections. Thus, for example,

$$\sigma_n = \sum_{l=0}^{\infty} \sigma_{n,l},$$

and

$$\sigma_r = \sum_{l=0}^{\infty} \sigma_{r,l}.$$

Arguments within parentheses will continue to be used to indicate differential cross sections. Thus, for example, $\sigma_n(\theta)$ is the differential elastic-scattering cross section measured, as always, in units of area per unit solid angle.

5.8 Conservation Laws in Nuclear Interactions

Certain physical and mathematical quantities have been found to be conserved in nuclear interactions. The conservation laws have somewhat different forms in relativistic and nonrelativistic mechanics. Since only interactions in the neutron energy range below 18 Mev will be considered, these conservation laws will be discussed in a nonrelativistic manner.

For nuclear interactions, the initial and final constellations of an isolated system are found to have the same (a) *total number of nucleons*, (b) *total electric charge*, (c) *total mass-energy*, (d) *total linear momentum*, (e) *total angular momentum*, (f) *total parity*, and (g) *total isospin*. The term "nuclear interaction" means an interaction in which nuclear forces are operative. Specifically excluded is β decay for which the responsible force is the "weak interaction" and in which parity and isospin are not necessarily conserved.

(a) *Total number of nucleons:* The number of nucleons in the final constellation equals the number of nucleons in the initial constellation,

$$\sum_i A_i = \sum_f A_f, \tag{5.62}$$

where the subscripts i and f refer to initial and final constellations, respectively.

(b) *Total electric charge:* The total electric charge in the final constellation equals the total electric charge in the initial constellation,

$$\sum_i Z_i = \sum_f Z_f. \tag{5.63}$$

It follows from the preceding two equations that the number of neutrons is also conserved,

$$\sum_i N_i = \sum_f N_f. \tag{5.64}$$

In beta decay the total number of nucleons and the total charge are conserved, but the total number of protons or neutrons is not.

(c) *Total mass–energy:* The sum of the kinetic energies and rest-mass energies Mc^2 of the particles in the initial and final constellations are equal,

$$\sum_i (Mc^2 + T)_i = \sum_f (Mc^2 + T)_f. \tag{5.65}$$

Note that the mass M includes any extra mass resulting from internal excitation energy. A nucleus having rest mass M_0 in its ground state has rest mass $M_0 + E^*/c^2$ in a state with excitation energy E^*.

(d) *Total linear momentum:* The components of the total linear momentum vector of the final constellation equal those of the initial constellation,

$$\sum_i \mathbf{P}_i = \sum_f \mathbf{P}_f. \tag{5.66}$$

In the C-system, the total linear momentum is always zero.

(e) *Total angular momentum:* The total angular momentum, intrinsic plus orbital, of a system of particles is conserved. Classically, all three components of the total angular-momentum vector $\mathbf{J}$ are conserved. Quantum mechanically, only the total magnitude of $\mathbf{J}$ and one component, say J_z, are conserved.

The operators J^2 and J_z have eigenvalues

$$J^2 = J(J + 1)\hbar^2 \qquad (J = \text{integer or half-integer}),$$

and

$$J_z = m_J \hbar \qquad (m_J = J, J - 1, \ldots, -J).$$

The quantum numbers J and m_J are always conserved in an interaction. The same is not true, in general, for the quantum numbers associated with the orbital angular-momentum operators L^2 and L_z and the spin operators s^2 and s_z. Total angular-momentum conservation may be formally expressed by the equations

$$J = \text{constant},$$

$$m_J = \text{constant}. \tag{5.67}$$

(f) *Total parity:* The total parity of the final constellation has been found to be equal to the total parity of the initial constellation in all nuclear interactions. Parity conservation for interactions whose initial and final constellations each consist of two bodies can be expressed by the equation

$$(-1)^l \Pi_1 \Pi_2 = (-1)^{l'} \Pi_3 \Pi_4, \tag{5.68}$$

where Π_1 and Π_2 are the intrinsic parities of the particle in the initial constellation, Π_3 and Π_4 are the intrinsic parities of the particles in the final constellation, and l and l' are the initial and final orbital angular-momentum quantum numbers. The value of Π_j is $+1$ for even parity and -1 for odd parity. If, for example, the product $\Pi_1 \Pi_2$ is even and the product $\Pi_3 \Pi_4$ is odd, the angular-momentum quantum numbers l and l' must differ in character; if l is even, l' must be odd and vice versa.

(g) *Total isospin:* The total isospin of the final constellation has been found to be equal to the total isospin of the initial constellation in all nuclear reactions,

$$\sum_i T_i = \sum_f T_f. \tag{5.69}$$

5.9 Mass–Energy Balance in Nuclear Reactions

A nuclear reaction with initial constellation $A + a$ and final constellation $B + b$ may be written formally as

$$A + a \rightarrow B + b + Q, \tag{5.70}$$

where Q represents the energy corresponding to the difference in atomic rest masses of the initial and final constellations.

$$Q = 931 \frac{\text{Mev}}{\text{amu}} [M(A) + M(a) - M(B) - M(b)]. \tag{5.71}$$

The atomic rest masses of charged particles, and not their nuclear masses, must be employed in the above equation to account for the masses of the orbital electrons. For example, if particle b is a proton, the mass of the atom ^{1}H must be used for $M(b)$; if b is an alpha particle, the mass of ^{4}He must be used.

If the Q value of the reaction is positive or zero, the reaction is exoergic and can proceed even if the particles in the initial constellation are at rest relative to one another. "Can proceed" is used rather than "will proceed" because it is necessary that a reaction be exoergic if it is to proceed at zero energy, but it is not sufficient. In subsequent chapters, several reactions that are exoergic but do not proceed because of other limitations will be examined. It will be found, for example, that charged-particle-out reactions in some nuclei are exoergic but, nevertheless, exhibit a thresholdlike behavior because of the existence of a Coulomb force between the outgoing particle and the residual nucleus.

If the Q of a reaction is negative, the reaction is endoergic and cannot proceed unless the available kinetic energy in the initial constellation exceeds the absolute value of Q. If nucleus A is at rest in the laboratory and particle a has kinetic energy $\tilde{E}$ in the laboratory, the available energy in the C-system is given by

$$E = \frac{M(A)}{M(A) + M(a)} \tilde{E}, \tag{5.72}$$

where $M(A)$ and $M(a)$ are the masses of A and a, respectively. The *threshold energy* of a reaction is defined as the least laboratory kinetic energy of the projectile particle at which the reaction can possibly proceed. Since the C-system energy E must equal $-Q$ at threshold, the threshold energy $\tilde{E}_t$ is given by

$$\tilde{E}_t = \frac{M(A) + M(a)}{M(A)}(-Q), \tag{5.73}$$

where $Q < 0$. For a neutron-induced reaction of the form

$$A + n \rightarrow B + b + Q,$$

the Q value is given by

$$Q = 931 \frac{\text{Mev}}{\text{amu}} [M(A) + m_n - M(B) - M(b)]; \tag{5.74}$$

and the threshold of an endoergic-neutron-induced reaction is given by

$$\tilde{E}_t = -\frac{A + 1}{A} Q \tag{5.75}$$

for $Q < 0$, where A is the rest mass of particle A expressed in neutron mass units, that is, where $A \equiv M(A)/m_n$.

5.10 The Reciprocity Relation

Denote the cross section for the reaction

$$A + a + E_1 \rightarrow B + b + E_2 \tag{5.76}$$

by $\sigma_{ab}(E_1)$. The quantities E_1 and E_2 are the total C-system kinetic energies of the initial and final constellations, respectively; they are not independent but are joined by mass–energy conservation since

$$E_2 = E_1 + Q, \tag{5.77}$$

where $Q = 931$ Mev/amu $[M(A) + M(a) - M(B) - M(b)]$. Denote the cross section for the inverse reaction

$$B + b + E_2 \rightarrow A + a + E_1 \tag{5.78}$$

by $\sigma_{ba}(E_2)$. We will now show, on the basis of a very general thermo-dynamic argument, that, if the cross sections are independent of the relative orientations of the spin vectors, then the cross sections are related by the reciprocity relation:

$$g_{Aa}p^2_{Aa}\sigma_{ab}(E_1) = g_{Bb}p^2_{Bb}\sigma_{ba}(E_2), \tag{5.79}$$

where g_{Aa} and g_{Bb} are the statistical weights of the $A + a$ and $B + b$ constellations, i.e., where

$$g_{Aa} = (2I_A + 1)(2I_a + 1), \tag{5.80}$$

$$g_{Bb} = (2I_B + 1)(2I_b + 1), \tag{5.81}$$

and where p_{Aa} and p_{Bb} are the total C-system momenta of the $A + a$ and $B + b$ constellations, respectively. These momenta are given by $p_{Aa} = (2m_1E_1)^{1/2}$ and $p_{Bb} = (2m_2E_2)^{1/2}$, where m_1 and m_2 are the reduced masses of the initial and final constellations, respectively. If one of the particles is a photon, its statistical weight is 2, corresponding to its two possible directions of polarization, and its momentum is E/c.

Consider a box containing large numbers of particles A, a, B, and b. At thermal equilibrium transitions occur in both directions ($A + a \leftrightarrows B + b$) at equal rates; therefore the number of transitions in energy interval dE about E_1 or E_2 is

$$v_{Aa}\sigma_{ab}(E_1)[A][a]\,dE = v_{Bb}\sigma_{ba}(E_2)[B][b]\,dE, \tag{5.82}$$

where the brackets denote number densities and the velocities v_{Aa} and v_{Bb} are the relative velocities of the particles in the constellations $A + a$ and $B + b$ corresponding to energies E_1 and E_2, respectively. The ratio of cross sections is therefore given by

$$\frac{\sigma_{ab}(E_1)}{\sigma_{ba}(E_2)} = \frac{[B][b]}{[A][a]} \frac{v_{Bb}}{v_{Aa}}. \tag{5.83}$$

The number density ratio, or what chemists call the equilibrium constant, is given by

$$K = \frac{[B][b]}{[A][a]} \tag{5.84}$$

and can be evaluated as follows:

Consider a spinless particle bound in a very large (practically infinite) box of volume Ω, which for simplicity will be assumed to be in the form of a cube of side L. The potential inside the box will be taken to be zero; the potential at the walls and everywhere outside is infinite. Schrödinger's equation for the particle within the box is

$$-\left(\frac{d^2}{dx^2} + \frac{d^2}{dy^2} + \frac{d^2}{dz^2}\right)\psi(x,y,z) = \frac{2mE}{\hbar^2}\psi(x,y,z)$$

$$= \frac{p_x^2 + p_y^2 + p_z^2}{\hbar^2}\psi(x,y,z),$$

and the boundary conditions are $\psi(x,y,z) = 0$ at $(x,y,z) = 0$ and $(x,y,z) = L$. Separating variables, we find that the following eigenfunctions satisfy the boundary conditions:

$$\psi(x,y,z) = \sin\left(\frac{p_x}{\hbar}x\right)\sin\left(\frac{p_y}{\hbar}y\right)\sin\left(\frac{p_z}{\hbar}z\right),$$

where $p_x/\hbar$ is restricted to the values $n_x\pi/L$ with n_x an integer and, similarly, where $p_y/\hbar$ and $p_z/\hbar$ are restricted to $n_y\pi/L$ and $n_z\pi/L$, respectively. The number of states, $N(p)$, of momentum less than p equals the number of combinations of p_x, p_y, and p_z such that $p_x^2 + p_y^2 + p_z^2 < p^2$ or, in terms of the quantum numbers n_x, n_y, and n_z, such that

$$n_x^2 + n_y^2 + n_z^2 < \frac{p^2L^2}{\pi^2\hbar^2}.$$

The number of sets (n_x,n_y,n_z) satisfying the above condition equals one-eighth the number of points in a cubical lattice enclosed by the sphere of radius $pL/\pi\hbar$. Thus

$$N(p) = \frac{1}{8}\frac{4\pi}{3}\left(\frac{pL}{\pi\hbar}\right)^3. \tag{5.85}$$

The factor $1/8$ comes from restricting n_x, n_y, and n_z to be greater than zero since negative values give no new independent values of p. Only the positive-value octant of the x-y-z coordinate system is considered. The number of states having momentum between p and $p + dp$, which will be called $n(p)\,dp$, is the differential of Equation 5.85 with respect to p,

$$n(p)\,dp = \frac{\Omega}{2\pi^2\hbar^3}p^2\,dp. \tag{5.86}$$

If the particle has spin I, the number of states is multiplied by the possible number of orientations of the spin vector, namely, $2I + 1$,

$$n(p)\,dp = (2I + 1)\frac{\Omega}{2\pi^2\hbar^3}p^2\,dp. \tag{5.87}$$

Now let us return to the proof of the reciprocity relation between cross sections. At thermal equilibrium the number of reaction pairs ($A + a$ or $B + b$) per momentum range dp about p is, according to the above equation, proportional to $gp^2\, dp$. Thus in the energy interval dE the number of pairs is proportional to $(gp^2/v)\, dE$. Hence the equilibrium constant is given by

$$\frac{[B][b]}{[A][a]} = \frac{g_{Bb}p_{Bb}^2}{g_{Aa}p_{Aa}^2}\frac{v_{Aa}}{v_{Bb}}. \tag{5.88}$$

Substituting this expression into Equation 5.83, we immediately have the reciprocity relation (Equation 5.79). Since kinematically the angular distributions of the reaction products in the C-system are identical for the direct and inverse reactions, the reciprocity relation is equally valid for differential cross sections and total cross sections.

The reciprocity relation has many applications, as will become apparent in subsequent chapters. Generally, it is used to infer an unknown cross section from its inverse known one. Thus, for example, the cross section for the photodisintegration of the deuteron in the reaction $^2H(\gamma,n)^1H$ can be inferred from the cross section for the capture of neutrons by protons in the reciprocal reaction $^1H(n,\gamma)^2H$.

The reciprocity relation, derived as it is on the basis of very general thermodynamic and quantum-mechanical arguments, is presumed to be exact. No exception to it has been experimentally discovered nor are any exceptions expected to be found.

SELECTED REFERENCES

1. Robley D. Evans, *The Atomic Nucleus*, McGraw-Hill Book Company, Inc., New York, 1955. Chapters 1 to 4 contain thorough discussions of the experimental evidence for nuclear properties such as charge, mass, radius, angular momentum, and electric and magnetic moments. Chapter 11 is an excellent introductory treatise on nuclear models.

2. John M. Blatt and Victor F. Weisskopf, *Theoretical Nuclear Physics*, John Wiley & Sons, Inc., New York, 1952. This has been the standard reference book in nuclear physics since its publication in 1952. The student is urged to read Chapter 1, "General Properties of the Nucleus," to reinforce the presentation in this chapter.

3. Sergio DeBenedetti, *Nuclear Interactions*, John Wiley & Sons, Inc., New York, 1964. A thorough and lucid account of nuclear interaction theory and basic experiments is found in this book written for the student who has had one or two courses in quantum mechanics. Chapter 2, "Nuclear Models," contains a particularly valuable account of the isobaric properties of light nuclei.

4. Maria Goeppert Mayer and J. Hans D. Jensen, *Elementary Theory of Nuclear Shell Structure*, John Wiley & Sons, Inc., New York, 1955. The experimental evidence for and theoretical development of the nuclear shell model are clearly explained by the two persons most responsible for the development of this model.

5. P. E. Nemirovskii, *Contemporary Models of the Atomic Nucleus*, The Macmillan Company, New York, 1963. A relatively brief, easily read account of the properties of atomic nuclei and the models built to explain these properties. Particularly valuable for its discussion of the unified or collective nuclear model.

EXERCISES

1. Using a table of atomic masses, (a) calculate the neutron and proton separation energies for the nuclides, $^{15}_{7}N$, $^{16}_{8}O$, $^{17}_{8}O$, and $^{18}_{9}F$ and (b) explain the results qualitatively on the basis of the shell model of the nucleus.

2. Derive the expression $Q_c = (q/5)(\alpha^2 - \beta^2)$ for the z component of the classical quadrupole moment of an ellipsoidal charge distribution.

3. Calculate the order of magnitude of the energy spacing between the two levels of an electronic spin doublet using Equation 5.51. If the nuclear spin-orbit force were given by the same equation as the electronic spin-orbit force,

$$V_{so}(r) = \frac{e}{2m^2c^2} \frac{1}{r} \frac{dV}{dr} \mathbf{L \cdot s},$$

with V and m interpreted as the nuclear potential and nucleon mass, what would be the order of magnitude of the spacing between a nuclear spin doublet?

4. Explain the following table of experimentally determined ground-state spins and parities on the basis of the shell model of the nucleus. The ground state of $^{11}_{5}B$ is the $1p^-_{3/2}$ state. What is the ground state of each of the other nuclides?

Nuclide	Measured spin	Measured parity
$^{11}_{5}B$	3/2	−
$^{13}_{6}C$	1/2	−
$^{17}_{8}O$	5/2	+
$^{39}_{19}K$	3/2	+
$^{207}_{82}Pb$	1/2	−
$^{209}_{83}Bi$	9/2	−

5. Determine the Q value and the threshold energy for each of the following neutron-induced reactions. Which are endoergic and which exoergic?

(a) $^{16}O(n,p)^{16}N$,
(b) $^{23}Na(n,p)^{23}Ne$,
(c) $^{23}Na(n,\alpha)^{20}F$,
(d) $^{23}Na(n,\gamma)^{24}Na$,

(e) $^{41}K(n,p)^{41}A$,
(f) $^{41}K(n,\alpha)^{38}Cl$,
(g) $^{87}Rb(n,p)^{87}Kr$,
(h) $^{85}Rb(n,p)^{85}Kr$.

6. Assume that the cross section for the reaction $B(\alpha,n)A$ is known over the α-particle laboratory energy range from $\tilde{E}_i$ to $\tilde{E}_f$. Exactly what can be inferred from the reciprocity relation about the cross section for the reaction $A(n,\alpha)B$, which you may assume has a negative Q value? Derive the equation by which you would infer $\sigma_{n\alpha}(\tilde{E}_1)$ from the known $\sigma_{\alpha n}(\tilde{\tilde{E}}_2)$.

7. Show that the excitation energy E^* of a compound nucleus is given in general by

$$E^* = \frac{M}{M + m} \tilde{E}_0 + B_i,$$

where m and M are the masses of the incident particle and target nucleus, respectively, $\tilde{E}_0$ is the laboratory kinetic energy of the incident particle, and B_i is the separation energy of the incident particle from the compound nucleus.

8. The highest thermal-neutron flux ever reported in a nuclear reactor is less than 10^{16} neutrons/cm²/sec. (a) Estimate the order of magnitude of the density of neutron–neutron collisions in this flux. (b) Show that neutron–neutron collisions are extremely rare compared with neutron–nucleus collisions in any nuclear reactor.

6 THE GENERAL THEORY OF NEUTRON–NUCLEUS INTERACTIONS

In examining a compilation of measured cross sections such as that in the "Barn Book,"[*] one is forcibly struck by the presence of sharp maxima, called *resonances*, in the measured cross sections. As will become increasingly evident, the formulation of the general theory describing nuclear interactions depends very strongly on the resonances in the measured cross sections. In light nuclei ($A < 25$), the resonances begin, on the average, to occur in the million-electron-volt region and are separated by gaps of the order of 1 Mev. In medium-weight nuclei ($25 \leq A \leq 70$), the resonances begin in the kilo-electron-volt region and are separated by energies of the order of 1 kev. In heavy nuclei ($A > 70$), the resonances begin in the electron-volt region and are separated by energies of the order of electron volts.

As the neutron energy is increased into the Mev region, the resonances in medium and heavy nuclei seem to disappear. An examination of the resolution triangles, whose information content is a vital part of any such data plot, shows that at least part of this leveling off of the cross section is due to lack of experimental energy resolution. With increased resolving power, we would find that the apparently smooth cross-section curve is actually made up of resonances more closely spaced than at the lower energies but still distinct. However, as the neutron energy is increased still further, we would expect eventually to reach an energy at which the resonances could not be resolved, even with infinite resolving power, simply because the resonance spacings have become smaller than the

[*] D. J. Hughes and R. B. Schwartz, *Neutron Cross Sections*, USAEC Report BNL-325, 2nd ed., Brookhaven National Laboratory, 1958.

resonance widths causing the resonances to overlap and merge into a continuum.

The ultimate goal of this chapter is to derive the equations that describe neutron elastic-scattering and reaction cross sections as functions of energy and angle. In the first section, the numbers and distributions of energy levels at high excitation energies are discussed, and the significance of the widths of energy levels is explored. Next, the partial-wave formalism for spinless particles is generalized to include the possibility of reactions. The isolated resonance cross-section equations that apply to spinless particles are then derived as a special case.

After spin, spin wave functions, and spin operators have been introduced, the partial-wave analysis is further generalized to include the spins of the interacting particles. The isolated resonance cross sections are again derived as a special case.

In Section 6.10, the highest generality is achieved with the introduction of the U-matrix treatment of nuclear interactions. In the next, and final, section of the chapter, the R-matrix theory is introduced, and how the unknown elements of the U matrix may be expressed in terms of the elements of the R matrix is shown. Hopefully, it will have become apparent to the reader by the end of the section that all nuclear interaction theory is intimately associated with the existence of discrete eigenstates within the nucleus.

6.1 Nuclear Energy Levels at High Excitation Energies

The existence of neutron resonances can be explained by considering the virtual energy levels of the compound nucleus. Note: It is the energy levels of the compound nucleus, not of the target nucleus, that are involved. The energy eigenvalues of a neutron in a potential determined by A other nucleons are the neutron energy levels in the nucleus with $A + 1$ nucleons. The situation is depicted schematically in Figure 6.1, which shows the neutron energy levels in the nucleus Z^{A+1}. Two distinct energy scales are employed: The excitation energy E^* of the nucleus Z^{A+1} has its origin at the ground state of the compound nucleus. The neutron's available kinetic energy E has its origin at an excitation energy equal to the separation energy S_n^c of a neutron from the compound nucleus.

Energy levels below $E^* = S_n^c$ are discrete and are indicated by lines (Figure 6.1). Above S_n^c, the energy eigenvalues actually are continuous; however, there are certain of these energies at which the wave function inside the nucleus is much larger than at nearby energies. These *virtual levels* are indicated by lines superimposed on the background continuum.

Only the levels above the separation energy S_n^c can be excited by a neutron incident on a target nucleus Z^A. A virtual level at excitation energy $E^* = S_n^c + E$ will manifest itself in a resonance at neutron energy E. Thus neutron resonances provide us with direct information only about those levels in the compound nucleus that are above S_n^c.

In addition to its energy, each level in the compound nucleus is characterized by a total angular-momentum quantum number J and a parity Π. The total angular-momentum quantum number, commonly called the spin quantum number, takes on either integral or half-integral values, as explained in Chapter 5. The parity is either even ($\Pi = +1$) or odd ($\Pi = -1$).

Level Widths and Lifetimes

To avoid clutter, we show the energy levels in Figure 6.1 as lines. Actually all the levels of a nucleus, with the possible exception of the ground state,

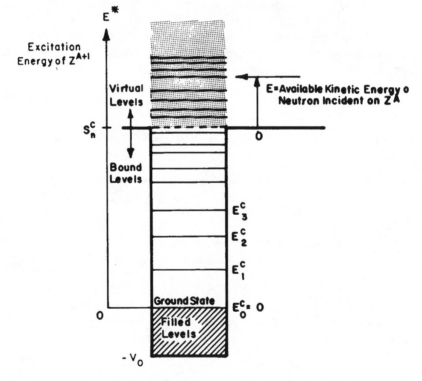

Figure 6.1—Neutron energy levels in the nucleus Z^{A+1} showing the filled, bound, and virtual levels and the two energy scales E^* and E. (This figure is highly schematic; the actual number of levels in most nuclei is much greater than that shown.)

have finite widths because they have finite probabilities of decay. The ground state E_0^c has no width if the nucleus Z^{A+1} is stable against radioactive decay; if the nucleus is unstable, the ground state also has finite width.

All the excited states, both bound and virtual, may decay by gamma emission. In addition, depending on the energetics of the reactions, it may be possible for virtual states and some of the higher bound states to decay by proton or alpha emission or, in some few heavy nuclei, by fission. The virtual states can, of course, always decay by neutron emission, i.e., compound elastic scattering. At sufficiently high energies, inelastic scattering is also allowed. Thus, associated with each of the excited states are one or more *decay probabilities* per unit time, $\Lambda_y(E^*)$, $\Lambda_n(E^*)$, $\Lambda_{n'}(E^*)$, $\Lambda_p(E^*)$, $\Lambda_\alpha(E^*)$, or $\Lambda_f(E^*)$. The reciprocal of a decay probability $\tau_j = 1/\Lambda_j$ is called the *mean life* of the state with respect to decay by mode j.

It is customary to express each decay probability in energy units, multiplying it by $\hbar$ to get

$$\Gamma_j = \hbar\Lambda_j = \frac{\hbar}{\tau_j}. \tag{6.1}$$

The quantity Γ_j is the *partial width* of the level toward decay by mode j. Note the correspondence between Equation 6.1 when written in the form $\Gamma_j\tau_j = \hbar$ and the Heisenberg uncertainty relationship for energy and time, $\Delta E \, \Delta t \simeq \hbar$. Because the state has a finite lifetime, the uncertainty in its time of decay is not infinite, and, therefore, the uncertainty in its energy must be finite.

Since a particular level may be able to decay by several modes, the total decay probability of an energy level is the sum of its partial decay probabilities,

$$\Lambda = \sum_j \Lambda_j$$

or

$$\Gamma = \sum_j \Gamma_j. \tag{6.2}$$

For bound states the *total width* Γ is the sum of Γ_y plus, possibly, Γ_p, Γ_α, and, in the fissile nuclides, Γ_f. For virtual levels Γ is the sum $\Gamma_n + \Gamma_y$ plus, possibly, $\Gamma_{n'}$, Γ_p, Γ_α, and, in the fissionable nuclides, Γ_f.

The mean life of a level is given by $\hbar/\Gamma$ or, numerically, by

$$\tau(\text{sec}) = \frac{6.6 \times 10^{-16}}{\Gamma(\text{ev})}. \tag{6.3}$$

We can see that a level with a width of the order of electron volts will have a mean life of the order of 10^{-16} sec; a level with a width of the order of a million electron volts will have a mean life of some 10^{-22} sec. Although times as short as these are not directly measurable at present, they are nevertheless believed to obtain, because there is no reason to doubt the universal validity of the uncertainty relationship between energy and time.

The width of an energy level is a quantum-mechanical concept that follows directly from the uncertainty relationship between energy and time. It is easy to show how the wave function describing a state may be modified to include the probability of decay of that state. Consider a situation in which a large number, N_0, of nuclei are created at time $t = 0$ in identical excited states with excitation energy E^*. At any subsequent time t, the number, $N(t)$, remaining in the excited state will be given by the well-known exponential decay law

$$N(t) = N_0 e^{-\Lambda t},$$

where Λ is the total decay probability per unit time. The quantum-mechanical statement of the exponential decay law is

$$|\Psi|^2 = |\Psi_0|^2 e^{-\Lambda t}.$$

The stationary wave Ψ_0 has the form

$$\Psi_0(\mathbf{r},t) = \psi_0(\mathbf{r})e^{-iE^* t/\hbar},$$

and, therefore, Ψ has the form

$$\Psi(\mathbf{r},t) = \psi_0(\mathbf{r})e^{-iE^* t/\hbar} e^{-\Lambda t/2}$$

$$= \psi_0(\mathbf{r})e^{-i(E^* - i\Lambda\hbar/2)t/\hbar},$$

which becomes, after we substitute $\Lambda\hbar = \Gamma$,

$$\Psi(\mathbf{r},t) = \psi_0(\mathbf{r})e^{-i(E^* - i\Gamma/2)t/\hbar}$$

Thus we can see that the wave function describing a nonstationary state is different from that for a stationary state in that the energy E^* is replaced by the complex energy $[E^* - (i\Gamma/2)]$. The square of the wave function for an unstable state decays exponentially; therefore, as time goes on, it becomes less and less likely that the nucleus described by this decaying wave function will still be in its initial state.

Distribution of Energy Levels

The density of energy levels (expressed as the number of levels per unit energy) at excitation energy E^* is given approximately by the semiempirical equation*

$$\rho(E^*) = C \exp(2\sqrt{aE^*}) \tag{6.4}$$

where the constants C and a are adjusted to agree with experimentally determined level densities.

Values of a and C that provide approximate agreement with experiment for odd-A nuclides are listed in Table 6.1. Level densities in odd-odd nuclides have been found to be about twice those in odd-A nuclides, and densities in even-even nuclides are about one-fifth those in odd-A nuclides. Thus, approximately,

$$C(\text{odd-odd}) = 2C(\text{odd-}A)$$

$$C(\text{even-even}) = \tfrac{1}{5}C(\text{odd-}A)$$

Table 6.1—Level Density Constants and Average Level Spacings

	Level density constants for odd-A nuclides, Mev^{-1}		Level spacings $[D(E^*)]$ in odd-A nuclides	
A	a	C	$E^* = 8$ Mev	$E^* = 18$ Mev
27	0.45	0.5	50 kev	7 kev
63	2	0.3	1 kev	20 ev
115	8	0.02	5 ev	2×10^{-3} ev
181	10	0.01	2 ev	2×10^{-4} ev
231	12	0.005	0.6 ev	3×10^{-5} ev

The empirical constant a is independent of the character of the mass number A. Also listed in Table 6.1 are the average distances between levels at an excitation energy of 8 Mev (corresponding roughly to the neutron separation energy, and, hence, to a neutron with zero kinetic energy) and at an excitation energy of 18 Mev (corresponding roughly to a neutron kinetic energy of 10 Mev). The average distance between levels, $D(E^*)$, is just the inverse of $\rho(E^*)$,

$$D(E^*) = \frac{1}{\rho(E^*)}. \tag{6.5}$$

A form of Equation 6.5 for the level density was derived by Hans Bethe in 1937.† In his derivation, Bethe introduced the first statistical

*See J. M. Blatt and V. F. Weisskopf, *Theoretical Nuclear Physics* (Reference 1 at the end of this chapter), pp. 365–372, for the derivation of this equation.
†H. A. Bethe, *Rev. Mod. Phys.*, **9**: 69 (1937).

model of the nucleus. The nucleus was pictured as a gas of nucleons confined in a box. The distribution of possible energies of these nucleons was determined by the methods of quantum statistical mechanics. The nucleus was assumed to be characterized by a temperature θ. The probability of occupation of a state of excitation energy E^* is proportional to $\exp(-E^*/\theta)$, where θ is expressed in energy units, i.e., where θ is the actual temperature times Boltzmann's constant. Bethe showed that the dependence of the level density ρ on excitation energy would be

$$\rho(E^*) \propto \frac{1}{(E^*)^2} \exp(2\sqrt{aE^*}) \tag{6.6}$$

where the constant a is related to the nuclear temperature through the equation $E^* = a\theta^2$. The nuclear temperature and, hence, the constant a are treated as parameters to be determined experimentally. The proportionality constant that converts expression 6.6 to an equation is likewise treated as an empirical parameter.

On the basis of similar statistical methods, the density of levels with given total angular-momentum quantum number J has been found to be given by

$$\rho(E^*,J) \propto (2J + 1)\exp\left[-\frac{J(J + 1)}{2\sigma^2}\right]\rho(E^*), \tag{6.7}$$

where σ^2 is an energy-dependent parameter that is rather large compared to unity at low E^* and that decreases slowly with energy. Empirical values of $2\sigma^2$ are of the order of 5 to 10 for excitation energies in the 8- to 14-Mev range. Thus, for small values of J, the spin dependence of the level density of the virtual levels goes approximately as

$$\rho(E^*,J) \propto (2J + 1)\rho(E^*). \tag{6.8}$$

The J values of the energy levels in a given nucleus are either all integral or all half-integral. If they are integral, Equation 6.8 tells us, for example, that there will be, on the average, about three times as many $J = 1$ as $J = 0$ levels. If the levels have half-integral J, then we can expect about twice as many $J = 3/2$ as $J = 1/2$ levels. The table of resonance parameters of light nuclei in the "Barn Book" confirms these expectations, at least approximately. Very large J values, on the other hand, are expected to be relatively rare. The decreasing exponential term in Equation 6.7 begins to be significant compared to $2J + 1$ for J values of the order of 3 or 4.

The Wigner Distribution of Level Spacings

The distribution of energy gaps between adjacent levels has been calculated exactly on the basis of the assumption that the resonance energies of the compound nucleus are randomly distributed.* The exact expression, which is quite complicated, can be closely approximated by a relatively simple expression previously derived by E. P. Wigner. In Wigner's approximation, the probability $P(D) \, dD$ that the spacing between adjacent levels lies in the range dD about D is given by

$$P(D) \, dD = \frac{\pi D}{2\langle D \rangle^2} \exp\left(-\frac{\pi D}{4\langle D \rangle^2}\right) dD, \tag{6.9}$$

where $\langle D \rangle$ is the average spacing between levels. Wigner's distribution has been found to be in reasonably close agreement with the observed spacings between resonances in low-energy neutron–nucleus cross sections.

The Porter–Thomas Distribution of Level Widths

Starting with the assumption that the wave functions describing the various compound nuclear states have random values and signs at the nuclear surface, Porter and Thomas found that the distribution of level widths for each type of interaction follows what statisticians call a chi-square distribution.† Specifically, $P(\Gamma_j) \, d\Gamma_j$, the probability that the width for the jth type interaction will fall in the range $d\Gamma_j$ about Γ_j is given by

$$P(\Gamma_j) \, d\Gamma_j = \frac{n_j}{2G(n_j/2)} (n_j x/2)^{(n_j/2)-1} \exp\left(-n_j x/2\right) dx, \tag{6.10}$$

where $x \equiv \Gamma_j/\langle \Gamma_j \rangle$ and G is the mathematical gamma function having the properties $G(y) = (y - 1)G(y - 1)$ and $G(1/2) = \pi^{1/2}$. The quantity n_j is an integer, called "the number of degrees of freedom," which reflects the number of distinct ways the compound nucleus can decay when emitting a particle of type j.

At low neutron energies the only channel open to neutron emission by the compound nucleus is the entrance channel. Hence there is but one degree of freedom, and as expected, the observed widths for neutron emission have a Porter–Thomas distribution with $n_j = n_n = 1$. The distribution of (n,γ) resonance widths is distinctly different. Since the compound nucleus can decay by gamma emission to any one of a large number of different final states, there are many degrees of freedoms, and the observed distribution of widths follows a Porter–Thomas distribution with $n_j = n_\gamma \gg 1$. In heavy nuclei n_γ often exceeds 50. The fission widths have

---

*M. L. Mehta and M. Gaudin, *Nucl. Phys.*, **18**: 420 (1960).

†C. E. Porter and R. G. Thomas, *Phys. Rev.*, **104**: 483 (1956).

been observed to follow a Porter–Thomas distribution with a small number (2 to 3) of degrees of freedom, indicating that there are only a few channels open to fission.

The mean square deviation of $\Gamma_j/\langle\Gamma_j\rangle$ from unity is $2/n_j$. Consequently, if the number of degrees of freedom is small, as it is in neutron and fission decay, the variation of the corresponding level width from resonance to resonance is great. But if the number of degrees of freedom is large, as in gamma decay, the corresponding level width is practically constant from resonance to resonance.

6.2 Partial-Wave Analysis of Interactions Between Spinless Particles

In Chapter 4 the method of partial waves was developed and applied to the problem of determining the elastic-scattering cross section. Processes that absorb neutrons were not considered in the analysis. Because neutron–nucleus reactions do in fact occur, it will be necessary to add some additional complexity to the quantum-mechanical formalism thus far developed. We seek expressions for the reaction cross section σ_r and the elastic-scattering cross section $\sigma_n(\theta)$ in the presence of processes that absorb neutrons. For the present, we will assume that the neutron and the nucleus are spinless. In a later section, the formalism will be generalized to include the spin.

In Section 4.7, the following expression for the asymptotic wave function for the case of elastic scattering of spinless particles was derived:

$$\psi(r,\theta) \sim \sum_{l=0}^{\infty} \frac{2l+1}{2kr} i^{l+1} \left\{ \exp\left[-i\left(kr - \frac{l\pi}{2} \right) \right] \right.$$
$$\left. - \exp\left(2i\delta_l \right) \exp\left[i\left(kr - \frac{l\pi}{2} \right) \right] \right\} P_l(\cos\theta). \tag{6.11}$$

In pure elastic scattering the *reflection factors* $\exp\left(2i\delta_l\right)$ must have modulus unity since they represent the relative amplitudes of the outgoing and incoming waves. Thus δ_l, the phase shift of the lth partial wave, must be real in pure elastic scattering.

Now the presence of reactions, such as (n,γ) and (n,p), that remove neutrons from the outgoing wave can be formally represented by allowing the phase shifts to be complex numbers with moduli less than unity. If the l-wave phase shift is written in the form $\delta_l = \alpha_l + i\beta_l$, where α_l and β_l are real numbers, then the l-wave reflection factor, which is given by

$$U_l \equiv e^{2i\delta_l} = e^{-2\beta_l} e^{2i\alpha_l}, \tag{6.12}$$

has *modulus* $e^{-2\beta_l}$ and *phase factor* $e^{2i\alpha_l}$. The modulus will be less than or equal to unity provided β_l is not negative. Hence a complex phase shift with a positive imaginary part describes an outgoing wave that has a smaller amplitude than the incoming wave, corresponding to part of the incoming flux being absorbed by the nucleus.

The partial-wave expression for the differential elastic-scattering cross section (Equation 4.120),

$$\sigma_n(\theta) = \lambda^2 \left| \sum_{l=0}^{\infty} (2l + 1)e^{i\delta_l}(\sin \delta_l)P_l(\cos \theta) \right|^2$$

can be expressed in terms of the reflection factors by writing $\sin \delta_l = (e^{i\delta_l} - e^{-i\delta_l})/2i$. Thus

Differential elastic: $$\sigma_n(\theta) = \frac{\lambda^2}{4} \left| \sum_{l=0}^{\infty} (2l + 1)P_l(\cos \theta)(1 - U_l) \right|^2. \qquad (6.13)$$

When this differential elastic-scattering cross section is integrated over all solid angles,

$$\sigma_n = \frac{\lambda^2}{4} \int_0^{\pi} \sum_l \sum_{l'} (2l + 1)(2l' + 1)P_l P_{l'}(1 - U_l)(1 - \bar{U}_{l'})2\pi \sin \theta \, d\theta,$$

the total elastic-scattering cross section is obtained in the form

Total elastic: $$\sigma_n = \pi\lambda^2 \sum_l (2l + 1)|1 - U_l|^2. \qquad (6.14)$$

The contribution to the elastic-scattering cross section from the *l*th partial wave is seen to be

l-wave elastic: $$\sigma_{n,l} = \pi\lambda^2(2l + 1)|1 - U_l|^2. \qquad (6.15)$$

The reaction cross section, defined as the total cross section for all nonelastic-scattering processes, is obtained as follows: The number of *l*-wave reactions per second per target nucleus is given by $N_{r,l} = \sigma_{r,l}\phi_{inc}$. As explained in Section 4.6, the incident flux has been normalized to v, the speed of the incident particles, and therefore $\sigma_{r,l} = N_{r,l}/v$. Now the number of reactions per second, $N_{r,l}$, is equal to the negative of the *net current* through a large gedanken sphere of radius R centered on the scatterer. Thus, using Equation 4.94 for the net current density in the radial direction and integrating over the sphere, we have

$$\sigma_{r,l} = \frac{N_{r,l}}{v} = -\frac{\hbar}{2miv} \int_{4\pi} \left(\frac{\partial \psi_l}{\partial r} \bar{\psi}_l - \psi_l \frac{\partial \bar{\psi}_l}{\partial r} \right) \bigg|_{r=R} R^2 \, d\Omega.$$

When the lth partial-wave function far from the scattering center, i.e.,

$$\psi_l(r,\theta) \sim \frac{2l + 1}{2kr} \, i^{l+1} \left\{ \exp\left[-i\left(kr - \frac{l\pi}{2} \right) \right] \right.$$

$$\left. - \, \mathsf{U}_l \exp\left[i\left(kr - \frac{l\pi}{2} \right) \right] \right\} P_l(\cos\theta),$$

is inserted into the above expression for $\sigma_{r,l}$, there results

l-wave reaction: $\qquad \sigma_{r,l} = \pi\lambda^2(2l + 1)(1 - |\mathsf{U}_l|^2).$ $\qquad$ (6.16)

An important conclusion can be drawn immediately from the form of Equations 6.15 and 6.16, namely: *No reaction can occur unless, at the same energy, there is also elastic scattering.* The proof is obvious, for, when $\sigma_{r,l}$ is nonzero, the amplitude of the reflection factor U_l is not unity and $\sigma_{n,l}$, which is proportional to $|1 - \mathsf{U}_l|^2$, has a value other than zero. However, these equations do not imply the converse, that all elastic scattering must be accompanied by a reaction, since $(1 - |\mathsf{U}_l|^2)$ can be zero when $|1 - \mathsf{U}_l|^2$ is not zero. In fact, exactly this situation will prevail at any energy at which β_l is zero and α_l is not zero. Thus we see that elastic scattering can occur in the absence of all reactions.

The total cross section for the lth partial wave is simply the sum of $\sigma_{n,l}$ and $\sigma_{r,l}$,

$$\sigma_{T,l} = \pi\lambda^2(2l + 1)[|1 - \mathsf{U}|^2 + (1 - |\mathsf{U}_l|^2)],$$

but, since $|1 - \mathsf{U}_l|^2 + (1 - |\mathsf{U}_l|^2) = 2 - \mathsf{U}_l - \bar{\mathsf{U}}_l = 2 - 2\,\mathrm{Re}\,\mathsf{U}_l,$

l-wave total: $\qquad \sigma_{T,l} = 2\pi\lambda^2(2l + 1)(1 - \mathrm{Re}\,\mathsf{U}_l).$ $\qquad$ (6.17)

Maximum Possible Cross Sections

It is interesting and instructive to estimate the maximum possible neutron cross sections on the basis of the equations derived above. The square of the reduced de Broglie wavelength of a neutron, which appears as a factor in each of the cross-section expressions, is given by $\lambda^2 = \hbar^2/2mE$, where m and E are the reduced mass of the neutron and its available kinetic energy, respectively. For all but the lightest of nuclei, the reduced mass is practically the same as the mass of the neutron. With this approximation, we find that

$$\lambda^2(\text{barns}) = \frac{2.1 \times 10^5}{E(\text{ev})}. \qquad (6.18)$$

We immediately see that neutron cross sections, which are proportional to λ^2, may obtain values of the order of 10^5 barns in the electron-volt region, 10^2 barns in the kilo-electron-volt region, and a few barns in the million-electron-volt region.

Whether or not the cross sections will reach their maximum possible values at a given energy depends on the reflection factor U_l at that energy. The reflection factor, in turn, depends on the complex phase shift. Substituting $U_l = \exp(-2\beta_l + 2i\alpha_l)$ into the equations for l-wave elastic-scattering and reaction cross sections, we obtain

$$\sigma_{n,l} = (2l + 1)\pi\lambda^2[1 + e^{-4\beta_l} - e^{-2\beta_l}(2\cos 2\alpha_l)],$$

$$\sigma_{r,l} = (2l + 1)\pi\lambda^2(1 - e^{-4\beta_l}). \tag{6.19}$$

The amplitude factor β_l plays a major role in determining the magnitude of the cross sections as summarized in the following table.

β_l	$\sigma_{r,l}$	$\sigma_{n,l}$
$\gg 1$	$(2l + 1)\pi\lambda^2$	$(2l + 1)\pi\lambda^2$
$\ll 1$	$(2l + 1)\pi\lambda^2(4\beta_l)$	$(2l + 1)\pi\lambda^2(4\sin^2\alpha_l + 4\beta_l\cos 2\alpha_l)$
0	0	$(2l + 1)\pi\lambda^2(4\sin^2\alpha_l)$

When $\beta_l \gg 1$, the l-wave reaction cross section assumes its maximum possible value, $(2l + 1)\pi\lambda^2$, and the elastic-scattering cross section assumes the same value. When $\beta_l \ll 1$, the reaction cross section is directly proportional to β_l, and the elastic-scattering cross section is determined by the relative sizes of α_l and β_l. If the phase shift α_l is an odd-integral multiple of $\pi/2$, the l-wave elastic cross section assumes its maximum possible value, $(2l + 1)4\pi\lambda^2$. When $\beta_l = 0$, the l-wave reaction cross section is zero, and the l-wave scattering cross section depends on the phase shift α_l. It will assume its maximum possible value at energies for which α_l is an odd-integral multiple of $\pi/2$.

Low Energy Cross Sections

At low neutron energies, measured elastic-scattering cross sections are constant and very roughly equal to $4\pi R^2$, where R is the nuclear radius (except at isolated resonances where σ_n may assume very large values). This can be explained by assuming that, except on the resonances, β_l is so small that the elastic-scattering cross section is given by

$$\sigma_n = \sum_l (2l + 1)\pi\lambda^2(4\sin^2\alpha_l).$$

A vanishingly small β_l means that the neutron does not penetrate into the nucleus and initiate a reaction. Thus, when $\beta_l \simeq 0$, the nucleus may be assumed to behave as an impenetrable sphere. In Section 4.7, we showed that at low energies ($kR \ll 1$) only the $l = 0$ phase shift need be considered for an impenetrable sphere. The $l = 0$ phase shift was given by $\delta_0 = -kb$, which is, in the notation of this section, $\alpha_0 = -kR$. Since $kR \ll 1$, we can write $\sin^2 kR \simeq k^2 R^2 = R^2/\lambdabar^2$ and, finally, $\sigma_n = \pi\lambdabar^2(4 \sin^2 \alpha_0) = 4\pi R^2$, which is the observed energy-independent elastic-scattering cross section at low energies. This is the *shape elastic-scattering* cross section, so named because the scattering takes place without formation of a compound nucleus.

A pure elastic-scattering resonance occurs at an energy at which β_l is vanishingly small and at which α_l takes on a value near $\pi/2$. At this energy the nuclear barrier is practically transparent because the neutron energy and angular momentum of the l-wave correspond to those of one of the possible virtual levels in the compound nucleus. The neutron enters the nucleus freely, spends an appreciable time within the nuclear potential, and then exits with the same energy (in the C-system) it had when it entered. This is *compound elastic scattering*. It will subsequently be shown that in a pure scattering resonance the compound elastic cross section for spinless particles actually assumes its maximum possible value, i.e., $(2l + 1)4\pi\lambdabar^2$.

High-Energy Cross Sections

At very high neutron energies (order of 100 Mev), measured total cross sections of most nuclei vary erratically with energy; but, averaged over an energy interval which contains many fluctuations, the cross sections equal approximately $2\pi R^2$. This average total cross section can be explained by assuming that at these very high energies the nucleus absorbs all neutrons that strike it. The nucleus is pictured as a black sphere. All reflection factors are zero, and β_l is much greater than 1. In this case, the l-wave reaction and elastic-scattering cross sections both equal $(2l + 1)\pi\lambdabar^2$. As shown in Chapter 4, the total cross section is given approximately by

$$\sigma_T \simeq 2\pi\lambdabar^2 \sum_{l=0}^{kR} (2l + 1) \simeq \frac{2\pi}{k^2} \int_0^{kR} 2l \, dl = 2\pi R^2,$$

which agrees with the average total cross sections obtained experimentally. This bit of evidence does not in itself prove that nuclei are black toward high-energy neutrons, but it is suggestive. This subject will be taken up again in Chapter 7 where we will show that the black-nucleus model is actually too severe to account for all the known features of the high-energy scattering cross sections and that the nucleus is much more accurately pictured as a sphere that partially absorbs and partially refracts neutrons.

Figure 6.2 shows the extent to which these simple theories agree with measured total cross sections at low, high, and very high energies. At low energies (1 to 100 ev), the total cross sections of those elements which happen to be free of resonances in this energy region are constant. They show a considerable scatter about the line $\sigma_T = 4\pi R^2 = 4\pi(1.5 \times 10^{-13} A^{1/3})^2$. This scatter is not interpreted as a defect in the equation $\sigma_T = 4\pi R^2$ but rather as a defect in the assumption that the effective nuclear radius is given by $R = 1.5 \times 10^{-13} A^{1/3}$. The low-energy data on σ_T are used to obtain an empirical measure of the effective nuclear radius R' by setting $\sigma_T = 4\pi(R')^2$ and solving for R' for each element.

As explained above, at very high energies (order of 100 Mev), the total cross sections fluctuate with energy. The data in Figure 6.2 were obtained by averaging these cross sections in the region 90 to 110 Mev. Note that σ_T in this case is represented rather well by $2\pi R^2$ for elements in the range $A < 100$ and reasonably well for those in the range $A > 100$.

The average total cross section in the energy region around 10 Mev is seen to be generally about 2 to $3\pi R^2$. In this energy region the cross sections are roughly approximated by $\sigma_T = 2\pi(R + \lambda)^2$.

Calculation of Reflection Factors

The reflection factors U_l, which substantially determine the cross sections, are obtained from the solutions to the radial Schrödinger equations. At any radius outside the nuclear potential, which is assumed to extend from $r = 0$ to $r = R$, the solution to the radial l-wave Schrödinger equation may be written in the form

$$W_l(r) = C_l[W_l^{(-)}(r) - U_l W_l^{(+)}(r)] \qquad \text{for } r \geq R, \qquad (6.20)$$

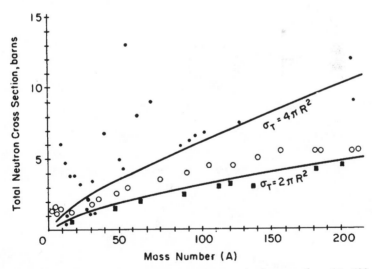

Figure 6.2—Measured total neutron cross sections: ●, in the range from 1 to 100 ev; ○, averaged around 10 Mev; and ■, averaged around 100 Mev.

where $W_l(r) \equiv rR_l(r)$. The functions $W_l^{(-)}(r)$ and $W_l^{(+)}(r)$, which have the form of incoming and outgoing wave functions for a free particle, are, for a neutron, the usual combinations of spherical Bessel and Neumann functions,

$$W_l^{(+)}(r) = kr[n_l(kr) + ij_l(kr)] \equiv A_l(r) + iB_l(r),$$
$$W_l^{(-)}(r) = kr[n_l(kr) - ij_l(kr)] \equiv A_l(r) - iB_l(r). \tag{6.21}$$

For a charged particle, $W_l^{(-)}$ and $W_l^{(+)}$ would be Coulomb wave functions. The functions A_l and B_l are introduced merely to shorten the equations which follow. A complex quantity, D_l, is now defined by the equation

$$D_l \equiv R\left[\frac{dW_l/dr}{W_l}\right]_{r=R}. \tag{6.22}$$

The quantity D_l will be called, somewhat inaccurately, "the logarithmic derivative of the radial wave function at the nuclear boundary." Actually D_l is not the logarithmic derivative of $R_l(r)$ but R times the logarithmic derivative of $W_l(r)$. Furthermore, R is not the "nuclear boundary" per se since realistic nuclear potentials do not have a sharp boundary but, actually, some radius beyond which the nuclear potential is so weak that it has only a negligible effect on the form of the wave function.

Substituting W_l from Equation 6.20 into Equation 6.22 and solving for the reflection factor, we obtain a most important basic expression:

$$U_l = \left\{\frac{D_l - R\left[\dfrac{dW_l^{(-)}/dr}{W_l^{(-)}}\right]}{D_l - R\left[\dfrac{dW_l^{(+)}/dr}{W_l^{(+)}}\right]} \frac{W_l^{(-)}}{W_l^{(+)}}\right\}_{r=R} \tag{6.23}$$

It should be noted that this expression for the reflection factor contains only one unknown, the logarithmic derivative D_l. The real numbers a_l and b_l are now defined by the equation

$$R\left[\frac{dW_l^{(+)}/dr}{W_l^{(+)}}\right]_{r=R} \equiv a_l + ib_l. \tag{6.24a}$$

It follows from Equation 6.21 that

$$R\left[\frac{dW_l^{(-)}/dr}{W_l^{(-)}}\right]_{r=R} \equiv a_l - ib_l. \tag{6.24b}$$

Using Equation 6.21 again in conjunction with Equations 6.24, we find the following expressions for a_l and b_l:

$$a_l = R \left[\frac{A_l(dA_l/dr) + B_l(dB_l/dr)}{A_l^2 + B_l^2} \right]_{r=R} \tag{6.25}$$

and

$$b_l = R \left[\frac{k}{A_l^2 + B_l^2} \right]_{r=R} \tag{6.26}$$

The numerator of Equation 6.26 follows from the readily verifiable relation $A_l(dB_l/dr) - B_l(dA_l/dr) = k$. The functions a_l and b_l are called the *shift factor* and the *penetration factor*, respectively, for reasons that are explained in Section 6.3.

Explicit expressions for a_l and b_l are obtained by straightforward calculation using Equations 6.25, 6.26, and 6.21. The first three a_l and b_l and the asymptotic forms are displayed in Table 6.2. The double factorial symbol (!!) employed in this table is defined as

$$n!! \equiv \begin{cases} n \cdot (n - 2) \cdot (n - 4) \cdots 5 \cdot 3 \cdot 1 & (n \text{ odd}) \\ n \cdot (n - 2) \cdot (n - 4) \cdots 6 \cdot 4 \cdot 2 & (n \text{ even}). \end{cases} \tag{6.27}$$

Table 6.2—The Shift Factor $a_l(\rho)$ and the Penetration Factor $b_l(\rho)$ at the Nuclear Surface*

l	a_l	b_l
0	0	ρ
1	$\dfrac{-1}{\rho^2 + 1}$	$\dfrac{\rho^3}{\rho^2 + 1}$
2	$\dfrac{-3(\rho^2 + 6)}{\rho^4 + 3\rho^2 + 9}$	$\dfrac{\rho^5}{\rho^4 + 3\rho^2 + 9}$
3	$\dfrac{-(6\rho^4 + 90\rho^2 + 675)}{\rho^6 + 6\rho^4 + 45\rho^2 + 225}$	$\dfrac{\rho^7}{\rho^6 + 6\rho^4 + 45\rho^2 + 225}$
$l \gg \rho$ or $\rho \ll 1$	$-l$	$\dfrac{\rho^{2l+1}}{[(2l - 1)!!]^2}$

*Where $\rho \equiv kR$.

The ratio $W_l^{(-)}/W_l^{(+)}$ that appears in Equation 6.23 will now be written in the form

$$\left[\frac{W_l^{(-)}}{W_l^{(+)}} \right]_{r=R} = e^{2i\Delta_l}. \tag{6.28}$$

The phase constant Δ_l, defined by the above equation, is a real number since the absolute value of $W_l^{(-)}/W_l^{(+)}$ is unity. In fact, Δ_l is the familiar phase shift for an impenetrable sphere of radius R. This is most easily proved by setting $W_l(R) = 0$ in Equation 6.20, a condition corresponding to the impenetrable-sphere boundary condition, and then solving for U_l. This U_l is seen to be identically equal to the left-hand side of Equation 6.28. Hence Δ_l is indeed the phase shift for the impenetrable sphere. The Δ_l are known functions of kR, given by Equation 4.127.

The final expression for the reflection factors is obtained by substituting Equations 6.24 and 6.28 into Equation 6.23:

$$U_l = \frac{D_l - a_l + ib_l}{D_l - a_l - ib_l} e^{2i\Delta_l}. \tag{6.29}$$

The only unknown in this equation is the logarithmic derivative D_l. The logarithmic derivatives eventually must be determined by solution of the radial Schrödinger equations for a neutron in an assumed nuclear potential. Note that, if D_l is real, $|U_l|^2 = 1$, so that the reaction cross section is zero and the interaction is pure elastic scattering.

The l-wave scattering and reaction cross sections are obtained as functions of the unknown logarithmic derivatives by substituting Equation 6.29 into Equations 6.15 and 6.16:

$$\sigma_{n,l} = (2l + 1)\pi\lambda^2 \tag{6.30}$$

$$\times \left| e^{2i\Delta_l} \left[(e^{-2i\Delta_l} - 1) + \frac{-2ib_l}{(\operatorname{Re} D_l - a_l) + i(\operatorname{Im} D_l - b_l)} \right] \right|^2,$$

$$\sigma_{r,l} = (2l + 1)\pi\lambda^2 \frac{-4b_l \operatorname{Im} D_l}{(\operatorname{Re} D_l - a_l)^2 + (\operatorname{Im} D_l - b_l)^2}. \tag{6.31}$$

In these equations the logarithmic derivative has been expressed explicitly in terms of its real and imaginary parts, $D_l \equiv \operatorname{Re} D_l + i \operatorname{Im} D_l$.

Several interesting conclusions can be drawn from examination of Equations 6.30 and 6.31:

1. Since $\sigma_{r,l}$ must be nonnegative and b_l is always positive, the imaginary part of the logarithmic derivative D_l must be nonpositive.

2. If the imaginary part of D_l is zero, the reaction cross section $\sigma_{r,l}$ is zero, and the interaction is pure elastic scattering.

3. If the second term within the brackets in Equation 6.30 is very small relative to the first term, the elastic-scattering cross section is simply

$$\sigma_{n,l} = (2l + 1)\pi\lambda^2 |e^{2i\Delta_l}(e^{-2i\Delta_l} - 1)|^2 = (2l + 1)4\pi\lambda^2 \sin^2 \Delta_l,$$

which will be recognized as the scattering cross section of an impenetrable sphere, the "potential scattering cross section." The first term within the brackets in Equation 6.30 is therefore identified as the amplitude for potential (shape elastic, external) scattering,

$$A_{se}^l \equiv e^{-2i\Delta_l} - 1. \tag{6.32}$$

The remaining term within the brackets (Equation 6.30) is identified as the amplitude for resonance (compound elastic, internal) scattering,

$$A_{ce}^l \equiv \frac{-2ib_l}{(\text{Re } D_l - a_l) + i(\text{Im } D_l - b_l)}. \tag{6.33}$$

The total l-wave elastic-scattering cross section now can be expressed in terms of the amplitudes for shape elastic and compound elastic scattering,

$$\sigma_{n,l} = (2l + 1)\pi\lambda^2 |A_{se}^l + A_{ce}^l|^2. \tag{6.34}$$

Specific attention is directed to the fact that shape elastic and compound elastic scattering add coherently. It is necessary to add the amplitudes, as in Equation 6.34, before taking the square of the absolute value. We will show in the next chapter that this coherence is responsible for the distinctive shape of elastic-scattering resonances. The elastic-scattering differential cross section follows from substitution of $1 - U_l = (\exp 2i\Delta_l)(A_{se}^l + A_{ce}^l)$ into Equation 6.13:

$$\sigma_n(\theta) = \frac{\lambda^2}{4} \left| \sum_l (2l + 1)e^{2i\Delta_l}(A_{se}^l + A_{ce}^l)P_l(\cos\theta) \right|^2. \tag{6.35}$$

The l-wave reaction cross section (Equation 6.31) can be written in terms of the amplitude for compound elastic scattering in the form

$$\sigma_{r,l} = (2l + 1)\pi\lambda^2 |A_{ce}^l|^2 \left(\frac{-\text{Im } D_l}{b_l} \right). \tag{6.36}$$

The total elastic-scattering and reaction cross sections are then, respectively,

$$\sigma_n = \sum_{l=0}^{\infty} \sigma_{n,l}$$

and $\hspace{9cm}$ (6.37)

$$\sigma_r = \sum_{l=0}^{\infty} \sigma_{r,l}.$$

Equations 6.30 to 6.37 are exact expressions for the elastic-scattering and reaction cross sections of spinless neutrons and nuclei.

The 1/v Law of Neutron Absorption

At low neutron energies, particularly in the absence of nearby reaction resonances, the cross sections for such reactions as (n,γ) and (n,α) are experimentally observed to be inversely proportional to the velocity of the neutron. This empirical fact can be explained on the basis of the natural assumption that only $l = 0$ neutrons contribute to the reaction cross section at low energies ($kR \ll 1$). Then Equation 6.31 takes the form

$$\sigma_r = \pi \lambdabar^2 \frac{-4b_0 \, \text{Im} \, D_0}{(\text{Re} \, D_0 - a_0)^2 + (\text{Im} \, D_0 - b_0)^2}. \tag{6.38}$$

From Table 6.2, a_0 is equal to 0 and b_0 equals kR. Therefore b_0 is proportional to $\sqrt{E}$ and, since $\lambdabar^2$ is proportional to $1/E$,

$$\sigma_r \propto \frac{1}{\sqrt{E}} \left[\frac{-\text{Im} \, D_0}{(\text{Re} \, D_0)^2 + (\text{Im} \, D_0 - kR)^2} \right]. \tag{6.39}$$

If we now assume that the logarithmic derivative is insensitive to energy and that $|D_0| \gg kR$, the expression within the brackets in Equation 6.39 is constant, and

$$\sigma_r \propto \frac{1}{\sqrt{E}} \propto \frac{1}{v}, \tag{6.40}$$

which is the well-known 1/v *law of neutron absorption*. The word "law" is very loosely used in this context to describe a cross-section variation that is common but not universal. The $1/v$ behavior prevails only under the conditions that (1) s-wave neutrons are responsible for the reaction, (2) the logarithmic derivative D_0 is energy independent and much greater than kR in absolute value, and (3) there are no nearby reaction resonances.

The 1/v law is often derived as a special case of the Breit–Wigner single-level resonance formula. Actually, as we demonstrated above, the $1/v$ variation in σ_r exists even in the absence of resonances. There are empirical examples of this predicted behavior.

It will be left as an exercise for the reader to show that, if l-wave neutrons are responsible for a reaction cross section near $E = 0$, then, in general,

$$\sigma_{r,l} \propto E^{l - 1/2}$$

as E approaches 0 so that the 1/v variation is the special case when $l = 0$.

6.3 Breit–Wigner Isolated-Resonance Formulas for Spinless Particles

An *isolated resonance* is, by definition, a resonance whose total width is much smaller than the distance (in energy units) to the nearest resonance on either side that has the same J value. The total width, Γ, is defined as the full width of the resonance in the total cross section at one-half the maximum height of the cross section; that is,

$$\Gamma \equiv E_2 - E_1,$$

where $\sigma_T(E_2) = \sigma_T(E_1) = \frac{1}{2}(\sigma_T)_{max}$.

The original theoretical derivation of the experimentally observed variation of cross section in the vicinity of an isolated resonance was made by Breit and Wigner.* The equations are called *Breit–Wigner isolated-resonance formulas*. The following derivation of the formulas bears little resemblance to the original derivation; instead, it is an approximation to the exact expressions for the elastic-scattering and reaction cross sections that were derived in the preceding section.

Note that the entire development of the preceding section was deliberately accomplished without reference to the form of the wave function within the nucleus. A prudent procedure since the specific form of the inner wave function is unknown. The inner wave function will continue to be ignored in the development of the Breit–Wigner isolated-resonance equations that follow.

All quantities in Equations 6.30 to 6.37 are known except the logarithmic derivatives at the nuclear surface, D_l. These logarithmic derivatives depend on the specific form of the (unknown) inner wave functions. We will now show that the theoretical scattering and reaction cross sections will exhibit resonances with the experimentally observed shapes if two *ad hoc* assumptions are made relative to the behavior of $D_l(E)$. Assume that:

1. the real part of the *l*-wave logarithmic derivative $D_l(E)$ becomes zero at some neutron energy $E = E_0'$, which shall be called the *formal resonance energy*,

$$\text{Re } D_l(E_0') = 0; \tag{6.41}$$

2. the imaginary part of $D_l(E)$ is either a constant or a slowly varying function of energy in the vicinity of the formal resonance energy E_0',

$$\text{Im } D_l(E) \simeq \text{constant}, \tag{6.42}$$

---

*G. Breit and E. Wigner, Capture of Slow Neutrons, *Phys. Rev.,* **49**: 519 (1936).

where E is nearly E_0'. The formal resonance energy E_0', defined by Equation 6.41, will be shown subsequently to correspond to the neutron energy at the experimentally observed resonance peak in the case of $l = 0$ resonances. For $l > 0$ resonances, E_0' is close to, but not exactly equal to, the energy of the observed resonance peak.

The real part of the logarithmic derivative is now expanded in a Taylor series about $E = E_0'$, and only the leading term is retained,

$$\text{Re } D_l(E) \simeq (E - E_0')\left(\frac{d \text{ Re } D_l}{dE}\right)_0. \qquad (6.43)$$

The derivative is evaluated at the formal resonance energy $E = E_0'$.

The expansion (Equation 6.43) is now substituted into the expression A_{ce}^l (Equation 6.33), and the resulting equation is divided by $(d \text{ Re } D_l/dE)_0$:

$$A_{ce}^l = \frac{-i\,\dfrac{2b_l}{(d \text{ Re } D_l/dE)_0}}{\left[E - E_0' - \dfrac{a_l}{(d \text{ Re } D_l/dE)_0}\right] + i\left[\dfrac{\text{Im } D_l}{(d \text{ Re } D_l/dE)_0} - \dfrac{b_l}{(d \text{ Re } D_l/dE)_0}\right]}. \qquad (6.44)$$

To simplify this expression, we define the *neutron width*:

$$\Gamma_n \equiv \frac{-2b_l}{(d \text{ Re } D_l/dE)_0}, \qquad (6.45)$$

the *reaction width*:

$$\Gamma_r \equiv \frac{2 \text{ Im } D_l}{(d \text{ Re } D_l/dE)_0}, \qquad (6.46)$$

the *total width*:

$$\Gamma \equiv \Gamma_n + \Gamma_r, \qquad (6.47)$$

and the *observed resonance energy*:

$$E_0 \equiv E_0' + \frac{a_l}{(d \text{ Re } D_l/dE)_0}. \qquad (6.48)$$

Note that the observed resonance energy is shifted relative to the formal resonance energy by an amount proportional to a_l, the shift factor, and that the cross section for compound-nucleus formation is proportional to b_l, the penetration factor. The subscript l, which should be attached to Γ_n, Γ_r, Γ, and E_0, has been omitted for notational convenience. When these expressions are substituted into Equation 6.44, the resonance scattering amplitude takes on the simple form

$$A_{ce}^l = \frac{i\Gamma_n}{(E - E_0) + i(\Gamma/2)}. \tag{6.49}$$

The various widths that were so blithely introduced into the development will be investigated in subsequent chapters. When we substitute A_{ce}^l from Equation 6.49 into Equation 6.34, we obtain the l-wave elastic-scattering cross section in the vicinity of an isolated l-wave resonance,

$$\sigma_{n,l} = (2l + 1)\pi\lambda^2 \left| e^{2i\Delta_l} \left[A_{se}^l + \frac{i\Gamma_n}{(E - E_0) + i(\Gamma/2)} \right] \right|^2. \tag{6.50}$$

Substituting Equation 6.49 and the ratio of Equation 6.46 to Equation 6.45 into Equation 6.36, we obtain the l-wave reaction cross section in the vicinity of an isolated l-wave resonance,

$$\sigma_{r,l} = (2l + 1)\pi\lambda^2 \frac{\Gamma_n\Gamma_r}{(E - E_0)^2 + (\Gamma/2)^2}. \tag{6.51}$$

The l-wave compound elastic-scattering cross section is given by $(2l + 1)\pi\lambda^2|A_{ce}^l|^2$, which becomes, after using Equation 6.49 for A_{ce}^l,

$$\sigma_{ce,l} = (2l + 1)\pi\lambda^2 \frac{\Gamma_n\Gamma_n}{(E - E_0)^2 + (\Gamma/2)^2}. \tag{6.52}$$

To the extent that the Bohr theory of the compound nucleus is correct, all interactions except potential scattering take place through the formation and subsequent independent decay of the compound nucleus and can therefore be written in the form

$$\sigma_{j,l} = \sigma_{c,l} \frac{\Gamma_j}{\Gamma}, \tag{6.53}$$

where σ_c is the cross section for compound nucleus formation and Γ_j/Γ is the probability for decay through emission of j, which may be either a massive particle or a photon. Equations 6.51 and 6.52 have indeed the form predicted by Bohr,

$$\sigma_{r,l} = \sigma_{c,l} \frac{\Gamma_r}{\Gamma},$$

$$\sigma_{ce,l} = \sigma_{c,l} \frac{\Gamma_n}{\Gamma}, \tag{6.54}$$

provided we identify the cross section for compound nucleus formation as

$$\sigma_{c,l} = (2l + 1)\pi\lambda^2 \frac{\Gamma_n\Gamma}{(E - E_0)^2 + (\Gamma/2)^2}. \tag{6.55}$$

Note the paramount role played by Γ_n in the cross section for compound nucleus formation. The compound nucleus cannot be formed by an incident neutron unless the width for neutron emission by the compound nucleus is nonzero. This conclusion may be related directly to the reciprocity relation. If the cross section for the reaction $(Z^{A+1})^* \rightarrow Z^A + n$ is zero, then certainly the cross section for the inverse reaction $Z^A + n \rightarrow (Z^{A+1})^*$ is also zero. Note that Γ_n has this unique role in determining the magnitude of σ_c only because the development has been restricted to neutron-induced interactions. We would replace Γ_n in Equation 6.55 by Γ_p if the incident particle were a proton, by Γ_α if the incident particle were an alpha particle, etc. In any case, the level width of the incident particle determines the magnitude of σ_c. Note, however, that the cross section for a specific reaction is given by an expression completely symmetrical in the level widths of the incoming and outgoing particles. For example, $\sigma_{n,p}$ is proportional to $\Gamma_n \Gamma_p$.

The width, and therefore the probability, for emission of a neutron of energy E by a compound nucleus will most certainly be zero if the neutron cannot exist within the compound nucleus for an appreciable time with energy near $E^* = E + S_n^c$. Hence, there must exist a virtual level for a neutron in the compound nucleus that has an energy approximately equal to $E^* = E + S_n^c$ for the compound nucleus to exhibit a resonance at neutron energy E. This argument leads to the important conclusion that an observed neutron resonance at neutron energy E is the result of a virtual level in the compound nucleus with energy near $E^* = E + S_n^c$ above the ground level. The word "near" is used because the observed resonance energy of an $l > 0$ resonance differs from the formal resonance energy by an amount proportional to a_l, as shown in Equation 6.48.

6.4 Intrinsic Angular Momentum in Quantum Mechanics

Thus far the wave-mechanical description of scattering phenomena has been restricted to spinless particles. But the three fundamental building blocks of atoms—the neutron, the proton, and the electron—all have intrinsic angular momentum; each acts as though it were spinning about an axis through its center.

Spin was introduced into quantum mechanics some seven years before the neutron was discovered. At that time, the energy-level structure of the hydrogen atom, as shown by optical spectrograms, had a fine structure that quantum mechanics did not predict. Uhlenbeck and Goudsmit* proposed that each electron spins on its axis while revolving around the

---

*G. E. Uhlenbeck and S. Goudsmit, *Naturwiss.*, **13**: 953 (1925); *Nature*, **117**: 264 (1926).

nucleus, much as the earth spins on its axis while circling the sun, and that this spinning electrical charge produces a magnetic moment. This intrinsic magnetic moment couples with the magnetic moment generated by the orbital motion of the electron and so introduces an additional potential term into the Schrödinger equation. This proposal was remarkably successful in explaining away the hitherto puzzling spectral data.

Uhlenbeck and Goudsmit hypothesized that (a) every electron has a spin angular momentum s whose component in any direction can have only one of the two values $+\hbar/2$ or $-\hbar/2$, i.e., whose component

$$s_z = \pm\tfrac{1}{2}\hbar, \tag{6.56}$$

and (b) every electron has a spin magnetic moment, μ, that is related to its spin angular momentum through the equation

$$\mu = \frac{e}{m}\,s, \tag{6.57}$$

where e and m are the electron charge and mass, respectively.

The first hypothesis, that the possible components of s were $\pm\hbar/2$, seemed strange. There was much experimental evidence at the time, and a great deal more now, proving that the components of L, the orbital angular momentum, are always integral multiples of $\hbar$, not halves. Only after the development of a fully relativistic theory of the electron by Dirac was the strange behavior of spin fully understood, only then could the empirical Uhlenbeck–Goudsmit relation between μ and s be theoretically derived. Indeed the spin angular momentum and the orbital angular momentum, though very closely related in many of their formal mathematical properties, are not on exactly the same footing. Spin angular momentum is a relativistic quantum-mechanical property.

Pauli's Theory of Spin Angular Momentum

Shortly after publication of Uhlenbeck and Goudsmit's hypothesis, Wolfgang Pauli* developed a mathematical formalism for including spin within the framework of quantum mechanics. Although developed for the electron, this theory is equally applicable to any other particle of spin 1/2, e.g., the neutron or the proton.

Two things are required if spin is to be included within the formalism of wave mechanics: (1) a wave function that depends not only on the spatial coordinates but also on the spin direction and (2) wave-mechanical operators for the components (s_x, s_y, s_z) of the spin vector.

* W. Pauli, Z. Physik, **32**: 601 (1927).

Spin Wave Functions for Particles of Spin 1/2

The wave function at a point (x,y,z) must be capable of telling us how many neutrons (or other particles of spin 1/2) are in a volume element about this point with their spins parallel or antiparallel to some axis. The terms *spin-parallel* and *spin-antiparallel* indicate spin components along the chosen axis equal to $+\hbar/2$ and $-\hbar/2$, respectively. Figure 6.3 is a pictorial representation of the two possible spin states of a particle of spin 1/2. In a state in which s^2 and s_z are known, the other two components of s, namely, s_x and s_y, are undetermined. The spin vector s may be pictured as lying in some undetermined position along the cones, or it may be assumed to be sweeping out the conical surfaces shown in this figure. Both interpretations have equal validity since the Heisenberg uncertainty principle stands mute on both. Stated in terms of angular momentum and

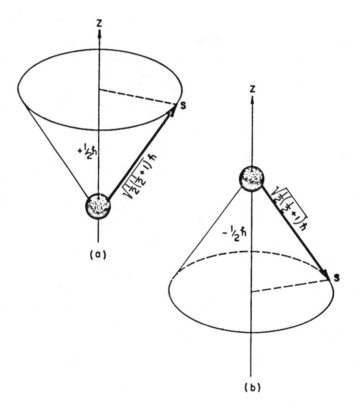

Figure 6.3—Pictorial representation of the spin vector s in the spin-parallel and spin-antiparallel states of a particle with spin 1/2.

angular position uncertainty, the Heisenberg principle is of the form $\Delta s \, \Delta\phi \geq \hbar/2$, where s is the spin angular momentum component conjugate to the angle ϕ.

The most straightforward way of describing the spin state is to replace the scalar wave function of the ordinary Schrödinger equation with a vector wave function having two components. The *Pauli wave function* has the form of a two-component column vector,

$$\psi(x,y,z) = \begin{pmatrix} \psi^p \\ \psi^a \end{pmatrix}, \tag{6.58}$$

where $\psi^p = \psi^p(x,y,z)$ is the spin-parallel wave function and $\psi^a = \psi^a(x,y,z)$ is the spin-antiparallel wave function. The complex conjugate of the Pauli ψ is written as a row symbol,

$$\bar{\psi} = (\bar{\psi}^p, \bar{\psi}^a). \tag{6.59}$$

The product $\bar{\psi}\psi$ then follows from the ordinary rule of matrix multiplication,

$$\bar{\psi}\psi = (\bar{\psi}^p, \bar{\psi}^a) \begin{pmatrix} \psi^p \\ \psi^a \end{pmatrix} = \bar{\psi}^p\psi^p + \bar{\psi}^a\psi^a. \tag{6.60}$$

As usual, the integral of $\bar{\psi}\psi$ over all space is required to equal unity,

$$\int_{\text{all space}} [\bar{\psi}^p\psi^p + \bar{\psi}^a\psi^a] \, d\tau = 1. \tag{6.61}$$

We interpret $\bar{\psi}^p\psi^p \, d\tau$ and $\bar{\psi}^a\psi^a \, d\tau$ as the probabilities of finding spin-parallel and spin-antiparallel neutrons in volume element $d\tau$ and $\bar{\psi}\psi \, d\tau$ as the total probability of finding a neutron, regardless of spin direction, in $d\tau$.

Since the spin under consideration may have only one of two orientations relative to a given axis, the spin eigenfunctions that describe the parallel and antiparallel states may be defined as:

$$\chi^p \equiv \begin{pmatrix} 1 \\ 0 \end{pmatrix},$$
$$\chi^a \equiv \begin{pmatrix} 0 \\ 1 \end{pmatrix}. \tag{6.62}$$

The components of the Pauli wave function may then be written:

$$\psi^p = \psi(x,y,z)\chi^p,$$
$$\psi^a = \psi(x,y,z)\chi^a. \tag{6.63}$$

Spin Operators for Particles of Spin 1/2

We noted in Chapter 4 that the components of the orbital angular-momentum operator L obey the commutation relations

$$[L_x,L_y] = i\hbar L_z,$$
$$[L_y,L_z] = i\hbar L_x, \tag{4.75}$$
$$[L_z,L_x] = i\hbar L_y.$$

Pauli assumed that the components of spin angular momentum would obey the identical commutation relations

$$[s_x,s_y] = i\hbar s_z,$$
$$[s_y,s_z] = i\hbar s_x, \tag{6.64}$$
$$[s_z,s_x] = i\hbar s_y.$$

This was a pure assumption, an inspired guess, that subsequently proved its validity in the way in which all physical theories are ultimately judged, by agreement of theoretical predictions with experiment.

To rid the notation of the clumsy factor $\hbar/2$, Pauli defined a vector operator σ related to the spin operator s by the equation

$$\mathbf{s} \equiv \frac{\hbar}{2}\boldsymbol{\sigma}. \tag{6.65}$$

This vector equation is shorthand for the three scalar equations

$$s_x = \frac{\hbar}{2}\sigma_x,$$
$$s_y = \frac{\hbar}{2}\sigma_y, \tag{6.66}$$
$$s_z = \frac{\hbar}{2}\sigma_z.$$

According to Equations 6.64 and 6.66, the components of the operator σ obey the commutation rules

$$[\sigma_x,\sigma_y] = 2i\sigma_z,$$
$$[\sigma_y,\sigma_z] = 2i\sigma_x, \tag{6.67}$$
$$[\sigma_z,\sigma_x] = 2i\sigma_y.$$

Furthermore, since the eigenvalues of the components of s are $+\hbar/2$ and $-\hbar/2$, the eigenvalues of the components of σ are 1 and -1. Thus σ_x^2, σ_y^2, and σ_z^2 have only the eigenvalue 1,

$$\sigma_x\sigma_x = \sigma_y\sigma_y = \sigma_z\sigma_z = 1. \tag{6.68}$$

It can be easily proved using Equations 6.67 and 6.68 that the operators σ_x, σ_y, and σ_z anticommute among themselves. That is,

$$\sigma_x\sigma_y + \sigma_y\sigma_x = 0,$$
$$\sigma_y\sigma_z + \sigma_z\sigma_y = 0, \tag{6.69}$$
$$\sigma_z\sigma_x + \sigma_x\sigma_z = 0.$$

Finally, it follows from Equations 6.67 and 6.69 that

$$\sigma_x\sigma_y = i\sigma_z,$$
$$\sigma_y\sigma_z = i\sigma_x, \tag{6.70}$$
$$\sigma_z\sigma_x = i\sigma_y.$$

Equations 6.67 through 6.70 are the defining properties of the σ operators. Since each σ has two eigenvalues and operates on a two-component wave function, σ may be represented by a 2-by-2 matrix. There are numerous ways of representing the σ's by 2-by-2 matrices that obey all the relations of Equations 6.67 through 6.70. Pauli began by associating with σ_z the simplest 2-by-2 matrix having eigenvalues 1 and -1, namely,

$$\sigma_z = \begin{pmatrix} 1 & 0 \\ 0 & -1 \end{pmatrix}.$$

Note that σ_z indeed does have the eigenvalues 1 and -1 since $\sigma_z\chi^p = \chi^p$ and $\sigma_z\chi^a = -\chi^a$, that is,

$$\begin{pmatrix} 1 & 0 \\ 0 & -1 \end{pmatrix}\begin{pmatrix} 1 \\ 0 \end{pmatrix} = \begin{pmatrix} 1 \\ 0 \end{pmatrix},$$

$$\begin{pmatrix} 1 & 0 \\ 0 & -1 \end{pmatrix}\begin{pmatrix} 0 \\ 1 \end{pmatrix} = -\begin{pmatrix} 0 \\ 1 \end{pmatrix}.$$

With simple algebra based on the properties shown in Equations 6.67 through 6.70, it is easy to prove that the complete set of *Pauli spin matrices* is

$$\sigma_x = \begin{pmatrix} 0 & 1 \\ 1 & 0 \end{pmatrix},$$

$$\sigma_y = \begin{pmatrix} 0 & -i \\ i & 0 \end{pmatrix},$$

$$\sigma_z = \begin{pmatrix} 1 & 0 \\ 0 & -1 \end{pmatrix}, \tag{6.71}$$

$$\boldsymbol{\sigma}\cdot\boldsymbol{\sigma} = \begin{pmatrix} 3 & 0 \\ 0 & 3 \end{pmatrix}.$$

It will be left as an exercise for the reader to derive these matrices and to verify that they do indeed obey all the rules of Equations 6.67 to 6.70.

Pauli Form of the Schrödinger Equation

Pauli wave functions are two-component vectors. All operators associated with these wave functions must be in the form of 2-by-2 matrices. To cast a scalar Schrödinger operator into matrix form, we simply multiply it by the unit matrix

$$I = \begin{pmatrix} 1 & 0 \\ 0 & 1 \end{pmatrix}.$$

Thus, for example, if Ω is a Schrödinger operator, then its Pauli form is given by ΩI, where

$$\Omega I = \Omega \begin{pmatrix} 1 & 0 \\ 0 & 1 \end{pmatrix} = \begin{pmatrix} \Omega & 0 \\ 0 & \Omega \end{pmatrix}. \tag{6.72}$$

The Schrödinger equation for a single particle moving under the influence of a central potential has the form

$$\left[-\frac{\hbar^2}{2m} \nabla^2 + V(r) \right] \psi(r) = E\psi(r).$$

The corresponding Pauli equation is

$$\begin{pmatrix} [-(\hbar^2/2m)\nabla^2 + V(r)] & 0 \\ 0 & [-(\hbar^2/2m)\nabla^2 + V(r)] \end{pmatrix} \begin{pmatrix} \psi^p(r) \\ \psi^a(r) \end{pmatrix}$$
$$= \begin{pmatrix} E & 0 \\ 0 & E \end{pmatrix} \begin{pmatrix} \psi^p(r) \\ \psi^a(r) \end{pmatrix}, \tag{6.73}$$

which can be written explicitly as two equations, one for the spin-parallel and one for the spin-antiparallel wave function,

$$\left[-\frac{\hbar^2}{2m} \nabla^2 + V(r) \right] \psi^p(r) = E\psi^p(r),$$

$$\left[-\frac{\hbar^2}{2m} \nabla^2 + V(r) \right] \psi^a(r) = E\psi^a(r). \tag{6.74}$$

The equations in this case are identical; so the solutions for the spin-parallel and spin-antiparallel wave functions are identical except possibly in amplitude, as indeed we might have expected since none of the operators is spin dependent. In the absence of a spin-dependent potential, the Pauli equations are seen to be uncoupled.

Suppose the potential contains a spin-orbit term of the form $f(r)\boldsymbol{\sigma} \cdot \mathbf{L}$.

The Schrödinger equation in this case is

$$\left[-\frac{\hbar^2}{2m}\nabla^2 + V(r) + f(r)\boldsymbol{\sigma}\cdot\mathbf{L}\right]\psi(r) = E\psi(r),\tag{6.75}$$

where

$$\boldsymbol{\sigma}\cdot\mathbf{L} = \sigma_x L_x + \sigma_y L_y + \sigma_z L_z$$

$$= \begin{pmatrix} 0 & 1 \\ 1 & 0 \end{pmatrix}\begin{pmatrix} L_x & 0 \\ 0 & L_x \end{pmatrix} + \begin{pmatrix} 0 & -i \\ i & 0 \end{pmatrix}\begin{pmatrix} L_y & 0 \\ 0 & L_y \end{pmatrix} + \begin{pmatrix} 1 & 0 \\ 0 & -1 \end{pmatrix}\begin{pmatrix} L_z & 0 \\ 0 & L_z \end{pmatrix}.$$

Performing the indicated multiplications and adding, we get

$$\boldsymbol{\sigma}\cdot\mathbf{L} = \begin{pmatrix} L_z & L_x - iL_y \\ L_x + iL_y & -L_z \end{pmatrix},\tag{6.76}$$

therefore the Pauli equation corresponding to Equation 6.75 is

$$\begin{pmatrix} -(\hbar^2/2m)\nabla^2 + V(r) + f(r)L_z & f(r)(L_x - iL_y) \\ f(r)(L_x + iL_y) & -(\hbar^2/2m)\nabla^2 + V(r) - f(r)L_z \end{pmatrix}$$

$$\times \begin{pmatrix} \psi^p(r) \\ \psi^a(r) \end{pmatrix} = \begin{pmatrix} E & 0 \\ 0 & E \end{pmatrix}\begin{pmatrix} \psi^p(r) \\ \psi^a(r) \end{pmatrix},$$

or, written in the form of two equations,

$$\left[-\frac{\hbar^2}{2m}\nabla^2 + V(r) + f(r)L_z\right]\psi^p + f(r)[L_x - iL_y]\psi^a = E\psi^p,$$

$$\left[-\frac{\hbar^2}{2m}\nabla^2 + V(r) - f(r)L_z\right]\psi^a + f(r)[L_x + iL_y]\psi^p = E\psi^a.\tag{6.77}$$

The parallel and antiparallel wave functions seem to be coupled by these two differential equations. We will show in Chapter 7 that these equations actually cease to be coupled when we take into account the conservation of total (orbital plus spin) angular momentum and that the operator $\boldsymbol{\sigma}\cdot\mathbf{L}$ takes on the simple form

$$\boldsymbol{\sigma}\cdot\mathbf{L} = \hbar \begin{pmatrix} l & 0 \\ 0 & -(l+1) \end{pmatrix},$$

in coordinates r, θ, and ϕ; so the radial Pauli equations become

$$\left[-\frac{\hbar^2}{2m}\nabla_r^2 + V(r) + f(r)\hbar l\right]R_l^p(r) = ER_l^p(r),$$

$$\left[-\frac{\hbar^2}{2m}\nabla_r^2 + V(r) - f(r)\hbar(l+1)\right]R_l^a(r) = ER_l^a(r),\tag{6.78}$$

where ∇_r^2 is the radial part of the Laplacian operator in spherical coordinates that includes the centrifugal potential term

$$\nabla_r^2 = \frac{1}{r^2}\frac{d}{dr}\left(r^2\frac{d}{dr}\right) - \frac{l(l+1)}{r^2}$$

and where $R_l(r)$ is the radial part of the wave function for l-wave neutrons. The form (Equation 6.78) of the radial Pauli equations is obtained by applying the technique of separation of variables to the matrix equation (Equation 6.75).

Spin Wave Functions and Operators for Spins Greater Than 1/2

It proves convenient to view the target nucleus in a nuclear interaction as a particle with an intrinsic spin **I**. The "intrinsic spin of the nucleus" is the vector sum of the actual intrinsic spins and the orbital angular momenta of the nucleons it contains, and thus **I** is not rigorously a spin. Nevertheless, to avoid constant circumlocution, we shall refer to **I** as the spin of the nucleus. Furthermore, we shall refer to the nucleus as a particle where the word "particle" is used in its broadest sense, not in its restricted sense of elementary particle.

The Pauli spin wave functions and spin operators are equally applicable to all particles with spin 1/2: the electron, the neutron, the proton, and all nuclei with $I = 1/2$. The extension of the spin wave-function and operator formalism to particles with spins greater than 1/2 is straightforward. The spin wave function for a particle with spin I is a column vector with $2I + 1$ components. Each component is labeled by a value of the spin projection, m_I, where m_I ranges from $+I$ to $-I$. The spin operators are $(2I + 1)$-by-$(2I + 1)$ matrices with properties identical, except for numerical factors, to those of the Pauli spin operators as given in Equations 6.67 to 6.70.

In general, we first define the spin operators σ^I by the equation

$$\mathbf{I} \equiv I\hbar\sigma^I, \tag{6.79}$$

which is the generalization of Equation 6.65. The commutation rules that the spin operators follow are then obtained from the corresponding commutation rules on the spins:

$$[I_i, I_j] = i\hbar I_k,$$

thus

$$[\sigma_i^I, \sigma_j^I] = \frac{i}{I}\sigma_k^I, \tag{6.80}$$

where $i, j,$ and k represent any cyclic permutation of the three coordinates

x, y, and z, i.e., where $(ijk) = (xyz)$ or (yzx) or (zxy). If we assume that the state to be described is one in which the square of the total angular momentum $\mathbf{I} \cdot \mathbf{I}$ and the z component of the angular momentum I_z are simultaneously observable, then

$$[\mathbf{I} \cdot \mathbf{I}, I_z] = 0$$

and

$$[\boldsymbol{\sigma}^I \cdot \boldsymbol{\sigma}^I, \sigma_z^I] = 0. \tag{6.81}$$

The simultaneous eigenfunctions of the total spin and the z component of the spin will be designated $\chi_I^{m_I}$. These eigenfunctions must obey the eigenvalue equations

$$\mathbf{I} \cdot \mathbf{I} \chi_I^{m_I} = I(I + 1)\hbar^2 \chi_I^{m_I},$$

from which we get

$$\boldsymbol{\sigma}^I \cdot \boldsymbol{\sigma}^I \chi_I^{m_I} = \frac{I(I + 1)}{I^2} \chi_I^{m_I}, \tag{6.82}$$

and

$$I_z \chi_I^{m_I} = m_I \hbar \chi_I^{m_I},$$

from which we get

$$\sigma_z^I \chi_I^{m_I} = \frac{m_I}{I} \chi_I^{m_I}, \tag{6.83}$$

where m_I takes on the $2I + 1$ values $I, I - 1, \ldots, -I$.

The $2I + 1$ eigenfunctions $\chi_I^{m_I}$ are now identified as column vectors, each of which has $2I + 1$ components according to the following scheme:

$$\chi_I^{m_I} = \chi_I^{I} = \begin{pmatrix} 1 \\ 0 \\ \vdots \\ 0 \end{pmatrix},$$

$$\chi_I^{m_I} = \chi_I^{I-1} = \begin{pmatrix} 0 \\ 1 \\ 0 \\ \vdots \end{pmatrix},$$

$$\vdots$$

$$\chi_I^{m_I} = \chi_I^{-I} = \begin{pmatrix} 0 \\ 0 \\ \vdots \\ 1 \end{pmatrix}. \tag{6.84}$$

The position of the nonzero component in the vector $\chi_I^{m_I}$ indicates immediately the value of m_I, the projection of $\mathbf{I} \cdot \mathbf{I}$ on the z-axis. If the number in the nth position is unity, then $m_I = I - n + 1$.

The matrices associated with the spin operators are now determined from their properties as expressed in Equations 6.80 through 6.83. The matrices for σ^2 and σ_z can be written down immediately, and from these σ_x and σ_y can be found. The simple algebraic details of this procedure will be left as an exercise for the reader. As an example of the results, the spin wave functions and operators associated with a particle having spin $I = 1$ are given by

$$\chi_1^1 = \begin{pmatrix} 1 \\ 0 \\ 0 \end{pmatrix}, \quad \chi_1^0 = \begin{pmatrix} 0 \\ 1 \\ 0 \end{pmatrix}, \quad \chi_1^{-1} = \begin{pmatrix} 0 \\ 0 \\ 1 \end{pmatrix},$$

$$\sigma_x^1 = \frac{1}{\sqrt{2}} \begin{pmatrix} 0 & 1 & 0 \\ 1 & 0 & 1 \\ 0 & 1 & 0 \end{pmatrix}, \quad \boldsymbol{\sigma}^1 \cdot \boldsymbol{\sigma}^1 = 2 \begin{pmatrix} 1 & 0 & 0 \\ 0 & 1 & 0 \\ 0 & 0 & 1 \end{pmatrix}, \tag{6.85}$$

$$\sigma_y^1 = \frac{i}{\sqrt{2}} \begin{pmatrix} 0 & -1 & 0 \\ 1 & 0 & -1 \\ 0 & 1 & 0 \end{pmatrix}, \quad \sigma_z^1 = \begin{pmatrix} 1 & 0 & 0 \\ 0 & 0 & 0 \\ 0 & 0 & -1 \end{pmatrix}.$$

We can verify without difficulty that these operators and vectors satisfy all the conditions in Equations 6.80 to 6.83. It is easily shown that in this representation none of the spin wave functions is an eigenfunction of σ_x or σ_y.

The complex conjugate of a spin wave function is written as a row vector, e.g., $\bar{\chi}_1^1 = (1\ 0\ 0)$, and the product $\bar{\chi}\chi$ is obtained by the ordinary role of matrix multiplication. It is readily verified that the spin wave functions are orthonormal, i.e., that

$$\bar{\chi}_I^\alpha \chi_I^\beta = \delta_{\alpha\beta}.$$

As an exercise, the reader may wish to obtain the spin wave functions and operators for the next highest spin, namely, $I = 3/2$.

6.5 Algebraic Properties and Coupling of Angular Momenta

The following algebraic properties of quantum-mechanical angular momentum will be employed repeatedly in subsequent sections. We suggest that these rules, which are presented here without proof, be committed to memory. The proofs may be found in almost any quantum mechanics textbook. For further information, in depth, about the properties of angular momentum, the reader may wish to consult the books by M. E. Rose and A. R. Edmonds that are listed at the end of this chapter.

The following rules apply to any quantum-mechanical angular momentum $\mathbf{j}$, whether an orbital angular momentum, a spin angular momentum, or a sum of both types:

1. The square of the operator $\mathbf{j}$ and one of its components, say j_z, commute and are therefore simultaneously measureable. The other two components of $\mathbf{j}$ do not commute with j_z.

2. The eigenvalues of $\mathbf{j} \cdot \mathbf{j}$ and j_z are given by

$$\mathbf{j} \cdot \mathbf{j} = j(j + 1)\hbar^2 \tag{6.86}$$

and

$$j_z = m_j\hbar, \tag{6.87}$$

where j can be either integral or half-integral (except that, when $\mathbf{j} = \mathbf{L}$, the orbital angular-momentum operator, then $j = l$ must be integral).

3. The values of the magnetic (azimuthal) quantum number m_j lie between and include $-j$ and $+j$, and are separated from one another by unity, i.e.,

$$-j \leq m_j \leq +j$$

or

$$m_j = j, j - 1, \ldots, -j + 1, -j. \tag{6.88}$$

There are therefore $2j + 1$ values of m_j. Every value of m_j has the same numerical character as j; that is, m_j is integral if j is integral, half-integral if j is half-integral.

The following rules apply to the addition or subtraction of any two angular momentum operators $\mathbf{j}'$ and $\mathbf{j}''$ that commute with one another. These may be, for example, the orbital and spin angular momentums of a single particle,* or the spin angular momentums of two different particles. (In the case of greatest interest to collision theory, $\mathbf{j}'$ is the orbital angular momentum of a two-particle constellation, and $\mathbf{j}''$ is the vector sum of the spins of the two particles.) The sum or difference of $\mathbf{j}'$ and $\mathbf{j}''$,

$$\mathbf{j} = \mathbf{j}' \pm \mathbf{j}'',$$

---

* It should be obvious from the previous discussion that the addition of an orbital angular-momentum operator to a spin angular-momentum operator is accomplished by first multiplying the orbital operator by a unit matrix (of the same size as the matrices associated with the spin operator) and then adding the two matrices.

is an angular-momentum operator with the following properties:

1. The eigenvalues of $\mathbf{j} \cdot \mathbf{j}$ are given by

$$\mathbf{j} \cdot \mathbf{j} = j(j + 1)\hbar^2, \tag{6.89}$$

where the quantum number j lies in the range

$$|j' - j''| \leq j \leq (j' + j''). \tag{6.90}$$

The possible values of j are given by

$$j = |j' - j''|, |j' - j''| + 1, |j' - j''| + 2, \ldots, (j' + j''). \tag{6.91}$$

2. The eigenvalues of the operator j_z associated with one of the specific values of j are given by Rules 2 and 3 preceding.

Angular-Momentum Coupling

In any nuclear interaction that begins with the formation of a single compound state,

$$A + a \to C^*,$$

the total angular-momentum quantum number, J, of the compound state is fixed beforehand by the dynamics of the compound nucleus and the energy of the initial constellation. The initial constellation must have the same J as the compound state since total angular momentum is conserved. There are five distinct angular-momentum vectors involved in this simple interaction:

$\mathbf{S}_A \equiv$ the true intrinsic spin of target nucleus A, the vector sum of the spins of all nucleons in A.

$\mathbf{L}_A \equiv$ the internal orbital angular momentum of A, the vector sum of the orbital angular momenta of all nucleons in A.

$\mathbf{s} \equiv$ the intrinsic spin of particle a, which we will assume is either a neutron, a proton, or an alpha particle, none of which has internal orbital angular momentum.

$\mathbf{L} \equiv$ the orbital angular momentum of particle a relative to nucleus A.

$\mathbf{J} \equiv$ the total angular momentum of the compound state, C^*, and the total angular momentum of the initial constellation.

Clearly the operator $\mathbf{J}$ must be the vector sum of all four operators: $\mathbf{J} = \mathbf{S}_A + \mathbf{L}_A + \mathbf{s} + \mathbf{L}$, but the order in which these operators are to be added is not usually known a priori. Three extreme coupling schemes are distinguished below on the basis of the order in which the four pertinent vectors are added to obtain $\mathbf{J}$.

Coupling Scheme	Order of Addition of Angular Momentums	
$L-S$	$\mathbf{J} = (\mathbf{S}_A + \mathbf{s}) + (\mathbf{L}_A + \mathbf{L}) \equiv \mathbf{S}_T + \mathbf{L}_T$	(6.92)
$j-j$	$\mathbf{J} = (\mathbf{S}_A + \mathbf{L}_A) + (\mathbf{s} + \mathbf{L}) \equiv \mathbf{j}_A + \mathbf{j}_a$	(6.93)
$L-j$	$\mathbf{J} = [(\mathbf{S}_A + \mathbf{L}_A) + \mathbf{s}] + \mathbf{L} \equiv \mathbf{j} + \mathbf{L}$	(6.94)

L–S Coupling: In this scheme which is also called *Russell–Saunders coupling*, all the spins are added together, all the orbital angular momenta are added together, and then these two sums are added. This scheme assumes that the operators corresponding to the simultaneous observables are $\mathbf{J}$, $\mathbf{S}_T$, $\mathbf{L}_T$, and the Hamiltonian. It has been found that *L–S* coupling does indeed predict the electronic energy levels in the light atoms.

j–j Coupling: In this scheme the spin and the orbital angular momentum of each particle are first added to obtain the total angular momentum of the particle, then the resulting total angular momentums are added. This scheme, which is also called *spin-orbit coupling*, was the one employed by Mayer and Jensen in their development of the nuclear shell theory. It was previously employed to calculate the virtual electronic energy levels of heavy atoms. (In heavy atoms, the valance electrons are sufficiently screened from the nucleus that the spin-orbit force becomes appreciable compared with the Coulomb force.) The *j–j* coupling scheme assumes that $\mathbf{j}_A$, $\mathbf{j}_a$, $\mathbf{J}$, and H commute in pairs and are therefore simultaneously observable.

L–j Coupling: This coupling scheme, which is commonly called the *channel spin formalism*, is the one most employed in nuclear interaction theory. The intrinsic spin and the intrinsic orbital angular momentum of the target nucleus are first added to obtain what is commonly called the nuclear spin $\mathbf{I}_A \equiv \mathbf{S}_A + \mathbf{L}_A$. The nuclear spin is then added to the intrinsic spin of the projectile particle to obtain the "channel spin" $\mathbf{j} \equiv \mathbf{I}_A + \mathbf{s}$. Finally, the channel spin and the relative orbital angular momentum are added to obtain the total angular momentum $\mathbf{J} = \mathbf{j} + \mathbf{L}$. The assumptions underlying the channel spin formalism will be further clarified in the next section.

In practice, we often find that none of these extreme coupling schemes predicts what is experimentally observed and that some intermediate mixture of the extreme schemes must be employed. The proper mixture is usually determined empirically by introducing mixing fractions whose values are ultimately adjusted to obtain the best fit to observations.

6.6 Elementary Consequences of the Conservation of J, Π, and M

In the derivation of the cross-section equations in Sections 6.2 and 6.3, we assumed that the interacting particles were spinless. The only angular momentum was that due to the relative motion of the particles; consequently, we assumed that L was conserved in the interaction. Actually, almost all of the particles with which the nuclear engineer must deal have intrinsic spin; so L is not the total angular momentum and need not be conserved. The quantum numbers representing quantities that must be conserved in any nuclear interaction are: the total angular momentum, J; the z component of the total angular momentum, M; and the total parity, Π. In this section, some of the consequences that follow from an elementary consideration of these laws will be developed. In particular, we will show that the Breit–Wigner isolated-resonance formulas must be modified to properly reflect the conservation of J and Π. The exercise in quantum-mechanical combinatorial analysis that follows is an example of "angular momentum bookkeeping," so named for reasons that will soon become apparent.

Consider any neutron-induced interaction that takes place through the excitation and subsequent decay of a single virtual level in the compound nucleus,

$$A + n \to C^* \to B + b.$$

Some very stringent limitations are imposed on the initial and final constellations in this interaction by the fact that the excited virtual level has predetermined J^Π. In effect, this level of the compound nucleus acts as a traffic controller. Only the fraction of the incident neutrons that have proper J and Π in conjunction with the target nucleus can form the compound nucleus. All other neutrons are turned back at the nuclear surface (in the low-energy impenetrable-sphere model) or are refracted by the nuclear potential without forming a compound nucleus (in the higher energy optical model).

The total angular-momentum quantum number J and the total parity Π are conserved in nuclear interactions. Therefore, in any nuclear interaction which proceeds through a compound state, i.e., a virtual level in the compound nucleus which has a definite J^Π, the initial and final constellations must have the same J and Π as the compound state. The projection of J on the z-axis ($M\hbar$) is also conserved in any interaction. The term M may have any one of the $2J + 1$ allowed values between $-J$ and $+J$. The energy of the compound state is assumed to be independent of M; nevertheless, the initial constellation, compound state, and final constellation must have the same value of M. For example, if $J = 3/2$, then the initial constellation, compound state, and final constellation must all have the same M, but M may be any one of the allowed values 3/2, 1/2, $-1/2$, and $-3/2$.

To illustrate the very stringent limitations that these conservation laws impose on the sequence of events in a nuclear interaction, we shall consider first the formation of the compound nucleus and then its decay. The various quantum numbers that are involved are listed in Table 6.3 for continuous reference.

Formation of the Compound Nucleus

The preinteraction configuration consists of an effectively infinite beam of monoenergetic particles with spin s incident on target nuclei with ground-state spin I. The particles and nuclei are unpolarized; i.e., all

Table 6.3—Quantum Numbers Associated with Compound-Nucleus Formation and Decay

---

Constants of Motion Associated with the Virtual Level of the Compound Nucleus, the Initial Constellation, and the Final Constellation

$J \equiv$ the total angular momentum quantum number
$M \equiv$ the azimuthal quantum number ($M \equiv m_J = J, J - 1, \ldots -J$)
$\Pi = \Pi_c \equiv$ the total parity

Parameters that Characterize the Initial Constellation ($A + n$)

$l \equiv$ the orbital angular momentum quantum number
$I \equiv$ the ground-state spin quantum number of the target nucleus
$s \equiv$ the spin quantum number of the neutron
$j \equiv$ the channel spin quantum number or numbers
$m_l, m_I, m_s, m_j \equiv$ the azimuthal quantum numbers associated with l, I, s, and j, respectively
$\Pi_n \equiv$ the intrinsic parity of the neutron
$\Pi_A \equiv$ the intrinsic parity of the target nucleus

Parameters that Characterize the Final Constellation ($B + b$)

$l' \equiv$ the orbital angular momentum quantum number (not necessarily equal to l)
$I' \equiv$ the spin of the residual nucleus (not necessarily in its ground state)
$s' \equiv$ the spin quantum number of the emergent particle
$j' \equiv$ the channel spin quantum number or numbers
$m_l', m_I', m_s', m_j' \equiv$ the azimuthal quantum numbers associated with l', I', s' and j'
$\Pi_b \equiv$ the intrinsic parity of particle b
$\Pi_B \equiv$ the intrinsic parity of the residual nucleus

$2s + 1$ projections of the vector **s** are equally represented in the beam, and all $2I + 1$ projections of the vector **I** are equally represented in the target nuclei. The incident beam consists of particles having all allowed values of angular momentum ($l = 0, 1, 2, \ldots$) relative to the target nuclei. The z-axis is defined as the direction of the incident beam, therefore the projection of the orbital angular-momentum vector **L** along the z-axis is zero; i.e., $L_z = m_l \hbar = 0$ for all values of l.

Now suppose that at energy $E^* = S_n^c + E$, where E is the available energy in the initial constellation, there is a virtual energy level in the compound nucleus that is characterized by a total angular-momentum quantum number J and parity Π_c. Assume that the level is degenerate with respect to the projection of **J**. The first question to be answered is: Which particles in the incident beam can enter the nucleus and form the compound state J^{Π_c}? The answer in its most general, and least useful, form is readily given: Any incident particle can form the compound state provided the sum of its orbital and spin vectors plus the intrinsic spin vector of the target nucleus add vectorially to the total angular momentum vector **J**, i.e., provided

$$\mathbf{J} = \mathbf{L} + \mathbf{s} + \mathbf{I} \tag{6.95}$$

and provided that the total parity in the initial constellation equals Π_c.

To proceed to a more useful formulation of this rule, we introduce the channel spin $\mathbf{j} \equiv \mathbf{I} + \mathbf{s}$. The quantum number j associated with the channel spin can have, according to Equation 6.90, the values

$$|I - s| \le j \le (I + s), \tag{6.96}$$

subject to the condition that successive values of j differ by unity. Thus, for example, if $I = 2$ and $s = 1/2$, then j can take on the values $3/2$ and $5/2$. If $I = 2$ and $s = 1$, then j can take on the values $1, 2,$ and 3.

Substituting $\mathbf{I} + \mathbf{s} = \mathbf{j}$ into Equation 6.95, we find that

$$\mathbf{L} = \mathbf{J} - \mathbf{j}. \tag{6.97}$$

Since the possible values of j are fixed beforehand by the spins of the target and projectile particles and since J is fixed by the level in the compound nucleus, Equation 6.97 imposes a constraint on the possible values of l that can contribute to the formation of the compound nucleus. As an example, consider a beam of neutrons incident on a target nucleus with spin and parity $I^{\Pi_A} = (3/2)^-$, and suppose the J^{Π_c} of the compound nucleus is 2^+. The neutron's spin ($s = 1/2$) must lie either parallel to or antiparallel to the spin of the target nucleus. Hence the channel spin j can assume the value 2 or 1. Now according to Equations 6.97 and 6.90, the value of l must lie in the range

$$|J - j| \le l \le (J + j).\tag{6.98}$$

In our example, therefore, if the channel spin $j = 2$, then l can assume the values 0, 1, 2, 3, and 4. If the channel spin $j = 1$, then l can assume the values 1, 2, and 3. On the basis of conservation of J alone, all partial waves with $l \ge 5$ have been ruled out as possible contributors to the interaction.

Let us now consider conservation of parity. The parity of the compound nucleus in the example under consideration is assumed to be even, hence the initial constellation must have even parity. The intrinsic parity of the ground state of the target nucleus is assumed to be -1, and the intrinsic parity of the neutron is $+1$. The parity of the wave function describing the relative motion of target and neutron must therefore be negative. The orbital parity is given by $(-1)^l$. Hence l must be an odd integer to conserve parity. The only values of l that conserve both J and Π_c are 1 and 3. Thus only the p and f partial waves can possibly contribute to the formation of the compound nucleus in this specific example. The relative contribution of each of these partial waves to the formation of the compound nucleus, which is given by $\sigma_{c,l}$, is directly proportional to the neutron penetrability which, in turn, depends on the neutron's energy. At low neutron energies, the neutron penetration factors b_l go as $(kR)^{2l+1}$; so we would expect only the lowest allowed l value to contribute appreciably. (In the example, $l = 1$ only.) At high neutron energies the penetration factors approach and exceed unity, and all allowed waves contribute to the formation of the compound nucleus. (In the example, $l = 1$ and 3.)

The Statistical Factor in Compound Nucleus Formation

The cross section for single level compound nucleus formation by the lth partial wave has been written in the form

$$\sigma_{c,l} = (2l + 1)\pi \lambda^2 \frac{\Gamma_n \Gamma}{(E - E_0)^2 + (\Gamma/2)^2}.\tag{6.99}$$

This form was derived for spinless particles incident on spinless target nuclei. It cannot be applied to the physically real situation in which the particles have spin because it assumes that all l-wave particles can form a compound nucleus. It does not take into account the conservation of J. All particles incident with a given l and s value cannot, in fact, form a compound nucleus with a given J value. Hence Equation o.99 must be multiplied by the probability that the vectors $\mathbf{s}$ and $\mathbf{I}$ and $\mathbf{L}$ combine to form a vector of fixed length $[J(J + 1)]^{1/2}\hbar$.

There are $2s + 1$, $2I + 1$, and $2l + 1$ equally probable orientations of the vectors $\mathbf{s}$, $\mathbf{I}$, and $\mathbf{L}$. Thus there is a total of $(2s + 1)(2I + 1)(2l + 1)$

possible ways in which the spins and orbital angular momentum of the
initial constellation can be oriented. If at least one of these orientations
adds up to J (otherwise the compound nucleus cannot be formed and
$\sigma_{c,l}$ will equal 0), then $2J + 1$ of the orientations add up to J. (The proof
of this bit of "bookkeeping" will be left as an exercise for the reader.)
Thus, the right-hand side of Equation 6.99 must be multiplied by the
ratio of *favorable orientations*, i.e., orientations in which $\mathbf{J} = \mathbf{L} + \mathbf{I} + \mathbf{s}$,
to *total orientations*

$$\frac{\text{Favorable orientations}}{\text{Total orientations}} = \frac{2J + 1}{(2s + 1)(2I + 1)(2l + 1)},$$ (6.100)

so that

$$\sigma_{c,l} = \frac{2J + 1}{(2s + 1)(2I + 1)} \, \pi \lambda^2 \frac{\Gamma_n \Gamma}{(E - E_0)^2 + (\Gamma/2)^2}.$$ (6.101)

The ratio $(2J + 1)/(2s + 1)(2I + 1)$ is commonly called the *statistical
factor* and is given the symbol $g(J)$,

$$g(J) \equiv \frac{2J + 1}{(2s + 1)(2I + 1)}.$$ (6.102)

All the previous equations for compound elastic-scattering and reaction
cross sections must be corrected by multiplication by the statistical factor.
We shall do this shortly, but first let us investigate the effects of the conservation laws of J, M, and Π on the possible modes of decay of the compound nucleus.

Compound Nucleus Decay

The decay of the compound nucleus is, of course, also governed by all
the conservation laws. In particular the J, Π, and M of the final constellation must equal the J, Π, and M of the virtual level of the compound
nucleus that has decayed.

The intrinsic spin of the emergent particle ($\mathbf{s}'$) and of the residual
nucleus ($\mathbf{I}'$) and the orbital angular momentum of the final constellation
($\mathbf{L}'$) must be related to the total angular momentum of the compound
nucleus ($\mathbf{J}$) by an equation exactly like Equation 6.95

$$\mathbf{J} = \mathbf{L}' + \mathbf{I}' + \mathbf{s}'.$$ (6.103)

(Note that I' is the spin of the residual nucleus B after emitting particle b.
It does not necessarily correspond to the ground-state spin of B.) Once
again defining the channel spin of the final constellation by

$$\mathbf{j}' \equiv \mathbf{I}' + \mathbf{s}',$$ (6.104)

we obtain the condition that **L′** must satisfy to conserve **J**,

$$\mathbf{L'} = \mathbf{J} - \mathbf{j'}. \tag{6.105}$$

That is, the value of l', must lie in the interval

$$|J - j'| \leq l' \leq (J + j'). \tag{6.106}$$

Again, as in the initial constellation, some of these values of l' can be eliminated because they do not conserve parity.

As a specific example, suppose J^{Π} of the compound nucleus is 2^{+}, and suppose the compound nucleus decays by alpha-particle emission to a state with $I' = 1$ and even parity. The intrinsic spin of the α particle is 0; hence the final channel spin $j' = I' = 1$. According to Equation 6.106, the orbital angular momentum of the final constellation can be $l' = 1, 2,$ or 3. Now parity conservation must be considered. Since the parity of the residual nucleus is even and the parity of the α particle is even, l' must also be even for the even parity of the compound state to be conserved. Thus the α particle in this specific case must emerge with $l' = 2$.

Consider any elastic scattering. Since the intrinsic parities of the particles in the initial and final constellations are identical, total parity will be conserved if and only if l and l' have the same character, both even or both odd. This elementary argument leads to an important selection rule for elastic scattering, namely,

$$(l' - l) \equiv \Delta l = 0, \pm 2, \pm 4, \ldots . \tag{6.107}$$

Because of this selection rule, neutron elastic scattering with a change in l is unlikely. At low energies, the neutron penetration factors b_l go as $(kR)^{2l+1}$, where $kR \ll 1$, so that only $l = 0$ waves contribute appreciably to the formation and decay of the compound nucleus. Compound states created by $l = 1$ neutrons are not likely to decay by emitting $l' = 3$ neutrons. At high energies, the competition from other modes of decay of the compound nucleus is so great that elastic scattering with a change of at least 2 units of l may be assumed to be a relatively rare phenomenon.

Formalization of Parity Conservation
In order to formalize parity conservation it is convenient to introduce the following notation:

$\Pi_c \equiv$ the parity of the compound nucleus,
$\Pi_a \equiv$ the intrinsic parity of the projectile particle,
$\Pi_b \equiv$ the intrinsic parity of the emergent particle,
$\Pi_A \equiv$ the parity of the target nucleus,
$\Pi_B \equiv$ the parity of the residual nucleus.

Each of the above symbols can have one of the two values $+1$ and -1. For even parity, $\Pi = +1$; for odd parity, $\Pi = -1$. Since

$$\begin{aligned}\text{Parity of initial constellation} &= (-1)^l \Pi_a \Pi_A, \\ \text{Parity of final constellation} &= (-1)^{l'} \Pi_b \Pi_B,\end{aligned} \qquad (6.108)$$

parity conservation is expressed by the equation

$$(-1)^l \Pi_a \Pi_A = \Pi_c = (-1)^{l'} \Pi_b \Pi_B. \qquad (6.109)$$

It proves convenient to provide for the conservation of parity by introducing the quantity $\omega_l(\Pi_i \Pi_j, \Pi_k)$, which is defined as

$$\omega_l(\Pi_i \Pi_j, \Pi_k) \equiv \begin{cases} 1 & \text{if } [(-1)^l \Pi_i \Pi_j = \Pi_k] \\ 0 & \text{if } [(-1)^l \Pi_i \Pi_j = -\Pi_k]. \end{cases} \qquad (6.110)$$

If this notation is used, the cross section for compound-nucleus formation by l-wave neutrons, Equation 6.101, can be written in the form

$$\sigma_{c,l} = \frac{2J + 1}{(2s + 1)(2I + 1)} \, \omega_l(\Pi_n \Pi_A, \Pi_c) \pi \lambda^2 \, \frac{\Gamma_n \Gamma}{(E - E_0)^2 + (\Gamma/2)^2}, \qquad (6.111)$$

where l is one of the values between $|J - j|$ and $(J + j)$, otherwise $\sigma_{c,l} = 0$. The quantity ω_l automatically ensures the conservation of parity since $\omega_l = +1$ if parity is conserved and $\omega_l = 0$ if parity is not conserved.

Table 6.4 is a list of the spins and parities of all the particles of interest in neutron–nucleus interactions. The ground-state parity of an even-even nucleus, in which all nucleons are paired, is even. The ground-state parity of an odd-A nucleus is determined by the l value of the orbit occupied by the single unpaired nucleon, $\Pi = (-1)^l$. The ground-state parity of an odd-odd nucleus is dependent on the orbits of both unpaired nucleons $\Pi = (-1)^l (-1)^{l'}$. The parities of excited states are dependent on the orbits of the unpaired nucleon or nucleons, which are even or odd depending on products of terms of the form $(-1)^l (-1)^{l'} (-1)^{l''} \ldots$.

6.7 Eigenfunctions of the Angular Schrödinger Equation in the Channel Spin Formalism

The major assumption underlying the channel spin formalism is that the eigenvalues of the operators L^2, j^2, J^2, and J_z are good quantum numbers, i.e., that the operators all commute. To proceed further with the theory, we must find a set of functions that are indeed simultaneous eigenfunctions of all four operators. Specifically, we must find a function Υ_{lj}^{JM} that satisfies

Table 6.4—Spins and Parities of Particles and Nuclei

	Body	Intrinsic total angular momentum quantum number	Intrinsic parity
Elementary particles	n	$s = 1/2$	$+1$
	p	$s = 1/2$	$+1$
	e	$s = 1/2$	$+1$
Simple composite particles	d	$s = 1$	$+1$
	α	$s = 0$	$+1$
Nuclear ground states	Even-even	$I = 0$	$+1$
	Odd-A	$I = $ half-integer	± 1
	Odd-odd	$I = $ integer > 0	± 1
Nuclear excited states	Even-even	$J = $ integer	± 1
	Odd-A	$J = $ half-integer	± 1
	Odd-odd	$J = $ integer	± 1

all the following eigenvalue equations:

$$L^2 \Upsilon_{lj}^{JM} = l(l + 1)\hbar^2 \Upsilon_{lj}^{JM}, \tag{6.112a}$$

$$j^2 \Upsilon_{lj}^{JM} = j(j + 1)\hbar^2 \Upsilon_{lj}^{JM}, \tag{6.112b}$$

$$J^2 \Upsilon_{lj}^{JM} = J(J + 1)\hbar^2 \Upsilon_{lj}^{JM}, \tag{6.112c}$$

$$J_z \Upsilon_{lj}^{JM} = M\hbar \Upsilon_{lj}^{JM}, \tag{6.112d}$$

where the operators J^2 and J_z are given by

$$J^2 \equiv (\mathbf{L} + \mathbf{j}) \cdot (\mathbf{L} + \mathbf{j}) = (L_x + j_x)^2 + (L_y + j_y)^2 + (L_z + j_z)^2,$$

$$J_z \equiv L_z + j_z.$$

(The unit matrix that multiplies each of the components of $\mathbf{L}$ in the above equations is again taken for granted.)

Each of the spherical harmonics $Y_l^{m_l}(\theta, \phi)$ is a solution to Equation 6.112a; each of the spin vectors $\chi_j^{m_j}$ is a solution to Equation 6.112b. It is a trivial exercise to prove that any product of these functions, $Y_l^{m_l}\chi_j^{m_j}$, with $m_l + m_j = M$ is a simultaneous solution to the first two as well as to the last eigenvalue equation. However, this product wave function will not, in general, satisfy Equation 6.112c; i.e., it will not, in general, be an eigenfunction of the operator J^2 with predetermined eigenvalue $J(J + 1)\hbar^2$. But a linear combination of product wave functions can be formed and forced to be a simultaneous eigenfunction of all four operators. This combination takes the form of a sum over the index m_j, the azimuthal quantum number associated with the channel spin,

$$\Upsilon_{lj}^{JM} = \sum_{m_j=-j}^{j} C_{lj}^{JM}(m_j) Y_l^{m_l} \chi_j^{m_j}.$$ (6.113)

Note that the quantum number m_l is not a free index in this summation. Its value is determined by the condition that $M = m_l + m_j$. Since M is fixed beforehand and since we are summing over m_j, once m_j is chosen, m_l is constrained to the value

$$m_l = M - m_j.$$ (6.114)

Let us now examine each of the four quantities in Equation 6.113. From right to left these are the spin functions $\chi_j^{m_j}$, the spherical harmonics $Y_l^{m_l}(\theta,\phi)$, the Clebsch–Gordon coefficients $C_{lj}^{JM}(m_j)$, and, finally, the channel spin wave functions Υ_{lj}^{JM}.

The Spin Functions $\chi_j^{m_j}$

The spin functions $\chi_j^{m_j}$ are the eigenfunctions of the spin operators j^2 and j_z, which are square matrices with $2j + 1$ rows and columns. These functions and operators were examined in Section 6.4. We found that $\chi_j^{m_j}$ is represented by a column vector with $2j + 1$ components, all of which are zero except one, whose value is unity. The component with value 1 appears in the $(j + 1 - m_j)$ position, counting from the top.

The Spherical Harmonics $Y_l^m(\theta,\phi)$

The spherical harmonic Y_l^m, which is the simultaneous eigenfunction of the quantum-mechanical operators L^2 and L_z with eigenvalues $l(l + 1)\hbar^2$ and $m\hbar$, respectively, is given by

$$Y_l^m(\theta,\phi) = \frac{(-1)^{(m+|m|)/2}}{2^l l!} \sqrt{\frac{(2l + 1)(l - |m|)!}{4\pi(l + |m|)!}} (\sin \theta)^{|m|}$$
$$\times \frac{d^{l+|m|}}{d(\cos \theta)^{l+|m|}} (\sin \theta)^{2l} e^{im\phi},$$ (6.115)

where $m \equiv m_l$ can take on any integral value between $-l$ and $+l$. The spherical harmonics are orthonormal over the full 4π solid angle,

$$\int_{4\pi} \overline{Y}_l^m Y_{l'}^{m'} d\Omega \equiv \int_0^{2\pi} d\phi \int_0^{\pi} \overline{Y}_l^m Y_{l'}^{m'} \sin \theta \, d\theta = \delta_{ll'}\delta_{mm'},$$ (6.116)

where $\overline{Y}_l^m$ is the complex conjugate of Y_l^m.

The spherical harmonics satisfy the following readily proved relations:

$$\overline{Y}_l^m = (-1)^m Y_l^{-m},$$ (6.117)

$$(L_x + iL_y)Y_l^m = \hbar \sqrt{(l - m)(l + m + 1)} \, Y_l^{m+1},$$ (6.118a)

$$(L_x - iL_y)Y_l^m = \hbar \sqrt{(l + m)(l - m + 1)} \; Y_l^{m-1}, \tag{6.118b}$$

where L_x and L_y are the operators whose spherical coordinate forms are given by Equation 4.77.

The spherical harmonic Y_l^m is connected with the associated Legendre function P_l^m and the Legendre polynomial P_l by the equations

$$Y_l^m(\theta,\phi) = (-1)^{(m+|m|)/2} \sqrt{\frac{(2l + 1)(l - |m|)!}{4\pi(l + |m|)!}} \; P_l^m(\cos\theta)e^{im\phi}, \tag{6.119a}$$

and

$$Y_l^0(\theta) = \sqrt{\frac{2l + 1}{4\pi}} \; P_l(\cos\theta). \tag{6.119b}$$

The defining equations for the functions P_l^m and P_l are given in Chapter 4.

The Clebsch–Gordon Coefficients $C_{lj}^{JM}(m_j)$

The numerical coefficients $C_{lj}^{JM}(m_j)$ in Equation 6.113 are called Clebsch–Gordon, Wigner, or vector addition coefficients. These are simply mixing fractions and normalization coefficients combined in a single number. Their values are determined in such a way that two conditions are forced on the channel spin wave function, namely,

$$J^2 \Upsilon_{lj}^{JM} = J(J + 1)\hbar^2 \Upsilon_{lj}^{JM} \tag{6.120}$$

and

$$\int_{4\pi} \overline{\Upsilon}_{lj}^{JM} \Upsilon_{lj}^{JM} \, d\Omega = 1. \tag{6.121}$$

The Clebsch–Gordon coefficients may be obtained as follows: We substitute the expression for Υ_{lj}^{JM} from Equation 6.113 into Equation 6.120. The Clebsch–Gordon coefficients are treated as undetermined constants. Because χ_j is a vector with $2j + 1$ components and Y_l is a scalar, the term Υ_{lj}^{JM} is a vector with $2j + 1$ components. The quantity J^2 is a $(2j + 1)$ by $(2j + 1)$ matrix operator. Hence Equation 6.120 is really $2j + 1$ equations. These reduce to $2j + 1$ simultaneous algebraic equations for the $2j + 1$ unknown Clebsch–Gordon coefficients. Because the equations are homogeneous, they determine only the relative values of the Clebsch–Gordon coefficients. Their absolute values are finally determined by the normalization condition, Equation 6.121. In fact, since the spherical harmonics which appear in the Υ functions are orthonormal, Equation 6.121 is actually a normalization condition on the Clebsch–Gordon coefficients. Substituting Equation 6.113 in Equation 6.121, we find that

$$\sum_{m_j = -j}^{j} [C_{lj}^{JM}(m_j)]^2 = 1. \tag{6.122}$$

Actually, one rarely determines Clebsch–Gordon coefficients by the straightforward but laborious procedure outlined above. This procedure was introduced only to demonstrate what simple things the terrible-looking Clebsch–Gordon coefficients really are. They resemble characters of an ancient language because of their profusion of subscripts and superscripts, but behind this formidable facade there lurks a simple real number that, as a matter of fact, is always found to have the form of the square root of a fraction.

Using the methods of group theory, Wigner derived[*] the following closed-form expression for the Clebsch–Gordon coefficients:

$$
C_{lj}^{JM}(m_j) = \sqrt{\frac{(J + l - j)!(J - l + j)!(l + j - J)!(J + M)!(J - M)!(2J + 1)}{(J + l + j + 1)!(l - M + m_j)!(l + M - m_j)!(j - m_j)!(j + m_j)!}}
$$

$$
\times \sum_{\nu} \frac{(-1)^{\nu + j + m_j}(J + j + M - m_j - \nu)!(l - M + m_j + \nu)!}{(J - l + j - \nu)!(J + M - \nu)!(l - j - M + \nu)!\nu!}.
$$

$$(6.123)$$

In the summation, ν takes on all positive integral values, including zero, except those values which make any argument of a factorial negative.

The Clebsch–Gordon coefficients for the channel spins $j = 1/2$ and $j = 1$ are listed in Tables 6.5a and 6.5b. Tables for $j = 3/2$ and 2 may be found in Condon and Shortley's *The Theory of Atomic Spectra*.[†] Numerical values of the coefficients for all l, j, and J less than or equal to 9/2 have been tabulated by Simon.[‡]

The Channel Spin Wave Functions $\Upsilon_{lj}^{JM}(\theta, \phi)$

The most straightforward way to become acquainted with the channel spin wave functions is to consider some specific examples. Suppose we desire to obtain the angular part of the wave function that describes $l = 1$ neutrons incident on a nucleus with ground-state spin $I = 0$. This collision is to be characterized by total angular-momentum quantum number $J = 3/2$ with projection $M = 1/2$ and channel spin $j = 1/2$. (Note that this is but one of the large number of subwaves in the total wave which would be characterized by the four quantum numbers s, l, I, and J.) Setting $l = 1, j = 1/2, J = 3/2$, and $M = 1/2$ in Equation 6.113, we write

$$
\Upsilon_1^{3/2\ 1/2} = C_1^{3/2\ 1/2}(+\tfrac{1}{2})Y_1^0 \begin{pmatrix} 1 \\ 0 \end{pmatrix} + C_1^{3/2\ 1/2}(-\tfrac{1}{2})Y_1^1 \begin{pmatrix} 0 \\ 1 \end{pmatrix}.
$$

[*] E. P. Wigner, *Gruppentheorie und ihre Anwendung auf die Quantenmechanik der Atomspektren*, Friedrich Vieweg und Sohn, Brunswick, 1931.

[†] E. U. Condon and G. H. Shortley, *The Theory of Atomic Spectra*, Cambridge University Press, New York, 1935.

[‡] A. Simon, *Numerical Tables of Clebsch–Gordon Coefficients*, USAEC Report ORNL-1718, Oak Ridge National Laboratory, 1954.

The Clebsch–Gordon coefficients are found, using the first line of Table 6.5a, to be $(2/3)^{1/2}$ and $(1/3)^{1/2}$, respectively. Hence

$$\Upsilon_{1\ 1/2}^{3/2\ 1/2} = \sqrt{\tfrac{2}{3}}Y_1^0 \begin{pmatrix} 1 \\ 0 \end{pmatrix} + \sqrt{\tfrac{1}{3}}Y_1^1 \begin{pmatrix} 0 \\ 1 \end{pmatrix} = \begin{pmatrix} \sqrt{\tfrac{2}{3}}Y_1^0(\theta,\phi) \\ \sqrt{\tfrac{1}{3}}Y_1^1(\theta,\phi) \end{pmatrix}. \tag{6.124}$$

The wave function is seen to be a column vector, each component of which is a spherical harmonic weighted by the appropriate Clebsch–Gordon coefficient. It will be left as an exercise for the reader to prove that this wave function actually satisfies all the conditions imposed by Equations 6.112a to 6.112d.

The number of components in the column vector that represents the channel spin wave function is obviously the same as the number of components in the spin functions $\chi_j^{m_j}$, namely, $2j + 1$. Some of the components may, however, be zero. To illustrate this, consider the following three Υ_{lj}^{JM}, which are characterized by the same set of l, j, and J but different M values.

Table 6.5—Clebsch-Gordon Coefficients $C_{lj}^{JM}(m_j)$

(a) Channel Spin $j = 1/2$

	$m_j = 1/2$	$m_j = -1/2$
$J = l + 1/2$	$\left[\dfrac{l + M + 1/2}{2l + 1}\right]^{1/2}$	$\left[\dfrac{l - M + 1/2}{2l + 1}\right]^{1/2}$
$J = l - 1/2$	$-\left[\dfrac{l - M + 1/2}{2l + 1}\right]^{1/2}$	$\left[\dfrac{l + M + 1/2}{2l + 1}\right]^{1/2}$

(b) Channel Spin $j = 1$

	$m_j = 1$	$m_j = 0$	$m_j = -1$
$J = l + 1$	$\left[\dfrac{(l+M)(l+M+1)}{(2l+1)(2l+2)}\right]^{1/2}$	$\left[\dfrac{(l-M+1)(l+M+1)}{(2l+1)(l+1)}\right]^{1/2}$	$\left[\dfrac{(l-M)(l-M+1)}{(2l+1)(2l+2)}\right]^{1/2}$
$J = l$	$-\left[\dfrac{(l+M)(l-M+1)}{2l(l+1)}\right]^{1/2}$	$\left[\dfrac{M}{[l(l+1)]^{1/2}}\right]$	$\left[\dfrac{(l-M)(l+M+1)}{2l(l+1)}\right]^{1/2}$
$J = l - 1$	$\left[\dfrac{(l-M)(l-M+1)}{2l(2l+1)}\right]^{1/2}$	$-\left[\dfrac{(l-M)(l+M)}{l(2l+1)}\right]^{1/2}$	$\left[\dfrac{(l+M+1)(l+M)}{2l(2l+1)}\right]^{1/2}$

$l = 1, j = 1, J = 1$:

$$(M = -1) \quad \Upsilon^1_{1\,1}{}^{-1} = \begin{pmatrix} 0 \\ -\sqrt{\tfrac{1}{2}}\; Y_1^{-1}(\theta,\phi) \\ \sqrt{\tfrac{1}{2}}\; Y_1^{0}\;(\theta,\phi) \end{pmatrix},$$

$$(M = 0) \quad \Upsilon^1_{1\,1}{}^{0} = \begin{pmatrix} -\sqrt{\tfrac{1}{2}}\; Y_1^{-1}(\theta,\phi) \\ 0 \\ \sqrt{\tfrac{1}{2}}\; Y_1^{1}\;(\theta,\phi) \end{pmatrix},$$

$$(M = +1) \quad \Upsilon^1_{1\,1}{}^{1} = \begin{pmatrix} -\sqrt{\tfrac{1}{2}}\; Y_1^{0}\;(\theta,\phi) \\ \sqrt{\tfrac{1}{2}}\; Y_1^{1}\;(\theta,\phi) \\ 0 \end{pmatrix}.$$

The complex conjugate of a channel spin wave function is a row vector whose components are the complex conjugates of the components in the column vector. For example, the complex conjugate of the wave function immediately preceding is

$$\overline{\Upsilon}^1_{1\,1}{}^{1} = (-\sqrt{\tfrac{1}{2}}\, \overline{Y}_1^{0}, \sqrt{\tfrac{1}{2}}\, \overline{Y}_1^{1}, 0).$$

The product of a channel spin wave function and its complex conjugate is a scalar function; for example,

$$\overline{\Upsilon}^1_{1\,1}{}^{1}\Upsilon^1_{1\,1}{}^{1} = \tfrac{1}{2}\, \overline{Y}_1^{0}Y_1^{0} + \tfrac{1}{2}\, \overline{Y}_1^{1}Y_1^{1}.$$

It is easy to verify, using Equation 6.116, that this product is indeed normalized over the sphere

$$\int_{4\pi} \overline{\Upsilon}^1_{1\,1}{}^{1}\Upsilon^1_{1\,1}{}^{1}\, d\Omega = \tfrac{1}{2}\int_{4\pi} \overline{Y}_1^{0}Y_1^{0}\, d\Omega + \tfrac{1}{2}\int_{4\pi} \overline{Y}_1^{1}Y_1^{1}\, d\Omega = \tfrac{1}{2} + \tfrac{1}{2} = 1,$$

as, of course, will always be the case since the Clebsch–Gordon coefficients were deliberately defined to ensure this normalization.

Generalization
Equation 6.113, which defines the channel spin wave functions, is a special case of a more general relation. The simultaneous eigenfunctions of three angular-momentum operators related by the equation $\mathbf{J} = \mathbf{j}' \pm \mathbf{j}''$ are

$$\Upsilon^{JM}_{j'j''} = \sum_{m''=-j''}^{j''} C^{JM}_{j'j''}(m'')A_{j'}^{m'}A_{j''}^{m''}, \tag{6.125}$$

where the functions $A_{j'}^{m'}$ and $A_{j''}^{m''}$ are either spherical harmonics or spin wave functions depending on whether their respective operators $\mathbf{j}'$ and $\mathbf{j}''$ are orbital angular-momentum or spin angular-momentum operators. As in the special case previously considered, $m' = M - m''$. Equation 6.125 allows us to find the simultaneous eigenfunctions of any triplet of commuting angular momenta related by $\mathbf{J} = \mathbf{j}' \pm \mathbf{j}''$ regardless of whether $\mathbf{j}'$ and $\mathbf{j}''$ are spin operators, orbital operators, or a mixture of both types. The same Clebsch–Gordon coefficients apply in all cases. Equation 6.125 can be used, for example, to set up wave functions in the j-j and L-S coupling schemes or, indeed, in any other coupling scheme that seems appropriate.

6.8 Channel Spin Analysis of Neutron–Nucleus Interactions

The method of partial waves will now be employed to obtain the elastic-scattering and reaction cross sections in the channel spin formalism. The procedure for handling particles with spin differs from the partial-wave analysis of spinless particles in that the partial waves are now characterized by the four quantum numbers l, j, J, and M instead of only l. This substantially increases the tediousness of the angular-momentum bookkeeping without, however, changing the basic procedure.

We can identify eight steps in the procedure for finding the cross sections for particles that have spin. The initial steps, which are obvious generalizations of analogous steps in the case of spinless particles, will merely be outlined. For details, the reader is referred to Blatt and Weisskopf.*

> *Step 1.* Write the general expression for the total (incoming and outgoing) partial-wave function that corresponds to an incoming partial wave of neutrons labeled by the quantum numbers l, j, J, and M:
>
> $$r\psi \equiv \Phi(ljJM, \mathbf{r})$$
>
> $$= A\left[e^{-i[kr - (l\pi/2)]}\Upsilon_{lj}^{JM} \sum_{l'=|J-j|}^{J+j} \mathsf{U}_{ll'}(jJ)e^{i[kr - (l'\pi/2)]}\Upsilon_{l'j}^{JM} \right]$$

It will be noted that the reflection factors U_l are replaced in this more general case by the elements $\mathsf{U}_{ll'}$ of the collision matrix $\mathbf{U}$. The complex number $\mathsf{U}_{ll'}(jJ)$ is the coefficient of the outgoing partial wave

---

*John M. Blatt and Victor F. Weisskopf, *Theoretical Nuclear Physics* (Reference 1 at the end of this chapter).

that is characterized by the quantum numbers l', j, J. There are also outgoing waves with channel spin $j' \neq j$. Since, however, the contributions of neutrons with distinct channel spins add incoherently, these neutrons are most easily treated as reaction products. The contribution of these neutrons to the elastic-scattering cross section must eventually be included since the measured elastic-scattering cross section does not distinguish between neutrons with various channel spins.

Step 2. Expand the incident plane wave in eigenfunctions of the operators $\mathbf{L}$, $\mathbf{j}$, and $\mathbf{J}$. Noting that $m_j = M - m_l = M$ since the neutrons are incident along the z-axis, we find that

$$\Phi_{\text{inc}}(jM, \mathbf{r}) = re^{ikz}\chi_j^M$$

$$= \frac{\sqrt{\pi}}{k} \sum_{l=0}^{\infty} \sum_{J=|l-j|}^{l+j} i^{l+1}\sqrt{2l+1}$$

$$\times C_{lj}^{JM}(M)\{e^{-i[kr-(l\pi/2)]} - e^{i[kr-(l\pi/2)]}\}\Upsilon_{lj}^{JM}.$$

Step 3. Compare the incoming parts of these two expressions to determine the value of the coefficient A,

$$A = \frac{\sqrt{\pi}}{k}\sqrt{2l+1}\,i^{l+1}C_{lj}^{JM}(M).$$

Step 4. Subtract the incident partial wave from the total partial wave to obtain the equation for the scattered partial wave:

$$\Phi_{\text{sc}}(ljJM, \mathbf{r})$$

$$= \frac{\sqrt{\pi}}{k}\sqrt{2l+1}\,i^{l+1}C_{lj}^{JM}(M) \sum_{l'=|J-j|}^{J+j} [\delta_{ll'} \quad \mathsf{U}_{ll'}]e^{i[kr-(l'\pi/2)]}\Upsilon_{l'j}^{JM}.$$

Step 5. Add the scattered partial waves corresponding to all allowed l values to get

$$\Phi_{\text{sc}}(jJM, \mathbf{r}) \equiv \sum_{l=|J-j|}^{J+j} \Phi_{\text{sc}}(ljJM, \mathbf{r}).$$

Step 6. The coefficient of e^{ikr} in the expression for $\Phi_{\text{sc}}(jJM,\mathbf{r})$ is the elastic-scattering amplitude for the jJM partial wave, $f(jJM,\theta\phi)$. The partial differential-scattering cross section is given by the familiar expression

$$\sigma_n(jJM,\theta\phi) = |f(jJM,\theta\phi)|^2.$$

Step 7. The partial elastic-scattering cross section is given by the integration of the differential cross section over all solid angles:

$$\sigma_n(jJM) = \int_{4\pi} |f(jJM,\theta\phi)|^2 \, d\Omega,$$

$$\sigma_n(jJM) = \pi\lambda^2 \sum_{l'=|J-j|}^{J+j} \left| \sum_{l=|J-j|}^{J+j} i^{l+1}\sqrt{2l+1} \, C_{lj}^{JM}(M)[\delta_{ll'} - U_{ll'}(jJ)] \right|^2 .$$

$$(6.126)$$

Step 8. The partial reaction cross section is obtained by dividing the net incoming current by the speed of the neutrons as in the derivation of Equation 6.16. We find that

$$\sigma_r(jJM) = \pi\lambda^2 \left\{ \sum_{l=|J-j|}^{J+j} (2l+1)[C_{lj}^{JM}(M)]^2 \right.$$

$$\left. \sum_{l'=|J-j|}^{J+j} \left| \sum_{l=|J-j|}^{J+j} i^{l+1}\sqrt{2l+1} \, C_{lj}^{JM}(M)U_{ll'}(jJ) \right|^2 \right\}. \quad (6.127)$$

These very general results are now specialized to the situation of greatest practical importance, namely, when $l' = l$, since, as explained in Section 6.5, elastic scattering with a change in l is a relatively improbable event. Thus we assume that $U_{ll'} = U_l \, \delta_{ll'}$ so that

$$\sigma_n(jJM) = \pi\lambda^2 \sum_{l=|J-j|}^{J+j} (2l+1)[C_{lj}^{JM}(M)]^2 |1 - U_l(jJ)|^2 \qquad (6.128)$$

and

$$\sigma_r(jJM) = \pi\lambda^2 \sum_{l=|J-j|}^{J+j} (2l+1)[C_{lj}^{JM}(M)]^2 [1 - |U_l(jJ)|^2]. \qquad (6.129)$$

These equations, which are the generalizations of Equations 6.15 and 6.16, reduce identically to the latter equations if the channel spin j equals zero, for in this case the summation has but one term, $l = J$, and the Clebsch–Gordon coefficient for $j = 0$ is identically unity. Hence

$$\sigma_{n,l} = \sigma_n(0lM) = \pi\lambda^2(2l+1)|1 - U_l|^2$$

and

$$\sigma_{r,l} = \sigma_r(0lM) = \pi\lambda^2(2l+1)(1 - |U_l|^2).$$

Equations 6.126 and 6.127 are simplified by averaging the cross sections over the possible orientation $M = m_j$ of the incident wave, which is assumed to be unpolarized,

$$\sigma_{n,r}(jJ) = \frac{1}{2j+1} \sum_{M=-j}^{j} \sigma_{n,r}(jJM). \qquad (6.130)$$

This averaging process eliminates the Clebsch–Gordon coefficients, which obey the relation

$$\sum_{M=-j}^{j} [C_{lj}^{JM}(M)]^2 = \frac{2J + 1}{2I + 1}.$$

(6.131)

Thus,

$$\sigma_n(jJ) = \frac{2J + 1}{2j + 1}\,\pi\lambda^2 \sum_{l=|J-j|}^{J+j} |1 - U_l(jJ)|^2,$$

$$\sigma_r(jJ) = \frac{2J + 1}{2j + 1}\,\pi\lambda^2 \sum_{l=|J-j|}^{J+j} [1 - |U_l(jJ)|^2].$$

(6.132)

The partial cross sections must now be averaged over all the channel spins that can contribute to the formation of the compound nucleus in state J^{π_c}. The statistical weight of a channel spin j is the ratio of the number of orientations of $\mathbf{j}$ to the total number of orientations of $\mathbf{I}$ and $\mathbf{s}$,

$$g(j) = \frac{2j + 1}{(2s + 1)(2I + 1)}.$$

(6.133)

We force parity conservation by inserting the previously defined quantity $\omega_l(\Pi_n\Pi_A,\Pi_c)$ into the equation, where $\Pi_n = +1$ is the parity of the neutron, Π_A is the parity of the ground state of the target nucleus, and Π_c is the parity of the compound nuclear state. When we insert $g(j)$ and ω_l and average, the result is

$$\sigma_n(J) = \pi\lambda^2 \frac{2J + 1}{(2s + 1)(2I + 1)} \sum_{j=|I-s|}^{I+s} \sum_{l=|J-j|}^{J+j} \omega_l(\Pi_n\Pi_A,\Pi_c)|1 - U_l(jJ)|^2,$$

(6.134)

$$\sigma_r(J) = \pi\lambda^2 \frac{2J + 1}{(2s + 1)(2I + 1)} \sum_{j=|I-s|}^{I+s} \sum_{l=|J-j|}^{J+j} \omega_l(\Pi_n\Pi_A,\Pi_c)[1 - |U_l(jJ)|^2].$$

(6.135)

These are the final expressions for the elastic-scattering and reaction cross sections for an interaction that takes place through the formation and decay of a compound-nucleus state of total angular momentum J and parity Π_c. These equations lack complete generality only in the assumption that there is no change in l in the elastic scattering. This assumption is generally justified by the relative rarity of an elastic scattering in which l changes. When it is not, the general forms, Equations 6.126 and 6.127, must be employed to determine the cross sections.

6.9 Breit–Wigner Isolated-Resonance Formulas for Neutron–Nucleus Interactions

The Breit–Wigner isolated resonance equations are obtained by substituting the proper form of the reflection factors $U_{ll'}(jJ)$ into Equations 6.126 and 6.127. The procedure for obtaining the reflection factors for particles that have spin is essentially identical to that employed in Sections 6.2 and 6.3 for spinless particles and will not be repeated here.

The cross section for the reaction $A + n \rightarrow (J^{\Pi_c}) \rightarrow B + b$ in the vicinity of an isolated resonance takes the form

$$\sigma_b(E) = \pi \lambda^2 g(J)$$

$$\times \frac{\left[\sum_j \sum_l \omega_l (\Pi_n \Pi_A, \Pi_c) \Gamma_n(lj) \right] \cdot \left[\sum_{j'} \sum_{l'} \omega_{l'} (\Pi_b \Pi_B, \Pi_c) \Gamma_b(l'j') \right]}{(E - E_0)^2 + (\Gamma/2)^2} ,$$

$$(6.136)$$

where $g(J) = (2J + 1)/(2s + 1)(2I + 1)$. The summation over the channel spin j contains only the two terms $j = I \pm 1/2$ if $I > 0$, and only one term $j = 1/2$ if $I = 0$. The outgoing channel spin quantum number j' takes on all values between $|I' - s'|$ and $(I' + s')$ by integers. The sums over l and l' include all values consistent with the total angular-momentum quantum number J, i.e.,

$$|J - j| \leq l \leq (J + j),$$

$$|J - j'| \leq l' \leq (J + j').$$

The functions ω_l and $\omega_{l'}$ eliminate from the summation all values of l and l' that are inconsistent with the conservation of parity.

The elastic-scattering cross section in the vicinity of an isolated resonance is rather complicated. It may be expressed as the sum of four terms, each of which has a definite source.

$$\sigma_n(E) = P(E) + R(E) + O(E) + S(E). \qquad (6.137a)$$

The first term is simply the cross section for the potential scattering which would be present if there were no resonance near the energy E,

$$P(E) = \pi \lambda^2 \sum_{l=0}^{\infty} (2l + 1) |A_{se}^l|^2. \qquad (6.137b)$$

The second term is the cross section for resonance scattering without change of orbital angular momentum or channel spin. This term shows the usual interference between compound elastic and shape elastic scattering,

$$R(E) = \pi \lambda^2 g(J) \sum_j \sum_l \omega_l (\Pi_n \Pi_A, \Pi_c)$$

$$\times \left[\left| \frac{i\Gamma_n(lj)}{E - E_0 + i\Gamma/2} + A_{se}^l \right|^2 - |A_{se}^l|^2 \right].$$

(6.137c)

It is readily verified that $R(E)$ vanishes far from the resonance energy E_0. The summations over j and l contain the usual terms: $j = I \pm 1/2$, $|J - j| \le l \le (J + j)$.

The third term is the cross section for elastic scattering in which the orbital angular momentum changes but the channel spin does not.

$$O(E) = \pi \lambda^2 g(J) \sum_j \sum_l \sum_{l'} \omega_l (\Pi_n \Pi_A, \Pi_c) \omega_{l'} (\Pi_n \Pi_A, \Pi_c) \frac{\Gamma_n(lj)\Gamma_n(l'j)}{(E - E_0)^2 + (\Gamma/2)^2}.$$

(6.137d)

In these summations j, l, and l' range over the usual values except that terms with $l = l'$ are omitted.

The final term is the cross section for elastic scattering in which the channel spin changes from $I - 1/2$ to $I + 1/2$ or vice versa.

$$S(E) = \pi \lambda^2 g(J) \sum_j \sum_{j'} \sum_l \sum_{l'}$$

$$\times \omega_l (\Pi_n \Pi_A, \Pi_c) \omega_{l'} (\Pi_n \Pi_A, \Pi_c) \frac{\Gamma_n(lj)\Gamma_n(l'j')}{(E - E_0)^2 + (\Gamma/2)^2}.$$

(6.137e)

In the summations over j and j', the terms with $j = j'$ are omitted. The sums over l and l' cover the usual ranges.

Equations 6.137 contain all possible contributions to the elastic-scattering cross section in the isolated resonance region. Usually this cross section is expressed as the sum of $P(E)$ and $R(E)$ only. The contribution of $O(E)$ to the cross section is very small because elastic scattering with a change of l is unlikely. The term $S(E)$, which is the scattering cross section with change of the channel spin, vanishes identically if the target nucleus is even-even, for in this most common case $I = I' = 0$ and therefore $j = j' = 1/2$. It also vanishes for s-wave ($l = 0$) resonances because only the channel spin $j = j' = J$ can contribute. Finally, it vanishes if the spin of the compound state is zero or 1/2, because in this case the allowed values of l and l' are j and $j + 1$, but l must change by multiples of two units in elastic scattering. Hence l cannot change, and therefore j cannot change. Thus spin-flipping elastic scattering can occur only if I and l are greater than zero and $J > 1/2$.

The cross section assumes its maximum value at the peak of an $l = 0$ pure elastic-scattering resonance.

$$\sigma_n(E_0) = \frac{2J + 1}{(2s + 1)(2I + 1)} 4\pi\lambda_0^2.$$

Hence an experimental determination of E_0 and $\sigma_n(E_0)$ provides a measure of the angular momentum J of the resonance level of the compound nucleus, assuming, of course, that the spin I of the target nucleus is known. The fact that $l = 0$ neutrons are responsible for a particular resonance can be inferred from the isotropic angular distribution of the scattered neutrons.

The Breit–Wigner equations give reliable predictions of isolated resonance cross sections provided *empirical* values of the level widths and observed resonance energy are used. Nuclear theory in its present state can give only order-of-magnitude estimates of these parameters. The actual values of the constants E_0, Γ_m, Γ, and so forth, are so chosen that when they are substituted into the Breit–Wigner equations, the equations do indeed give the experimentally observed cross sections in the vicinity of an isolated resonance.

6.10 The General Theory of Nuclear Interactions

In this section attention will be directed to the general formal theory for the cross sections for all nuclear interactions that proceed from an initial constellation $a + A$ to any of a set of possible final constellations.

$$a + A \rightarrow \begin{cases} A + a \\ A^* + a' \\ B + b \\ C + c \\ \vdots \end{cases}$$

The development will be restricted to interactions in which all members of the initial and final constellations are massive particles. Inasmuch as photons have zero rest mass, reactions that involve gamma rays are excluded. In Chapter 9, gamma rays are considered, and the theory advances one step further in generality. Although the theory can be readily generalized to three or more particles, consideration is restricted here to reactions in which both constellations consist of two particles. With these exceptions, the theory which is presented in this section is completely general; it applies equally well to direct reactions and to reactions that proceed through a compound nuclear state. Furthermore, because the nucleus is treated as a black box, the theory applies regardless of the particular model used to describe the nucleus.

Insofar as possible, we shall employ the notation of Lane and Thomas, whose review article[†] constitutes a most thorough and readable exposition of both the general theory of nuclear reactions and R-matrix theory. The reader is urged to supplement the material in this section and the next by a careful reading of Lane and Thomas.

Definitions

The totality of quantum numbers that describe one particular partial wave in the initial constellation is called an *entrance channel* and is designated by the symbol c. The symbol c implies any exhaustive set of quantities that describe one particular partial wave. In practice, the two sets with which we shall be concerned are

$$c = \{\alpha l j m \nu\}$$

and

$$c = \{\alpha l j J M\}.$$

The symbol α designates the internal variables of the pair of particles in the initial constellation. Thus α includes the types of particles, their degrees of internal excitation, if any, and their spins; i.e., $\alpha = \{AaE_A^* E_a^* I_A I_a\}$. The quantum numbers l, j, J, and M have been defined previously. The symbols m and ν are used in place of the previously defined m_l and m_j, respectively, in order to avoid subscripted subscripts.

The *exit channels* are similarly designated by exhaustive sets of quantum numbers:

$$c' = \{\alpha' l' j' m' \nu'\}$$

or

$$c' = \{\alpha' l' j' J' M'\}.$$

We will assume that for any channel c there exists a finite radial distance of separation, a_c, beyond which the nuclear forces are zero. For distances greater than the *channel radius* a_c, we assume that no force is operative on neutrons and that only the long-range Coulomb force is operative in the case of charged particles. A commonly used prescription for a_c is the sum of the radii of the particles in the constellation,

$$a_c \equiv a_\alpha = R_0(A_1^{1/3} + A_2^{1/3}), \tag{6.138}$$

where A_1 and A_2 are the mass numbers of the two reactants, and R_0 is taken to be approximately 1.5 fm.

[†] A. M. Lane and R. G. Thomas, R-Matrix Theory of Nuclear Reactions, *Rev. Mod. Phys.*, 30: 257 (1958).

In addition, we define the following channel quantities, all of which are quite familiar from preceding chapters:

The reduced mass of the pair of particles in channel c is

$$M_c \equiv M_\alpha \equiv \frac{M(A)M(a)}{M(A) + M(a)}.$$

The energy of relative motion is defined as

$$E_c \equiv E_\alpha.$$

The wave number is given by

$$k_c \equiv k_\alpha \equiv \frac{\sqrt{2M_\alpha|E_\alpha|}}{\hbar}.$$

The relative velocity is

$$v_c \equiv v_\alpha = \frac{\hbar k_\alpha}{M_\alpha}.$$

The distance measured in reduced de Broglie wavelengths is

$$\rho_c \equiv \rho_\alpha \equiv k_\alpha r_\alpha.$$

Channel Wave Functions

The complete channel wave functions corresponding to incoming ($\mathscr{I}$) and outgoing ($\mathcal{O}$) waves normalized to one particle per second through any sphere larger than a_c centered at the origin are given in the $(\alpha l j m v)$ scheme by

$$\mathscr{I}_{\alpha l j m v} = (i^l Y_l^m) \chi_{\alpha j}^v \frac{I_{\alpha l}}{r_\alpha \sqrt{v_\alpha}} \tag{6.139a}$$

and

$$\mathcal{O}_{\alpha l j m v} = (i^l Y_l^m) \chi_{\alpha j}^v \frac{O_{\alpha l}}{r_\alpha \sqrt{v_\alpha}}, \tag{6.139b}$$

where I and O are the radial parts of the incoming and outgoing wave functions, respectively. The channel wave functions in the $(\alpha l j J M)$ scheme are related to those in the $(\alpha l j m v)$ scheme by

$$\mathscr{I}_{\alpha l j J M} = \sum_{v+m=M} C_{lj}^{JM}(M) \mathscr{I}_{\alpha l j m v} \tag{6.140a}$$

and

$$\mathcal{O}_{\alpha l j J M} = \sum_{v+m=M} C_{lj}^{JM}(M) \mathcal{O}_{\alpha l j m v}. \tag{6.140b}$$

Radial Wave Functions

The radial wave functions I and O are the two linearly independent solutions to the radial Schrödinger equation at points external to the nucleus. In the case of charged particles the relevant differential equation is Equation 4.69,

$$W''_{\alpha l}(\rho_\alpha) + \left[1 - \frac{2\eta_\alpha}{\rho_\alpha} - \frac{l(l+1)}{\rho_\alpha^2} \right] W_{\alpha l}(\rho_\alpha) = 0, \tag{6.141}$$

where the prime indicates differentiation with respect to the dimensionless variable $\rho_\alpha \equiv k_\alpha r_\alpha$. The Coulomb field parameter η_α is given by

$$\eta_\alpha = \frac{(Z_A Z_a e^2) k_\alpha}{2 E_\alpha} = \frac{Z_A Z_a e^2}{\hbar v_\alpha}. \tag{6.142}$$

For positive-energy channels, that is, for positive values of E_α, the asymptotic forms of the solutions of Equation 6.141 may be written in terms of the *regular* (F) and *irregular* (G) Coulomb wave functions, whose asymptotic forms for large ρ_α are

$$F_c \equiv F_{\alpha l} \sim \sin \left[\rho_\alpha - \eta_\alpha \ln 2\rho_\alpha - (l\pi/2) + \zeta_{\alpha l} \right], \tag{6.143a}$$

$$G_c \equiv G_{\alpha l} \sim \cos \left[\rho_\alpha - \eta_\alpha \ln 2\rho_\alpha - (l\pi/2) + \zeta_{\alpha l} \right]. \tag{6.143b}$$

We can readily verify that the following combinations of solutions represent incoming and outgoing radial waves:

$$I_c \equiv I_{\alpha l} \sim (G_c - iF_c)e^{i\omega_c}, \tag{6.144a}$$

$$O_c \equiv O_{\alpha l} \sim (G_c + iF_c)e^{i\omega_c}, \tag{6.144b}$$

where ω_c, the *Coulomb phase shift*, which determines the magnitude of the purely electrostatic scattering, is given by

$$\omega_c \equiv \omega_{\alpha l} = \zeta_{\alpha l} - \zeta_{\alpha 0} = \sum_{n=1}^{l} \tan^{-1}(\eta_\alpha/n). \tag{6.145}$$

In the absence of a Coulomb field (for example, if one of the particles in the constellation α is a neutron), the Coulomb field parameter η_α equals 0, and the I and O functions are the usual combination of spherical Bessel and Neumann functions multiplied by ρ_α since the "inverse square" falloff of the probability density has already been taken explicitly into account in the expressions for $\mathscr{I}$ and $\mathscr{O}$. Thus, we may write, for neutrons,

$$I_c \equiv I_{\alpha l} = -i\rho_\alpha[j_l(\rho_\alpha) - in_l(\rho_\alpha)] \tag{6.146a}$$

and

$$O_c \equiv O_{\alpha l} = -i\rho_\alpha[j_l(\rho_\alpha) + in_l(\rho_\alpha)]. \tag{6.146b}$$

The Collision Matrix

The total wave function in the region outside the nucleus can be expressed, very generally, as a linear combination of incoming and outgoing wave functions in all possible entrance and exit channels:

$$\psi = \sum_c y_c \mathcal{I}_c + \sum_{c'} x_{c'} \mathcal{O}_{c'}. \tag{6.147}$$

The numbers y_c and $x_{c'}$ are the amplitudes of the incoming waves $\mathcal{I}_c$ and the outgoing waves $\mathcal{O}_{c'}$, respectively. For any given system of interacting particles, the numbers y_c are known and the numbers $x_{c'}$ must be determined by the nature of the forces between the interacting particles.

The *collision matrix* **U** is defined as the matrix of numbers that gives an expression for the unknown $x_{c'}$ in terms of the known y_c. By convention, **U** is introduced through the definition

$$x_{c'} \equiv -\sum_c U_{c'c} y_c. \tag{6.148}$$

Substituting this expression for $x_{c'}$ into Equation 6.147, we find the general expression for the total wave function:

$$\psi = \sum_c y_c \left[\mathcal{I}_c - \sum_{c'} U_{c'c} \mathcal{O}_{c'} \right]. \tag{6.149}$$

The preceding three equations can be expressed succinctly in matrix notation, as follows:

$$\psi = \mathbf{y}\mathcal{I} + \mathbf{x}\mathcal{O}, \tag{6.147'}$$

$$\mathbf{x} \equiv -\mathbf{U}\mathbf{y}, \tag{6.148'}$$

$$\psi = \mathbf{y}[\mathcal{I} - \mathbf{U}\mathcal{O}]. \tag{6.149'}$$

The Outgoing Amplitudes

The outgoing amplitudes are obtained by the usual procedure of subtracting the incoming wave from the total wave in the asymptotic region so as to obtain the outgoing wave. Only the results are presented here; the details may be found in Lane and Thomas' article. The scattering or reaction amplitude is found to be given by

$$f_{\alpha'j'v',\alpha jv}(\Omega_{\alpha'}) = \sqrt{\pi}\,\lambda_\alpha \left\{ -C_{\alpha'}(\theta_{\alpha'})\,\delta_{\alpha'j'v',\alpha jv} + i \sum_{l'm'l} \sqrt{2l+1} \right. \tag{6.150}$$

$$\left. \times [e^{2i\omega_{\alpha'l'}}\,\delta_{\alpha'j'l'v'm',\alpha jlvo} - U_{\alpha'j'l'v'm',\alpha jlvo}]Y_{l'}^{m'}(\Omega_{\alpha'}) \right\}.$$

The function $C_{\alpha'}(\theta_{\alpha'})$ is the amplitude of the Coulomb scattered wave, if it exists. It is given by

$$C_{\alpha'}(\theta_{\alpha'}) = \frac{\eta_{\alpha}}{\sqrt{4\pi}} \operatorname{cosec}^2\left(\frac{\theta_{\alpha'}}{2}\right) \exp\left[-2i\eta_{\alpha'} \ln \sin\left(\frac{\theta_{\alpha'}}{2}\right)\right] \tag{6.151}$$

The symbol $\delta_{a'b'c'\cdots,abc\cdots}$ has the meaning

$$\delta_{a'b'c'\cdots,abc\cdots} \equiv \delta_{a'a}\,\delta_{b'b}\,\delta_{c'c}\cdots, \tag{6.152}$$

where each of the doubly subscripted deltas is a Kronecker delta. The symbol $(\Omega_{\alpha'})$ means $(\theta_{\alpha'},\phi_{\alpha'})$, the C-system polar and azimuthal angles of the final constellation.

For a neutron-induced interaction, Equation 6.150 reduces to

$$f_{\alpha'j'v',\alpha jv}(\Omega_{\alpha'}) = \pi\lambda_{\alpha}i \sum_{l'm'l} \sqrt{2l+1}$$

$$\times [\delta_{\alpha'j'l'v'm',\alpha jlvo} - U_{\alpha'j'l'v'm',\alpha jlvo}]Y_{l'}^{m'}(\Omega_{\alpha'}). \tag{6.153}$$

Differential Cross Sections

The differential cross section for the interaction $\alpha jv \to \alpha'j'v'$ is, by definition, the absolute square of the outgoing amplitude,

$$\sigma_{\alpha jv,\alpha'j'v'}(\Omega_{\alpha'}) = |f_{\alpha'j'v',\alpha jv}(\Omega_{\alpha'})|^2. \tag{6.154}$$

For unpolarized particles in the initial constellation, the preceding cross section may be summed over v' and averaged with respect to v to obtain the differential cross section for the interaction $\alpha j \to \alpha'j'$,

$$\sigma_{\alpha j,\alpha'j'}(\Omega_{\alpha'}) = \frac{1}{2j+1}\sum_{vv'} |f_{\alpha'j'v',\alpha jv}(\Omega_{\alpha'})|^2. \tag{6.155}$$

In the same way, Equation 6.155 may be summed over j' and averaged with respect to j to obtain the observed differential cross section for the reaction $\alpha \to \alpha'$,

$$\sigma_{\alpha\alpha'}(\Omega_{\alpha'}) = \frac{1}{(2I+1)(2s+1)}\sum_{jj'vv'} |f_{\alpha'j'v',\alpha jv}(\Omega_{\alpha'})|^2, \tag{6.156}$$

where I and s are the spins of the nucleus and the incident particle in the initial constellation, respectively.

To perform the absolute squaring operation indicated in Equation 6.156, we transform U from the $(\alpha jlvm)$ scheme to the $(\alpha jlJM)$ scheme and then replace the absolute square by a double sum over the indices $(J_1M_1l_1l'_1m'_1)$ and $(J_2M_2l_2l'_2m'_2)$. The result is that Equation 6.155

becomes*

$$\sigma_{\alpha j, \alpha' j'}(\Omega_{\alpha'}) = \pi \lambda_\alpha^2 \left\{ |C_{\alpha'}(\theta_{\alpha'})|^2 \, \delta_{\alpha' j', \alpha j} + \frac{1}{\pi} \sum_L \frac{B_L(\alpha' j', \alpha j)}{(2j + 1)} \, P_L(\cos \theta_{\alpha'}) \right.$$

$$\left. + \frac{1}{\sqrt{4\pi}} \sum_{Jl} \frac{(2J + 1)}{(2j + 1)} 2 \, \mathrm{Re} \left[iT^J_{\alpha' j' l', \alpha j l} C_{\alpha'}(\theta_{\alpha'}) P_l(\cos \theta_{\alpha'}) \right] \right\}, \tag{6.157}$$

where, by definition,

$$T^J_{\alpha' j' l', \alpha j l} \equiv e^{2i\omega_{\alpha'} l'} \, \delta_{\alpha' j' l', \alpha j l} - U^J_{\alpha' j' l', \alpha j l} \tag{6.158}$$

and

$$B_L(\alpha' j', \alpha j) = \frac{(-1)^{j - j'}}{4} \sum_{J_1 J_2 l_1 l_2 l_1' l_2'} \tilde{Z}(l_1 J_1 l_2 J_2, jL)$$

$$\times \, \tilde{Z}(l_1' J_1 l_2' J_2, j'L) T^{J_1}_{\alpha' j' l_1', \alpha j l_1} \overline{T}^{J_2}_{\alpha' j' l_2', \alpha j l_2}. \tag{6.159}$$

The $\tilde{Z}$ coefficients are related to the tabulated Z coefficients of Blatt and Biedenharn† by the equation

$$\tilde{Z}(l_1 J_1 l_2 J_2, jL) = i^{l_1 - l_2 - L} Z(l_1 J_1 l_2 J_2, jL). \tag{6.160}$$

The matrix elements of U in the $(\alpha j l v m)$ and the $(\alpha j l J M)$ schemes are related by

$$U_{\alpha' j' l' v' m', \alpha j l v m} = \sum_{JM} C^{JM}_{lj}(v + m) C^{JM}_{j'l'}(v' + m') U^J_{\alpha' j' l', \alpha j l}. \tag{6.161}$$

Integrated Cross Sections

The integrated cross section for the reaction $aj \to \alpha' j'$ is obtained by integrating the differential cross section (Equation 6.157) over the solid angle $\Omega_{\alpha'}$. Setting $C_{\alpha'} = 0$, in order to eliminate the physically unrealistic infinite contribution to the integrated elastic scattering from the Coulomb field, we can write the integral as

$$\sigma_{\alpha j, \alpha' j'} = \frac{\lambda_\alpha^2}{(2j + 1)} \int_0^\pi \sum_L B_L(\alpha' j', \alpha j) P_L(\cos \theta_{\alpha'}) 2\pi \sin \theta_{\alpha'} \, d\theta_{\alpha'}.$$

Because of the orthogonality of the Legendre polynomials, all contributions to the integral are zero except that of the term containing P_0. Using the relation

$$\tilde{Z}(l_1 J_1 l_2 J_2, j0) = (-1)^{J_1 - j} \sqrt{2J_1 + 1} \, \delta_{l_1 l_2} \delta_{J_1 J_2}, \tag{6.162}$$

---

*See A. M. Lane and R. G. Thomas, R-Matrix Theory of Nuclear Interactions, *Rev. Mod. Phys.*, **30**: 292 (1958).

†J. M. Blatt and L. C. Biedenharn, *Rev. Mod. Phys.*, **24**: 258 (1952).

we get the very important result

$$\sigma_{\alpha j, \alpha' j'} = \frac{\pi \lambda_\alpha^2}{(2j + 1)} \sum_{Jll'} (2J + 1) |T^J_{\alpha' j'l', \alpha jl}|^2. \tag{6.163}$$

The observed cross section for the reaction $\alpha \to \alpha'$ (assuming unpolarized particle and nuclei in the initial constellation) is obtained by summing over j' and averaging over j:

$$\sigma_{\alpha \alpha'} = \pi \lambda_\alpha^2 \sum_{Jll'jj'} g(J) |T^J_{\alpha' j'l', \alpha jl}|^2, \tag{6.164}$$

where $g(J)$ is the usual statistical factor (see Equation 6.102).

Finally, the observed total cross section is obtained by summing the above equation over all possible α', including $\alpha' = \alpha$. The result is

$$\sigma_{\alpha T} = 2\pi \lambda_\alpha^2 \sum_J g(J) \sum_{jl} (1 - \text{Re } U^J_{\alpha jl, \alpha jl}). \tag{6.165}$$

The derivation of Equation 6.165 depends on the fact that the collision matrix $\mathbf{U}$ is unitary, i.e., its complex conjugate equals its reciprocal $\bar{\mathbf{U}} = \mathbf{U}^{-1}$; so $\bar{\mathbf{U}}\mathbf{U} = 1$, or

$$\sum_{c'} |U_{c'c}|^2 = 1. \tag{6.166}$$

Channel-to-Channel Integrated Cross Sections

It is instructive to examine the forms of the integrated cross sections for neutron-induced interactions before the various sums and averages over quantum numbers were performed.

The cross section for a collision leading from channel c to channel c' is

$$\sigma_{cc'} = \pi \lambda_\alpha^2 |\delta_{c'c} - U_{c'c}|^2. \tag{6.167}$$

We shall now define an interaction that has identical entrance and exit channels as *channel elastic scattering*. Any interaction that is not channel elastic scattering will be called a *channel reaction*. Note that channel elastic scattering is only a part of the experimentally observed elastic scattering since elastic scattering with a change in orbital or spin quantum number is considered here as a channel reaction.

The cross section for channel elastic scattering is obtained by setting $c' = c$ in Equation 6.167:

$$\sigma_{cc} = \pi \lambda_\alpha^2 |1 - U_{cc}|^2. \tag{6.168}$$

The channel reaction cross section is obtained by summing Equation 6.167 over c' with $c' \neq c$, and using the unitary property of $\mathbf{U}$,

$$\sigma_{cr} = \pi \lambda_\alpha^2 (1 - |U_{cc}|^2). \tag{6.169}$$

The total cross section for a given entrance channel c is obtained by summing Equation 6.167 over all exit channels, or by adding σ_{cc} and σ_{cr},

$$\sigma_{cT} = 2\pi\lambda_\alpha^2(1 - \text{Re } U_{cc}). \tag{6.170}$$

Equations 6.167 to 6.170 are the generalizations of the equations derived in Section 6.2.

The Symmetry of the Collision Matrix and Reciprocity

If a channel c is designated by the quantum numbers $\{\alpha ljmv\}$, then the channel $-c$ is defined as that channel which has quantum numbers $\{\alpha lj-m-v\}$. The channel $-c$ is therefore identical to the channel c except for reversed angular momenta directions.

Now it can be shown by an argument based on the invariance of the Schrödinger equation to time reversal,[*] that the collision matrix has the following fundamental symmetry property

$$U_{cc'} = U_{-c', -c},$$

that is, the element of the collision matrix for the reaction $c \rightarrow c'$ is identical to that for the inverse reaction $c' \rightarrow c$ with the orbital and spin angular-momentum vectors reversed.

Using Equation 6.167, we can write

$$\sigma_{cc'} = \pi\lambda_\alpha^2|\delta_{c'c} - U_{c'c}|^2$$

and

$$\sigma_{-c', -c} = \pi\lambda_{\alpha'}^2|\delta_{-c, -c'} - U_{-c, -c'}|^2.$$

But $U_{c'c} = U_{-c, -c'}$; therefore

$$\frac{\sigma_{cc'}}{\lambda_\alpha^2} = \frac{\sigma_{-c', -c}}{\lambda_{\alpha'}^2}$$

or, in terms of momenta,

$$p_\alpha^2\sigma_{cc'} = p_{\alpha'}^2\sigma_{-c', -c}.$$

This is the fundamental channel-to-channel reciprocity relation. For unpolarized particles, summing each side of this equation results in the previously derived Equation 5.79,

$$g_\alpha p_\alpha^2\sigma_{\alpha\alpha'}(E_1) = g_{\alpha'} p_{\alpha'}^2\sigma_{\alpha'\alpha}(E_2),$$

where, for example, $g_\alpha = (2I_A + 1)(2I_a + 1)$.

---

[*]See pp. 474–477 of M. A. Preston, *Physics of the Nucleus* (Reference 4 at the end of this chapter).

6.11 The R-Matrix Theory

In the preceding section, general equations were developed for the differential and total cross sections for interactions that proceeded from initial constellation α to final constellation α'. The unknowns in these equations are the elements of the collision matrix **U**. The aim of the R-matrix theory is to determine the form of the elements of **U**.

The R-matrix theory, which was introduced by Wigner and Eisenbud[*] in 1947, is rigorous in both its quantum-mechanical and its mathematical structure. Furthermore, its application is not restricted to interactions that proceed via a compound nucleus; it can be applied to all nuclear interactions, both direct and indirect. Interaction models, such as the black-nucleus model and the cloudy-crystal-ball model, are encompassed as special cases within the R-matrix theory. The Breit–Wigner single-level resonance formulas and their multilevel multichannel generalizations are also derivable from the R-matrix theory.

Despite its great generality, the R-matrix theory suffers from the deficiency that is common to all nuclear interaction theories: it does not explicitly present ways to calculate wave functions for the nucleons within the nucleus. It treats the nucleus as a black box. The unknown internal wave functions and their derivatives at the nuclear boundary appear as part of the elements in the R matrix. These elements contain the parameters E_λ and $\gamma_{\lambda c}$ whose values are not determined by R-matrix theory *per se*. These unknown parameters can be determined to some extent by examination of measured cross sections, and it is hoped that eventually they will be determined by a theory of nuclear forces and nuclear structure.

Given all the elements of the R matrix for a particular initial constellation, all cross sections could be computed with absolute accuracy. Actually, the elements of the R matrix are not known a priori. What has been determined by R-matrix theory is that the energy dependence of the elements must have the simple form

$$R_{cc'} = \sum_\lambda \frac{\gamma_{\lambda c}\gamma_{\lambda c'}}{E_\lambda - E},$$

where the E_λ's are the complete set of energy eigenvalues of the energy eigenstates within the nucleus with $A_1 + A_2$ nucleons and the $\gamma_{\lambda c}$ are the *reduced width amplitudes* for states λ and channels c. The E_λ and $\gamma_{\lambda c}$ are energy-independent unknown parameters. In the isolated resonance region, these parameters are closely related to the *formal resonance energy* E'_0 and the *resonance width* Γ_c of the Breit–Wigner treatment.

[*] E. P. Wigner and L. Eisenbud, *Phys. Rev.*, **72**: 29 (1947).

R-Matrix Treatment of the Elastic Scattering of Spinless Particles

The general theory of the R-matrix must, of necessity, be presented in terms of complicated mathematical expressions. These expressions unfortunately tend to obscure the relatively simple basis of the theory. We begin therefore with the elementary case of the elastic scattering of spinless particles. This special case, though mathematically simple, contains all the essential elements of the general theory. Since the interacting particles are spinless and elastic scattering is assumed, the orbital angular-momentum number l is the only symbol required to completely designate the channel.

We begin by noting that the crucial step in the solution to any scattering problem is equating the logarithmic derivatives of the inner and outer wave functions at the surface of the potential, i.e., at $r = a$. It is from this step that the phase shifts, or the elements of the U matrix are obtained. To proceed to the solution, we therefore must first obtain an expression for the inner radial wave function. If the potential were known, the inner wave function could be obtained by a straightforward solution, probably numerical, of the Schrödinger equation. Since the nuclear potential is unknown, we must proceed in a less direct manner. The inner wave function W_l at any energy E is expanded in the eigenfunctions of the energy levels, bound and virtual, in the compound nucleus:

$$W_l(E, r) = \sum_\lambda A_{l\lambda} W_l(E_\lambda, r), \tag{6.171}$$

where the E_λ values are the energy eigenvalues and the corresponding functions $W_l(E_\lambda, r)$ are the eigenfunctions that satisfy the zero derivative boundary condition

$$\left[\frac{dW_l(E_\lambda, r)}{dr} \right]_{r=a} = 0. \tag{6.172}$$

Since these are eigenfunctions of a real Hamiltonian, they are orthogonal and may be assumed to be normalized,

$$\int_0^a W_l(E_\lambda, r) W_l(E_{\lambda'}, r) \, dr = \delta_{\lambda\lambda'}. \tag{6.173}$$

The next objective is to find an expression for the expansion coefficients, the $A_{l\lambda}$. This expression may be obtained by multiplying Equation 6.171 by $W_l(E_\lambda, r)$ and integrating over r from 0 to a. Because of the orthogonality of the eigenfunctions, we find that

$$A_{l\lambda} = \int_0^a W_l(E_\lambda, r) W_l(E, r) \, dr. \tag{6.174}$$

The wave functions in the preceding expression are solutions of the radial Schrödinger equations

$$\left\{\frac{d^2}{dr^2} + \frac{2m}{\hbar^2}\left[E - V(r) - \frac{l(l + 1)\hbar^2}{2mr^2}\right]\right\}W_l(E, r) = 0 \tag{6.175a}$$

and

$$\left\{\frac{d^2}{dr^2} + \frac{2m}{\hbar^2}\left[E_\lambda - V(r) - \frac{l(l + 1)\hbar^2}{2mr^2}\right]\right\}W_l(E_\lambda, r) = 0. \tag{6.175b}$$

An expression for the right-hand side of Equation 6.174 can be obtained by multiplying Equation 6.175a by $W_l(E_\lambda, r)$ and Equation 6.175b by $W_l(E, r)$, subtracting the resulting equations and integrating from 0 to a:

$$\frac{2m}{\hbar^2}(E - E_\lambda)\int_0^a W_l(E_\lambda, r)W_l(E, r)\, dr = \int_0^a (W_\lambda W_E'' - W_E W_\lambda'')\, dr$$

$$= W_\lambda W_E' - W_E W_\lambda'\Big|_0^a \tag{6.176}$$

$$= W_\lambda(a)\left[\frac{dW_E}{dr}\right]_{r=a}.$$

In the above equations, for notational convenience, $W_l(E, r)$ and $W_l(E_\lambda, r)$ have been written as W_E and W_λ, respectively. The primes indicate differentiation with respect to r. The last step of Equation 6.176 follows from the condition that all wave functions must vanish at $r = 0$ and W_λ' vanishes at $r = a$. From Equations 6.174 and 6.176, we obtain the expression for $A_{l\lambda}$,

$$A_{l\lambda} = \frac{\hbar^2}{2m}(E_\lambda - E)^{-1}W_l(E_\lambda, a)\left[\frac{dW_l(E, r)}{dr}\right]_{r=a}. \tag{6.177}$$

When this expression is substituted into Equation 6.171, we obtain for the wave function at the surface of the nucleus

$$W_l(E, a) = \frac{\hbar^2}{2ma}\sum_\lambda\left[\frac{W_l(E_\lambda, a)W_l(E_\lambda, a)}{E_\lambda - E}\right]\cdot\left[r\,\frac{dW_l(E, r)}{dr}\right]_{r=a}. \tag{6.178}$$

This is the desired expression that relates the value of the inner wave function to its derivative at the nuclear surface. The R function, which in the general many-channel case will be replaced by the R matrix, is defined as the coefficient of the derivative term; i.e.,

$$R_l \equiv \frac{\hbar^2}{2ma}\sum_\lambda\frac{[W_l(E_\lambda, a)]^2}{E_\lambda - E} = \sum_\lambda\frac{\gamma_{\lambda l}^2}{E_\lambda - E}, \tag{6.179}$$

where, by definition, the *reduced-width amplitude* for state λ and channel l is given by

$$\gamma_{\lambda l} \equiv \sqrt{\frac{\hbar^2}{2ma}}\, W_l(E_\lambda, a). \tag{6.180}$$

The reduced-width amplitudes, which depend on the values of the inner wave functions at the nuclear surface, are unknown energy-independent constants. The fact that they are energy-independent is most noteworthy because it means that the entire energy dependence of the R function appears in its resonance-type denominator.

It only remains to relate the R function to the U function. First, let us note that Equation 6.178 can be written in the form

$$W_l(E, a) = R_l \left[r\, \frac{dW_l(E, r)}{dr} \right]_{r=a} \tag{6.181}$$

or

$$R_l = \left[\frac{W_l(E, r)}{r\, dW_l(E, r)/dr} \right]_{r=a} \tag{6.182}$$

in which it is clear that the R function is simply the inverse of the logarithmic derivative of the inner wave function at the nuclear boundary. Equating R_l to the inverse of the logarithmic derivative of the outer wave function at the nuclear boundary,

$$R_l = \frac{1}{a} \left\{ \frac{W_l^{(-)} - U_l W_l^{(+)}}{[dW_l^{(-)}/dr] - U_l[dW_l^{(+)}/dr]} \right\}_{r=a}, \tag{6.183}$$

and solving for U_l, we obtain the final expression

$$U_l = e^{2i\Delta_l} \left[\frac{1 - (a_l - ib_l)R_l}{1 - (a_l + ib_l)R_l} \right] \tag{6.184}$$

In deriving this expression we used Equations 6.24 and 6.28. We now shall use this expression to examine several special cases of particular interest.

Example 6.1   The Isolated Resonance
In the region of an isolated l-wave resonance, we may assume that E is so close to E_λ that only one term in the R function is significant, namely the term

$$R_l = \frac{\gamma_{\lambda l}^2}{E_\lambda - E}.$$

Upon substituting the preceding expression for R_l into Equation 6.184 and the resulting expression for U_l into the familiar equation

$$\sigma_{n,l} = \pi \lambdabar^2 (2l + 1)|1 - U_l|^2, \tag{6.185}$$

we obtain the previously derived Breit–Wigner expression

$$\sigma_{n,l} = (2l + 1)\pi \lambdabar^2 \left| e^{2i\Delta_l} \left[e^{-2i\Delta_l} - 1 + \frac{i\Gamma_{n,l}}{(E - E_0) + (i\Gamma_{n,l}/2)} \right] \right|^2.$$

The width for elastic scattering was defined as a function of the reduced width by the equation

$$\Gamma_{n,l} = 2\gamma_{\lambda l}^2 b_l, \tag{6.186}$$

and the observed resonance energy E_0 was related to the formal resonance energy E_λ by

$$E_0 = E_\lambda - \gamma_{\lambda l}^2 a_l. \tag{6.187}$$

Note the correspondence between Equations 6.186 and 6.45 and between Equations 6.187 and 6.48. The term $(-d \, \mathrm{Re} \, D_l/dE)_{E=E_0}$ which appeared in the earlier *ad hoc* derivation has been replaced in the rigorous R matrix derivation by the reduced width $\gamma_{\lambda l}^2$. Once again, it is clear that the observed resonance energy is not the energy eigenvalue E_λ but is shifted by an amount proportional to a_l.

Example 6.2 The Two-Level Approximation
Next consider a situation in which the energy E of the initial constellation is such that only two terms of the R function for a particular l wave have appreciable magnitude; so R_l can be written as

$$R_l = \frac{\gamma_{1l}^2}{E_1 - E} + \frac{\gamma_{2l}^2}{E_2 - E}. \tag{6.188}$$

When this expression for R_l is substituted into Equation 6.184 and the resulting expression for U_l is substituted into Equation 6.185, there results

$$\sigma_{n,l} = \pi \lambdabar^2 (2l + 1) \left| e^{2i\Delta_l} \left[(e^{-2i\Delta_l} - 1) \right. \right.$$
$$\left. \left. + \frac{2ib_l B}{(E_1 - E)(E_2 - E) - (a_l + b_l)B} \right] \right|^2 \tag{6.189}$$

where, by definition, $B \equiv \gamma_{1l}(E_2 - E) + \gamma_{2l}(E_1 - E)$.

This case was chosen to illustrate one important point: The cross section that results from two or more levels of the same l value (or J value in the case of particles with spin) cannot be expressed as the simple sum of single-level Breit–Wigner equations. It is true, however, that in the vicinity of a maximum, say, when $E = E_1$, the one-level expression is an accurate description provided the widths of the neighboring resonances are much less than their distance in energy units from E_1.

General R-Matrix Theory

The generalization of the preceding analysis to include the spin of the particles and the possibility of many exit channels is straightforward but algebraically cumbersome. Only the results will be displayed here; the derivation of these results may be found in the Lane and Thomas article and in Preston's *Physics of the Nucleus.**

In the general multilevel multichannel theory, the cc' element of the R matrix takes the form

$$R_{cc'} = \sum_{\lambda} \frac{\gamma_{\lambda c} \gamma_{\lambda c'}}{E_{\lambda} - E}, \qquad (6.190)$$

where the reduced width amplitude $\gamma_{\lambda c}$ is a real constant given by

$$\gamma_{\lambda c} \equiv \sqrt{\frac{\hbar^2}{2 M_c a_c}}\, W_c(E_{\lambda}, a_c). \qquad (6.191)$$

The relation between the U matrix and the R matrix in the general theory is given by

$$\mathbf{U} = \rho^{1/2}\mathbf{O}^{-1}[\mathbf{I} - \mathbf{R}(\mathbf{L} - \mathbf{D})]^{-1}[\mathbf{I} - \mathbf{R}(\overline{\mathbf{L}} - \mathbf{D})]\mathbf{I}\,\rho^{-1/2}, \qquad (6.192)$$

where $\mathbf{I}$ is the unit matrix, $\mathbf{L}$ is the diagonal matrix whose elements are $a_l + ib_l$, $\overline{\mathbf{L}}$ is the diagonal matrix whose elements are $a_l - ib_l$, $\mathbf{D}$ is the diagonal matrix whose elements are the logarithmic derivatives of the external wave functions at the nuclear boundary, and $\mathbf{O}$ and $\mathbf{I}$ are the diagonal matrices whose elements are the external outgoing and incoming wave functions O_c and I_c evaluated at the nuclear boundary. The matrix $\rho^{1/2}$ is the diagonal matrix whose elements are $(k_c a_c)^{1/2}$.

All matrices on the right-hand side of Equation 6.192 are diagonal except the R matrix. Furthermore, all these matrices except $\mathbf{R}$ are known, and the only unknowns in $\mathbf{R}$ are the reduced-width amplitudes and the formal resonance energies. Equation 6.192 is an exact expression for the collision matrix.

Two fundamental difficulties hinder the use of Equation 6.192 in the determination of $\mathbf{U}$ and hence of all cross sections. One of these difficulties

---

*M. A. Preston, *Physics of the Nucleus* (Reference 4 at the end of this chapter).

is physical, the other mathematical. The physical problem is the determination of the parameters $\gamma_{\lambda c}$ and E_λ, which depend on the inner structure of the nucleus. Even if all the $\gamma_{\lambda c}$ and E_λ were known, the mathematical problem of inverting the matrix $[\mathbf{I} - \mathbf{R}(\mathbf{L} - \mathbf{D})]$ would be formidable. For these reasons various approximations are employed to examine special cases. For example, the single- and double-level Breit–Wigner elastic-scattering and reaction cross-section equations can be obtained by making simplifying assumptions relative to magnitudes of the elements in the $\mathbf{R}$ matrix. In the region of overlapping resonances, cross-section expressions can be obtained by assuming that the signs of the reduced-width amplitudes are random, uncorrelated, etc. To pursue these many ramifications of $\mathbf{R}$-matrix theory would take us far beyond the intended scope of this book. The reader who plans to devote his professional life to interaction theory should consider this but a small antipasto to his studies, the first significant portion of which should consist of digesting the article by Lane and Thomas.

SELECTED REFERENCES

1. John M. Blatt and Victor F. Weisskopf, *Theoretical Nuclear Physics*, John Wiley & Sons, Inc., New York, 1952. For a more thorough exposition of the material in this chapter, read Chapter VIII, "Nuclear Reactions: General Theory."

2. M. E. Rose, *Elementary Theory of Angular Momentum*, John Wiley & Sons, Inc., New York, 1957. The formal theory of angular momentum in quantum mechanics is presented in a lucid, but not quite "elementary," form. The algebraic properties presented without proof in Section 6.5 of this book are proved in Chapters II and III of Rose's book. The properties of the Clebsch–Gordon coefficients are also presented in Chapter III. Almost one-half the book consists of instructive applications of the general theory.

3. A. R. Edmonds, *Angular Momentum in Quantum Mechanics*, 2nd ed., Princeton University Press, Princeton, N.J., 1960. This book gives a somewhat more elementary treatment of angular momentum than that of Rose. Edmonds' book is particularly valuable for its tables of formulas and for its references to compilations of the various coefficients which enter into angular-momentum bookkeeping.

4. M. A. Preston, *Physics of the Nucleus*, Addison-Wesley Publishing Company, Inc., Reading, Mass., 1962. Chapter 16 includes descriptions of basic reaction theory in the collision matrix formalism and R-matrix theory. The unitarity and symmetry of the U matrix are proved.

5. J. E. Lynn, *The Theory of Neutron Resonance Reactions*, Clarendon Press, Oxford, England, 1968. A lucid, up-to-date, and nearly exhaustive account of neutron resonance theory. Includes a considerable amount of experimental data and their theoretical interpretation. Highly recommended.

EXERCISES

1. Using the semiempirical level density equation, estimate the average spacing between energy levels near the neutron separation energy in aluminum, cadmium, and uranium. Compare your results with the measured cross sections in USAEC

Report BNL-325 (the "Barn Book") and draw some conclusions relative to the degree of validity of the level density equation.

2. Prove that the reaction cross section near $E = 0$ would be proportional to $E^l/\sqrt{E}$ if l-wave neutrons were responsible for the reaction.

3. Prove that the channel spin wave function is an eigenfunction of the angular part of the Hamiltonian for a single reduced-mass particle in a central potential.

4. Show by direct substitution that the *index raising* and *lowering operators* $(L_x + iL_y)$ and $(L_x - iL_y)$ have the following effects on a spherical harmonic:

$$(L_x + iL_y)Y_l^m = \hbar\sqrt{(l - m)(l + m + 1)}\, Y_l^{m+1},$$

$$(L_x - iL_y)Y_l^m = \hbar\sqrt{(l + m)(l - m + 1)}\, Y_l^{m-1}.$$

Show that the *spin-flip operators* $(\sigma_x + i\sigma_y)$ and $(\sigma_x - i\sigma_y)$ affect the spin functions of particles of spin 1/2 as follows:

$$(\sigma_x + i\sigma_y)\chi^p = 0, \qquad (\sigma_x - i\sigma_y)\chi^p = 2\chi^a,$$

$$(\sigma_x + i\sigma_y)\chi^a = 2\chi^p, \qquad (\sigma_x - i\sigma_y)\chi^a = 0.$$

Show that $(\sigma_x^1 + i\sigma_y^1)$ and $(\sigma_x^1 - i\sigma_y^1)$ are *spin raising* and *lowering operators* for particles of spin 1.

5. Determine the spin wave functions and spin operators for a nucleus with spin $I = 3/2$.

6. Find the channel spin wave function Υ_{lj}^{JM} with $l = 1, j = 1/2, J = 3/2$, and $M = 1/2$. Prove by direct substitution that this wave function is an eigenfunction of the operators

(a) L^2 with eigenvalue $l(l + 1)\hbar^2 = 2\hbar^2$,
(b) j^2 with eigenvalue $j(j + 1)\hbar^2 = (3/4)\hbar^2$,
(c) J_z with eigenvalue $M\hbar = (1/2)\hbar$,
(d) J^2 with eigenvalue $J(J + 1)\hbar^2 = (15/4)\hbar^2$.

7. Consider a nucleus with spin $I = 1$ and a neutron with spin $s = 1/2$. Assume the spins are added to obtain a total spin $\mathbf{j} = \mathbf{I} + \mathbf{s}$. (a) Determine the simultaneous spin eigenfunctions of the operators $\mathbf{I}\cdot\mathbf{I}$, $\mathbf{s}\cdot\mathbf{s}$, $\mathbf{j}\cdot\mathbf{j}$, and j_z. (b) Develop a formalism in which all the operators are 5-by-5 matrices and all the eigenfunctions are five-component column vectors.

8. Prove the following: In any elastic scattering (or reaction) that proceeds through a single level of the compound nucleus, the differential cross section is symmetric about $\theta = \pi/2$; hence, only Legendre polynomials with even indices ($P_0, P_2, P_4, \ldots$) are required to express the differential cross section.

9. Cadmium-113 has a giant (n, γ) resonance at $E_0 = 0.178$ ev. The measured parameters of this resonance are $I = 1/2, J = 1, l = 0, \Gamma_n = 0.065 \times 10^{-3}$ ev, and $\Gamma_\gamma = 0.113$ ev. The isotope ^{113}Cd comprises 12.26% of the element cadmium. Based on these data, calculate the total cross section of the element cadmium in the energy range $0.01 \le E \le 1.0$ ev and compare your results with the measured cross section. in the Barn Book. (The only interactions in this energy range are elastic scattering in all isotopes of cadmium and radiative capture in ^{113}Cd.)

10. Choose an isolated pure-scattering resonance from the Barn Book. Specialize the general Breit–Wigner equation (Equation 6.137) to conform to the tabulated values of I, J, l, and $\Gamma_{n,l}$ for this resonance and plot the Breit–Wigner

scattering cross section $\sigma_n(E)$. Compare the theoretical and experimental graphs and explain any discrepancies.

11. Under what conditions might one expect the Breit–Wigner equation to yield the correct reaction cross section near neutron energy $E = 0$, even though the isolated resonance to which it is applied has its center at several thousands to tens of thousands of electron volts? (Since the Breit–Wigner equations are expected to be valid only in the near vicinity of an isolated resonance, you must explain in what respect $E = 0$ is near $E = 10^4$ or 10^5 ev.)

12. Prove that the collision matrix $\mathbf{U}$ is symmetric and unitary. (See, for example, the article by Lane and Thomas or the book by Preston.)

13. Prove that the cross sections for channel elastic scattering and channel reactions are given by

$$\sigma_{cc} = \pi \lambda_\alpha^2 |1 - U_{cc}|^2,$$
$$\sigma_{cr} = \pi \lambda_\alpha^2 (1 - |U_{cc}|^2).$$

ELASTIC
SCATTERING 7

Neutron–nucleus elastic scattering is distinguished from other types of interactions by two conditions: (1) the initial and final constellations consist of identical particles,

$$Z^A + n \rightarrow Z^A + n;$$

and (2) the total kinetic energy is conserved. In the C-system, the scattered neutron will emerge with a kinetic energy identical to its initial kinetic energy. In the L-system, the scattered neutron will emerge with a kinetic energy that depends on the scattering angle. However, in either system the sums of the kinetic energies of the neutron and the nucleus before and after the interaction are identical.

Markedly different procedures are employed for determining cross sections in the low-energy region of isolated resonances and in the high-energy region of overlapping resonances. In the low-energy region, the Breit–Wigner formalism is employed to a considerable extent. In this energy region the nucleus is treated as an impenetrable sphere except at those neutron energies which correspond to virtual levels in the compound nucleus. At those energies the empirically determined resonance parameters are employed in conjunction with the Breit–Wigner equations to find the cross sections. Elastic-scattering cross sections at high energies are calculated by partial-wave analysis using an optical-model potential. The bulk of this chapter is devoted to descriptions of these calculations and their results. Before getting into these descriptions, we shall examine the kinematics of elastic scattering of neutrons off nuclei that are at rest before the interaction. The discussion of elastic interactions between neutrons and moving nuclei is delayed until Chapter 12. To the extent that the initial kinetic energy of the neutron is appreciably greater than that of the nucleus (kT), the assumption of stationary nuclei is a reasonably good one. The exception is noted in Chapter 12 in reference to the Doppler effect.

7.1 Kinematics of Elastic Scattering of Neutrons Off Nuclei

Neutron elastic scattering, and indeed all neutron interactions, in the energy range below 20 Mev (the highest energy of interest to nuclear engineers) can be treated nonrelativistically. The speed of a 20-Mev neutron is 6.2×10^9 cm/sec. Particle speeds enter relativistic equations in the ratio $(v/c)^2$, where c is the velocity of light, 3×10^{10} cm/sec. If the ratio $(v/c)^2$ is much less than unity, relativistic kinematic analysis is unnecessary. For a 20-Mev neutron, the quantity $(v/c)^2 = 0.04$. Thus, with an error of the order of 4%, one may ignore relativistic effects and use all the nonrelativistic kinematic formulas derived in Chapter 2. The solutions found in Chapter 2 will now be restated in a more appropriate notation. The masses of all particles will be stated in *neutron masses*, which are very closely, but not exactly, equal to atomic mass units. In these units the neutron mass is identically unity, and the mass of the nucleus is A neutron masses. The initial kinetic energy of the neutron in the L-system will be called $\tilde{E}$; the total energy in the C-system, E. Final energies will be indicated by primes. The neutron scattering angle in the L-system will be called $\tilde{\vartheta}$; and in the C-system, θ.

Consider a neutron of kinetic energy $\tilde{E}$ incident on a target nucleus of mass A that is at rest in the laboratory system before the collision. After the elastic scattering, the neutron moves off at angle $\tilde{\vartheta}$ relative to its initial direction. The final laboratory kinetic energy of the neutron is given by Equation 2.44. In the new notation, it is

$$\tilde{E}' = \tilde{E} \left\{ \frac{1}{A+1} \cos \tilde{\vartheta} + \left[\frac{A-1}{A+1} + \left(\frac{1}{A+1} \right)^2 \cos^2 \tilde{\vartheta} \right]^{1/2} \right\}^2. \quad (7.1)$$

The nucleus will be given a laboratory kinetic energy $\tilde{E}'_A$, which is readily determined from conservation of kinetic energy,

$$\tilde{E}'_A = \tilde{E} - \tilde{E}'. \quad (7.2)$$

As we proved in Chapter 2, the nucleus must scatter in the forward direction. Its laboratory angle of scattering $\tilde{\vartheta}_A$ is given by Equation 2.50, which may be written in the form

$$\sin \tilde{\vartheta}_A = \sqrt{\frac{\tilde{E}'}{A\tilde{E}'_A}} \sin \tilde{\vartheta}. \quad (7.3)$$

The center-of-mass scattering angle of the neutron, θ, is implicitly given by Equation 2.46

$$\tan \tilde{\theta} = \frac{\sin \theta}{(1/A) + \cos \theta}. \tag{7.4}$$

Finally, the center-of-mass scattering angle of the nucleus is given by

$$\theta_A = \pi - \theta. \tag{7.5}$$

Upon replacing the laboratory angle $\tilde{\theta}$ in Equation 7.1 by the appropriate expression for the center-of-mass angle θ, we obtain another very useful relation,

$$\frac{\tilde{E}'}{\tilde{E}} = \frac{A^2 + 1 + 2A \cos \theta}{(A + 1)^2}. \tag{7.6}$$

Equations 7.1 to 7.6 constitute the complete solution to the kinematics of neutron–nucleus nonrelativistic elastic scattering. The reader will no doubt find several of these relations useful in accomplishing the suggested exercises at the end of this chapter.

Several cases of these general relations are of particular interest: If the neutron scatters directly forward, $\cos \tilde{\theta} = 1$, and Equation 7.1 tells us that $\tilde{E}' = \tilde{E}$; the neutron loses none of its initial energy. The greater the angle of scattering, the greater the energy loss the neutron suffers. If the mass of the nucleus A is greater than one neutron mass, the neutron may scatter through any laboratory angle between 0 and π radians. If it scatters through an angle of π radians, it transfers the maximum possible energy to the nucleus; the neutron's final energy is then given by

$$(\tilde{E}')_{\min} = \left[\frac{A - 1}{A + 1}\right]^2 \tilde{E}. \tag{7.7}$$

If the mass of the target nucleus were identically equal to the mass of the neutron then, as explained in Chapter 2, the maximum laboratory scattering angle of the neutron is $\pi/2$ radians. This angle corresponds to a head-on collision in which the entire kinetic energy of the neutron is given to the target nucleus. It should be noted, however, that there are actually no nuclei with exactly the mass of the neutron. The proton has a mass that is very slightly less than the mass of the neutron, actually 0.998622 neutron masses. It will be left as an exercise for the student to prove that this slight mass difference has the consequences that (a) the maximum n–p laboratory scattering angle is 86° 59.5′, not the usually assumed value of 90°, (b) a neutron scattered at an angle very close to π in the C-system scatters through a very small angle in the L-system, and

(c) for any given value of the laboratory angle $\bar{\theta}$ there are two possible values of the center-of-mass angle θ but for each value of the center-of-mass angle there is one unique value of the laboratory angle.

7.2 Measured Elastic-Scattering Cross Sections

Perhaps a few general remarks describing some of the more salient tendencies and characteristics of elastic-scattering cross sections would be of value here. Elastic-scattering cross sections, both total and differential, have been measured for almost all elements, as well as for a few separated isotopes, generally at widely separated energies in the range up to at least 20 Mev.

At the very low end of the neutron energy scale, the elastic-scattering cross section for individual neutron–nucleus collisions consists of a constant shape elastic cross section (very roughly equal to $4\pi R^2$) upon which isolated resonances are superimposed. The spacing of these resonances varies with the atomic weight of the scatterer so that resonances appear at lower and lower energies as the scattering nucleus becomes more and more massive. This is a direct result of the energy-level spacing in the various nuclei. For example, the first resonance in carbon is in the million-electron-volt region whereas uranium has resonances in the electron-volt

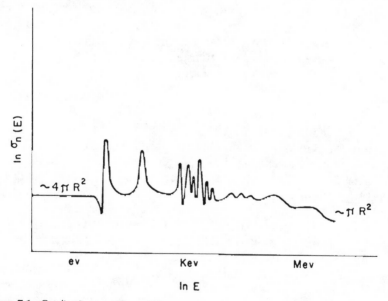

Figure 7.1—Qualitative variation of the neutron elastic-scattering cross section for a typical medium or heavy nuclide.

region. As pointed out in Chapter 6, all resonances, even those which are primarily reaction resonances, must have an elastic-scattering component.

As the neutron energy is increased still further, individual resonances are no longer seen, partly because of experimental lack of resolution and partly because the individual resonance levels are overlapping and merging into a continuum. The onset of the continuum region occurs in the kilo-electron-volt region for heavy scatterers and in the million-electron-volt region for medium-weight scatterers and for the heavy magic elements like bismuth and lead.

Once in the continuum region, the cross section is relatively constant with energy but does exhibit a slow wavelike variation. The measured elastic-scattering cross section, which does not include the highly forward peaked shadow scattering, finally approaches πR^2 at very high energies. Figure 7.1 illustrates, qualitatively, the usual behavior of the elastic cross section as a function of energy.

Angular distributions of elastically scattered neutrons are functions of the atomic weight of the scatterer, its level structure, and the energy of the incident neutrons. At low energies the angular distributions are isotropic in the C-system except in the vicinity of isolated $l > 0$ resonances. At higher energies the angular distributions become increasingly more forward directed as p, d, and higher order partial waves begin to add to the cross section. At still higher energies the angular distributions exhibit strong forward peaks and wavelike structure. This behavior is illustrated in Figure 7.2, which shows the measured differential cross section of iron in the range 0.22 Mev to 14 Mev. Below about 0.2 Mev elastic scattering from iron is isotropic; above this energy the differential cross section exhibits a variation that is rather typical of its behavior for all but the lightest elements. In particular, we note an increase in forward scattering and in the number of maxima with increasing energy.

The behavior of the ^{1}H elastic-scattering cross section is an important exception to the general behavior described above. The n–p scattering cross section is almost constant at 20.4 barns from 0 up to about 1000 ev, at which energy it has decreased to about 20.0 barns. It then begins to decrease monotonically to about 0.5 barns at 20 Mev. The differential scattering cross section of ^{1}H is isotropic in the C-system up to approximately 13 Mev. Between 14 and 20 Mev, it exhibits some slight anisotropy, being roughly 10% greater at 180° than at 90° in this energy region.

7.3 Isolated Scattering Resonances

Most isolated resonances are characterized by a single value of l, and, for this case, the general Breit–Wigner equation (Equation 6.137) can be written in the more simple form

$$\sigma_n(E) = \pi\lambda^2 \sum_{l=0}^{\infty} (2l + 1)|A_{se}^l|^2$$

$$+ \pi\lambda^2 g(J)[|e^{2i\Delta_{l^*}}(A_{ce}^{l^*} + A_{se}^{l^*})|^2 - |A_{se}^{l^*}|^2], \quad (7.8)$$

where l^* is the angular-momentum quantum number of the neutrons that

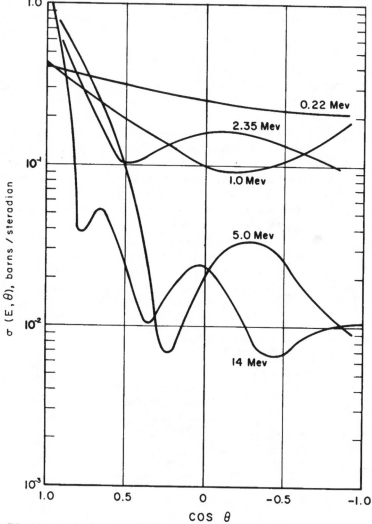

Figure 7.2—Measured differential elastic-scattering cross section of iron. (From USAEC Report BNL-400, Brookhaven National Laboratory, October 1962).

are involved in the resonance scattering. The amplitudes for shape elastic and compound elastic scattering are given by Equations 6.32 and 6.49, respectively:

$$A_{se}^l = e^{-2i\Delta_l} - 1$$

and

$$A_{ce}^l = \frac{i\Gamma_{n,l}}{(E - E_0) + i\Gamma/2},$$

where Δ_l is the l-wave phase shift of the impenetrable sphere.

The Neutron Width $\Gamma_{n,l}$

Let us now investigate the behavior of the level width for neutron emission, $\Gamma_{n,l}$, as a function of energy. The neutron width and the reaction width, which were introduced into the *ad hoc* derivation of the Breit–Wigner formulas by means of Equations 6.45 and 6.46, are first of all positive quantities since (1) b_l is positive, (2) Im D_l is negative, and (3) the derivative $(d$ Re $D_l/dE)$ is negative. The proofs of (1) and (2) have already been given; the proof of (3) will be left as an exercise for the reader.

The neutron width $\Gamma_{n,l}$ is defined by Equation 6.45

$$\Gamma_{n,l} \equiv \frac{-2b_l}{(d \text{ Re } D_l/dE)_0}.$$

To advance the investigation of $\Gamma_{n,l}$, we must examine the behavior of the derivative of Re D_l, which, in turn, requires some assumptions, however crude, about the radial wave function within the nucleus and near its boundary. This wave function will be assumed to have the form of a simple incoming and outgoing wave,

$$W_l(r) \propto e^{-iKr} + U_l e^{iKr}, \tag{7.9}$$

for $r < R$, where U_l is the reflection factor $e^{-2\beta_l}e^{2i\alpha_l}$ and K is the wave number near the nuclear boundary. If the nucleus were, for example, a spherical square well of depth V_0, then K would be given by

$$K = \frac{\sqrt{2m(E + V_0)}}{\hbar}. \tag{7.10}$$

Regardless of the actual form of the nuclear potential, there is every reason to expect that its depth is such that the inner wave number K is much greater than the outer wave number k for all neutron energies below several million electron volts. The inner wave function can be written in the equivalent form

$$W_l(r) \propto \cos(Kr + \alpha_l + i\beta_l). \tag{7.11}$$

The logarithmic derivative of the inner wave function at the nuclear boundary is

$$D_l \equiv R\left[\frac{dW_l/dr}{W_l}\right]_{r=R} = -KR \tan[KR + \alpha_l(E) + i\beta_l(E)], \tag{7.12}$$

where particular attention has been called to the fact that α_l and β_l are functions of the neutron energy E.

Now consider a pure scattering resonance or any resonance that is predominantly scattering. For such a resonance, β_l is much less than α_l, and therefore

$$\text{Re } D_l \simeq -KR \tan[KR + \alpha_l(E)] \equiv -KR \tan Z(E). \tag{7.13}$$

The function $Z(E)$ is defined by the preceding equation. Now we recall that the formal resonance energies E_0' were defined as those energies at which $\text{Re } D_l = 0$. From the above equation it is clear that this corresponds to those energies at which

$$Z(E_0') = n\pi, \tag{7.14}$$

where n is an integer. The derivative of $\text{Re } D_l$ is

$$\left(\frac{d\,\text{Re } D_l}{dE}\right)_{E=E_0'} = -\left[\frac{d(KR)}{dE}\tan Z(E) - KR\sec^2 Z\frac{dZ}{dE}\right]_{E=E_0'}.$$

But $\tan Z(E_0') = 0$, and $\sec Z(E_0') = 1$; therefore

$$\left(\frac{d\,\text{Re } D_l}{dE}\right)_0 = \left[-KR\frac{dZ}{dE}\right]_{E=E_0'}. \tag{7.15}$$

The actual magnitude of dZ/dE is unknown; however, its order of magnitude may be estimated as follows: According to Equation 7.14, $Z(E)$ must change by π in going from one l-wave resonance energy to the next. If these levels are separated in energy by an average spacing D, then, to order of magnitude, the derivative is

$$\left(\frac{dZ}{dE}\right)_0 \stackrel{\circ}{=} \frac{\pi}{D}. \tag{7.16}$$

Substituting this approximation into Equation 7.15 and the resulting expression into Equation 6.45, we find that

$$\Gamma_{n,l} \stackrel{\circ}{=} (2b_l)\left(\frac{D}{\pi KR}\right). \tag{7.17}$$

This equation is the final result. The neutron width is seen to be the product of two factors. The first factor, $2b_l$, is a function of the wave number k outside the nucleus. The second factor $(D/\pi KR)$ is a function of the wave number K just inside the nuclear boundary.

The reader will recall that the neutron width expression (Equation 6.186) that was derived from the rigorous R-matrix theory was

$$\Gamma_{n,l} = 2b_l\gamma_{\lambda l}^2,$$

where $\gamma_{\lambda l}^2$ is energy independent. The preceding two expressions can be combined to obtain the expected order of magnitude of the reduced width, namely,

$$\gamma_{\lambda l}^2 \overset{\circ}{=} \frac{D}{\pi KR}. \tag{7.18}$$

Now let us specifically restrict our attention to the low-energy range defined by the condition that $kR \ll 1$. In terms of energy and mass number

$$kR = \frac{\sqrt{2mE}}{\hbar}[1.5 \times 10^{-13}A^{1/3}] = 3.3 \times 10^{-4}\sqrt{E(\text{ev})}\,A^{1/3}. \tag{7.19}$$

Since A never exceeds about 250, we see that any neutron energy below about 100 ev can be considered to be low energy for even the heaviest of nuclei. From the data on b_l in Table 6.2, we can verify that at low energies

$$b_l \simeq (kR)^{2l+1} \propto E^{l+1/2} \qquad \text{(for } kR \ll 1\text{).}$$

Hence,

$$\Gamma_{n,l} \propto E^{l+1/2} \qquad \text{(for } kR \ll 1\text{).} \tag{7.20}$$

The proportionality can also be derived from Equation 7.17 and from the fact that the inner wave number K and the level spacing D are practically constant in the neutron-energy range below 100 ev.

Many low-energy resonances in light nuclei and most low-energy resonances in heavy nuclei are $l = 0$ resonances. For these resonances,

$$\Gamma_{n,0} \propto \sqrt{E}. \tag{7.21}$$

It is common practice to isolate the energy-independent part of the neutron level width for $l = 0$ resonances by defining a "reduced neutron width" $\Gamma_{n,0}^0$ through the equation

$$\Gamma_{n,0} \equiv \sqrt{\frac{E_0}{1 \text{ ev}}}\,\Gamma_{n,0}^0, \tag{7.22}$$

where E_0, $\Gamma_{n,0}$, and $\Gamma^0_{n,0}$ are all expressed in energy units, i.e., electron volts. The width actually measured in the laboratory is $\Gamma_{n,0}$. Attention is called to the unfortunate fact that the name "reduced neutron width" is widely used in the literature for the two related, but not identical, quantities $\gamma^2_{\lambda l}$ and $\Gamma^0_{n,0}$. In this text, $\gamma^2_{\lambda l}$ is called the reduced neutron width, and $\Gamma^0_{n,0}$ is called the "reduced neutron width" (in quotation marks).

Measured values of "reduced neutron widths" range from about 1 Mev in the light elements to fractions of a milli-electron volt in the heavy elements. Most "reduced neutron widths" in the light elements are in the range from 1 to 10 kev. In the heavy elements, most are in the range 10^{-1} to 10 milli-electron volts. For more details, the reader should consult a tabulation of resonance parameters such as that in the "Barn Book."

Low-Energy s-Wave Isolated Resonances

At low energies ($kR \ll 1$), the s-wave phase shift for the impenetrable sphere is given by $\Delta_0 = -kR$; hence the Breit–Wigner equation (Equation 7.8) for the special case $l^* = 0$ becomes

$$\sigma_n(E) = 4\pi R^2 + 4\pi \lambdabar^2 g(J)$$

$$\times \left[\left| \frac{\Gamma_{n,0}/2}{(E - E_0) + (i\Gamma/2)} + \frac{R}{\lambdabar} \right|^2 - \frac{R^2}{\lambdabar^2} \right]. \tag{7.23}$$

At the peak of a pure scattering resonance, the cross section takes on a particularly simple form. Setting $E = E_0$ and $\Gamma = \Gamma_{n,0}$ in Equation 7.23, we find

$$\sigma_n(E_0) = 4\pi R^2 + 4\pi \lambdabar^2 g(J) \simeq 4\pi \lambdabar^2 g(J), \tag{7.24}$$

since $4\pi \lambdabar^2 g(J) \gg 4\pi R^2$. Equation 7.24 may be used to infer the statistical factor $g(J)$ for a particular s-wave resonance from a measurement of its peak cross section $\sigma_n(E_0)$. If, in addition, the spin I of the target nucleus is known, the J of the resonance level may be determined. Note that, if $I = 0$, the statistical factor is unity and the scattering cross section actually assumes its maximum allowed value, $4\pi \lambdabar^2$, at $E = E_0$. If $I > 0$, then $g(J)$ is less than unity and the cross section remains below the $4\pi \lambdabar^2$ value, as pictured in Figure 7.3.

Low-energy s-wave scattering resonances exhibit a characteristic minimum at an energy just below the resonance energy. This minimum is a direct result of the interference between shape elastic and compound elastic scattering, which is introduced through the term $A^0_{ce} + A^0_{se}$ in Equation 7.8.

P and Higher Order Isolated Resonances

At low energies, the major difference between an s-wave resonance and a resonance excited by a higher order wave is that the shape elastic amplitude A_{se}^l, which goes as $(kR)^{2l+1}$, is much smaller for the higher order waves. The interference between shape elastic and compound elastic scattering is much less severe; hence the dip in the cross section before the peak is less pronounced and may even disappear. At high energies, however, A_{se}^l may be quite large for higher order waves, and the dip, therefore, quite pronounced.

Differential Cross Sections in Isolated Resonances

The differential cross section in an isolated resonance that is excited by a wave of order $l*$ will contain a large component of the Legendre polynomial of order $l*$, as can be seen by inspection of Equation 6.35:

$$\sigma_n(E,\theta) = \frac{\lambda^2}{4} \left| \sum_l (2l + 1)e^{2i\Delta_l}(A_{se}^l + A_{ce}^{l^*})P_l(\cos \theta) \right|^2$$

If the resonance peak lies well above the shape elastic background, the differential cross section at the peak will be essentially proportional to $[P_{l^*}(\cos \theta)]^2$. The differential cross section therefore furnishes an experimental method for determining the $l*$ value, or values, of an isolated resonance.

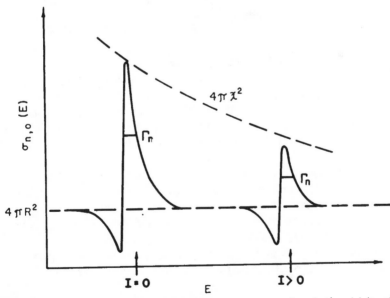

Figure 7.3—Variation of low-energy elastic-scattering cross sections in the vicinity of pure isolated s-wave scattering resonances.

7.4 The Optical Model of Neutron Scattering

The original black-nucleus or strong-coupling model* was remarkably successful in describing neutron–nucleus cross sections in the region of isolated resonances. But at high energies this model predicted that total cross sections averaged over resonances should be monotonically decreasing functions of energy at a fixed mass number A and slowly increasing functions of mass number at a fixed energy. These predictions were not verified experimentally. As can be seen in Figure 7.4a, measured total cross sections are not monotonic with energy, nor do they increase monotonically with increasing A.

These difficulties with the strong-coupling model led Fernbach, Serber, and Taylor† to propose a model in which the nucleus was not considered to be totally absorbing, as in the black-nucleus model, but only partially absorbing. The nucleus is pictured as a cloudy crystal ball that can partially refract and partially absorb the wave associated with the incident neutron. Under this reduced interaction, the incident neutron may penetrate into the nucleus, move within it, and escape from it without forming a compound nucleus; in this event, the target acts as a potential well. The formation of a compound state, which occurs with probability less than unity, is in effect an absorption of the incident neutron. This absorption can be accounted for in the wave-mechanical description of the interaction by assuming that the nuclear potential has the form of a complex well so that the total outgoing probability current is less than the total incoming current.

Such an oversimplified model cannot be expected to reproduce fine resonance features in the cross sections since these depend on the details of the quantum states in the compound nucleus that are not even considered in the optical model. At best, this model will describe the average behavior of the elastic-scattering, compound-nucleus-formation, and total cross sections as functions of energy and atomic weight.

Theory of Average Cross Sections

Neutron cross sections in the resonance region are not readily derivable from a one-particle potential because of their rapid fluctuations with energy. We will now show that the averages of the cross sections taken over an energy interval ΔE, which includes many resonances, will be the cross sections belonging to a new scattering problem, the *gross-structure* problem. This gross-structure problem is described by the optical model.

---

*H. Feshbach and V. F. Weisskopf, *Phys. Rev.*, **76**: 1550 (1949).
†S. Fernbach, R. Serber, and T. Taylor, *Phys. Rev.*, **75**: 1352 (1949).

In the development of the partial-wave analysis for spinless particles, the elastic-scattering, reaction, and total cross sections for the lth partial wave were found to be given by

$$\sigma_{n,l} = \pi\lambda^2(2l + 1)|1 - U_l|^2, \tag{7.25a}$$

$$\sigma_{r,l} = \pi\lambda^2(2l + 1)(1 - |U_l|^2), \tag{7.25b}$$

and

$$\sigma_{T,l} = 2\pi\lambda^2(2l + 1)(1 - \mathrm{Re}\, U_l), \tag{7.25c}$$

where the reflection factor $U_l = \exp(2i\delta_l)$, with δ_l the complex phase shift $\alpha_l + i\beta_l$.

For the gross-structure problem, we assume that we can average over the cross-section fluctuations. An *average reflection factor* at energy E is defined as

$$\langle U_l(E) \rangle \equiv \frac{1}{\Delta E} \int_{E - \Delta E/2}^{E + \Delta E/2} U_l(E')dE'. \tag{7.26}$$

Average cross sections are defined in the same manner. Hence the optical-model l-wave elastic-scattering, reaction, and total cross sections are given by the averages of Equations 7.25:

$$(\sigma_{n,l})_{\mathrm{opt}} = \pi\lambda^2(2l + 1)\langle|1 - U_l|^2\rangle, \tag{7.27a}$$

$$(\sigma_{r,l})_{\mathrm{opt}} = \pi\lambda^2(2l + 1)(1 - \langle|U_l|^2\rangle), \tag{7.27b}$$

$$(\sigma_{T,l})_{\mathrm{opt}} = 2\pi\lambda^2(2l + 1)(1 - \mathrm{Re}\langle U_l\rangle), \tag{7.27c}$$

where all averages are over an interval ΔE, which is assumed to be so much smaller than E that λ^2 need not be averaged. After expanding the expression within the absolute-value sign in Equation 7.27a and averaging, we obtain for the elastic-scattering cross section

$$(\sigma_{n,l})_{\mathrm{opt}} = \pi\lambda^2(2l+1)(|1 - \langle U_l\rangle|^2) + \pi\lambda^2(2l+1)(\langle|U_l|^2\rangle - |\langle U_l\rangle|^2), \tag{7.28}$$

and we note that the sum of the optical-model elastic-scattering and reaction cross sections is consistent with the expression for the optical-model total cross section, i.e.,

$$(\sigma_{T,l})_{\mathrm{opt}} = (\sigma_{n,l})_{\mathrm{opt}} + (\sigma_{r,l})_{\mathrm{opt}} = 2\pi\lambda^2(2l + 1)(1 - \mathrm{Re}\langle U_l\rangle). \tag{7.29}$$

We note further that the reaction cross section can be written in the form

$$(\sigma_{r,l})_{\mathrm{opt}} = \pi\lambda^2(2l + 1)(1 - \langle|U_l|^2\rangle) \tag{7.30}$$

$$= \pi\lambda^2(2l + 1)(1 - |\langle U_l\rangle|^2) - \pi\lambda^2(2l + 1)(\langle|U_l|^2\rangle - |\langle U_l\rangle|^2).$$

Now the term $(\langle|U_l|^2\rangle - |\langle U_l\rangle|^2)$ which appears in the optical-model expression for both the elastic-scattering and reaction cross sections is just the mean-square fluctuation of the coefficient of the outgoing wave in the interval ΔE. We therefore define the l-wave fluctuation cross section as

$$\sigma_{fl,l} \equiv \pi\lambda^2(2l + 1)(\langle|U_l|^2\rangle - |\langle U_l\rangle|^2). \tag{7.31}$$

Note that, when we have determined $\langle U_l\rangle$, all cross sections except this fluctuation cross section can be evaluated. The elastic-scattering and reaction cross sections may now be written

$$(\sigma_{n,l})_{\text{opt}} = \pi\lambda^2(2l + 1)(|1 - \langle U_l\rangle|^2) + \sigma_{fl,l}, \tag{7.32a}$$

$$(\sigma_{r,l})_{\text{opt}} = \pi\lambda^2(2l + 1)(1 - |\langle U_l\rangle|^2) - \sigma_{fl,l}. \tag{7.32b}$$

Now we assume that the coefficient of the outgoing wave U_l that we obtain by solving the appropriate Schrödinger equation with an optical potential is to be identified with the average value $\langle U_l\rangle$. That is, we assume

$$U_l^{\text{opt}} = \langle U_l\rangle. \tag{7.33}$$

The elastic-scattering cross section as given by Equation 7.32a consists of the sum of a smoothly varying cross section and a fluctuating cross section. The smoothly varying part is identified as the shape elastic cross section

$$(\sigma_{se,l})_{\text{opt}} = \pi\lambda^2(2l + 1)(|1 - \langle U_l\rangle|^2). \tag{7.34}$$

The rest of the elastic-scattering cross section is identified as compound elastic scattering. It is therefore assumed that

$$\sigma_{ce,l} = \sigma_{fl,l}. \tag{7.35}$$

Finally, the cross section for compound-nucleus formation is given by Equations 7.32b and 7.35 in the form

$$(\sigma_{c,l})_{\text{opt}} = (\sigma_{r,l})_{\text{opt}} + \sigma_{ce,l} = \pi\lambda^2(2l + 1)(1 - |\langle U_l\rangle|^2). \tag{7.36}$$

Differential Elastic-Scattering Cross Sections

The optical-model differential elastic-scattering cross sections are obtained by averaging the differential cross section expression over the range ΔE about E

$$[\sigma(\theta)]_{\text{opt}} = \left\langle \left|\frac{\lambda^2}{4}\right| \sum_{l=0}^{\infty} (2l + 1)P_l(\cos\theta)[1 - U_l(E)]\right|^2\right\rangle.$$

Only the terms containing $U_l(E)$ need be averaged; they have the form

$$\langle(1 - \bar{U}_l)(1 - U_{l'})\rangle = [(1 - \langle\bar{U}_l\rangle)(1 - \langle U_{l'}\rangle)] + [\langle\bar{U}_l U_{l'}\rangle - \langle\bar{U}_l\rangle\langle U_{l'}\rangle].$$

The first bracketed term can be seen by comparison with Equation 7.34 to be due to shape elastic scattering. Therefore, we can write

$$[\sigma_{se}(\theta)]_{opt} = \frac{\lambdabar^2}{4} \left| \sum_{l=0}^{\infty} (2l + 1)P_l(\cos \theta)(1 - \langle U_l \rangle) \right|^2 . \tag{7.37}$$

The second bracketed term is ascribed to compound elastic scattering

$$[\sigma_{ce}(\theta)]_{opt} = \frac{\lambdabar^2}{4} \sum_{l=0}^{\infty} \sum_{l'=0}^{\infty} (2l + 1)(2l' + 1)$$

$$\times P_l(\cos \theta)P_{l'}(\cos \theta)[\langle \bar{U}_l U_{l'} \rangle - \langle \bar{U}_l \rangle \langle U_{l'} \rangle]. \tag{7.38}$$

The total differential elastic-scattering cross section is then given by the sum $\sigma_{se}(\theta) + \sigma_{ce}(\theta)$.

Penetrabilities

In the analysis of such reactions as (n, n'), (n, p), and (n, α), the most significant parameters obtained from the optical model are the penetrabilities, sometimes called transmission coefficients. The penetrabilities, T_l, are defined by expressing the cross section for formation of the compound nucleus by the lth partial wave in the form

$$\sigma_{c,l}(E) = \pi\lambdabar^2(2l + 1)T_l(E). \tag{7.39}$$

Comparing Equations 7.36 and 7.39, we see that the l-wave penetrability is defined as

$$T_l(E) \equiv 1 - |\langle U_l \rangle|^2 . \tag{7.40}$$

Penetrabilities can be obtained for charged particles, such as protons and alpha particles, as well as for neutrons by means of optical-model calculations that use appropriate complex potentials.

Let us now discuss what we can and cannot determine from the optical model. The solution of the radial Schrödinger equation with an optical-model (complex) potential will determine the logarithmic derivatives D_l, hence the reflection factors U_l^{opt}. Under the primary assumption that $\langle U_l \rangle = U_l^{opt}$, we can determine the total cross section, the shape elastic cross section, and the cross section for compound-nucleus formation, all of which depend only on $\langle U_l \rangle$. We cannot determine the compound elastic cross section (the fluctuation cross section) or the reaction cross section because these depend on the unknown $\langle |U_l|^2 \rangle$. However, if it is known that with a certain target nuclide at a certain energy reactions do not occur, then σ_{ce} can be determined, for in this event $\sigma_{ce} = \sigma_c$, which is obtained from the optical model.

The optical model is usually employed to obtain the shape elastic cross section at all energies, but it is used for the compound elastic cross section only in the nonreaction region. In the reaction region, the method of Wolfenstein, Hauser, and Feshbach, described in Chapter 8, is used in conjunction with penetrabilities obtained from the optical model to determine compound elastic-scattering and reaction cross sections.

7.5 The Complex Square Well

One of the earliest cloudy-crystal-ball calculations of neutron cross sections was made by Feshbach, Porter, and Weisskopf.* Their assumptions and results will be briefly described here both because of their intrinsic historical value and, more importantly, because they show quite clearly that even the simplest possible optical-model potential does remarkably well in predicting the gross structure of measured cross sections.

Feshbach, Porter, and Weisskopf chose the most elementary possible potential to describe the neutron–nucleus interaction, a complex spherical square well,

$$V(r) = \begin{cases} -V_0(1 + i\zeta) & \text{(for } r < R), \\ 0 & \text{(for } r > R), \end{cases}$$

where V_0 and ζ are positive constants and R is the nuclear radius. The well depth V_0 and the *absorption constant* ζ were varied to obtain the best overall fit to the measured cross sections. The values $V_0 = 42$ Mev and $\zeta = 0.03$ gave the best overall fit. The nuclear radius was chosen to be $R = R_0 A^{1/3} = 1.45 A^{1/3}$ fm where A is the nuclear mass number.

The researchers found that the calculated cross sections were quite insensitive to V_0 as long as the well strength, $V_0 R_0^2$, was kept constant. The shapes of the cross sections, on the other hand, were sensitive to the value of the absorption constant ζ. A change in ζ of as much as 0.01 was found to result in significantly worse agreement between theoretical and observed cross sections.

Comparison with Observations

Figure 7.4b shows the total cross section calculated by Feshbach, Porter, and Weisskopf as a function of neutron energy and mass number. The energy is expressed in terms of the dimensionless parameter $x^2 = (R/\lambda)^2 = (kR)^2$. In more common units,

$$E(\text{Mev}) = 1.66 \, x^2 \quad \text{(for } A = 16),$$

$$E(\text{Mev}) = 0.47 \, x^2 \quad \text{(for } A = 100),$$

$$E(\text{Mev}) = 0.29 \, x^2 \quad \text{(for } A = 200).$$

*H. Feshbach, C. E. Porter, and V. F. Weisskopf, *Phys. Rev.*, **96**: 448 (1954).

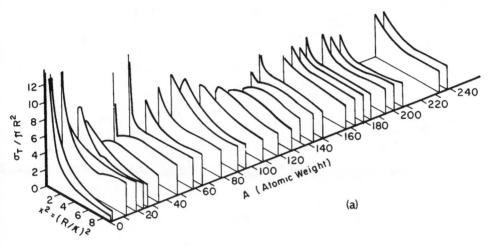

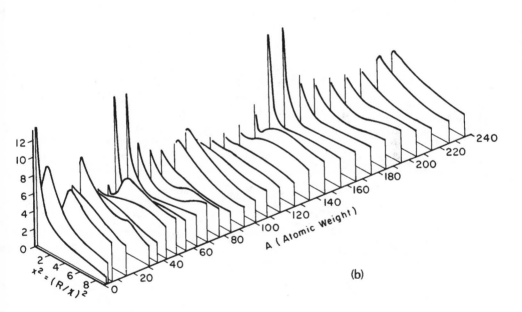

Figure 7.4—(a) Observed average neutron cross sections as a function of energy and mass number with $R = 1.45A^{1/3}$ fm. (b) Calculated total cross sections based on the optical model with $R = 1.45A^{1/3}$ fm, $V_0 = 42$ Mev, and $\zeta = 0.03$. [From H. Feshbach, C. E. Porter, and V. F. Weisskopf, *Phys. Rev.*, **96**: 448 (1954).]

The experimental curves in Fig. 7.4a are averages over resonances. Comparison of the figures indicates that the theory reproduces the gross structure rather well. Note in particular the representation of the drop in σ_T at low energies in the regions near $A = 40$ and between $A = 100$ and 140 and the representation of greatly increased σ_T in the regions near $A = 60$, 90, and 150.

In Figure 7.5a, experimental values of differential elastic-scattering cross sections are plotted as a function of angle in the C-system for 1-Mev neutrons. Certain characteristics are evident: (a) relatively flat distributions from $A = 50$ to 70, (b) strong forward and increasing backward peaking from $A = 100$ to 140, and (c) the development of a secondary maximum at 90° near $A = 180$.

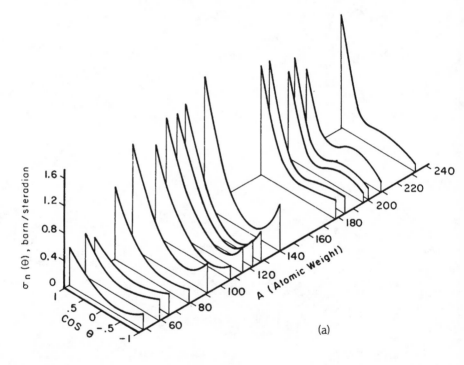

Figure 7.5—(a) Observed differential cross sections of 1-Mev neutrons as measured by Walt and Barschall. (b) Calculated shape elastic differential cross section of 1-Mev neutrons based on the optical model with $R = 1.45A^{1/3}$ fm, $V_0 = 42$ Mev, and $\zeta = 0.03$. (c) Calculated shape elastic-scattering cross section plus maximum compound elastic-scattering cross section based on the optical model with $R = 1.45A^{1/3}$ fm, $V_0 = 42$ Mev, and $\zeta = 0.03$. [From H. Feshbach, C. E. Porter, and V. F. Weisskopf, *Phys. Rev.*, **96**: 459 (1954).]

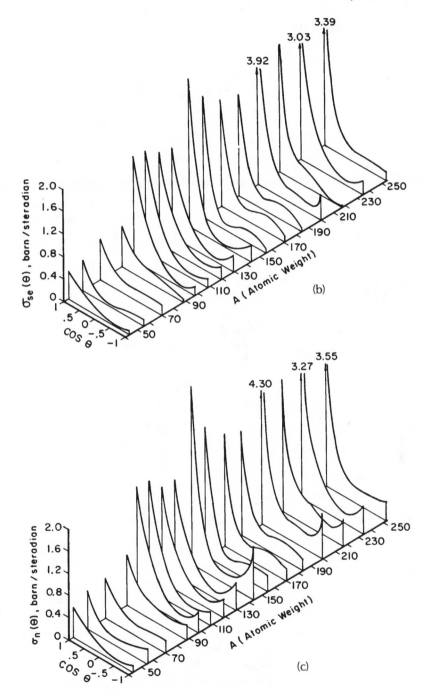

In Figure 7.5b, calculated values of the shape elastic-scattering cross section are plotted. Figure 7.5c shows the calculated shape elastic-scattering plus maximum compound elastic-scattering cross sections, assuming that the compound state decays exclusively via the entrance channel, i.e., by elastic scattering. These figures indicate clearly that the main features of the observed differential elastic-scattering cross sections are reproduced by the theory, except possibly above $A = 200$. The flatness of the cross sections in the region from $A = 50$ to 70 is explained theoretically by the fact that the p-wave contribution, which has strong angular dependence, is very weak in this region.

The theory of Feshbach, Porter, and Weisskopf did somewhat less well in predicting σ_c, the cross section for compound-nucleus formation. Although no direct measurement of σ_c is possible, they were able to make indirect comparisons with $\sigma_{n'}$ and σ_r, the inelastic and reaction cross sections; and they found that the observed values were of the predicted order of magnitude but did not agree in detail with the theoretical values.

In summary, we can state that the work of Feshbach, Porter, and Weisskopf demonstrated rather conclusively that the optical model, even with an extremely simple potential, was capable of reproducing both the gross structure of the total cross sections and the differential elastic-scattering cross sections for neutron–nuclei interactions.

7.6 Refinements in Optical-Model Potentials

The encouraging, though partial, success of Feshbach, Porter, and Weisskopf precipitated a rush of attempts to improve optical-model calculations. A great variety of potentials were tried, ranging from square wells whose depth V_0 and absorption coefficient ζ were allowed to vary with A and E, to trapezoidal wells, and to more realistic radial forms that accounted for the known decrease in density of the nucleons near the nuclear surface. Before the advent of high-speed digital computers, potentials were chosen for their amenability to analytical solutions; since the advent of computers, which are capable of calculating radial wave functions in fractions of minutes regardless of the potential, more realistic, and hence more complicated, potentials have been employed. These refinements have naturally led to more accurate predictions of cross sections.

The recent potentials have three things in common: (1) a real potential well that tapers off gradually to zero, (2) an imaginary potential well that is concentrated near the nuclear surface, and (3) a spin-orbit potential term that is also concentrated near the surface.

The Real Potential

The real or *refracting potential* is generally chosen to be approximately proportional to the density of nucleons in the nucleus. It has the form

$$V_c \rho(r), \tag{7.41}$$

where V_c is a negative constant and $\rho(r)$ is chosen to approximate the measured nuclear charge density (see Figure 5.2), which is assumed to be proportional to the nucleon density. The most common form is the *Saxon–Woods potential*

$$\rho(r) = \frac{1}{1 + \exp[(r - R)/a]}, \tag{7.42}$$

where R, the nuclear radius, is generally chosen to be $R = R_0 A^{1/3}$ and a is an adjustable parameter, called the *diffuseness*, that is usually varied around 0.55 fm to obtain optimum fit.

The Imaginary Potential

The imaginary or *absorptive potential* was initially chosen to be a volume potential like the real potential. However, the experimental cross-section data could be better reproduced by putting the absorption in a surface layer. One widely used analytical form for the surface absorption potential is the Gaussian

$$iW_s \exp\left[-\left(\frac{r - R}{b}\right)^2\right], \tag{7.43}$$

where R is the nuclear radius $R_0 A^{1/3}$, b is an adjustable parameter, usually set equal to approximately 1 fm, and W_s is an adjustable negative constant.

It is also possible to obtain a surface absorption potential by using the derivative of the real potential. If, for example, a Saxon–Woods potential is employed as the real potential, then the imaginary potential may be expressed as

$$iW_s a \frac{d\rho}{dr} = iW_s \rho^2 \exp\left(\frac{r - R}{a}\right). \tag{7.44}$$

This procedure has the advantage of involving one less arbitrary parameter in the potential since the parameter b is eliminated.

The Saxon–Woods potential $\rho(r)$ and its derivative for a nucleus with a radius of 7 fm and a diffuseness of 0.55 fm are shown in Figure 7.6. Since each of these curves is multiplied by a negative constant before being introduced into the Schrödinger equation, the actual potentials used in the

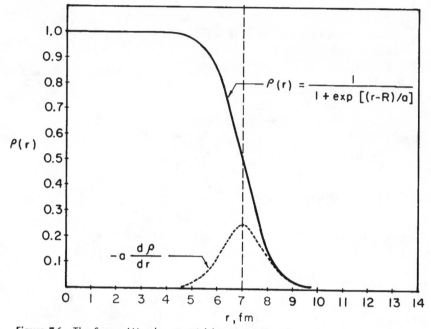

Figure 7.6—The Saxon–Woods potential function $\rho(r)$ and $-a\,d\rho/dr$ with parameters $R = 7$ fm and $a = 0.55$ fm.

optical model look like the mirror images of these curves. Note the rapid decrease in $\rho(r)$ beyond the nuclear surface and the concentration of the absorption in the immediate vicinity of the surface.

The Spin-Orbit Potential

For the optical model to be capable of predicting, as is observed, that scattered neutrons are partially polarized, it is necessary to include a spin-dependent term in the potential. Nuclear theoretical analysis indicates that this potential should have the form*

$$V_{\sigma \cdot l}(r) = -\alpha V_c \left(\frac{\hbar}{2mc}\right)^2 \frac{1}{r}\frac{d\rho}{dr}\,\boldsymbol{\sigma} \cdot \mathbf{l}, \tag{7.45}$$

where $\mathbf{l}$ and $\boldsymbol{\sigma}$ are the dimensionless operators related to the orbital angular momentum and spin angular momentum of the incident neutron, $\mathbf{L} \equiv \hbar \mathbf{l}$ and $\mathbf{s} \equiv \hbar \boldsymbol{\sigma}/2$. The function $\rho(r)$, which describes the radial variation of

---

*S. Fernbach, W. Heckrotte, and J. V. Lepore, *Phys. Rev.*, **97**: 1059 (1955); G. E. Brown, *Proc. Phys. Soc. A*, **70**: 351 (1957).

the nucleon density, is usually chosen as the Saxon–Woods function; $(\hbar/mc) = 0.21$ fm is the rationalized Compton wavelength of the neutron; V_c is the negative constant that appears in the real potential term; and α is an adjustable positive constant. The nuclear spin-orbit potential is identical to the electronic one (Equation 5.50) except for the minus sign and the empirical constant α.

The spin-orbit potential, like the absorptive potential, is concentrated at the nuclear surface. Unlike the absorptive potential, however, it is usually taken as real because making it complex has not been found to appreciably increase the agreement between theory and experiment.

Note, by the way, that the constants V_c and $(\hbar/2mc)^2$ are seemingly superfluous in the sense that both these constants could have been included in the adjustable parameter α. The Compton wavelength appears as a result of nuclear theory, and V_c is included primarily because it appears as a coefficient in the other real term in the potential; so α becomes a dimensionless parameter that measures the strength of the spin-orbit potential relative to the refracting potential. Since $(\hbar/2mc)^2$ is of the order of 10^{-2} fm^2, the quantity $(d\rho/dr)/r$ is of order 10^{-1} fm^{-2}, and $\mathbf{l}\cdot\mathbf{\sigma}$ is of order unity, the parameter α may take on values of order 10 to 10^2 and still represent a relatively small perturbation to the refracting potential.

Nonspherical Potentials

Ellipsoidal and spheroidal potentials have been employed in attempts to improve optical-model calculations for nuclei that are known to be nonspherical. These have met with considerable success* but not without additional analytical difficulties arising from the fact that the radial Schrödinger equations for different l values are coupled together by a nonspherical potential.

Nonlocal Potentials

Nuclear theory indicates that the force acting on a neutron whose center is at a point $\mathbf{r}$, within or near the nucleus, depends not only on $\mathbf{r}$ but also on the value of the wave function throughout all space. This might be expected since the uncertainty principle says that it is impossible to localize a neutron with fixed momentum. The force and the potential from which it is derived are therefore nonlocal, and the term $V(\mathbf{r})\,\psi(\mathbf{r})$ in the Schrödinger equation for the neutron–nucleus system should rigorously be replaced by

$$\int V(\mathbf{r}, \mathbf{r}')\,\psi(\mathbf{r}')\,d\mathbf{r}'. \tag{7.46}$$

*See, for example, D. M. Chase, L. Wilets, and A. R. Edmonds, *Phys. Rev.*, **110**: 1080 (1958).

Solutions of the optical model have been obtained using nonlocal potentials.* In general, they do seem to give slightly better results than those obtained with local potentials. At present, however, it is not at all clear whether this improvement is a direct result of the increased validity of the nonlocal potential or simply follows from the greater number of adjustable parameters that are included in the nonlocal potential.

Further Refinements

Additional refinements in optical potentials have been suggested from time to time. These include spin–spin potentials of the form $(\mathbf{I} \cdot \mathbf{\sigma})f(r)$, target–spin-orbit potentials, such as $(\mathbf{I} \cdot \mathbf{l})f(r)$, and combination potentials, such as $(\mathbf{I} \cdot \mathbf{\sigma})(\mathbf{l} \cdot \mathbf{I})f(r)$. The need for such forces to explain experimental cross sections has not been proved conclusively. Their effects on cross sections are likely to be of second order except, possibly, in the scattering of polarized nucleons from nuclei with large spins.

Theoretical Justification of Optical Potentials

The rather remarkable success of the optical model in predicting the gross structure of elastic-scattering and total cross sections has naturally stimulated attempts to derive the optical potentials from fundamental nuclear theory. This theory is of necessity very complex since it involves the many-body interaction of an incident nucleon with the nucleons in the nucleus; nevertheless, some success has been realized. It has been possible to estimate the depth of the potential, the radial variation of the imaginary potential, and the form of the spin-orbit potential. It has not been possible to fix exact values for the many parameters that enter an optical-model potential, but it has become quite clear that the optical model is a consequence of a deeper reality and not simply a fortuitous accident.

For further details of the theory of the optical model, the reader is encouraged to explore the short but stimulating books by P. E. Hodgson and I. Uhehla, L. Gomolcak, and Z. Pluhar listed at the end of this chapter.

7.7 A Typical Optical-Model Calculation

This section consists of a typical optical-model calculation that was performed by R. S. Caswell† to obtain the differential and total elastic-scattering cross sections of calcium in the neutron-energy range from 0.734 Mev up to 18 Mev. This particular calculation was chosen as an

---

*F. G. J. Perey and B. Buck, *Nucl. Phys.*, **32**: 353 (1962).

†R. S. Caswell, *J. Res. Nat. Bur. Std.*, **66**A: 389 (1962).

example from the hundreds in the literature primarily because the potential and the method of determining the phase shifts are typical of those employed in the majority of such calculations. All the significant steps are detailed; so the diligent reader should have sufficient information after studying this section to begin his own optical-model calculations of cross sections if the need should arise.

The potential has the form of a complex well $[V(r) + iW(r)]$ with surface absorption and a spin-orbit coupling term. The real and imaginary parts are given by

$$V(r) = V_c\rho(r) - \alpha V_c\left(\frac{\hbar}{2mc}\right)^2 \mathbf{l}\cdot\boldsymbol{\sigma}\frac{1}{r}\frac{d\rho}{dr},$$

$$W(r) = W_s\left(4a\frac{d\rho}{dr}\right) = W_s 4\rho^2 \exp\left(\frac{r-R}{a}\right),$$

(7.47)

where $\rho(r)$ is the Saxon–Woods potential and V_c, W_s, α, a, and R are adjustable parameters. The nuclear radius R and the spin-orbit parameter α were held constant at $1.25A^{1/3}$ fm and 35, respectively. Three parameters – V_c, W_s, and a – were allowed to vary with energy for optimum fit to the experimental data. The values of these parameters, as finally used, are shown in Figure 7.7.

The element calcium consists of 96.97% ^{40}Ca with 2.06% ^{44}Ca and small amounts of ^{42}Ca, ^{43}Ca, ^{46}Ca, and ^{48}Ca. In this calculation it was considered to be pure ^{40}Ca except that a slightly larger nuclear radius was used to approximate the average radius of the element calcium. All the isotopes except ^{43}Ca are even-even; so the spin of the nucleus was taken to be zero. The neglect of the nonzero spin of ^{43}Ca was justified on the basis of the fact that this isotope has an abundance of only 0.145%.

The Spin-Orbit Term

The spin-orbit interaction term $\mathbf{l}\cdot\boldsymbol{\sigma}$ is evaluated as follows: The orbital angular momentum of an l-wave neutron has square magnitude

$$|\mathbf{L}|^2 = l(l+1)\hbar^2.$$

The spin angular momentum has square magnitude

$$|\mathbf{s}|^2 = s(s+1)\hbar^2,$$

where $s = 1/2$. These two angular momenta add vectorially, as shown in Figure 7.8, to give the total squared angular momentum. Thus, from the fact that $|\mathbf{J}|^2 \equiv (\mathbf{L}+\mathbf{s})\cdot(\mathbf{L}+\mathbf{s})$ we find

$$|\mathbf{J}|^2 = J(J+1)\hbar^2 = |\mathbf{L}|^2 + |\mathbf{s}|^2 + 2\mathbf{L}\cdot\mathbf{s}.$$

Solving for $\mathbf{L}\cdot\mathbf{s}$, we get

$$\mathbf{L}\cdot\mathbf{s} = \frac{\hbar^2}{2}[J(J+1) - l(l+1) - s(s+1)].$$

If the spin and orbital angular momentum are "parallel," then $J = l + s$, in a "spin-antiparallel" state, $J = l - s$. Hence, from the above equation,

$$\textit{Parallel:} \quad \mathbf{L}\cdot\mathbf{s} = \frac{\hbar^2}{2}\,l,$$

$$\textit{Antiparallel:} \quad \mathbf{L}\cdot\mathbf{s} = -\frac{\hbar^2}{2}(l+1).$$

The vectors $\mathbf{L}$ and $\mathbf{l}$ are related by $\mathbf{L} \equiv \hbar\mathbf{l}$; the vector $\mathbf{s}$ and $\boldsymbol{\sigma}$, by $\mathbf{s} \equiv \hbar\boldsymbol{\sigma}/2$.

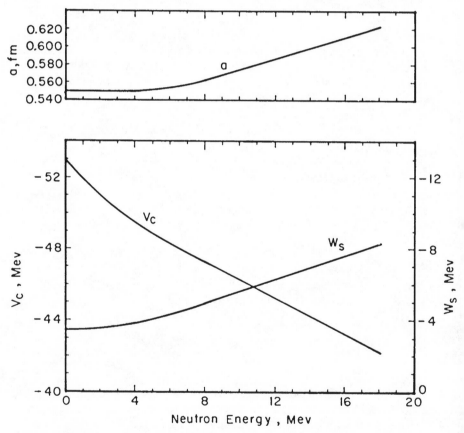

Figure 7.7—Optical-model parameters vs. energy [as adopted by Caswell in *J. Res. Nat. Bur. Stand.,* **66A**: 389 (1962)].

Hence

> *Parallel:* $\mathbf{l} \cdot \mathbf{\sigma} = l,$
>
> *Antiparallel:* $\mathbf{l} \cdot \mathbf{\sigma} = -(l + 1).$

(7.48)

These quantities are substituted into the potential $V(r)$ in place of the $\mathbf{l} \cdot \mathbf{\sigma}$ operator. More precisely, the operator $\mathbf{l} \cdot \mathbf{\sigma}$ is given by the diagonal matrix

$$\mathbf{l} \cdot \mathbf{\sigma} = \begin{pmatrix} l & 0 \\ 0 & -(l+1) \end{pmatrix},$$

so that $(\mathbf{l} \cdot \mathbf{\sigma})\chi^p = l\chi^p$ and $(\mathbf{l} \cdot \mathbf{\sigma})\chi^a = -(l+1)\chi^a$.

Solution of the Wave Equations

The calculation begins with the radial part of the Schrödinger equation including the complex potential

$$\frac{d^2 W_l(r)}{dr^2} + \frac{2m}{\hbar^2} \left[E - \frac{l(l+1)\hbar^2}{2mr^2} - V(r) - iW(r) \right] W_l(r) = 0, \quad (7.49)$$

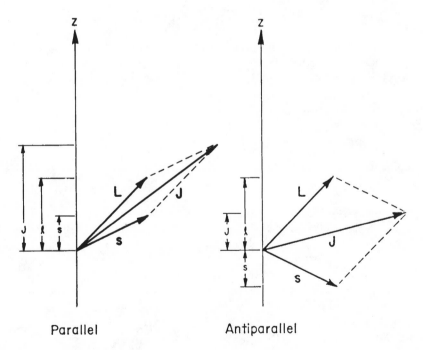

Parallel   Antiparallel

Figure 7.8—Vector models of the addition of orbital and spin angular momenta for parallel and antiparallel states.

where m is the reduced mass of the neutron, E is the total energy in the C-system, and $W_l(r) \equiv rR_l(r)$ with $R_l(r)$ the radial wave function. The reduced wave function $W_l(r)$ is now written in terms of its real and imaginary components:

$$W_l(r) \equiv X_l(r) + iY_l(r), \tag{7.50}$$

so that Equation 7.49 becomes

$$\frac{d^2(X_l + iY_l)}{dr^2} + \frac{2m}{\hbar^2}\left[E - \frac{l(l+1)\hbar^2}{2mr^2} - V - iW\right](X_l + iY_l) = 0. \tag{7.51}$$

This equation is now separated into real and imaginary parts,

$$\frac{d^2 X_l(r)}{dr^2} + \frac{2m}{\hbar^2}\left[E - \frac{l(l+1)\hbar^2}{2mr^2} - V(r)\right]X_l(r) + \frac{2m}{\hbar^2}W(r)Y_l(r) = 0,$$

$$\frac{d^2 Y_l(r)}{dr^2} + \frac{2m}{\hbar^2}\left[E - \frac{l(l+1)\hbar^2}{2mr^2} - V(r)\right]Y_l(r) - \frac{2m}{\hbar^2}W(r)X_l(r) = 0. \tag{7.52}$$

For a given value of l and a given spin direction (parallel or antiparallel), these two coupled differential equations are solved numerically through step-by-step outward integration. Caswell found that the time required on an IBM-7090 computer for 15 l values, both spin directions, and 200 integration steps was about 1 min per energy, including the calculation of all angular distributions. From the outward integrations, the values and slopes of X_l and Y_l are found at a radius R' at which the potential is effectively zero (3 to 4 nuclear radii), the logarithmic derivative then is found from the expression

$$D_l = R'\left[\frac{(dX_l/dr) + i(dY_l/dr)}{X_l + iY_l}\right]_{r=R'} \tag{7.53}$$

and the reflection factor U_l is obtained from Equation 6.23.

For a given l and E, the coupled wave equations must be solved twice because the $\boldsymbol{\sigma} \cdot \mathbf{l}$ term in the potential has two different forms, l if the spin is parallel to the orbital angular momentum and $-(l+1)$ if it is antiparallel. Thus for a given l and E, two logarithmic derivatives and reflection factors are found, U_l^p and U_l^a, where the superscripts p and a refer, as usual, to parallel and antiparallel spin. The coefficients of the outgoing waves, which are here designated U_l^p and U_l^a, are those which result from an optical-model potential. They are therefore identified as being equal to the average values of the underlying, rapidly fluctuating coefficients over an energy interval which contains many resonances; that is, we make the basic assumption of the optical model: $\langle U_l^p \rangle = U_l^p$ and $\langle U_l^a \rangle = U_l^a$. The only task which remains is that of relating the desired cross sections to the known reflection factors. The relevant equations will now be derived.

The incident neutrons, which have spin 1/2, are represented by a plane wave

$$\psi_{\text{inc}} = e^{ikz}\chi_{\text{inc}}, \tag{7.54}$$

where χ_{inc}, the spin wave function of the neutron, is

$$\chi_{\text{inc}} = A^p\chi^p \quad A^a\chi^a = A^p\begin{pmatrix}1\\0\end{pmatrix} + A^a\begin{pmatrix}0\\1\end{pmatrix},$$

where $|A^p|^2$ and $|A^a|^2$ are the relative fractions of incident neutrons with spin parallel to and antiparallel to their orbital angular momenta. Since $|\chi_{\text{inc}}|^2 = 1$, the quantity $|A^p|^2 + |A^a|^2$ equals 1. For completely unpolarized incident neutrons, the case of usual interest, A^p equals A^a equals $(1/2)^{1/2}$, indicating that the incident neutrons are equally likely to be in either state of polarization.

In addition to this notation, it is convenient to introduce the operators Π_l^p and Π_l^a, which select the states for parallel and antiparallel spin, respectively. These operators are

$$\Pi_l^p = \frac{l + 1 + \boldsymbol{\sigma} \cdot \mathbf{l}}{2l + 1}$$

and

$$\Pi_l^a = \frac{l - \boldsymbol{\sigma} \cdot \mathbf{l}}{2l + 1}. \tag{7.55}$$

It is readily verified that, if the spin is parallel, $\Pi_l^p = 1$ and $\Pi_l^a = 0$ and, if the spin is antiparallel, $\Pi_l^p = 0$ and $\Pi_l^a = 1$. By using these operators, we can express the total wave function, which is the usual sum of incoming and outgoing waves, far from the nucleus, in a very compact form:

$$\psi_{\text{tot}}(r, \theta) \sim \sum_{l=0}^{\infty} \frac{2l + 1}{2} i^l (\Pi_l^p + \Pi_l^a)[h_l^{(-)}(kr) + U_l h_l^{(+)}(kr)]P_l(\cos\theta)\chi_{\text{inc}}, \tag{7.56}$$

where $h_l^{(-)}$ and $h_l^{(+)}$ are the previously defined spherical Hankel functions. The incident-wave function is given by

$$\psi_{\text{inc}} \sim \sum_{l=0}^{\infty} \frac{2l + 1}{2} i^l [h_l^{(-)}(kr) + h_l^{(+)}(kr)]P_l(\cos\theta)\chi_{\text{inc}}. \tag{7.57}$$

The asymptotic forms of $h_l^{(+)}$ and $h_l^{(-)}$,

$$h_l^{(+)}(kr) \sim \frac{1}{kr} e^{i[kr-(l+1)\pi/2]}$$

and

$$h_l^{(-)}(kr) \sim \frac{1}{kr} e^{-i[kr-(l+1)\pi/2]}, \tag{7.58}$$

are substituted into Equations 7.56 and 7.57, and the incident-wave function is subtracted from the total-wave function. The difference is the asymptotic scattered-wave function

$$\psi_{sc} \sim \frac{e^{ikr}}{r} \left\{ \sum_{l=0}^{\infty} \frac{1}{2ik} [(l+1)(U_l^p - 1) + l(U_l^a - 1) \right.$$
$$\left. + (U_l^p - U_l^a)\boldsymbol{\sigma}\cdot\mathbf{l}]P_l(\cos\theta) \right\} \chi_{inc}. \tag{7.59}$$

In obtaining this equation, we used the expressions in Equation 7.55 for the operators and the fact that, when both U_l^p and U_l^a are unity, ψ_{sc} must equal zero.

We can argue on very basic grounds that the asymptotic form of the scattered Pauli wave function for a central potential must be given by

$$\psi_{sc} \sim \frac{e^{ikr}}{r} F(\theta, \phi)\chi_{inc}. \tag{7.60}$$

This equation, when written out in full, has the form

$$\begin{pmatrix} \psi_{sc}^p \\ \psi_{sc}^a \end{pmatrix} \sim \frac{e^{ikr}}{r} \begin{pmatrix} F_{11}F_{12} \\ F_{21}F_{22} \end{pmatrix} \begin{pmatrix} A^p \\ A^a \end{pmatrix}. \tag{7.61}$$

In this equation, F_{11} is the amplitude for the scattering of initially spin-parallel neutrons into spin-parallel neutrons, in brief, $p \to p$ scattering. Similarly, the elements F_{12}, F_{21}, and F_{22} are the amplitudes for $a \to p$, $p \to a$, and $a \to a$ scattering, respectively. Comparing Equations 7.59 and 7.60, we note that the scattering matrix $F(\theta,\phi)$ is given by

$$F(\theta,\phi) = \sum_{l=0}^{\infty} \frac{1}{2ik} [(l+1)(U_l^p - 1)\mathbf{I} + l(U_l^a - 1)\mathbf{I}$$
$$+ (U_l^p - U_l^a)\boldsymbol{\sigma}\cdot\mathbf{l}]P_l(\cos\theta), \tag{7.62}$$

where the unit matrix $\mathbf{I}$ has been indicated explicitly. Note that the diagonal elements F_{11} and F_{22} are functions of θ only, but the off diagonal elements F_{12} and F_{21} are functions of both θ and ϕ. The dependence is introduced in the operator $\boldsymbol{\sigma}\cdot\mathbf{l}$ which has functions of ϕ in its off diagonal elements. The scattering matrix will now be expressed in the form

$$F(\theta,\phi) = A(\theta) + B(\theta)\boldsymbol{\sigma}\cdot\mathbf{n}, \tag{7.63}$$

where $A(\theta)$ and $B(\theta)$ are diagonal matrices and $\mathbf{n}$ is a unit vector normal to the plane of scattering. The unit vector $\mathbf{n}$ is defined by

$$\mathbf{n} = \frac{\mathbf{v}_i \times \mathbf{v}_f}{|\mathbf{v}_i \times \mathbf{v}_f|}, \tag{7.64}$$

where $\mathbf{v}_i$ and $\mathbf{v}_f$ are the neutron's velocity vectors before and after scattering. The orbital angular-momentum vector $\mathbf{L} = \mathbf{r} \times \mathbf{p}$ is also normal to the plane of scattering. After scattering, the neutron's spin will be directed either parallel to or antiparallel to $\mathbf{n}$. By convention, a beam of neutrons with spins predominantly parallel to the $\mathbf{n}$ direction is said to be positively polarized.

If the incident beam is unpolarized, then the polarization of the emergent beam is introduced through the spin-orbit operator $\boldsymbol{\sigma}\cdot\mathbf{l}$ operating on $P_l(\cos\theta)$:

$$(\boldsymbol{\sigma}\cdot\mathbf{l})P_l(\cos\theta) = \boldsymbol{\sigma}\cdot\left[\mathbf{n}\left(-i\frac{\partial}{\partial\theta}\right)P_l(\cos\theta)\right].$$

However, $\dfrac{\partial}{\partial\theta} = -\sin\theta\dfrac{\partial}{\partial(\cos\theta)}$; so

$$(\boldsymbol{\sigma}\cdot\mathbf{l})P_l(\cos\theta) = (\boldsymbol{\sigma}\cdot\mathbf{n})\left[i\sin\theta\frac{\partial}{\partial(\cos\theta)}P_l(\cos\theta)\right]$$

or, finally,

$$(\boldsymbol{\sigma}\cdot\mathbf{l})P_l(\cos\theta) = (\boldsymbol{\sigma}\cdot\mathbf{n})iP_l^1(\cos\theta). \tag{7.65}$$

The last step follows directly from the definitions of the Legendre polynomials $P_l(\mu)$ and associated Legendre functions $P_l^m(\mu)$ given in Equations 4.39 and 4.38, respectively. Upon substituting Equation 7.65 into Equation 7.62, we find that the scattering matrix can indeed be expressed in the form $F(\theta,\phi) = A(\theta) + B(\theta)\boldsymbol{\sigma}\cdot\mathbf{n}$, where $A(\theta)$ and $B(\theta)$ are the 2-by-2 diagonal matrices

$$A(\theta) = \frac{1}{2ik}\sum_{l=0}^{\infty}\left[(l+1)(U_l^p - 1) + l(U_l^a - 1)\right]P_l(\cos\theta)\mathbf{I},$$

and

$$B(\theta) = \frac{1}{2k}\sum_{l=0}^{\infty}(U_l^p - U_l^a)P_l^1(\cos\theta)\mathbf{I}, \tag{7.66}$$

where I is the unit matrix. The differential shape elastic-scattering cross section for an unpolarized beam is obtained by averaging the parallel and antiparallel cross sections.

$$\sigma_{se}(\theta) = \tfrac{1}{2}(|F_{11}|^2 + |F_{12}|^2 + |F_{21}|^2 + |F_{22}|^2)$$
$$= |A|^2 + |B|^2, \tag{7.67}$$

and we note that the differential cross section for an unpolarized beam is independent of the azimuthal angle.

The shape elastic, compound-nucleus-formation, and total cross sections are given by

$$\sigma_{se} = \pi \lambda^2 \sum_{l=0}^{\infty} [(l + 1)|1 - U_l^p|^2 + l|1 - U_l^q|^2], \tag{7.68}$$

$$\sigma_c = \pi \lambda^2 \sum_{l=0}^{\infty} [(l + 1)(1 - |U_l^p|^2) + l(1 - |U_l^q|^2)], \tag{7.69}$$

$$\sigma_T = 2\pi \lambda^2 \sum_{l=0}^{\infty} [(l + 1) \, \mathrm{Re} \, (1 - U_l^p) + l \, \mathrm{Re} \, (1 - U_l^q)]. \tag{7.70}$$

When we compare these equations with the ones that were derived neglecting spin-orbit coupling, Equations 7.34, 7.36, and 7.29, we note that they are practically identical except that the coefficient $(2l + 1)$ has been broken into two parts, $(l + 1)$ and l, which multiply, respectively, the spin-parallel and spin-antiparallel reflection-factor terms.

It has been recognized for some time that the optical model does not yield accurate compound elastic-scattering cross sections in the energy region above the first excited state of the target nucleus, which is at 3.35 Mev in ^{40}Ca. For this reason, Caswell followed the common procedure of obtaining $\sigma_{ce}(\theta)$ above 3.35 Mev by using the method of Wolfenstein, Hauser, and Feshbach described in Chapter 8. The shape elastic cross section was obtained at all energies from the optical model, as was the compound elastic cross section below 3.35 Mev.

The compound elastic (fluctuation) differential cross section in the absence of inelastic scattering and other reactions may be obtained directly from the optical model.* When the target nucleus has zero spin $(I = 0)$ and the incident particle has spin 1/2, the fluctuation cross section is given by

$$\sigma_{ce}(\theta) = |A_{fl}(\theta)|^2 + |B_{fl}(\theta)|^2, \tag{7.71}$$

---

*H. Feshbach, Chaps. V.A. and VI.D. of *Nuclear Spectroscopy, Part B*, Fåy Ajzenberg-Selove (Ed.), Reference 3 at the end of this chapter.

where

$$|A_{\rm fi}|^2 = \frac{\lambda^2}{4} \sum_{l=0}^{\infty} [(l+1)^2(1-|U_l^p|^2) + l^2(1-|U_l^a|^2)] \cdot [P_l(\cos\theta)]^2$$

$$(7.72)$$

and

$$|B_{\rm fi}|^2 = \frac{\lambda^2}{4} \sum_{l=0}^{\infty} [(1-|U_l^p|^2) + (1-|U_l^a|^2)] \cdot [P_l^1(\cos\theta)]^2, \qquad (7.73)$$

where

$$P_l^1(\cos\theta) = \sin\theta \, \frac{d}{d(\cos\theta)} P_l(\cos\theta).$$

Note that the compound elastic cross section is not isotropic in this energy region. In the next chapter, we will show that there are theoretical reasons for believing that it does become practically isotropic in the energy regions where the levels of the target nucleus are close enough to be practically continuous. This region is between about 5 and 6 Mev in calcium. At higher energies (above 6 Mev in calcium), there are so many open exit channels for the decay of the compound nucleus that compound elastic scattering becomes highly improbable, and its cross section is therefore assumed to be zero.

Polarization

The polarization **P** of spin-1/2 particles, such as neutrons and protons, is defined as the expectation value of the Pauli spin operator σ,

$$\mathbf{P} \equiv \langle\sigma\rangle = \frac{\bar{\psi}^p\sigma\psi^p + \bar{\psi}^a\sigma\psi^a}{\bar{\psi}^p\psi^p + \bar{\psi}^a\psi^a}. \tag{7.74}$$

For an unpolarized incident beam we find, by direct evaluation of this equation using the scattered wave functions of Equation 7.61, that

$$\langle\sigma_x\rangle = -P(\theta)\sin\phi, \langle\sigma_y\rangle = P(\theta)\cos\phi, \text{ and } \langle\sigma_z\rangle = 0,$$

where $P(\theta)$, the scalar polarization, is given by

$$P(\theta) \equiv \frac{2\,\mathrm{Re}(\bar{A}B)}{|A|^2 + |B|^2}.$$

Noting that the components of the vector **n** are $(-\sin\phi, \cos\phi, 0)$, we see that **P** is in fact directed along **n**. Those particles scattered in any direction $\mathbf{v}_f$ have a non-vanishing expectation value of the spin only for the component parallel to **n**, that is, normal to both the initial and final velocity vectors.

If the incident beam is polarized in the direction of the unit vector $\mathbf{c}$, the differential scattering cross section for spin-zero target nuclei takes the form

$$\sigma_{se}(\theta, \phi) = [|A|^2 + |B|^2][1 + P(\theta)\mathbf{n}\cdot\mathbf{c}] = \sigma_{se}(\theta)[1 + P(\theta)\mathbf{n}\cdot\mathbf{c}]. \quad (7.75)$$

Thus the differential cross section depends not only on θ, the angle between $\mathbf{v}_i$ and $\mathbf{v}_f$, but also on ϕ, the angle between $\mathbf{c}$ and $\mathbf{n}$.

Results

A sampling of the differential elastic-scattering cross sections obtained from the foregoing analysis is shown in Figures 7.9, 7.10, and 7.11, along with data from direct measurements of these cross sections. The solid

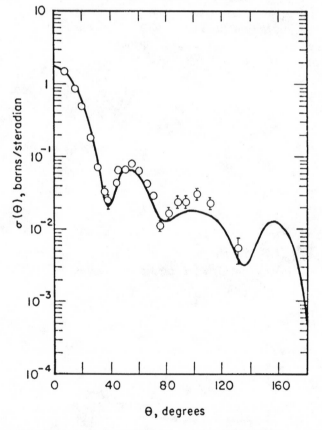

Figure 7.9—Elastic-scattering differential cross section of calcium at 14.6 Mev. [Experimental data are those of W. G. Cross and R. G. Jarvis, *Nucl. Phys.*, **15**: 160 (1960).]

lines are the cross sections determined by Caswell. The agreement between theory and experiment that is displayed here is rather typical of
that obtained from most modern optical-model calculations—encourag-
ingly good, but obviously not perfect.

Of course, the theoretical cross sections shown in these figures were
obtained by adjusting the parameters V_c, W_s, and a to obtain the best
overall fit to the measured cross sections. In view of this empirical adjust-
ment, the student may well wonder why lengthy optical-model calculations are performed to obtain cross sections that seem to be known experimentally. The answer depends in large part on whether you are a
nuclear engineer or a nuclear physicist.

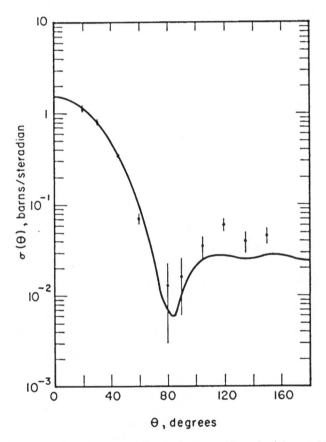

Figure 7.10—Elastic-scattering differential cross section of calcium at 6.0 Mev. [Experi-
mental data are those of J. D. Seagrave, L. Cranberg, and J. E. Simmons; see R. J. Howerton,
USAEC Report UCRL-5573, 1961.]

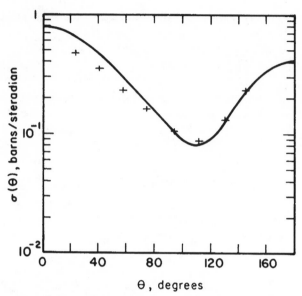

σ(Θ), barns/steradian

Θ, degrees

Figure 7.11—Elastic-scattering differential cross section of calcium at 1.57 Mev. [Experimental data are those of R. O. Lane, A. S. Langsdorf, Jr., J. B. Monahan, and A. J. Elwyn; see R. J. Howerton, USAEC Report UCRL-5573, 1961.]

The nuclear engineer must know the differential elastic-scattering cross section. If $\sigma(\theta)$ were known experimentally for all nuclides at all energies up to 18 Mev, he would obviously have no further need for optical-model calculations of $\sigma(\theta)$. Until that day arrives, however, it is necessary to use theory to interpolate and extrapolate measured differential cross sections. In practice, this is done by varying the adjustable parameters in the optical potential to obtain agreement between calculated and measured cross sections at the energies at which they have been measured and then smoothly varying the parameters between the measured energies to obtain the cross sections at energies where no measurements exist. We can also interpolate in the mass number A if no differential cross sections have been measured for a particular element. Furthermore, as has been pointed out by an outstanding experimentalist, "proper extrapolation of experimental values using the [optical] model may lead to a more accurate cross section than can be obtained experimentally." *

The nuclear physicist on the other hand, views the optical model as a window, albeit cloudy, into the mysteries of nuclear structure. His main concern is understanding and predicting the form of and numerical parameters in the optical potential. Even if $\sigma(\theta)$ had been measured at all

*A. B. Smith, USAEC Report ANL-7110, p. 13, Argonne National Laboratory, 1965.

energies for all elements, the physicist would still be interested in comparisons between optical-model calculations and experiments.

An Informal Note of Explanation

The sensitive reader is probably, at this point, more than a little bewildered and distressed by the course that this narrative has taken. He had ample reason to expect that, after the development of the rigorous U- and R-matrix expressions, the theory of nuclear cross sections was complete and that the calculation of any desired cross section would consist merely of the specialization of the general U- and R-matrix equations to the interaction of interest. Why, he may wonder, has the nonrigorous optical model been introduced?

Perhaps a simple analogy may help to clarify the situation: The combined U and R theory is much like a cookbook that details in every way except one the way to bake any cross-section cake. The ingredients are listed as well as the exact way in which to combine them. Unfortunately, the cookbook has one major deficiency. It fails to list the required quantities of each of the hundreds of ingredients (the values of the level parameters E_λ, J^π, and $\gamma_{\lambda c}$). These, we have reason to believe, may eventually be found in another, as yet incomplete, book, namely, nuclear theory.

The R-matrix method of determining the elements of the U matrix is exact but thus far has been found to be of practical value only in the region of isolated or nearly isolated resonances. In the high-energy region of overlapping resonances, other less-exact methods based on average behavior are employed to determine the cross sections. These latter methods can, however, be shown to be approximations to the rigorous theory and derivable from it.*

7.8 Legendre Polynomial Expansions of Differential Elastic-Scattering Cross Sections

Differential elastic-scattering cross sections are commonly recorded in the form of tabulations of Legendre polynomial expansion coefficients, f_l, in the center-of-mass system. Such data are more compact and practical than graphs of the differential cross section vs. the angle of scattering. Furthermore, data in this form are directly applicable in the spherical-harmonics method of solution of the neutron-transport equation.

The data for iron, Table 7.1, are typical of such a tabulation. At high energies many Legendre polynomials are required to fit the differential cross section; at lower energies fewer are required until, finally, below some energy (22 kev in the case of iron), the angular distribution becomes

---

*See, for example, Section XI of A. M. Lane and R. G. Thomas, R-Matrix Theory of Nuclear Reactions, *Rev. Mod. Phys.*, **30**: 257 (1958).

isotropic. The energy listed in Table 7.1 is the kinetic energy of the neutron in the laboratory system. The expansion coefficients are those which apply to the center-of-mass system. The differential cross section as a function of the C-system angle is given in terms of these expansion coefficients by

$$\sigma_n(\tilde{E}, \mu) = \sigma_n(\tilde{E}) \sum_{l=0}^{N} \frac{(2l + 1)}{4\pi} f_l(\tilde{E}) P_l(\mu), \tag{7.76}$$

where $\sigma_n(\tilde{E})$ is the total elastic-scattering cross section at lab energy $\tilde{E}$, $\mu \equiv \cos \theta$ with θ the center-of-mass scattering angle, and $P_l(\mu)$ is the lth Legendre polynomial in the variable μ.

Because of the way in which the differential cross section is expressed in Equation 7.76, the zero expansion coefficient f_0 is identically unity. This is readily proved by integrating both sides of Equation 7.76 over the entire sphere:

$$\int_{-1}^{+1} \sigma_n(\tilde{E}, \mu) 2\pi \, d\mu = \sigma_n(\tilde{E}) \sum_{l=0}^{N} \frac{2l + 1}{4\pi} f_l \int_{-1}^{+1} P_l(\mu) 2\pi \, d\mu.$$

The left-hand side is by definition $\sigma_n(\tilde{E})$. On the right-hand side, only the $l = 0$ integral is nonzero because $P_0 = 1$ and the Legendre polynomials are orthogonal over the range $-1 \leq \mu \leq +1$. The integral yields 4π; hence, $f_0 = 1$ identically.

Average Fractional Energy Loss

The average fractional energy loss suffered by a neutron in an elastic collision can be readily shown to be a function only of the first expansion coefficient, f_1, and not of the higher order expansion coefficients. The *fractional energy loss* is defined as $(\tilde{E} - \tilde{E}')/\tilde{E}$ where $\tilde{E}$ and $\tilde{E}'$ are the neutron's laboratory energies before and after the collision, respectively. From the kinematic expression

$$\frac{\tilde{E}'}{\tilde{E}} = \frac{A^2 + 1 + 2A\mu}{(A + 1)^2},$$

where A is the mass of the nucleus expressed in neutron masses, we obtain the fractional energy loss as a function of μ,

$$\frac{\tilde{E} - \tilde{E}'}{\tilde{E}} = \frac{2A(1 - \mu)}{(A + 1)^2}.$$

The average fractional energy loss is then defined by

$$\left\langle \frac{\tilde{E} - \tilde{E}'}{\tilde{E}} \right\rangle \equiv \frac{1}{\sigma_n(\tilde{E})} \int_{-1}^{+1} \left[\frac{\tilde{E} - \tilde{E}'}{\tilde{E}} \right] \sigma_n(\tilde{E}, \mu) 2\pi \, d\mu.$$

Substituting the preceding expression into the integrand in place of

Table 7.1—Legendre Expansion Coefficients for the Angular
Distribution of Neutrons Elastically Scattered by Iron*

$\tilde{E}$, Mev	$\sigma_n(\tilde{E})$, barns	f_1	f_2	f_3	f_4	f_5	f_6	f_7	f_8
18.0	1.05	0.807	0.690	0.580	0.487	0.384	0.291	0.195	0.1135
14.0	1.21	0.815	0.700	0.566	0.450	0.334	0.228	0.155	0.0820
10.9	1.51	0.819	0.717	0.550	0.405	0.277	0.168	0.108	0.0520
8.51	1.95	0.810	0.685	0.512	0.350	0.210	0.111	0.061	0.0220
6.63	2.21	0.790	0.640	0.450	0.282	0.141	0.067	0.025	0.0047
5.16	2.33	0.726	0.570	0.373	0.201	0.081	0.037	0.012	0.0005
4.02	2.38	0.584	0.455	0.290	0.147	0.046	0.018	0.0043	−0.0004
3.13	2.16	0.475	0.395	0.219	0.090	0.022	0.006	−0.0005	−0.0005
2.44	2.20	0.374	0.340	0.161	0.050	0.005	0.000	−0.0020	−0.0003
1.90	2.19	0.293	0.285	0.113	0.029	−0.0015	−0.0029	−0.0020	0.0000
1.48	2.20	0.244	0.276	0.006	0.026	−0.0035	−0.0030	−0.0015	0.000
1.15	1.93	0.240	0.236	0.034	0.022	−0.0020	−0.0020	−0.0010	0.000
0.897	2.15	0.277	0.176	0.017	0.010	−0.0017	−0.0010	−0.0005	0.000
0.699	2.40	0.108	0.124	0.009	0.010	−0.0008	−0.0005	0.000	0.000
0.544	2.70	0.197	0.094	0.004	−0.002	−0.0003	0.000	0.000	0.000
0.424	4.70	0.111	0.044	0.002	0.000	0.000	0.000	0.000	0.000
0.330	2.75	0.079	0.019	0.001	0.000	0.000	0.000	0.000	0.000
0.257	2.90	0.066	0.010	0.000	0.000	0.000	0.000	0.000	0.000
0.200	3.40	0.053	0.00505	0.000	0.000	0.000	0.000	0.000	0.000
0.155	4.70	0.0413	0.00307	0.000	0.000	0.000	0.000	0.000	0.000
0.121	2.35	0.0322	0.00186	0.000	0.000	0.000	0.000	0.000	0.000
0.0943	4.50	0.0250	0.00113	0.000	0.000	0.000	0.000	0.000	0.000
0.0734	1.95	0.0195	0.00068	0.000	0.000	0.000	0.000	0.000	0.000
0.0572	4.40	0.0151	0.00042	0.000	0.000	0.000	0.000	0.000	0.000
0.0445	5.50	0.0118	0.00025	0.000	0.000	0.000	0.000	0.000	0.000
0.0347	8.10	0.0092	0.00015	0.000	0.000	0.000	0.000	0.000	0.000
0.0270	24.0	0.0071	0.00010	0.000	0.000	0.000	0.000	0.000	0.000
0.0221	1.7	0.0058	0.00006	0.000	0.000	0.000	0.000	0.000	0.000

*From E. S. Troubetzkoy, USAEC Report NDA 2111-3, United Nuclear Corporation, 1959.

$(\tilde{E} - \tilde{E}')/\tilde{E}$ and the Legendre polynomial expansion in place of $\sigma_n(\tilde{E}, \mu)$, we find that, because of the orthogonality of the Legendre polynomials, only the terms involving f_0 and f_1 survive. Thus

$$\left\langle \frac{\tilde{E} - \tilde{E}'}{\tilde{E}} \right\rangle = \frac{2A}{(A + 1)^2} [1 - f_1(\tilde{E})]. \tag{7.77}$$

If the scattering is isotropic in the C-system, f_1 equals 0 and the average fractional energy loss is simply $2A/(A + 1)^2$.

Expansion of Experimental Differential Cross Sections

Differential cross-section data obtained from experiment are in the form of graphs of $\sigma_n(\tilde{E}, \tilde{\mu})$ vs. $\tilde{\mu}$, the cosine of the laboratory scattering angle. This data can readily be expressed in the form of a Legendre polynomial expansion in the laboratory cosine $\tilde{\mu}$,

$$\sigma_n(\tilde{E}, \tilde{\mu}) = \sigma_n(\tilde{E}) \sum_{k=0}^{N} \frac{2k + 1}{4\pi} \tilde{f}_k(\tilde{E}) P_k(\tilde{\mu}). \tag{7.78}$$

The kth expansion coefficient, $\tilde{f}_k$, is obtained by numerical integration of the kth Legendre polynomial, P_k, times the measured cross section. Multiplying the above equation through by $P_k(\tilde{\mu})$ and integrating both sides from $\tilde{\mu} = -1$ to $\tilde{\mu} = +1$, we find that

$$\tilde{f}_k(\tilde{E}) = \frac{2\pi}{\sigma_n(\tilde{E})} \int_{-1}^{+1} \sigma_n(\tilde{E}, \tilde{\mu}) P_k(\tilde{\mu}) \, d\tilde{\mu}. \tag{7.79}$$

The value of the integral may be obtained by numerical or graphical integration since both terms in its integrand are known.

The expansions of the differential cross sections are terminated at some order N that is consistent with the accuracy in the data. Unfortunately, such finite expansions can, and sometimes do, result in differential cross sections that assume negative values at one or more angles. Whether or not these unrealistic negative cross sections can be tolerated depends on the specific problem in which the cross sections are being used. The nuclear engineer should, in any event, be aware of this problem and check cross sections for negative values before using them.

Expansion of Theoretical Differential Cross Sections

We have seen how experimental differential cross sections can be expressed in Legendre polynomial expansions; let us now examine the way in which theoretical calculations of the differential cross section produce similar expansions.

The differential elastic-scattering cross section calculated by the method of partial wave results in an expression of the form

$$\sigma_n(E, \mu) = \lambda^2 \left| \sum_l (2l + 1) e^{i\delta_l} \sin \delta_l P_l(\mu) \right|^2.$$

where the phase shifts depend on the energy E and the details of the potential. The squared absolute value can be expressed as a double summation that can be written as

$$\sigma_n(E, \mu) = \sum_p \sum_q A_{pq} P_p P_q, \tag{7.80}$$

where all coefficients of the Legendre polynomial product have been lumped in the quantity A_{pq}. The value of A_{pq} depends only on E for a given potential and is not a function of μ.

The double summation of a product of Legendre polynomials may be expressed as a single summation of single Legendre polynomials,

$$\sum_p \sum_q A_{pq} P_p P_q = \sum_r B_r P_r, \tag{7.81}$$

since the coefficients B_r can be obtained by multiplying through by each Legendre polynomial P_r in turn and integrating over μ from -1 to $+1$. The right-hand side integrates to $2B_r/(2r + 1)$ because of the ortho-normality of the polynomials; hence,

$$B_r = \frac{2r + 1}{2} \int_{-1}^{+1} \sum_p \sum_q A_{pq} P_p P_q P_r \, d\mu. \tag{7.82}$$

An expression for integrals involving the products of three Legendre polynomials was obtained by J. C. Adams in 1878,* namely,

$$\int_{-1}^{+1} P_p P_q P_r \, d\mu = \frac{2}{p + q + r + 1} \times \frac{1\cdot 3 \cdots (p + q - r - 1)}{2\cdot 4 \cdots (p + q - r)}$$

$$\times \frac{1\cdot 3 \cdots (p + r - q - 1)}{2\cdot 4 \cdots (p + r - q)} \times \frac{1\cdot 3 \cdots (r + q - p - 1)}{2\cdot 4 \cdots (r + q - p)}$$

$$\times \frac{2\cdot 4 \cdots (p + q + r)}{1\cdot 3 \cdots (p + q + r - 1)}. \tag{7.83}$$

The integral is zero if the sum of any two of the integers p, q, and r is less than the third.

Thus we see that the B_r coefficients can be readily obtained as functions of the known A_{pq} coefficients and that $\sigma_n(E, \mu)$ can be expressed as a summation of single Legendre polynomials as in Equation 7.76.

Transformations of Expansions

Equations 7.76 and 7.78 are two alternative expressions for the differential elastic-scattering cross section. Of these two, the one in terms of the laboratory variable $\tilde{\mu}$ is generally the most useful to the nuclear engineer. Most tabulations are generated theoretically and so produce expansion

*J. C. Adams, *Proc. Roy. Soc. (London)*, **27**: 63 (1878).

coefficients in the center-of-mass variable μ. It is necessary, therefore, to have a procedure for transforming center-of-mass expansion coefficients to laboratory coefficients. The relevant equations have been developed by P. F. Zweifel and H. Hurwitz.* In the notation of this chapter, the transformation of the coefficients from the C-system to the L-system is given by

$$\tilde{f}_k = \sum_l G_{kl} \frac{2l + 1}{2k + 1} f_l \qquad \text{(for } k, l = 0, 1, 2, \ldots\text{).} \tag{7.84}$$

The inverse transformation is given by

$$f_l = \sum_k G_{lk}^{-1} \frac{2k + 1}{2l + 1} \tilde{f}_k \qquad \text{(for } k, l = 0, 1, 2, \ldots\text{).} \tag{7.85}$$

The elements of the transformation matrix G and the inverse matrix G^{-1} are functions of a parameter γ given by

$$\gamma = \left[\frac{m_1 m_3 (m_3 + m_4) E}{m_2 m_4 (m_1 + m_2)(E + Q)} \right]^{1/2}, \tag{7.86}$$

where m_1 and m_2 are the masses of the projectile and target particles, respectively, m_3 and m_4 are the masses of the reaction products, E is the total energy in the C-system, and Q is the reaction energy. In the special case of neutron elastic scattering, since $Q = 0$, $m_1 = m_3 = m$, and $m_2 = m_4 = M$, therefore $\gamma = m/M = 1/A$, where A is the mass of the target nucleus expressed in neutron masses.

The first five rows and columns of G and G^{-1} are displayed here. The elements in both these matrices are labelled in the standard way; the elements G_{ij} and G_{ij}^{-1} appear in the ith row and jth column of the appropriate matrix. These matrices are valid only if $\gamma > 1/2$. The reader who wishes information on the derivation of these matrices and their extension to higher order should refer to the original article by Zweifel and Hurwitz.

$$G = \begin{bmatrix} 1 & 0 & 0 & 0 & 0 & \cdots \\ 2\gamma & 1 - \frac{3}{5}\gamma^2 & -\frac{2}{5}\gamma & \frac{9}{35}\gamma^2 & 0 & \cdots \\ \gamma^2 & 2\gamma & 1 - \frac{11}{7}\gamma^2 & -\frac{6}{7}\gamma & \frac{16}{21}\gamma^2 & \cdots \\ 0 & \frac{8}{5}\gamma^2 & \frac{12}{5}\gamma & 1 - \frac{46}{15}\gamma^2 & -\frac{4}{3}\gamma & \cdots \\ 0 & 0 & \frac{18}{7}\gamma^2 & \frac{20}{7}\gamma & 1 - \frac{390}{77}\gamma^2 & \cdots \\ \vdots & \vdots & \vdots & \vdots & \vdots & \end{bmatrix}$$

*P. F. Zweifel and H. Hurwitz, *J. Appl. Phys.*, **25**: 1242 (1954).

$$
G^{-1} = \begin{bmatrix}
1 & 0 & 0 & 0 & 0 & \cdots \\
-2\gamma & 1 - \frac{1}{5}\gamma^2 & \frac{2}{5}\gamma & \frac{3}{35}\gamma^2 & 0 & \cdots \\
3\gamma^2 & -2\gamma & 1 - \frac{9}{7}\gamma^2 & \frac{6}{7}\gamma & \frac{8}{21}\gamma^2 & \cdots \\
0 & \frac{16}{5}\gamma^2 & -\frac{12}{5}\gamma & 1 - \frac{14}{5}\gamma^2 & \frac{4}{3}\gamma & \cdots \\
0 & 0 & \frac{30}{7}\gamma^2 & -\frac{20}{7}\gamma & 1 - \frac{370}{77}\gamma^2 & \cdots \\
\vdots & & & & &
\end{bmatrix}
$$

In the case of neutron elastic scattering from heavy elements, the parameter γ, which equals $1/A$, is much smaller than unity so all the off-diagonal elements in the transformation matrix are effectively zero. Since, in this case, the diagonal elements are very close to unity, the Legendre coefficients in the laboratory and the center-of-mass systems are practically identical.

Wick's Inequality

A very useful expression for setting a lower limit on the differential elastic-scattering cross section in the forward direction, where measurements of this cross section are practically impossible, was derived on the basis of a rigorous argument by G. C. Wick.* *Wick's inequality* has the form

$$
\sigma(\theta = 0) \geqq \left[\frac{\sigma_T}{4\pi\lambda} \right]^2, \tag{7.87}
$$

where $\sigma(\theta = 0)$ is the differential elastic-scattering cross section in the forward direction and σ_T is the total integrated cross section for all interactions. The nuclear engineer should check before using elastic-scattering data to see that they do indeed obey Wick's inequality. There are, unfortunately, many examples of theoretically derived, published, and widely used data that disagree with this inequality, presumably because of errors in the calculations.

SELECTED REFERENCES

1. P. E. Hodgson, *The Optical Model of Elastic Scattering*, Oxford University Press, American House, London, 1963. A clear and comprehensive survey of the voluminous literature on the optical model prior to 1962 is given in this book. The theoretical development has been limited to the minimum necessary to understand the results and conclusions.

2. Ivan Ulehla, Ladislav Gomolcak, and Zdenek Pluhar, *Optical Model of the Atomic Nucleus*, Academic Press Inc., New York, 1964. A good introduction to the

*G. C. Wick, *Phys. Rev.*, **75**: 1459 (1949).

theory of optical-model calculations, this book contains more of the mathematics of the optical model than Hodgson's book but less of the physics and fewer comparisons with experiments.

3. Fay Ajzenberg-Selove, Ed., *Nuclear Spectroscopy, Part B*, Academic Press Inc., New York, 1960. The two chapters by H. Feshbach constitute an admirably rigorous and clear account of the theories of the compound nucleus and the optical model.

EXERCISES

1. Verify that the maximum neutron–proton laboratory scattering angle is 86° 59.5'. Derive and plot the relation between the laboratory angle $\bar\vartheta$ and the center-of-mass angle θ for *n–p* scattering.

2. There is a pure scattering resonance in the interaction of ^{12}C and a neutron at $E_0 = 2.08$ Mev. According to the Barn Book, the parameters associated with this resonance are $I = 0$, $J = 5/2$, $l = 2$, $\Gamma_n = 7.0 \pm 0.3$ kev. Plot the expected Breit–Wigner behavior of the scattering cross section $\sigma_n(E)$ in the energy range from 10^{-2} Mev to 2.4 Mev. Compare your results with the curve in the Barn Book. Explain any differences. What is the value of the channel spin associated with this resonance?

3. Why may it be concluded from the shapes of the elastic-scattering resonances in sulphur at 111 kev and 585 kev that the first resonance is an $l = 0$ and the second is an $l > 0$ resonance?

4. In the region of isolated resonances, shape elastic and compound elastic scattering interfere with one another so that $\sigma_n(\theta) \neq \sigma_{se}(\theta) + \sigma_{ce}(\theta)$. Why then does the optical model predict (Equation 7.40) that there is no interference between shape elastic and compound elastic scattering in the gross-structure problem? (*Hint:* Consider the time scale of events and the numbers of exit channels in shape elastic scattering and compound elastic scattering at low and high energies.)

5. Prove that $(\boldsymbol{\sigma} \cdot \mathbf{l})P_l(\cos\theta) = (\boldsymbol{\sigma} \cdot \mathbf{n})iP_l^1(\cos\theta)$, as stated in the text.

6. Prove that $\sigma_{se}(\theta) = |A|^2 + |B|^2$ as stated in Equation 7.67.

7. Assume that, in an optical-model calculation of neutrons interacting with $I = 0$ target nuclei at a certain energy, the reflection factors are given by the list which follows.

l	U_l^p	U_l^a
0	$0.2 + 0.3i$	$0.1 + 0.2i$
1	$0.4 - 0.1i$	$0.3 + 0.4i$
≥ 2	1	1

Assume that the incoming beam is unpolarized. Calculate all possible cross sections, differential and otherwise, in units of $\pi\lambda^2$. Also calculate the polarization of the scattered beam as a function of θ.

8. In Section 7.7, the optical-model equations were developed for neutrons interacting with nuclei having zero spin. Develop the equivalent equations for the case of nuclei with spin 1/2 and spin 1.

9. Using the data in Table 7.1, determine the Legendre polynomial coefficients of the expansion of the iron differential cross section in the laboratory system at energy $\tilde{E} = 330$ kev.

10. What average fractional energy loss does a neutron suffer when it elastically scatters from an iron nucleus at initial laboratory energies 18 Mev, 4 Mev, 330 kev, 22 kev, and 10 ev?

11. The C-system Legendre polynomial coefficients for the elastic scattering of neutrons off lithium at $\tilde{E} = 0.2$ Mev are $f_1 = 0.289$, $f_2 = 0.044$, and $f_3 = 0.00015$. The elastic-scattering cross section is 1.78 barns. Using these data, calculate the L-system Legendre polynomial coefficients and plot the C-system and L-system differential cross sections of lithium at this energy.

12. (a) Approximate the neutron–proton potential by a spherical square well of depth V_0 and radius R. Find reasonable values of V_0 and R (note Figure 5.8) that yield best agreement with the experimental total scattering cross section of hydrogen for neutrons with laboratory energies in the range from 0 to 10 Mev.

(b) Investigate the possibility of obtaining better agreement by postulating two independent spherical square wells, one for the parallel spin state and one for the antiparallel spin state. [The scattering cross section in this case is given by $\sigma_n(E) = (3/4)\sigma_n^p(E) + (1/4)\sigma_n^a(E)$ since the probability of scattering with parallel spins is three times the probability of scattering with antiparallel spins.]

(c) At what energy is it necessary to take into account p-wave as well as s-wave scattering to ensure about 10% accuracy in the neutron-scattering cross section of hydrogen?

8 INELASTIC SCATTERING AND CASCADE REACTIONS

Neutron–nucleus inelastic scattering is defined by two conditions: (1) the initial and final constellations are identical except that the product nucleus is in an excited state,

$$Z^A + n \to (Z^A)^* + n',$$

and (2) the total kinetic energy of the final constellation is less than the total kinetic energy of the initial constellation by an amount equal to the excitation energy of the product nucleus.[†]

The product nucleus in an inelastic-scattering event usually emits one or more gamma rays to reach its ground state,

$$(Z^A)^* \to Z^A + \gamma.$$

Most gamma rays from inelastic scattering are emitted within an extremely short time, less than 10^{-10} sec, following the interaction. In some rare cases the product nucleus reaches a metastable state from which the gamma emission occurs with a much greater half-life.

In the energy region of interest to nuclear engineers, i.e., the region below 20 Mev, inelastic scattering takes place primarily through the formation and decay of a compound nucleus. Inelastic scattering is simply one of the possible channels of compound-nucleus decay, provided, of course, that the laws of conservation are not violated in this decay. The entire sequence of events, including the formation of the excited compound nucleus and its subsequent decay by neutron emission into an excited

---

[†] Specifically excluded from consideration here are inelastic interactions between low-energy neutrons and molecules or solids, in which the neutron may gain energy. Such interactions are considered in Chapter 12.

product nucleus that decays by gamma emission, may be symbolized as

$$Z^A + n \rightarrow (Z^{A+1})^* \rightarrow (Z^A)^* + n'$$
$$\qquad\qquad\qquad \hookrightarrow Z^A + \gamma.$$

Clearly compound elastic scattering may be viewed as a special case of inelastic scattering, one in which the product nucleus is left in its ground state rather than an excited state.

Inelastic scattering cannot occur unless the incident neutron has an energy high enough that the available energy is greater than the energy of the first excited state in the target nucleus. Remember that the available energy, E, is the total kinetic energy of the neutron and the nucleus in the C-system. From Equation 2.27 the available energy is

$$E = \frac{A}{A+1} \tilde{E}, \qquad (8.1)$$

where $\tilde{E}$ is the laboratory kinetic energy of the incident neutron and A is the mass of the target nucleus measured in neutron mass units.

Let us designate the energies of the excited states in the target nucleus (which is also the product nucleus) by E_1, E_2, E_3, etc., and the energy levels in the compound nucleus by E_1^c, E_2^c, E_3^c, etc. These energies are measured with respect to the ground states in the respective nuclei.

The *threshold energy* for any interaction is defined as the lowest possible neutron kinetic energy at which the interaction is kinematically possible. From the preceding discussion, it is clear that the laboratory threshold energy for inelastic scattering of a neutron off a nucleus whose lowest energy level is E_1 is given by

$$\tilde{E}_t = \frac{A+1}{A} E_1. \qquad (8.2)$$

When an incident neutron is absorbed by the target nucleus to form a compound nucleus, the compound nucleus is in a highly excited state, a state with energy

$$E_j^c = S_n^c + E, \qquad (8.3)$$

where S_n^c is the neutron's separation energy from the compound nucleus. On an absolute energy scale, shown in Figure 8.1, the neutron brings into the compound nucleus its rest energy, 931 Mev; its separation energy, S_n^c; and its available energy, E. However, the ground state of the compound nucleus is located at an absolute energy of 931 Mev $-$ S_n^c above the ground state of the target nucleus; therefore the net effect is to create a compound nucleus with excitation energy E_j^c given by Equation 8.3.

When a neutron is emitted by the compound nucleus, 931 Mev again appears as rest energy of the neutron. Since the rest energy of the neutron plays a neutral role in the kinematics of inelastic scattering, it is more

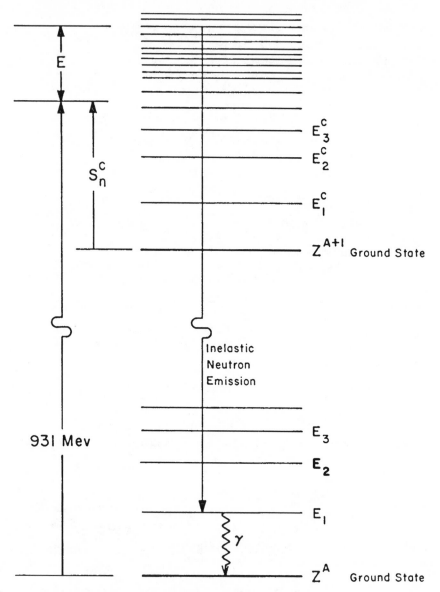

Figure 8.1—Inelastic scattering on an absolute energy scale.

convenient to picture the kinematics only in terms of the energy levels of the target (and product) nucleus, as in Figure 8.2. It must be understood, however, that it is into an energy level of the compound nucleus that the neutron is absorbed. Figure 8.2 pictures all possible neutron-emission reactions that can follow absorption of an incident neutron whose available energy is greater than E_2 but less than E_3. The emitted neutron may have one of three energies in the C-system which correspond to compound elastic scattering, to inelastic scattering in which the product nucleus is left in its first excited state, or to inelastic scattering in which the product nucleus is left in its second excited state. If we call the three energies E'_{n0}, E'_{n1}, and E'_{n2}, respectively, then, as will be shown shortly,

$$E'_{n0} = \frac{A}{A+1} E \quad \text{(compound elastic scattering)},$$

$$E'_{n1} = \frac{A}{A+1}(E - E_1) \quad \text{(inelastic group 1)}, \quad (8.4)$$

$$E'_{n2} = \frac{A}{A+1}(E - E_2) \quad \text{(inelastic group 2)}.$$

The gammas that follow these inelastic events are also shown in Figure 8.2. Inelastic neutron group 1 is accompanied by a single gamma of energy E_1; inelastic neutron group 2 is accompanied by either a single gamma of energy E_2 or by two gammas of energies $(E_2 - E_1)$ and E_1. Some of these gammas may not appear because the transition is forbidden by conservation laws. We will discuss this possibility later.

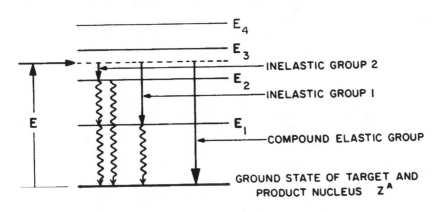

Figure 8.2—Inelastic scattering in terms of the energy levels of the target (and product) nucleus.

At much higher incident neutron energies, many levels in the product nucleus can be reached. The energy spectrum of inelastic neutrons is then practically continuous, and each neutron is accompanied by a cascade of gammas of various energies.

8.1 Kinematics of Inelastic Scattering

There are two distinct ways of examining the kinematics of inelastic scattering. We can analyze the interaction by considering only the incoming and outgoing particles and ignoring completely the existence of the compound nucleus as an intermediate state. Alternatively, we can analyze the interaction in two steps: the creation of the compound nucleus and its subsequent decay. The two methods yield identical results, but each points up certain features of the kinematics that the other hides. Let us examine the kinematics from first one viewpoint and then the other. Our major goal is to determine how much kinetic energy the inelastically scattered neutron will carry off at angle $\tilde{\theta}$ if the residual nucleus is left with an excitation energy E^*. The symbols we shall use are listed for reference in Table 8.1.

The One-Step Viewpoint

The one-step viewpoint is illustrated in Figure 8.3. A neutron of mass 1, speed $\tilde{v}$, and kinetic energy $\tilde{E}$ is scattered by a stationary nucleus of mass A. After the collision a neutron moves off at angle $\tilde{\theta}$ relative to the initial neutron direction with speed $\tilde{v}'$ and kinetic energy $\tilde{E}'$. The residual nucleus moves off at angle $\tilde{\theta}_A$ with speed $\tilde{V}$ and internal excitation energy E^*. All speeds, kinetic energies, and angles are measured in the laboratory system. The mass of the residual nucleus is very slightly greater than A because of its added internal energy. This slight mass

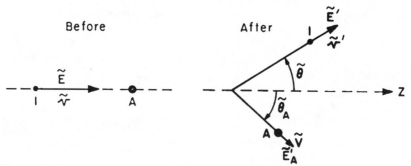

Figure 8.3—Before and after representation of an inelastic-scattering event in the L-system.

Table 8.1—Symbols for Inelastic Scattering

	L-System	C-System
Initial kinetic energy of neutron	$\tilde{E}$	E_n
Initial speed of neutron	$\tilde{v}$	v
Final kinetic energy of neutron	$\tilde{E}'$	E'_n
Final speed of neutron	$\tilde{v}'$	v'
Speed of product nucleus	$\tilde{V}$	V
Kinetic energy of product nucleus	$\tilde{E}'_A$	E'_A
Scattering angle of neutron	$\tilde{\theta}$	θ
Scattering angle of nucleus	$\tilde{\theta}_A$	θ_A

Energy levels of target nucleus	$E_1, E_2, E_3, \ldots, E_i, \ldots$
Energy levels of compound nucleus	$E^c_1, E^c_2, E^c_3, \ldots, E^c_j, \ldots$
Separation energy of neutron from compound nucleus	S^c_n
Mass of target nucleus in neutron masses	A
Excitation energy of product nucleus after emission of neutron	E^*
Threshold energy of interaction	$\tilde{E}_t$
Available energy in C-system	E
Kinetic energy of compound nucleus in L-system	$\tilde{E}_{cn}$
Speed of compound nucleus in L-system	$\tilde{V}_{cn}$

increase will be ignored (see Problem 1 at the end of the chapter for justification).

Conservation of linear momentum along and perpendicular to the initial neutron direction is expressed by

$$\tilde{v} = \tilde{v}' \cos \tilde{\theta} + A\tilde{V} \cos \tilde{\theta}_A, \tag{8.5}$$

$$\tilde{v}' \sin \tilde{\theta} = A\tilde{V} \sin \tilde{\theta}_A. \tag{8.6}$$

Conservation of energy is expressed by

$$\tilde{E} = \tilde{E}' + \tfrac{1}{2}A\tilde{V}^2 + E^*. \tag{8.7}$$

We now proceed to eliminate the unknown $\tilde{\theta}_A$ by rearranging, squaring, and adding Equations 8.5 and 8.6. The resulting expression for $\tilde{V}^2$ is substituted into Equation 8.7, yielding a quadratic in $(\tilde{E}'/\tilde{E})^{1/2}$ that is solved to give

$$\tilde{E}' = \tilde{E} \left\{ \frac{1}{A+1} \cos \tilde{\theta} \right.$$

$$\left. + \left[\left(\frac{1}{A+1} \cos \tilde{\theta} \right)^2 + \frac{A}{A+1} \left(1 - \frac{1}{A} - \frac{E^*}{\tilde{E}} \right) \right]^{1/2} \right\}^2 \tag{8.8}$$

This is the general expression for the final neutron energy $\tilde{E}'$ as a function of its initial energy $\tilde{E}$, the scattering angle $\tilde{\theta}$, and the excitation energy E^*. It can readily be shown to be identical to the previously derived Equation 2.65. Note that when $A \gg 1$ the energy of the emergent neutron is practically independent of angle and equal to $\tilde{E} - E^*$.

Let us now examine some interesting special cases of this general expression.

Case A: Suppose the scattering is elastic; E^* equals 0, and Equation 8.8 becomes identical to Equation 7.1, which was derived for elastic scattering.

Case B: Suppose the incident neutron transfers all its energy to the nucleus, part to excitation energy and part to kinetic energy. The neutron has zero energy after the interaction; so the left side of Equation 8.8 is zero. On the right-hand side $\cos \tilde{\theta}$ equals -1, and for the entire right-hand side to be zero, the term within the last brackets must be identically zero; that is,

$$E^* = \left(1 - \frac{1}{A}\right) \tilde{E}. \tag{8.9}$$

The remainder of the incident energy $\tilde{E}$ is given to the kinetic energy of the product nucleus:

$$\tilde{E}'_A = \tilde{E} \quad E^* = \frac{1}{A} \tilde{E}. \tag{8.10}$$

We note that the kinetic energy given the product nucleus is that which is required to conserve linear momentum. This is most easily seen by rearranging Equation 8.10 to read

$$\sqrt{2A\tilde{E}'_A} = \sqrt{2\tilde{E}}.$$

Case C: Suppose the emergent neutron goes along with the product nucleus. Both must move along the initial direction of the neutron to conserve momentum perpendicular to that direction. Their speeds are equal ($\tilde{v}' = \tilde{V}$), hence conservation of momentum is expressed by

$$\tilde{v} = \tilde{v}' + A\tilde{v}'$$

or

$$\tilde{v}' = \frac{1}{A+1} \tilde{v}.$$

Thus the final energy of the neutron is

$$\tilde{E}' = \tfrac{1}{2}(\tilde{v}')^2 = \left(\frac{1}{A+1}\right)^2 \tilde{E}. \tag{8.11}$$

Conservation of energy is expressed by

$$\tilde{E} = \tilde{E}' + \tfrac{1}{2}A(\tilde{v}')^2 + E^*$$

or

$$E^* = \tilde{E} - \tilde{E}' - A\tilde{E}' = \tilde{E} - \tilde{E}'(A + 1).$$

Upon substituting Equation 8.11 for $\tilde{E}'$, we obtain

$$E^* = \frac{A}{A + 1}\tilde{E}. \tag{8.12}$$

When we compare this expression for E^* with that in Equation 8.9, we come to an interesting conclusion: More excitation energy can be given to the nucleus in an inelastic interaction in which the emitted neutron travels along with the product nucleus than in an interaction in which the incident neutron gives up all its kinetic energy to the nucleus. This result seems strange from the one-step viewpoint but becomes entirely natural from the two-step viewpoint since it corresponds to decay of the compound nucleus by emission of a neutron of zero energy relative to the product nucleus.

The Two-Step Viewpoint
First consider the kinematics of the creation of the compound nucleus. Conservation of linear momentum is expressed by

$$\tilde{v} = (A + 1)\tilde{V}_{cn}, \tag{8.13}$$

where $\tilde{V}_{cn}$ is the speed of the compound nucleus in the L-system,

$$\tilde{V}_{cn} = \frac{1}{A + 1}\tilde{v}.$$

This speed is equal to the speed of the center of mass since in the intermediate (compound-nucleus) state the center of mass is coincident with the compound nucleus.

Now the kinetic energy of the compound nucleus in the L-system is

$$\tilde{E}_{cn} = \tfrac{1}{2}(A + 1)\tilde{V}_{cn}^2 = \tfrac{1}{2}(A + 1)\left(\frac{1}{A + 1}\tilde{v}\right)^2 = \frac{1}{A + 1}\tilde{E},$$

so its internal energy (exclusive of the separation energy of the neutron) must be

$$\tilde{E} - \tilde{E}_{cn} = \tilde{E}\left(1 - \frac{1}{A + 1}\right) = \frac{A}{A + 1}\tilde{E}. \tag{8.14}$$

Before

After

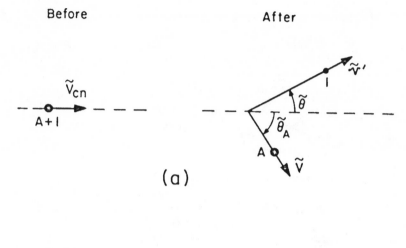

(a)

Before

After

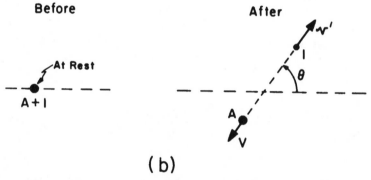

(b)

Figure 8.4—Decay of the compound nucleus by neutron emission. (a) L-system. (b) C-system.

We previously identified this quantity as the available energy, the total kinetic energy in the C-system.

Now let us consider the decay of the compound nucleus by neutron emission. We assume that the compound nucleus emits a neutron in its transition to an energy level in the product nucleus that has energy E^* above the ground state of the product nucleus. We assume that this neutron is emitted at angle θ in the C-system, which corresponds to angle $\tilde{\theta}$ in the L-system. The center-of-mass and laboratory views of the compound-nucleus decay are shown in Figure 8.4.

In the C-system, the kinematics are, as usual, extremely simple. Conservation of momentum requires that

$$v' = AV,$$

from which the kinetic energy of the product nucleus is given by

$$E'_A = \tfrac{1}{2}AV^2 = \tfrac{1}{2}A\left(\frac{1}{A}\,v'\right)^2 = \frac{1}{A}\,E'_n. \tag{8.15}$$

The total available energy in the C-system is now divided among the kinetic energies of the neutron and product nucleus and the internal energy of the product nucleus. Thus

$$E = E'_n + \frac{1}{A}\,E'_n + E^*$$

and, from this,

$$E'_n = \frac{A}{A+1}\,[E - E^*], \tag{8.16}$$

where

$$E = \frac{A}{A+1}\,\tilde{E}.$$

Equation 8.16 expresses the kinetic energy of the emitted neutron in the C-system. Note that the energy is independent of the angle of emission. Of course, the neutron's energy in the L-system is a function of its angle of emission because the velocity of the center of mass adds vectorially to the neutron's C-system velocity to give the neutron's L-system velocity. The interested student is invited to complete this two-step analysis by deriving Equation 8.8. We have gone far enough to examine once again the three special cases previously considered.

  Case A: Compound elastic scattering corresponds to $E^* = 0$; so Equation 8.16 gives us the well-known result that the final neutron's center-of-mass energy is equal to the initial neutron's center-of-mass energy, namely,

$$E'_n = \frac{A}{A+1}\,E. \tag{8.17}$$

  Case B: To obtain the physical situation corresponding to case B, we must imagine that the compound nucleus decays by emitting a neutron in the direction antiparallel to the incident neutron's velocity with speed equal to the speed of the compound nucleus,

$$v' = \tilde{V}_{cn} = \frac{1}{A+1}\,\tilde{v}.$$

In this event,

$$E'_n = \tfrac{1}{2}(v')^2 = \left(\frac{1}{A+1}\right)^2 \tilde{E}.$$

Substituting this expression for E'_n into Equation 8.16 and solving for E^*, we find

$$E^* = \left(1 - \frac{1}{A}\right) \tilde{E},$$ \hfill (8.18)

which duplicates, as it must, Equation 8.9.

Case C: If the neutron leaves the compound nucleus with zero kinetic energy in the C-system, Equation 8.16 tells us that the nucleus will be left with maximum excitation energy given by

$$E^* = E = \frac{A}{A+1} \tilde{E},$$ \hfill (8.19)

which duplicates Equation 8.12 of the one-step analysis.

8.2 General Behavior of Inelastic Cross Sections

Since inelastic scattering in the energy range below 20 Mev is assumed to proceed through the formation and decay of a compound nucleus (except as noted in Section 8.7), we may express its cross section, as usual, as the product of a formation cross section and a decay probability,

$$\sigma_{n'}(E) = \sigma_c(E) \frac{\Gamma_{n'}}{\Gamma},$$ \hfill (8.20)

where $\sigma_c(E)$ is the cross section for formation of the compound nucleus by neutrons of energy E and $\Gamma_{n'}/\Gamma$ is the probability that the compound nucleus will decay by emission of a neutron to one of the excited states of the product nucleus. The value of $\Gamma_{n'}/\Gamma$ depends on the excitation energy of the compound nucleus.

The energy of the lowest nuclear energy level is of the order of 0.5 to 5 Mev in light nuclides ($A < 25$), 0.1 to 0.2 Mev in medium-weight nuclides ($25 \leq A < 70$), and 0.05 to 0.1 Mev in the heaviest nuclides ($A \geq 70$). This means that inelastic scattering is a relatively high-energy, certainly greater than 50-kev, interaction. Since the cross section for compound-nucleus formation behaves approximately as $\pi \lambda^2$ times a transmission coefficient that is always smaller than unity, we can state with reasonable confidence that inelastic-scattering cross sections will never exceed a few barns in any element.

When the kinetic energy of the incident neutron (rigorously, its available energy) exceeds the energy of the second excited level in the target nucleus, the compound nucleus can decay inelastically in more than one mode. Corresponding to each neutron group that can be emitted are a partial width $\Gamma_{n'}(E_i)$ and a partial cross section for excitation of the E_ith level in the product nucleus. This cross section is written

$$\sigma_{n'}(E,E_i) = \sigma_c(E)\frac{\Gamma_{n'}(E_i)}{\Gamma}. \tag{8.21}$$

The total inelastic-scattering cross section is then the sum of these partial cross sections,

$$\sigma_{n'}(E) = \sum_{E_i} \sigma_{n'}(E,E_i). \tag{8.22}$$

In Figure 8.5 the typical behavior of the partial inelastic-scattering cross sections in the vicinity of the low-lying levels is shown. The partial cross section associated with a particular level is zero until the kinetic energy of the incident neutron reaches the threshold energy for excitation

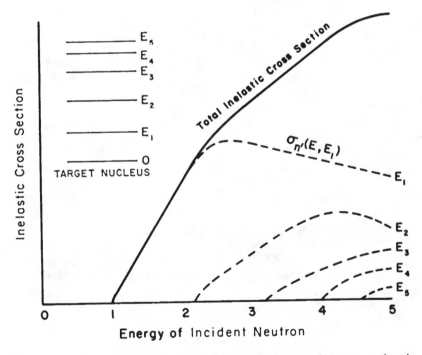

Figure 8.5—Typical behavior of partial and total inelastic-scattering cross sections in the region of resolved energy levels.

of that level. The cross section then increases until competition from other open channels forces it to decrease. At neutron energies of the order of a few million electron volts for most target nuclides, the only modes of compound-nucleus decay that compete with inelastic scattering are compound elastic scattering and photon emission (n,γ). As the energy increases, charged-particle-out reactions, particularly (n,p) and (n,α), begin to compete, and the total inelastic cross section eventually must decrease. The more pieces into which $\sigma_c(E)$ is cut, the smaller each piece must be.

When the kinetic energy of the incident neutron is so great that there are many channels open for the inelastic decay of the compound nucleus, then the energy distribution of the emitted neutrons becomes quite complex. With ideal energy resolution it might look like part a of Figure 8.6. The lines are not infinitely sharp but have a width at half-maximum given by $\Gamma_{n'} = \hbar/\tau_{n'}$, where $\tau_{n'}$ is the mean life for the emission of each neutron group from the compound nucleus. The lines at low energies are only a few electron volts apart (impossible to resolve with today's experimental techniques).

The same experiment performed with the sort of energy resolution available today (and probably for several decades into the future) produces quite a different picture, as shown in part b of Figure 8.6. The inelastically scattered neutrons appear to have a continuous spectrum ranging from zero to the incident energy. Only a few of the higher energy groups are resolved. It is no longer fruitful to interpret such data in terms of the properties of individual levels. Weisskopf[*] has developed a statistical theory for this situation that predicts an energy distribution of the inelastically scattered neutrons which has a Maxwellian form

$$\sigma_{n'}(E,E') = \sigma_{n'}(E)\frac{E'}{\tau^2}e^{-E'/\tau}\frac{\text{barns}}{\text{Mev}}. \tag{8.23}$$

The quantity $\sigma_{n'}(E,E')\,dE'$ is the cross section for the inelastic scattering of a neutron of initial energy E into the energy range dE' about E'. Thus $\sigma_{n'}(E,E')$ is a differential cross section of a type we are encountering for the first time; it is differential in energy, not in solid angle.

The quantity τ, the *nuclear temperature*, is a measure of the excitation energy of the product nucleus. It is also the most probable energy of the inelastically scattered neutrons, as can be readily proved by setting the derivative of $\sigma_{n'}(E,E')$ with respect to E' equal to zero and solving for E' at the maximum.

The exponential factor in Equation 8.23 results from the increased density of energy levels in the product nucleus at high excitation energies.

[*]V. F. Weisskopf, *Phys. Rev.*, 52: 295 (1937).

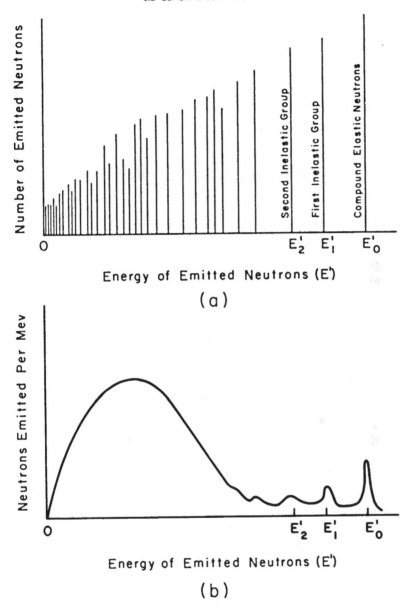

Figure 8.6—Spectrum of inelastic neutrons. (a) Perfect resolution. (b) Actual resolution.

The more energy levels, the more exit channels for neutron emission; thus low-energy neutron emission is favored when there are many channels open. At large excitation energies, the density of the energy levels in the product nucleus, as explained in Chapter 6, is of the form

$$\rho(E^*) = C \exp(2\sqrt{aE^*}) \text{ levels per Mev,} \tag{8.24}$$

where C and a are constants given in Table 6.1. Statistical mechanical reasoning leads to a relation between the nuclear temperature and the level density,[*]

$$\frac{1}{\tau(E^*)} = \frac{\partial}{\partial E^*} [\ln \rho(E^*)]. \tag{8.25}$$

Then the nuclear temperature is obtained from Equations 8.24 and 8.25 in the form

$$\tau(E^*) = \sqrt{E^*/a}. \tag{8.26}$$

The very rapid increase in level density with excitation energy, particularly in heavy nuclei, has the effect that most inelastically scattered neutrons have much lower energies than the incident neutron; therefore the energy of excitation of the product nucleus, E^*, is approximately the same as the incident-neutron energy, E. Thus, over most of the inelastic spectrum, E^* is approximately equal to E, and therefore τ is a constant for a fixed initial energy

$$\tau(E) \simeq \sqrt{E/a}. \tag{8.27}$$

At high incident-neutron energies the spectrum of inelastic neutrons can be obtained from Equations 8.23 and 8.27. It must be emphasized, however, that this is only a rough estimate of the spectrum. In the next section, a more accurate procedure based on Hauser–Feshbach theory is presented.

8.3 Hauser–Feshbach Cross-Section Theory

A 1951 paper by L. Wolfenstein[†] and its 1952 sequel by Walter Hauser and Herman Feshbach[‡] form the basis for the calculation of inelastic-scattering cross sections. Wolfenstein's development is not restricted to inelastic-scattering cross sections but considers how cross sections can be calculated for all reactions that proceed through the formation and decay of a compound nucleus. The only restriction is that the compound

---

[*] For a derivation of level density and nuclear-temperature equations, see John M. Blatt and Victor F. Weisskopf, *Theoretical Nuclear Physics* (John Wiley & Sons, Inc., New York, 1952), pp. 365–372.

[†] L. Wolfenstein, Conservation of Angular Momentum in the Statistical Theory of Nuclear Reactions, *Phys. Rev.*, **82**: 690 (1951).

[‡] Walter Hauser and Herman Feshbach, The Inelastic Scattering of Neutrons, *Phys. Rev.*, **87**: 366 (1952).

nucleus must have so large a density of levels at its excitation energy that it can be treated by the statistical theory.

The state of the compound nucleus, hence its mode of decay, are often said to be independent of the way in which the compound nucleus is formed. However, as pointed out in Section 6.6, this statement must be qualified by the conditions that any compound nuclear state must have the same total angular momentum J, z component of total angular momentum M, and parity Π as the initial state. The conservation of J and Π results in selection rules that forbid transitions between particular initial and final states to proceed through certain compound nuclear states. The conservation of M may be said to have as its consequence that the compound nucleus "remembers" the axis of approach of the initial constellation and decays accordingly. If the compound nucleus truly "forgot" its mode of formation, all inelastically scattered neutrons would be emitted isotropically (in the C-system). Experimentally, all are not; hence there is some "memory" of the mode of formation.

Hauser and Feshbach applied the procedure developed by Wolfenstein to the particular case of neutrons incident and emerging from the compound nucleus. With only minor changes, however, the formalism of Hauser and Feshbach is applicable to charged particles either incident or emerging. In particular, it is commonly employed in the calculation of (n,p) and (n,α) cross sections as well as of inelastic and compound elastic cross sections. To avoid circumlocution, we shall restrict the development at this point to neutron-in and -out interactions.

Hauser and Feshbach distinguish between two categories of inelastic scattering. In category I they place those scatterings in which the statistical model may be applied to the compound nucleus but not to the residual nucleus. In this category are interactions in which the incident-neutron energy is such that many levels are excited in the compound nucleus but only a few are excited in the residual nucleus. This situation will prevail at low neutron energies in medium and heavy nuclei. In category II they place high-energy neutron scatterings in which the statistical model may be applied to both the compound nucleus and the residual nucleus. They prove, as will shortly be shown, that the angular distribution of neutrons in category II is isotropic, whereas it is generally anisotropic in category I. The Hauser–Feshbach method does not apply to very light nuclei because statistical theory is inapplicable to nuclei whose energy levels are widely separated.

Notation

In this chapter the following notation is used:

I, I' = spins of target and residual nucleus, in units of $\hbar$

l, l' = orbital angular momenta of incoming and outgoing neutrons, in units of $\hbar$

E, E' = energy of incoming and outgoing neutrons (rigorously in C-system but practically in either C- or L-system since $A \gg 1$)

$j_1 = I + 1/2$ or $j_2 = I - 1/2$ = incoming channel spin

$j_1' = I' + 1/2$ or $j_2' = I' - 1/2$ = outgoing channel spin (When $I = 0$, j can have only the value $+1/2$; similarly when $I' = 0, j' = 1/2$)

J = spin of the excited level in the compound nucleus and also the total angular-momentum quantum number

As explained in Chapter 6, the spin of the compound nucleus $\mathbf{J}$ is formed by combining $\mathbf{l}$ and $\mathbf{j}$ or $\mathbf{l}'$ and $\mathbf{j}'$, $\mathbf{J} = \mathbf{l} + \mathbf{j} = \mathbf{l}' + \mathbf{j}'$. For a given J the values of l and l' that may contribute to the reaction are given by

$$|J - j| \leq l \leq (J + j), \quad |J - j'| \leq l' \leq (J + j'). \tag{8.28}$$

The z-axis is taken along the path of the incident neutron so that the z component of its orbital angular momentum is zero. The z component of the initial channel spin (which must be equal to the z component of $\mathbf{J}$) is called M. The z component of the total angular momentum of the emergent neutron is called m'; thus the z component of $\mathbf{j}'$ is $M - m'$.

In addition to J and M, parity is also conserved. The overall parity, which is either odd or even (-1 or $+1$), actually is not an independent quantum number but is determined by the value of l. It is noted separately because it is conserved whereas l is not. Changes in the parity of the nucleus are compensated for by appropriate changes in the neutron orbital angular momentum, from l to l'. If the parities of the initial and final nucleus are the same, $|l - l'|$ must be even; if the parities differ, $|l - l'|$ must be odd, i.e.,

No nuclear parity change:

$$|l - l'| = 0, 2, 4, \ldots, \tag{8.29a}$$

Nuclear parity change:

$$|l - l'| = 1, 3, 5, \ldots. \tag{8.29b}$$

Statistical Assumptions

It is assumed that the incident beam of neutrons is sufficiently broad in energy that many levels in the compound nucleus are excited. Note that the spread in the beam may be only 1% around an average energy of, say, 1 Mev; but this 10-kev region will contain, according to Equation 8.24, on the order of a hundred levels in a medium-weight compound nucleus and many more in a heavy nucleus. The wave functions corresponding to these various levels are assumed to have random phases with respect to one another so that when phase averages are performed all interference terms will vanish.

It is further assumed that all other processes by which the compound nucleus may decay, e.g., by gamma, proton, or alpha emission, have sufficiently small cross sections relative to $\sigma_{n'}$ in the energy region under consideration that they may be ignored in calculating the inelastic cross section. When this assumption is invalid, corrections can be made.

Development of Cross Sections

The cross section for a scattering in which the nuclear spin changes from I to I' may be written

$$\sigma(I,I') = \sum_{l,l'} \sigma(l,I|l',I'), \tag{8.30}$$

where $\sigma(l,I|l',I')$ is the cross section for the scattering from l to l' resulting in I'. This may be written

$$\sigma(l,I|l',I') = \frac{1}{2(2I+1)} \sum_{j,j'} \sigma(l,j|l',j'), \tag{8.31}$$

where $\sigma(l,j|l',j')$ is the cross section for the scattering from channel spin j to j' and from l to l'. As usual, we have averaged over initial states and summed over final states.

Now, of course, not all neutrons of orbital angular momentum l will form a compound nucleus. The cross section for compound-nucleus formation by l-wave neutrons is

$$\sigma_{c,l} = (2l + 1)\pi\lambda^2 T_l(E), \tag{8.32}$$

where the l-wave penetrability $T_l(E)$ is obtained from the optical-model reflection factor U_l in the form

$$T_l(E) = (1 - |U_l|^2). \tag{8.33}$$

(We are restricting the development at this point to potentials that do not include spin-orbit terms so that the penetrabilities involve l only and are independent of j and J.)

The cross section for formation of a compound nucleus of spin J by l-wave neutrons is given by $\sigma_{c,l}$ times the probability that the incident neutron and target form a system of spin J. This probability is equal to the square of the Clebsch–Gordon coefficient $C_{lj}^{JM}(M)$. Thus

$$\sigma_{c,l}[C_{lj}^{JM}(M)]^2 = (2l + 1)\pi\lambda^2 T_l(E)[C_{lj}^{JM}(M)]^2 \tag{8.34}$$

is the cross section for formation of a compound nucleus of spin J by incident l-wave neutrons. To obtain the cross section for a particular mode of decay, we must multiply Equation 8.34 by the relative probability of this particular process. Under the assumption that only neutron emission is possible, the relative probability for a particular mode of decay can be obtained by applying the reciprocity theorem to Equation 8.34. The cross section for the production of neutrons of energy E', angular momentum l', and channel spin j' moving in direction θ in the C-system is given by

$$\sigma(l,j|l',j'|\theta) = (2l + 1)\pi\lambda^2 T_l(E)\sum_J \frac{T_{l'}(E')}{\sum_{p;q,r} T_p(E'_q)} A_J(l,j|l',j'|\theta), \qquad (8.35)$$

where the index p refers to possible final neutron orbital angular momenta, the index q to possible final energies E'_q, and the index r to possible final channel spins. Since the coefficient $(2l + 1)\pi\lambda^2 T_l(E)$ is just the cross section for the formation of a compound nucleus by l-wave neutrons, the ratio of outgoing transmission coefficients

$$\sum_J \frac{T_{l'}}{\sum T_p} \qquad (8.36)$$

is equivalent to the decay probability $\Gamma_{l'}/\Gamma$ for this particular reaction. Note that the sum over J implies that the compound nucleus does indeed "remember" the initial constellation.

The factor A_J in Equation 8.35 carries the angular dependence of the cross section; it has the form of a sum of products of Clebsch–Gordon coefficients times a spherical harmonic

$$A_J(l,j|l',j'|\theta) = \sum_{M,m'} [C_{lj}^{JM}(M)]^2 [C_{l'j'}^{JM}(M - m')]^2 |Y_{l'}^{m'}(\theta,\phi)|^2. \qquad (8.37)$$

The absolute-value sign on the spherical harmonic wipes out the dependence of the scattering on the azimuthal angle ϕ. Blatt and Biedenharn* have shown that the A_J functions can be expressed in terms of the Legendre polynomials as

$$A_J(l,j|l',j'|\theta) = \frac{1}{4\pi(2l + 1)}\left|\sum_L Z(lJlJ;jL)Z(l'Jl'J;j'L)P_L(\cos\theta)\right|, \qquad (8.38)$$

where the coefficients $Z(abcd;ef)$ have been tabulated by Biedenharn and Simon†. From the general properties of the Z coefficients, the summation

---

*J. M. Blatt and L. C. Biedenharn, *Rev. Mod. Phys.*, 24: 258 (1952).

†L. C. Biedenharn, *Revised Z Tables of the Racah Coefficients*. USAEC Report ORNL-1501, Oak Ridge National Laboratory, May 1953; and L. C. Biedenharn and A. Simon, *Revised Z Tables of the Racah Coefficients*, USAEC Report ORNL-1501 (Suppl. 1), Oak Ridge National Laboratory, February 1954.

index L takes on only even values, and $L \leq$ min $(2l,2l',2J)$. Since L is always even, only even Legendre polynomials appear in the sum; so the angular distribution must be symmetric about $\theta = \pi/2$. Furthermore, if either l or l' is zero or if J is either zero or $1/2$, the function A_J will be independent of θ, and the inelastically scattered neutrons will be isotropically distributed.

Total inelastic-scattering cross sections can be obtained by first integrating the differential cross section over θ to obtain $\sigma(l,j|l',j')$ and then summing over j and j' as in Equation 8.31 and over l and l' as in Equation 8.30 to obtain $\sigma(I|I')$. The final result is

$$\sigma(I|I') = \frac{\pi \lambda^2}{2(2I + 1)} \sum_l T_l(E) \sum_J \frac{(2J + 1)\varepsilon(jlJ) \sum_{l',j'} \varepsilon(j'l'J)T_{l'}(E')}{\sum_{j'',l'',q} \varepsilon(j''l''J)T_{l''}(E_q')} , \quad (8.39)$$

where the symbol $\varepsilon(jlJ)$ takes on the values 2, 1, or 0 according to

$$\varepsilon(jlJ) = \begin{cases} 2, \text{ if both } j_1 \text{ and } j_2 \\ 1, \text{ if } j_1 \text{ or } j_2, \text{ not both} \\ 0, \text{ if neither } j_1 \text{ nor } j_2 \end{cases} \text{ satisfy } |J - l| \leq j_i \leq (J + l).$$

$$(8.40)$$

The summation in the denominator is, as before, over all nonforbidden exit channels of a fixed J. Equation 8.39 expresses the cross section for an inelastic event that leaves the residual nucleus in an energy level with spin I'. If several levels can be excited by neutrons of energy E, the total inelastic-scattering cross section is given by the sum of $\sigma(I,I')$ over I'.

Inelastic Scattering in Category II

Category II interactions, the reader will recall, are those in which the level density in the residual nucleus as well as in the compound nucleus is so great that the statistical model can be applied to both. The number of levels in the residual nucleus having an excitation energy between E^* and $E^* + dE^*$ is $\rho(E^*) \, dE^*$, or equivalently $\rho(E - E') \, d(E - E')$ since $E^* = E - E'$. When this is inserted into Equation 8.35, the cross section for inelastic scattering from E to dE' about E' in the direction $d\Omega$ about θ is obtained in the form

$$\sigma(E|E',\theta) \, dE' \, d\Omega = \frac{\lambda^2}{2(2I + 1)} \sum_{l,l',I',j} (2l + 1)T_l(E)T_{l'}(E')$$

$$\times \sum_{j',J} \frac{(2j' + 1)A_J(l,j|l',j'|\theta)\rho(E - E') \, dE' \, d\Omega}{\sum_{p,q,r} T_p(E_q')} ,$$

where the factor $2j' + 1$ is inserted to account for the degeneracy of each state with a given angular momentum j'. The summation over j' can be performed to yield

$$\sum_{j'} (2j' + 1)A_J(l,j|l',j'|\theta) \doteq \frac{(2J + 1)^2(2l' + 1)}{4\pi(2l + 1)},$$

which is independent of θ, thus proving that the category II inelastic-scattering cross section is isotropic in the C-system. Now the summations over j and J can be performed, and, since the integration over solid angle yields 4π, the final result is

$$\sigma(E|E') = \pi\lambda^2 \frac{\left[\sum_l (2l + 1)T_l(E)\right]\left[\sum_{l'} (2l' + 1)T_{l'}(E')\right]\rho(E - E')}{\sum_{l''} (2l'' + 1)\int_0^E T_{l''}(E'')\rho(E - E'')\,dE''}.$$

$$(8.41)$$

This is the desired formula for the differential (in energy) inelastic-scattering cross section for category II events. In place of the level density $\rho(E - E')$, we now substitute, from Equation 8.24

$$\rho(E - E') = C \exp\left[2\sqrt{a(E - E')}\,\right],$$

where the constants C and a are those appropriate for the residual nucleus. Since the penetrabilities $T_l(E)$ are obtainable from an optical-model calculation, all is known on the right-hand side of Equation 8.41.

Equation 8.41 can be rewritten in terms of the cross sections for compound-nucleus formation

$$\sigma_c = \pi\lambda^2 \sum_l (2l + 1)T_l(E)$$

as

$$\sigma(E|E') = \sigma_c(E) \frac{E'\sigma_c(E')\rho(E - E')}{\int_0^E E''\sigma_c(E'')\rho(E - E'')\,dE''},$$

$$(8.42)$$

which looks very much like the Bohr expression

$$\sigma(E,E') = \sigma_c(E) \frac{\Gamma(E')}{\Gamma}.$$

This implies that the compound nucleus, which is excited in a category II interaction, "forgets" its mode of excitation.

General Wolfenstein–Hauser–Feshbach
Cross-Section Expressions

For future use, we present here the Wolfenstein–Hauser–Feshbach expression for the differential and total reaction cross sections for any incident and emergent particles, not necessarily neutrons. The expressions are further generalized to include the possibility of spin-orbit potentials, for which the penetrabilities are functions of j and J.

The differential cross section is given by

$$\sigma(\alpha|\alpha',\theta) = \frac{\lambda^2}{4} \sum_{ljJl'j'L} \frac{(-1)^{j'-j}}{(2s+1)(2I+1)} \frac{T^J_{\alpha jl} T^J_{\alpha'j'l'}}{\sum\limits_{\alpha''j''l''} T^J_{\alpha''j''l''}}$$

$$\times Z(lJlJ;jL)Z(l'Jl'J;j'L)P_L(\cos\theta), \quad (8.43)$$

where α = channel symbol for the initial constellation; it includes type of incident particle (neutron, proton, etc.), energy of incident particle, and type of target nucleus,

α' = channel symbol for the final constellation; it includes type and energy of emergent particle and excitation energy of product nucleus,

I = target-nucleus spin,

s = incident-particle spin,

j = incident channel spin,

j' = emergent channel spin,

T = penetrability.

The cross section for the reaction integrated over all solid angles is

$$\sigma(\alpha|\alpha') = \pi\lambda^2 \sum_{ljJl'j'} \frac{(2J+1)}{(2s+1)(2I+1)} \frac{T^J_{\alpha jl} T^J_{\alpha'j'l'}}{\sum\limits_{\alpha''j''l''} T^J_{\alpha''j''l''}}. \quad (8.44)$$

Conservation of total angular momentum requires that only channel spins (j and j') that obey the usual coupling relations be included in the summations, namely,

$$|J-l| \le j \le (J+l), \qquad |J-l'| \le j' \le (J+l'),$$
$$|I-s| \le j \le (I+s), \qquad |I'-s'| \le j' \le (I'+s'). \quad (8.45)$$

The double-primed summations in Equations 8.43 and 8.44 are over all possible exit channels, subject as always to the conservation laws of total energy, total angular momentum, and parity. Included in the possible exit channels are emergent particles that are both the same and different in type from those whose cross section is being calculated.

These Wolfenstein–Hauser–Feshbach expressions were derived from purely kinematical arguments employing the conservation laws of angular momentum and parity. Kinetics (forces) appear in the form of the yet-to-be-evaluated penetrabilities. It is now common to calculate the penetrabilities for neutrons and charged particles using the complex potential-well model with spin-orbit coupling as described in Chapter 7. In the case of charged particles, a Coulomb term is added to the other terms in the potential to account for the strictly electrostatic forces. Photon penetrabilities are estimated from their radiation widths.*

8.4 A Typical Hauser–Feshbach Calculation

In Section 7.7 it was shown how the optical model could be used to obtain the shape elastic differential scattering cross sections of a typical element, calcium. In this section we will show how Hauser–Feshbach theory can be employed to find the compound elastic differential cross section and the inelastic-scattering cross sections of this same element. This specific element is examined for the sake of clarity; however, the method is quite general and can obviously be used to determine the cross sections of any element whose levels are such that Hauser–Feshbach theory is valid.

Energy Levels in ^{40}Ca and ^{41}Ca

The energies, spins, and parities of the low-lying energy levels in ^{40}Ca are shown in Figure 8.7. Below 5 Mev there are four excited levels. Above 5 Mev it is assumed that the levels form a continuum. Actually the level density formula (Equation 8.24) predicts approximately seven levels per Mev at 5 Mev in an even-even nuclide with $A = 40$. However, in the absence of specific experimental data on the energies, spins, and parities of these levels, it is not unreasonable to assume a continuum for the purpose of calculation.

The energy for neutron separation from the compound nucleus ^{41}Ca is given by

$$S_n^c = 931 \frac{\text{Mev}}{\text{amu}} [M(^{40}\text{Ca}) + m_n - M(^{41}\text{Ca})] = 8.36 \text{ Mev.}$$

The density of ^{41}Ca levels at an excitation energy of 8.36 Mev is, from Equation 8.24, approximately 100 per Mev. This density is assumed to be sufficiently great that the statistical model applies to the compound nucleus. In other words, we are stating that our results apply to a situation in which the energy resolution of the incident beam of neutrons is no better than about 0.1 Mev; so of the order of 10 levels or more are being

---

* See, for example, P. A. Moldauer, C. A. Engelbrecht, and G. J. Duffy, *NEARREX, a Computer Code for Nuclear Reaction Calculations*, USAEC Report ANL-6978, Argonne National Laboratory, December 1964.

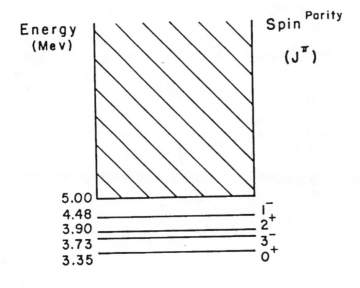

Energy
(Mev)   Spin   Parity

                                                          (J^π)

5.00
4.48 ——————————————————————— 1^-
3.90 ——————————————————————— 2^+
3.73 ——————————————————————— 3^-
3.35 ——————————————————————— 0^+

0 ——————————————————————— 0^+

Figure 8.7—Low-lying energy levels of ^{40}Ca. [From R. S. Caswell, *J. Res. Nat. Bur. Stand.*, **66**A: 389 (1962).]

excited in the compound nucleus. At higher neutron energies the density of levels increases; therefore the energy resolution may be better than 0.1 Mev without affecting our analysis.

Laboratory and Center-of-Mass Energies and Angles
We note that for an element as heavy as calcium ($A \simeq 40$) laboratory and center-of-mass energies and angles differ by only about 2%, which is far less than the usual experimental resolution in energy and angle. Thus for all practical purposes the slight difference between laboratory and center-of-mass variables may be ignored.

Compound Elastic Scattering
Compound elastic scattering from ^{40}Ca was treated by R. S. Caswell* in four different ways, depending on the incident-neutron energy:

1. Below the threshold for inelastic scattering at 3.35 Mev, $\sigma_{ce}(\theta)$ was obtained from the optical-model phase shifts, with a small correction for

*R. S. Caswell, *J. Res. Nat. Bur. Std.*, **66**A: 389 (1962).

the observed (n,p) cross section. The relevant equations are Equations 7.71, 7.72, and 7.73.

2. In the energy range from 3.35 to 5.0 Mev, a Hauser–Feshbach category-I calculation, which was based on the discrete energy levels of the target nucleus, was made. Equation 8.35 was specialized to the case of compound elastic scattering $(E = E', l = l', j = j')$ from a spin-zero nucleus so that $(I = I' = 0, j = j' = 1/2,$ and $J = l \pm 1/2)$. When the spin-dependent penetrabilities are taken into account, the l-wave contribution to the compound elastic differential cross section takes the form

$$\sigma_{ce,l}(E,\theta) = \frac{\pi \lambda^2}{2} \left[\frac{(l+1)T_l^p(E)T_l^p(E)A_{l+1/2}(l,\tfrac{1}{2}|l,\tfrac{1}{2}|\theta)}{\sum\limits_{n,q,r} T_n^p(E_q')} \right.$$

$$\left. + \frac{lT_l^a(E)T_l^a(E)A_{l-1/2}(l,\tfrac{1}{2}|l,\tfrac{1}{2}|\theta)}{\sum\limits_{n,q,r} T_n^a(E_q')} \right]. \quad (8.46)$$

The penetrabilities that appear in the denominator of Equation 8.46 include those for the experimentally observed (n,p) and (n,α) cross sections. The A functions are evaluated in terms of the tabulated Z coefficients of Blatt and Biedenharn,

$$A_J(l,j|l,j|\theta) = \frac{1}{4\pi(2l+1)} \sum_L [Z(lJlJ;jL)]^2 P_L(\cos\theta),$$

and the final cross-section expression is given by the simple summation of the partial-wave cross sections without phase factors,

$$\sigma_{ce}(E,\theta) = \sum_l \sigma_{ce,l}(E,\theta).$$

Note that the resulting cross section is symmetric about $\theta = \pi/2$ since the A functions are sums of even Legendre polynomials.

3. In the energy range from 5.0 to 6.0 Mev, a Hauser–Feshbach category-II calculation was made based on a statistical model of the residual nucleus. In this case compound elastic scattering is isotropic because of the assumed randomness of the spins and parities of the levels in the compound nucleus.

4. Above 6 Mev there is so much competition from (n,n'), (n,α), and (n,p) reactions that compound elastic scattering may be assumed to be negligible.

Inelastic Scattering, Category I

Neutrons incident with energy below 5.0 Mev can excite the four discrete levels in ^{40}Ca (see Figure 8.7). Interest therefore centers on the partial cross sections for excitation of each of these levels as a function of incident-neutron energy. These cross sections are given by Equation 8.39, which will now be explored in greater depth.

To avoid circumlocution, let us choose a very specific case. Suppose we seek to find the cross section for excitation of the ^{40}Ca level at 3.90 Mev by neutrons with energy E somewhere between 4.5 and 5.0 Mev. Neutrons of this energy will excite all four of the discrete levels (five, including the ground state), but we seek at this time only the cross section for excitation of the level at 3.90 Mev. This level has a spin and parity of 2^+, whereas the ground state has a spin and parity of 0^+. The transition in the target nucleus is from 0^+ to 2^+, and its cross section is therefore $\sigma(I|I') = \sigma(0|2)$.

The incoming channel spin is $j = 1/2$. The outgoing channel spins are $j'_1 = 3/2$ and $j'_2 = 5/2$. The conservation of total angular momentum, as expressed by Equation 8.28, together with the condition that $|l - l'|$ must equal 0 or a multiple of 2 to conserve parity, determines the possible values of l, l', and J. These are listed in Table 8.2 for all l and l' up to 3.

From this table, we see that for $l = 0$ the summation over J in Equation 8.39 includes only one term, namely $J = 1/2$, and the summation over l' and j' has only one term

$$\sum_{l',j'} \varepsilon(j'l'J)T_{l'}(E') = 2T_2(E'),$$

since l' has only one value for $l = 0$ and ε has the value 2 because both $j' = 3/2$ and $j' = 5/2$ satisfy inequality 8.40.

To write explicit expressions for the terms within the double-primed summation, we must first develop tables like Table 8.2 for each of the other allowed transitions $(0^+ \rightarrow 1^-)$ and $(0^+ \rightarrow 3^-)$. This "bookkeeping" will be left as an exercise for the reader.

Table 8.2—Values of l, l', and J^π that Enter Calculation of $\sigma(l, l')$ for a $0^+ \rightarrow 2^+$ Transition for $l, l' \leq 3$

l	l'	J^π	l	l'	J^π
0	2	$(1/2)^+$	2	0	$(3/2)^+$
			2	2	$(3/2)^+$
1	1	$(1/2)^-$	2	2	$(5/2)^+$
1	3	$(1/2)^-$			
1	1	$(3/2)^-$	3	1	$(5/2)^-$
1	3	$(3/2)^-$	3	3	$(5/2)^-$
			3	3	$(7/2)^-$

   The partial cross sections for excitation of the first four levels in
calcium, as calculated by Troubetzkoy et al.,* are shown in Table 8.3.
The magnitudes and general behavior of these inelastic cross sections are
reasonably typical of medium-weight even-A nuclei. In odd-A nuclei and

Table 8.3—Partial Cross Sections for Excitation of First Four Levels in
Calcium

	$\sigma_{n'}(E,E_i)$, barns			
E, Mev	$E_1 = 3.35$ Mev	$E_2 = 3.73$ Mev	$E_3 = 3.90$ Mev	$E_4 = 4.48$ Mev
4.91	0.046	0.140	0.092	0.036
4.67	0.040	0.150	0.084	0.008
4.44	0.036	0.136	0.074	0
4.23	0.032	0.113	0.058	0
4.02	0.029	0.069	0.015	0
3.82	0.024	0.000	0	0
3.64	0.019	0	0	0
3.46	0.008	0	0	0
3.29	0	0	0	0

heavier nuclei, the lower levels are closer together and the cross sections
are generally somewhat greater.
   The total inelastic cross section may be obtained if desired by summing the partial cross sections:

$$\sigma_{n'}(E) = \sum_i \sigma_{n'}(E,E_i).$$

Inelastic Scattering, Category II

It is assumed that above 5 Mev, inelastic scattering in calcium can be
treated as a category-II event. The emergent neutrons are distributed
continuously in energy with the differential inelastic-scattering cross
section given by Equation 8.41 or Equation 8.42.
   The results of such a calculation are shown in Table 8.4. The total
inelastic cross section can be obtained by graphical or numerical integration over E',

$$\sigma_{n'}(E) = \int \sigma_{n'}(E,E') \, dE'.$$

*E. S. Troubetzkoy, M. H. Kalos, H. Lustig, J. H. Ray, and B. H. Trupin, *Fast-Neutron
Cross Sections of Manganese, Calcium, Sulfur, and Sodium,* Report NDA-2133-4, Nuclear
Development Corp. of America, January 1961.

Table 8.4—Differential Cross Section for Production of Inelastic Neutrons of Energy E' by Neutrons of Energy E Incident on Calcium*

E, Mev	$\sigma_{n'}(E,E')$, barns/Mev for E' given in Mev								
	0.5	1	1.5	2	3	4	5	6	7
18.0	0.041	0.067	0.076	0.076	0.065	0.049	0.033	0.021	0.013
16.3	0.046	0.070	0.082	0.084	0.069	0.049	0.031	0.019	0.010
14.75	0.047	0.082	0.092	0.090	0.071	0.050	0.030	0.018	0.009
13.3	0.048	0.090	0.097	0.095	0.072	0.048	0.028	0.014	0.008
12.1	0.054	0.097	0.104	0.098	0.071	0.045	0.025	0.017	0.005
10.9	0.068	0.106	0.110	0.100	0.070	0.042	0.022	0.012	0.016
9.89	0.079	0.113	0.112	0.101	0.068	0.037	0.019	0.022	0.000
8.95	0.088	0.115	0.114	0.099	0.061	0.034	0.045	0.000	0.000
8.10	0.086	0.114	0.113	0.097	0.054	0.069	0.000	0.000	0.000
7.33	0.085	0.113	0.108	0.096	0.100	0.000	0.000	0.000	0.000
6.63	0.083	0.107	0.108	0.100	0.145	0.000	0.000	0.000	0.000
6.00	0.075	0.110	0.165	0.220	0.000	0.000	0.000	0.000	0.000
5.43	0.080	0.155	0.231	0.310	0.000	0.000	0.000	0.000	0.000
5.16	0.094	0.190	0.290	0.000	0.000	0.000	0.000	0.000	0.000

*Abstracted from E. S. Troubetzkoy et al., Report NDA-2133-4, Nuclear Development Corp. of America, January 1961.

8.5 Energy and Angular Distributions of Inelastic Gammas

Following emission of the inelastic neutron from the compound nucleus, the still-excited product nucleus decays by emission of one or more gamma rays until it finally reaches its ground state, assuming, of course, that the excitation is not sufficiently high to allow massive particle emission. Most of these gammas are emitted promptly, within times less than 10^{-10} sec after neutron emission.

There are, however, a number of nuclides that have a fairly low-lying level with a spin differing by several units from the spins of all lower-lying levels. The deexcitation of this metastable isomeric state of the product nucleus by gamma radiation may be delayed by seconds or even hours. Such metastable states are relatively rare when compared with the thousands of rapidly decaying states. Experimental nuclear physicists take great delight in them, however, because their gamma emission can be investigated free from the background problems attendant on neutron emission. After a target has been irradiated for several half-lives of the metastable state, the neutron beam can be shut off, and the gamma activity of the sample can be counted to obtain information on gamma energy, angular distribution, and level half-life.

Energy Spectrum of Inelastic Gammas

The energy spectrum of inelastic gammas, like the corresponding neutron spectrum, has one of two different forms depending on the energy of the bombarding neutrons. A category-I inelastic event excites at most a few levels in the target (product) nucleus, and the resulting gamma-ray spectrum consists of a few discrete lines, as in Figure 8.2. A category-II inelastic event leaves the product nucleus in a highly excited state, actually many highly excited states determined by the spectrum of emergent neutrons. Below these states there are scores of other states. Since decay can now take place via hundreds of distinct cascade paths, the resulting energy spectrum is practically continuous.

Category-I gamma-ray-production cross sections for calcium are shown in Table 8.5. Included in this table are not only inelastic gamma rays but also gamma rays from the $^{40}Ca(n,p\gamma)^{40}K$ and $^{40}Ca(n,\alpha\gamma)^{37}Ar$ reactions, which have appreciable probability even at energies below 5 Mev because of the relatively low Coulomb barrier in calcium. These charged-particle-out reactions are considered in detail in Chapter 10. We remark here only that ^{40}K has levels at 0.029, 0.797, and 0.885 Mev and ^{37}Ar has levels at 1.46 and 1.66 Mev; hence the emission of gamma rays of these energies.

Of immediate interest are the last four columns in Table 8.5, which list the partial cross sections for gamma-ray production following inelastic scattering in calcium. These cross sections are defined so that the number of gammas of energy E_γ produced per unit volume and unit time is given by

$$N(E_\gamma) = \sigma(E,E_\gamma)[\text{Ca}]\phi(E), \tag{8.47}$$

where [Ca] is the number density of calcium atoms and $\phi(E)$ is the flux of neutrons of energy E.

Examining the last four columns in this table, the alert reader undoubtedly will detect a puzzling anomaly; there is no gamma corresponding to the transition from the first excited level of ^{40}Ca, at 3.35 Mev, to the ground state, and, stranger still, there is a 0.511-Mev gamma despite the fact that there is no energy level at 0.511 Mev! Before reading the solution to this mystery, the student is invited to try to figure it out for himself. Clues are to be found in Figure 8.7 and in the relation between the cross section for excitation of the 3.35-Mev level in Table 8.3, and the cross section for production of the 0.511-Mev gammas in Table 8.5.

Because both the ground state and the first excited level of ^{40}Ca have spin zero and because a gamma must carry off at least one unit of angular

Table 8.5—Cross Section for Production of Discrete Gamma Rays by Neutrons of Energy E Incident on Calcium*

	$\sigma(E,E_\gamma)$, barns, for E_γ in Mev								
	${}^{40}Ca(n,p\gamma){}^{40}K$			${}^{40}Ca(n,\alpha\gamma){}^{37}Ar$		${}^{40}Ca(n,n'\gamma){}^{40}Ca$			
E, Mev	0.029	0.797	0.885	1.46	1.66	0.511	3.73	3.90	4.48
4.91	0.121	0.073	0.025	0.013	0.016	0.091	0.140	0.092	0.036
4.67	0.112	0.066	0.018	0.012	0.014	0.081	0.150	0.084	0.008
4.44	0.107	0.055	0.010	0.011	0.012	0.071	0.136	0.074	0.000
4.23	0.104	0.047	0.009	0.009	0.011	0.064	0.113	0.058	0.000
4.02	0.104	0.040	0.007	0.008	0.009	0.057	0.069	0.000	0.000
3.82	0.104	0.032	0.005	0.006	0.007	0.048	0.000	0.000	0.000
3.64	0.095	0.023	0.003	0.005	0.005	0.038	0.000	0.000	0.000
3.46	0.082	0.015	0.002	0.004	0.004	0.015	0.000	0.000	0.000
3.29	0.071	0.010	0.000	0.003	0.003	0.000	0.000	0.000	0.000
2.97	0.048	0.003	0.000	0.001	0.002	0.000	0.000	0.000	0.000
2.69	0.030	0.001	0.000	0.000	0.000	0.000	0.000	0.000	0.000
2.44	0.017	0.000	0.000	0.000	0.000	0.000	0.000	0.000	0.000
2.21	0.008	0.000	0.000	0.000	0.000	0.000	0.000	0.000	0.000
2.00	0.004	0.000	0.000	0.000	0.000	0.000	0.000	0.000	0.000
1.90	0.002	0.000	0.000	0.000	0.000	0.000	0.000	0.000	0.000
1.81	0.000	0.000	0.000	0.000	0.000	0.000	0.000	0.000	0.000

*Abstracted from E. S. Troubetzkoy et al., Report NDA-2133-4, Nuclear Development Corp. of America, January 1961.

momentum, the $0^+ \to 0^+$ gamma transition is absolutely forbidden by the law of conservation of total angular momentum.

There are two nonforbidden modes that an excited nucleus may employ to reach its ground state if gamma emission is impossible, namely, *internal conversion* or *pair emission*. In internal conversion the excess energy of the nucleus is given to one of the atom's orbital electrons, usually an electron in the low-lying K- or L-shells. This electron is ejected from the atom and carries off the nuclear excitation energy. If the excitation energy of the nucleus is above 1.02 Mev, it may emit a positron-electron pair to carry off its excitation energy. In either event the nucleus returns to its ground state.

It has been found experimentally that the first excited state of ${}^{40}Ca$ decays by pair emission, not by internal conversion. After its emission, the positron, upon coming to rest, is annihilated by an electron to produce two 0.511-Mev gammas. These *annihilation gammas* are, from a practical viewpoint, indistinguishable from the rest of the inelastic gammas; and, since they are produced as a byproduct of the (n,n') reaction, they are tabulated as inelastic gammas. Because two 0.511-Mev gammas are emitted per decay of ^{40}Ca by pair emission, the cross section for production of the 0.511-Mev gammas must be twice the cross section for excitation of the lowest level in ^{40}Ca, hence the relation between these cross sections in Tables 8.3 and 8.5.

Since, as previously stated, category-II gammas come off with a practically continuous distribution in energy, it is no longer feasible to catalog production cross sections for individual gamma energies. Instead, a differential cross section $\sigma(E,E_\gamma)$ is defined such that $\sigma(E,E_\gamma)\,dE_\gamma$ is the cross section for the emission of gammas in the energy range dE_γ about E_γ due to a reaction initiated by a neutron of energy E. A typical compilation of category-II cross sections is shown in Table 8.6 for the element calcium. The data shown here include gammas, not only from the $(n,n'\gamma)$ reaction but also from the $(n,p\gamma)$ and $(n,\alpha\gamma)$ reactions. Experimentally, there is no way of distinguishing between category-II gammas from these

Table 8.6—Differential Cross Section for Production of Gammas of Energy E_γ by Neutrons of Energy E Incident on Calcium*

	$\sigma(E,E_\gamma)$, barns/Mev for E_γ given in Mev								
E, Mev	0.5	1	1.5	2	3	4	5	6	7
18.0	0.323	0.455	0.569	0.617	0.516	0.269	0.164	0.117	0.066
16.3	0.311	0.438	0.541	0.585	0.486	0.249	0.151	0.106	0.058
14.75	0.303	0.421	0.517	0.556	0.455	0.230	0.138	0.094	0.051
13.3	0.290	0.396	0.485	0.516	0.415	0.207	0.122	0.082	0.042
12.1	0.278	0.377	0.457	0.483	0.382	0.187	0.109	0.070	0.029
10.9	0.264	0.354	0.422	0.444	0.345	0.165	0.093	0.058	0.028
9.89	0.251	0.333	0.396	0.412	0.313	0.145	0.080	0.047	0.020
8.95	0.236	0.308	0.363	0.375	0.277	0.124	0.065	0.036	0.012
8.10	0.222	0.285	0.331	0.339	0.242	0.105	0.053	0.026	0.007
7.33	0.206	0.262	0.304	0.304	0.211	0.087	0.039	0.016	0.000
6.63	0.195	0.243	0.275	0.274	0.181	0.068	0.028	0.006	0.000
6.00	0.180	0.221	0.244	0.240	0.150	0.052	0.016	0.000	0.000
5.43	0.162	0.196	0.212	0.205	0.120	0.035	0.006	0.000	0.000
5.16	0.153	0.183	0.197	0.188	0.105	0.028	0.000	0.000	0.000

*Abstracted from E. S. Troubetzkoy et al., Report NDA-2133-4, Nuclear Development Corp. of America, January 1961.

three reactions; and, of course, for most nuclear engineering use, it is the total gamma-ray production cross section that is desired, not the partial production from each of the three competing reactions. In the next chapter we show how one may theoretically determine the gamma spectrum from each separate reaction.

Data such as that displayed in Tables 8.3 to 8.6 are obtained by using theory to fill in gaps that have not been explored experimentally. In the present state of nuclear theory, experimental data, regardless of uncertainties, is almost always accepted in preference to data that is derived purely from theory. Perhaps, at some far distant time, nuclear theory will be so completely developed that uncertainties in calculated results will actually be less than experimental uncertainties. There are many areas of engineering that are still, hundreds of years after their initiation, primarily empirical. There is little reason to hope that nuclear theory will prove to be more tractable than, say, fluid heat transfer.

Angular Distributions of Inelastic Gammas

Angular distributions of inelastic gammas, like the corresponding neutron distributions, have different forms depending on whether the inelastic scattering giving rise to the gammas is a category-I or a category-II event. Angular distributions of decay gammas from low-lying levels directly excited by inelastic neutron emission have been found to be given by a summation of even Legendre polynomials in cos θ:

$$\sigma(E_\gamma, \theta) = \sigma(E_\gamma) \sum_{L=0}^{N} a_L P_L(\cos \theta),$$

where usually only the first three even polynomials P_0, P_2, P_4 are required. Angular distribution of category-II gammas have been found to be isotropic

$$\sigma(E_\gamma, \theta) \, dE_\gamma = \frac{\sigma(E_\gamma) \, dE_\gamma}{4\pi}.$$

The theoretical analysis of this particular subject is most logically treated as a special case of gamma emission in general and is so treated in Chapter 9.

8.6 Cascade Reactions, Particularly (*n*,2*n*)

A *cascade reaction* is defined as a reaction in which the final constellation consists of more than one massive particle* in addition to the final nucleus. Neutron-induced cascade reactions include (*n*,2*n*), (*n*,*np*), (*n*,2*α*),

---

*A *massive particle* is defined as one with a nonzero rest mass; hence reactions such as (*n*,*n'γ*) and (*n*,*pγ*) are not called cascade reactions.

(n,nt), etc. We shall consider only the $(n,2n)$ reaction at this point. Because neutrons have much greater penetrabilities than charged particles, the $(n,2n)$ reaction is, with few exceptions, the most probable cascade reaction in the energy range below 18 Mev, particularly from medium-heavy and heavy nuclides.

Threshold of $(n,2n)$ Reaction

Following formation of the compound nucleus by a highly energetic incident neutron, it may happen that a neutron is emitted with sufficiently low energy that the residual nucleus still has enough excitation energy to emit a second neutron. If it does, this is an $(n,2n)$ reaction,

$$Z^A + n \rightarrow (Z^{A+1})^* \rightarrow (Z^{A-1})^* + 2n + Q,$$

where the product nucleus if excited will subsequently decay by gamma emission.

The Q value of the $(n,2n)$ reaction is

$$Q(n,2n) = 931 \, \frac{\text{Mev}}{\text{amu}} \, [M(Z^A) + m_n - M(Z^{A-1}) - 2m_n]$$

$$= 931 \, \frac{\text{Mev}}{\text{amu}} \, [M(Z^A) - M(Z^{A-1}) - m_n], \tag{8.48}$$

which is just the negative of the separation energy of a neutron from the target nucleus (compare with Equation 5.9). Thus

$$Q(n,2n) = -S_n(Z^A). \tag{8.49}$$

The laboratory threshold energy for the $(n,2n)$ reaction is given by the same expression as that for all neutron-induced reactions, namely, $-(A+1)/A$ times Q:

$$\tilde{E}_t(n,2n) = -\frac{A+1}{A} \, Q(n,2n). \tag{8.50}$$

Since Q is negative, $\tilde{E}_t$ is positive.

Most $(n,2n)$ thresholds are obtained, in practice, by observing the threshold for the $Z^A(\gamma,n)Z^{A-1}$ reaction, which, starting as it does from the same target and leading to the same product nucleus, has the same Q value as the corresponding $(n,2n)$ reaction. Since the ratio of the photon's momentum to energy is very small, the threshold energy and the Q value of a (γ,n) reaction are practically identical in magnitude but opposite in sign; hence

$$\tilde{E}_t(n,2n) = \frac{A+1}{A} \, \tilde{E}_t(\gamma,n). \tag{8.51}$$

Table 8.7—Selected List of (γ,n) and Corresponding $(n,2n)$ Threshold Energies

Target nuclide	$\tilde{E}_t(\gamma,n)$, Mev	$\tilde{E}_t(n,2n)$, Mev
^{2}H	2.226	3.34
^{4}He	20.5	25.6
^{9}Be	1.666	1.85
^{12}C	18.7	20.3
^{14}N	10.54	11.3
^{23}Na	12.05	12.6
^{54}Fe	13.8	14.1
^{56}Fe	11.15	11.3
^{57}Fe	7.75	7.9
^{118}Sn	9.10	9.2
^{119}Sn	6.55	6.6
^{124}Sn	8.50	8.6
^{197}Au	8.05	8.1
^{206}Pb	8.25	8.3
^{207}Pb	6.88	6.9
^{208}Pb	7.40	7.4
^{238}U	5.97	6.0

A few selected (γ,n) and $(n,2n)$ thresholds are listed in Table 8.7. The unusually low thresholds of ^{2}H and ^{9}Be are of significance to the nuclear engineer because of the widespread use of heavy water and beryllium as reactor moderators. Except for these two special cases, the $(n,2n)$ reaction is a high-energy one. Most thresholds cluster around the 7- to 12-Mev range; the highest threshold being that of ^{4}He, about 26 Mev.

Cross Sections for $(n,2n)$ Reactions

In light nuclei the $2n$ mode of decay of the compound nucleus has competition from compound elastic scattering, inelastic scattering, photon emission, and charged-particle emission. In heavy nuclei the main competition comes from inelastic scattering since charged-particle-out reactions are generally strongly suppressed by large Coulomb barriers and compound elastic scattering is generally negligible compared to inelastic scattering at energies at which there are many inelastic channels open.

When (n,n') and $(n,2n)$ are indeed the only modes of decay having appreciable probability and if one assumes, as is commonly done, that a nucleus will emit a neutron rather than a photon whenever energetically possible, then the $(n,2n)$ cross section can be easily related to the (n,n') cross section: The first neutron emitted by the highly excited compound nucleus is an "inelastically" scattered neutron with an energy distribution given by a category-II Hauser–Feshbach calculation. If this neutron is emitted with an energy sufficiently low that the residual nucleus is left with an excitation energy greater than the neutron separation energy S_n, then the nucleus will emit another neutron. Hence the cross section for the $(n,2n)$ reaction is that part of the inelastic cross section that produces inelastic neutrons with energies E' less than $E - S_n$,

$$\sigma_{2n}(E) = \int_0^{E-S_n} \sigma_{n'}(E,E')\, dE'. \tag{8.52}$$

If the first neutron were emitted from the compound nucleus with an energy $E' \geq E - S_n$, then the excitation energy of the residual nucleus would be less than S_n, and the reaction would be pure inelastic scattering. The situation is depicted schematically in Figure 8.8, where the shaded area represents the integral in Equation 8.52. All neutrons emitted with energies below $E - S_n$ are assumed to be followed by a second neutron; those with energies above $E - S_n$ are followed only by gamma rays.

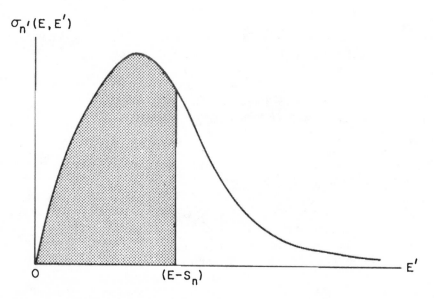

Figure 8.8—Typical differential inelastic-scattering cross section at high initial energy. The shaded area is σ_{2n}.

In any but the lightest elements, the emission of the first neutron is a category-II event; so the energy spectrum of first neutrons may be expected to be given by Equation 8.42,

$$\sigma_{n_1}(E,E') \propto E'\sigma_c(E')\rho(E - E'), \tag{8.53}$$

where $E' < E - S_n$ in order that emission of a second neutron be possible. The spectrum of second neutrons is rigorously proportional to

$$\sigma_{n_2}(E,E')\, dE' \propto dE' \int_{S_n + E'}^{E} P_1(E,E - E^*)P_2(E^*,E')\, dE^*, \tag{8.54}$$

where $P_1(E,E - E^*)\, dE^*$ is the probability that emission of the first neutron leaves the residual nucleus Z^A in an energy range dE^* about E^* and $P_2(E^*,E')\, dE'$ is the probability that the excited residual nucleus emits a neutron in an energy range dE' about E'.

Both P_1 and P_2 can be obtained from Hauser–Feshbach theory. In the special case that all energies involved are sufficiently great that category-II Hauser–Feshbach theory applies, one obtains from Equation 8.42

$$P_1(E,E - E^*) \propto (E - E^*)\sigma_c(E - E^*)\rho(E^*) \tag{8.55}$$

and

$$P_2(E^*,E') \propto E'\sigma_c(E')\rho(E^* - E' - S_n). \tag{8.56}$$

These can be substituted into Equation 8.54, and the spectrum can be determined from

$$\sigma_{n_2}(E,E') \propto E'\sigma_c(E') \int_{S_n + E'}^{E} (E - E^*)\sigma_c(E - E^*) \\ \times \rho(E^*)\rho(E^* - E - S_n)\, dE^*, \tag{8.57}$$

where the compound-nucleus formation cross sections are obtained, for example, from Hauser–Feshbach theory using optical-model penetrabilities and the level densities are known or approximated by some expression, such as Equation 8.24.

The Product Nucleus
The ground state of the product nucleus Z^{A-1} that is produced in the $Z^A(n,2n)Z^{A-1}$ reaction may or may not be stable against radioactive decay. If it is unstable and if it is a light- or medium-weight element (say $A < 70$), it will tend to decay by positron emission since it usually will have too few neutrons for stability. Some nuclei (e.g., ^{55}Fe) accomplish this same objective of increasing the neutron-to-proton ratio by capturing

an orbital electron. A unique exception is ^{8}Be, formed in the ^{9}Be$(n,2n)^8$Be reaction, which decays with a half-life of order 10^{-16} sec to two alpha particles.

Heavy radioactive product nuclides tend to decay by positron emission (or its equivalent, electron capture) if they are proton rich and by negatron emission if they are neutron rich. Thus, for example, $(n,2n)$ reactions in tin produce the radioactive isotopes ^{111}Sn, ^{113}Sn, ^{121}Sn, and ^{123}Sn. The lighter two isotopes decay by β^+ emission and/or electron capture; the heavier two isotopes decay by β^- emission.

8.7 Direct Reactions

There are observations that have led nuclear theorists to believe that at least some (n,n'), (n,p), and (n,α) reactions take place without the formation of a compound nucleus. These *direct reactions* are pictured as occurring at or near the nuclear surface. The incoming high-energy neutron knocks out a neutron, a proton, or an alpha particle from the surface of the target nucleus in the process of being absorbed itself.

The evidence for such reactions is primarily observed angular distributions of reaction products which differ from those predicted by the statistical theory, namely, symmetry about 90°. However, as pointed out by Goldstein,* even where the observed angular distributions show asymmetries about 90°, it is not safe to conclude the presence of direct interaction since an incomplete statistical mixture of states (i.e., too few states per unit energy interval) in the compound nucleus would lead to angular distributions with asymmetries about 90°.

At this time the evidence for direct reactions is rather conclusive. However, they probably constitute a small fraction of all interactions in the energy range below 18 Mev, even less below 2 Mev; thus they are of but slight significance to nuclear engineers. To pursue the theory of such reactions would lead us too far afield. The curious reader is referred to the article "Direct Reaction Theories," by N. Austern, which appears in *Fast Neutron Physics, Part II*, referenced below.

SELECTED REFERENCES

1. J. B. Marion and J. L. Fowler (Eds.), *Fast Neutron Physics, Part II, Experiments and Theory*, Interscience Publishers, Inc., New York, 1963. This book consists of about twenty review articles that describe the results obtained from experiments involving fast neutrons and the theoretical basis for the interpretation of these results.

*H. Goldstein, "Statistical Model Theory of Neutron Reactions and Scattering," in *Fast Neutron Physics, Part II*, J. B. Marion and J. L. Fowler (Eds.), Reference 1 above.

To supplement the material in this chapter, the reader might wish to read Chapters V. H. to V. K.

V. H. "*Neutron Nonelastic Collision Cross Sections*," by R. C. Allen, R. E. Carter, and H. Lyndon Taylor

V. I. "*Excitation Functions for Inelastic Scattering*," by J. B. Guernsey and D. A. Lind

V. J. "*Statistical Model Theory of Neutron Reactions and Scattering*," by H. Goldstein

V. K. "*Gamma Radiation from Neutron Inelastic Scattering*," by J. M. Freeman

Part I of *Fast Neutron Physics* is concerned with experimental techniques.

EXERCISES

1. Using Einstein's familiar relation $E = Mc^2$, calculate the difference in mass between a nucleus in its ground state and in an excited state E^*. Show that this mass difference may, for all practical purposes, be ignored in the kinematical analysis of inelastic scattering.

2. The momentum of a gamma ray of energy E_γ is E_γ/c, where c is the velocity of light in vacuo. (a) Prove that the recoil kinetic energy ΔE given to a nucleus of mass M as it emits a gamma ray is

$$\Delta E = \frac{1}{2M}\left(\frac{E_\gamma}{c}\right)^2.$$

(b) Show that this recoil energy is orders of magnitude less than the recoil energy that would result from the emission of a neutron of the same energy as the gamma.

3. (a) Show that under the assumption that the inelastic-scattering cross section is given by

$$\sigma_{n'}(E,E') = \sigma_{n'}(E)\frac{E'}{T^2}\,e^{-E'/T},$$

the $(n,2n)$ cross section is

$$\sigma_{2n}(E) = \sigma_{n'}(E)\left\{1 - \left[1 + \frac{E - E_t}{T}\right]e^{-(E-E_t)/T}\right\}.$$

(b) Show that near threshold, therefore, the $(n,2n)$ cross section behaves like

$$\sigma_{2n} = \frac{\sigma_c}{T^2}(E - E_t)^2.$$

This dependence on the square of the energy above threshold has been experimentally observed; see, for example, J. L. Fowler and J. M. Slye, Jr., *Phys. Rev.*, 77: 787 (1950).

4. Prepare a table similar to Table 8.2 that lists the values of l, l', and J^{11} which enter the calculation of the cross section for the transition of ^{40}Ca from its ground state to its second excited state ($0^+ \rightarrow 3^-$).

5. Using Equation 8.39 and Table 8.2, write in detail the equation for the inelastic cross section for excitation of the 2^+ level in ^{40}Ca by neutrons with $l \le 2$. Assume that all penetrabilities are known.

9 RADIATIVE CAPTURE

The radiative-capture reaction consists of the absorption of a neutron by a nucleus, and the emission, usually within an extremely short time ($\simeq 10^{-14}$ sec), of one or more gamma rays:

$$Z^A + n \rightarrow Z^{A+1} + \gamma.$$

The capture reaction is believed to proceed, like all other reactions in the neutron-energy region of primary interest to nuclear engineers ($E < 18$ Mev), by the formation and subsequent decay of a compound nucleus:

$$Z^A + n \rightarrow (Z^{A+1})^* \rightarrow Z^{A+1} + \gamma.$$

Radiative-capture (n,γ) reactions are particularly significant in reactor technology for the following reasons. (1) Over much of the energy range of a reactor neutron, they are the only energetically possible parasitic reactions; hence they strongly influence the neutron balance in a chain reactor. (2) Capture gamma rays from (n,γ) reactions often constitute the dominant sources in the radiation heating and bulk shielding of nuclear reactor vessels. (3) Often the ground state or a low-lying isomeric state of the product nucleus formed in an (n,γ) reaction is unstable against radioactive decay. It decays with some characteristic half-life, usually by the emission of a beta particle and decay gamma rays. These decay gamma rays engender shielding and maintenance problems during reactor operations and, more significantly, for times ranging up to years after reactor shutdown.

9.1 Kinematics of the Radiative-Capture Reaction

The kinematics of radiative capture are completely straightforward, indeed, almost trivial. A target nucleus Z^A absorbs a neutron to form a compound nucleus, which then decays by gamma emission

$$Z^A + n \rightarrow (Z^{A+1})^* \rightarrow Z^{A+1} + \gamma + Q. \tag{9.1}$$

Since the rest mass of a gamma ray is zero, the Q value is given by

$$Q(n,\gamma) = 931 \frac{\text{Mev}}{\text{amu}} [M(Z^A) + m_n - M(Z^{A+1})], \tag{9.2}$$

which is seen, by comparison with Equation 5.9, to be just the energy necessary to separate the neutron from the nuclide Z^{A+1}, the compound nucleus,

$$Q(n,\gamma) = S_n^c. \tag{9.3}$$

Since neutron separation energies are always positive, radiative capture is an exoergic reaction; its threshold energy is zero, and neutrons of any energy can be captured with gamma emission. In fact, the (n,γ) reaction is, for the vast majority of nuclides, the only reaction that can be triggered by thermal neutrons, the exceptions being a few pathological (n,p), (n,α), and (n,f) reactions, which are described in subsequent chapters.

Following the capture of a neutron, the compound nucleus is excited to a level with an energy given by the sum of the neutron separation energy and the available energy in the C-system (Figure 9.1):

$$E^* = S_n^c + E, \tag{9.4}$$

where $E = [A/(A + 1)]\tilde{E}$, provided the laboratory kinetic energy of the target nucleus is negligible compared with the laboratory kinetic energy $\tilde{E}$ of the neutron, which it will certainly be, except possibly where $\tilde{E} \lesssim 10$ ev. But in this case, $\tilde{E}$ is completely negligible compared with S_n^c. Thus, at all neutron energies the excitation energy of the compound nucleus is very closely approximated by

$$E^* = S_n^c + \frac{A}{A + 1} \tilde{E}. \tag{9.5}$$

The compound nucleus decays by the emission of one or more gamma rays to reach its ground state as depicted in Figure 9.1. The sum of the energies of the gammas is, for all practical purposes, equal to the excitation energy; i.e.,

$$E^* = \sum_i (E_\gamma)_i. \tag{9.6}$$

This equation is not rigorously correct because no account was taken of momentum conservation in the decay of the compound nucleus. As we will now show, it is nevertheless extremely close to being correct because the momentum carried off by a photon, namely, E_γ/c, imparts negligible recoil energy to a nucleus. Consider a nucleus of mass M that has emitted a photon of energy E_γ. Conservation of momentum is expressed by

$$\frac{E_\gamma}{c} = \sqrt{2M\,\Delta E}, \tag{9.7}$$

where ΔE is the recoil kinetic energy of the nucleus. Hence

$$\frac{\Delta E}{E_\gamma} = \frac{E_\gamma}{2Mc^2}. \tag{9.8}$$

Now Mc^2 is at least 2×931 Mev for the lightest gamma-emitting nucleus and more like 100×931 Mev on the average. On the other hand, E_γ is of the order of 1 to 10 Mev; hence the ratio $\Delta E/E_\gamma$ is smaller than 10^{-2} and usually of the order 10^{-5}. The value of ΔE is therefore negligible compared with the energy of the gamma. Since very little of the energy of the gamma ray must be expended in conserving momentum, the center-of-momentum and laboratory energies of the gamma ray will be practically identical.

Isomeric Transitions

Very often gamma transitions between the first or second excited level of a nucleus and the ground state of the nucleus are forbidden, and the

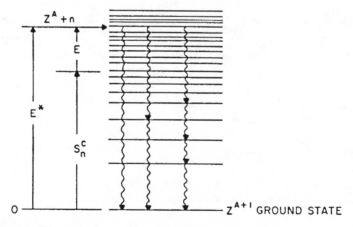

Figure 9.1—Schematic diagram showing the energy of excitation, E^*, of the compound nucleus formed by the capture of a neutron of energy E. Most nuclei exhibit many more modes of radiative deexcitation than are shown in this diagram.

nucleus must find some alternative mode of decay. We shall examine some of these alternative modes shortly. First, however, we shall define the word "forbidden." The word forbidden is applied in two senses to quantum-mechanical transitions. In the first, and most rigorous, sense, it means that a certain transition cannot occur because it violates some conservation law. In the second, loose but more common, usage, it means that a certain transition is highly unlikely; i.e., the transition has a half-life that is very long relative to other possible transitions. We shall distinguish between these two meanings by using the words "absolutely forbidden" to mean that the transition cannot occur.

In Chapter 8 we encountered an example of an absolutely forbidden transition. The transition between the first excited state and the ground state of ^{40}Ca by single gamma emission is absolutely forbidden by the law of conservation of total angular momentum because both states have $I = 0$, and a gamma ray must carry off at least one unit of angular momentum. In general, $0 \rightarrow 0$ single gamma transitions are absolutely forbidden. It is possible for a transition to occur with the simultaneous emission of two gammas, but this is extremely unlikely when compared with other modes of decay, such as internal conversion, pair emission, or beta decay.

Any gamma transition that involves a large change in angular momentum between the initial and final energy levels has a long half-life and is therefore forbidden in the second sense. Two examples are shown in Figure 9.2. The isomeric states ^{77}Ge and ^{77m}Ge are formed in the reaction ^{76}Ge$(n,\gamma)^{77}$Ge, ^{77m}Ge. With 0.025-ev neutrons, the cross sections for the formation of both isomers are approximately equal. This implies that the gamma cascades from the higher energy levels have equal probability of ending at the ground state or at the first excited state. Gamma deexcitation of ^{77m}Ge occurs but has a relatively long half-life because the isomeric transition $1/2^- \rightarrow 7/2^+$ requires the gamma to carry off at least three units of angular momentum. In addition to the isomeric transition,^{77m}Ge decays to the ground state and the first excited state of ^{77}As by β^- emission. The product nucleus ^{77}As also has an isomeric level at 0.473 Mev.

The isomeric states ^{115}Cd and ^{115m}Cd are formed in the reaction ^{114}Cd$(n,\gamma)^{115}$Cd, ^{115m}Cd with 0.025-ev cross sections of 1.1 and 0.14 barns, respectively. Because the isomeric transition $11/2^- \rightarrow 1/2^+$ would require the gamma ray to carry off at least five units of angular momentum, its half-life is of the order of 10^5 years. This transition has never been detected. Instead the much shorter half-life (43-day) beta transitions to the ground and excited states of ^{115}In, as pictured in Figure 9.2, are the preferred modes of deexcitation. It is interesting to note that the lowest two levels in the product nucleus ^{115}In also form an isomeric pair.

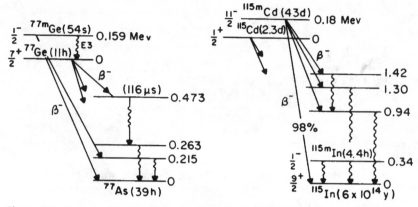

Figure 9.2—Decay schemes of the nuclear isomers of ^{77}Ge and ^{115}Cd. [From C. M. Lederer, J. M. Hollander, and I. Perlman, *Table of Isotopes*, 6th ed., John Wiley & Sons, Inc., New York, 1967.]

The existence of isomeric levels has several consequences of practical significance to the nuclear engineer. If the isomeric level has a long half-life, then the betas and gammas that emerge from the decay of this level may constitute appreciable decay-heat sources after reactor shutdown. On the other hand, if the isomeric level has a relatively short half-life, of the order of seconds, and decays primarily by gamma emission to the ground state, then this gamma should be included in the capture gamma spectrum during reactor operation and in the decay gamma spectrum at short times after shutdown.

9.2 Measured Capture Gamma Spectra

Capture gamma spectra resulting from the absorption of thermal neutrons have been measured for most elements and a few selected isotopes. Relatively few spectra have been measured for other than thermal-neutron captures, but the emphasis is shifting to the nonthermal area, and we may confidently expect sufficient data for nuclear engineering purposes within the next decade. Let us first examine some of the more salient features of thermal-neutron-capture gamma spectra.

Gamma Spectra from Thermal-Neutron Capture

The mode of decay of a compound nucleus after the capture of a thermal neutron is characteristic of that nucleus, but one can still form general classifications of the types of capture gamma-ray spectra yielded by various nuclei. There are three major types of capture gamma-ray spectra.*

*P. S. Mittleman and R. A. Liedtke, *Nucleonics*, **13** (5): 50–51 (1955).

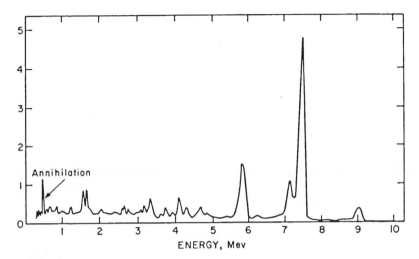

Figure 9.3—Type 1 capture gamma-ray spectrum for iron shows few gamma rays. Most of the deexcitation energy is carried by a single 6- to 8-Mev gamma ray. The 0.51-Mev line is the result of positron annihilation. (From L. V. Groshev et al., *Atlas of Gamma-Ray Spectra from Radiative Capture of Thermal Neutrons*, Pergamon Press, New York, 1959.)

These spectra are determined by the detailed character of the states of the residual nucleus and their relation to the capturing state.

Type 1 spectra have few gamma rays. The ground-state transition line dominates. Most of the energy is carried by a single 6- to 8-Mev gamma ray. This type of spectrum is illustrated in Figure 9.3. Iron-57 is a typical example since it emits about 35% of its gammas in a single transition from the excited state to the ground state. The gamma-ray transition diagram for ^{57}Fe is shown in Figure 9.4.

Type 2 capture gamma-ray spectra have many gamma rays, but distinct line structure is evident. This is typical of light- and medium-weight elements with fairly large spacing between levels where transitions between levels are as likely as transitions to the ground state. This type of spectrum is illustrated in Figure 9.5, for ^{52}V, which is a typical example of an isotope that yields a spectrum of this type. The corresponding transition diagram is given in Figure 9.6.

Type 3 capture gamma-ray spectra have many gamma rays and no evident line structure below 5 Mev. This is typical of the heavier elements; these have high level density, and the probability of transition to any level becomes a statistical quantity, i.e., transition to almost any energy is possible. Figure 9.7 shows this type of spectrum, and Figure 9.8 is the corresponding transition diagram.

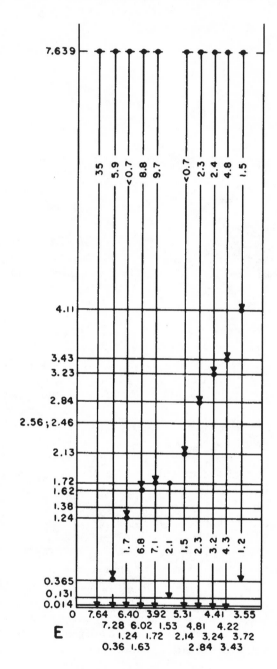

Figure 9.4—Gamma-ray transition diagram for ^{57}Fe formed by the capture of a thermal neutron in ^{56}Fe. The ground-state transition line is dominant. The numerals within the lines indicate the number of gammas per 100 thermal-neutron captures. (From L. V. Groshev et al., *Atlas of Gamma-Ray Spectra from Radiative Capture of Thermal Neutrons*, Pergamon Press, New York, 1959.)

Note that the number of gamma rays in the neutron-capture spectrum depends on the number of states below the initial state to which transition can occur. For instance, in the lightest product nuclei, ^{2}H and ^{3}H, only one such state, the ground state, is available, and the capture gamma spectrum consists of only one line. As the mass number increases, the number of energy levels below the neutron separation energy rapidly becomes larger, and the spectrum ultimately becomes very complex. There are a number of references that list information on gamma-ray spectra, but of these probably the most exhaustive is the *Atlas of Gamma-ray Spectra from Radiative Capture of Thermal Neutrons*, compiled by L. V. Groshev, V. N. Lutsenko, A. M. Demidov, and V. I. Pelekhov, Pergamon Press, New York, 1959.

A word of warning: You will find if you examine some of the compilations of thermal-neutron capture gamma spectra that the sum total of all the listed gamma energies multiplied by their individual frequencies does not add properly to the separation energy from the compound nucleus. In some cases the discrepancy is 20% or more. The discrepancies

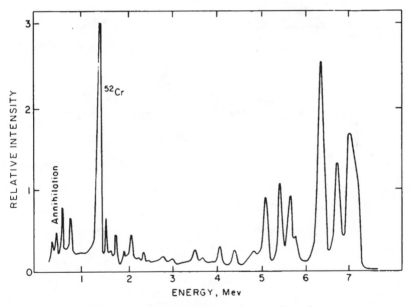

Figure 9.5—Type 2 capture gamma-ray spectrum for vanadium shows many gamma-rays. Distinct line structure is evident. This is typical of light- and medium-weight elements. The strong line at 1.4 Mev is a decay gamma from the reaction ^{52}V → ^{52}Cr + β^- + γ. (From L. V. Groshev et al., *Atlas of Gamma-Ray Spectra from Radiative Capture of Thermal Neutrons*, Pergamon Press, New York, 1959.)

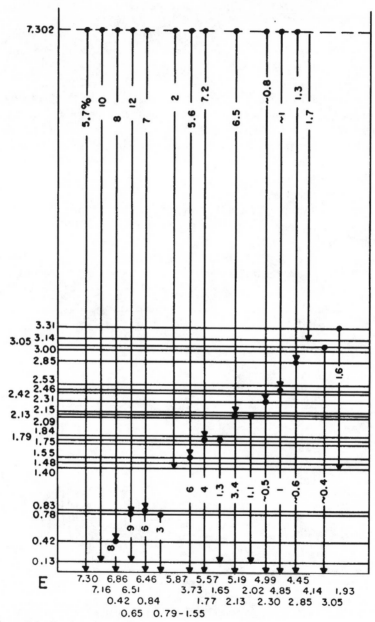

Figure 9.6—Gamma-ray transition diagram for type 2 spectrum for ⁵²V. Typical of elements with fairly large level spacing between levels and transitions between levels as likely as transitions to ground. (From L. V. Groshev et al., *Atlas of Gamma-Ray Spectra from Radiative Capture of Thermal Neutrons*, Pergamon Press, New York, 1959.)

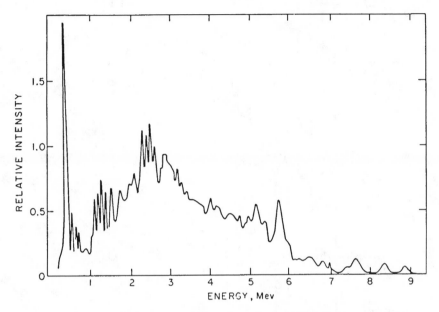

Figure 9.7—Type 3 capture gamma-ray spectrum for cadmium shows many gamma rays but no line structure below 5 Mev. This is typical of heavier elements. (From L. V. Groshev et al., *Atlas of Gamma-Ray Spectra from Radiative Capture of Thermal Neutrons,* Pergamon Press, New York, 1959.)

are the result of experimental difficulties, particularly in the measurement of the numbers of low-energy gammas. Since the total energy emitted is of great importance in some nuclear engineering applications, e.g., the calculation of heating rates, the data should be suitably renormalized so as to conserve total energy. Usually this can be done by referring to the original papers to find the method of measurement and its inherent uncertainties.

Gamma Spectra from Nonthermal-Neutron Capture

At the present time very few experimental data are available on the capture gamma spectra from nonthermal-neutron capture. But what is available indicates quite clearly that significant changes can take place in the spectra, particularly in discrete transitions from the capturing state to the vicinity of the ground state. Prominent lines in the thermal capture spectrum may be completely suppressed in capture at a resonant energy because of the operation of angular-momentum selection rules when the resonance capturing state has a different J^{Π} from the thermal capturing state. This matter is discussed further in Section 9.6.

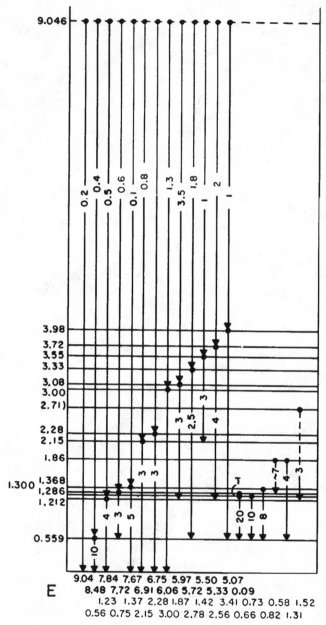

Figure 9.8—Gamma-ray transition diagram for type 3 spectrum for [114]Cd. This is typical of elements with a great density of levels near the capturing state. (From L. V. Groshev et al., *Atlas of Gamma-Ray Spectra from Radiative Capture of Thermal Neutrons,* Pergamon Press, New York, 1959.)

The nuclear engineer should be aware that the common practice of assuming that the capture spectrum is independent of neutron energy is generally invalid and may have serious consequences in such things as radiation heating and shielding calculations. The nuclear engineer should search out data on nonthermal capture spectra if a significant fraction of all captures occur in the nonthermal region, as for example in epithermal and fast reactors.

9.3 Radiative-Capture Cross Sections

In the present state of nuclear theory, the degree of accuracy in the determination or the estimation of σ_γ is very limited and yields results only to an order of magnitude. So we must turn to experimental data to study the general behavior of the (n,γ) cross section as a function of energy and atomic mass. Radiative-capture cross sections have been measured for almost all elements, as well as for a few separated isotopes, at thermal and epithermal energies and at a few widely separated energies (e.g., 1, 4.5, and 14 Mev) in the million-electron-volt region. As time-of-flight apparatus becomes more widespread, we may expect the gaps eventually to be completely filled in.

At the low end of the neutron-energy scale, well below the first capture resonance, σ_γ is found to be inversely proportional to the velocity of the neutron, or approximately so. The radiative-capture cross section assumes the typical Breit–Wigner resonance shape in the vicinity of isolated resonances. At still higher energies the resonances overlap and can no longer be distinguished. Furthermore, other reactions begin to compete with radiative capture for their share of the compound-nucleus cross section, hence σ_γ decreases to very small values at still higher energies. Figure 9.9 qualitatively illustrates the usual behavior of the radiative-capture cross section as a function of energy. The region of isolated resonances begins approximately in the million-electron-volt region in light nuclei, the kilo-electron-volt region in medium nuclei, and the electron-volt region in very heavy nonmagic nuclei.

The Region of Isolated Resonances

In medium and heavy nuclei, the isolated resonances are practically always found at such low energies that only $l = 0$ interactions need be considered. The (n,γ) cross section for a single isolated $l = 0$ resonance is given by the familiar Breit–Wigner dispersion formula

$$\sigma_\gamma(E) = \pi \lambda^2 \frac{2J + 1}{2(2I + 1)} \frac{\Gamma_n(E)\,\Gamma_\gamma}{(E - E_0)^2 + (\Gamma/2)^2}, \tag{9.9}$$

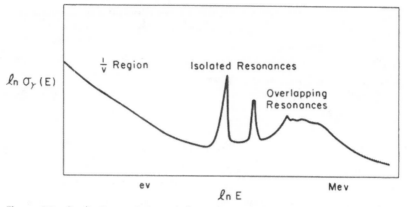

Figure 9.9—Qualitative variation of the radiative-capture cross section. In heavier nuclei there are many more isolated resonances than are shown here, and they occur at much lower energies.

where I is the spin of the target nucleus and J is the spin of the compound nucleus, either $I + 1/2$ or $I - 1/2$ depending on the spin of the level that happens to be available at the excitation energy. If $I = 0$, only $J = 1/2$ is possible.

The radiative-capture width, Γ_y, is generally of the order of 0.5 to 0.1 ev in the medium-mass nuclei and is somewhat smaller (0.1 to 0.03) in the heavy nuclei. It exhibits little variation from level to level in a given nuclide. These widths imply, through the uncertainty principle $\Gamma_y \tau_y \simeq \hbar$, mean radiative lifetimes of the order of 10^{-13} to 10^{-15} sec. These lifetimes are, with few exceptions, orders of magnitude smaller than those for other possible deexcitation processes, such as proton and alpha emission, which might follow slow-neutron capture. This is the principal reason why radiative capture is the most common reaction triggered by slow neutrons. Of course there are exceptions: In a few very light elements Γ_p and Γ_α are sometimes as large as, or larger than, Γ_y, and in the very heavy fissile elements the width for fission, Γ_f, often exceeds Γ_y.

The fact that radiative-capture resonances are found experimentally to have the symmetrical Breit–Wigner form leads one to the conclusion that there is little or no interference between levels. The basic reason for this lack of interference is the large number of exit channels. Each of the many possible gamma transitions from the capturing level to a lower level represents a distinct exit channel. For processes involving a large number of exit channels, the rigorous R-matrix multilevel formula reduces to the single-level Breit–Wigner form.

It is reasonable that Γ_y should be approximately constant from level to level within a given nuclide because radiation can lead from the compound nucleus to many hundreds of levels of lower energy. It would be most surprising if the widths of all these channels varied in the same direction simultaneously from one capturing level to another. This is not to say that there will not be great variations in the gamma-ray spectra from level to level; such variations have indeed been experimentally detected. Rather we are saying that the total radiation width, which is a sum over all the partial radiation widths,

$$\Gamma_\gamma = \sum_j \Gamma_{\gamma j},$$ (9.10)

is expected on simple statistical grounds to be approximately constant and is, indeed, found to be so.

The 1/v Variation of $\sigma_\gamma(E)$

In the very low neutron-energy region, the (n,γ) cross sections of most elements vary as the inverse of the neutron's velocity. This $1/v$ behavior obtains because the energy levels in the compound nucleus near the neutron separation energy S_n^c are widely separated compared with the widths Γ_γ of the levels. By "widely separated" we mean $D \gg \Gamma_\gamma$, but since Γ_γ is only a fraction of an electron volt, the average separation D need only be a few electron volts. With a separation of the order of 10 ev, it is highly unlikely that a level will occur within a fraction of an electron volt of the zero of neutron energy. The most striking and familiar exception to this expectation is the level at 0.178 ev in the target nucleus ^{113}Cd.

The $1/v$ behavior of the (n,γ) cross section can be readily derived as a special case of the Breit–Wigner single-level formula for an $l = 0$ resonance,

$$\sigma_\gamma(E) = \pi \lambda^2 g(J) \frac{\Gamma_n(E)\Gamma_\gamma}{(E - E_0)^2 + (\Gamma/2)^2}.$$

Under the assumption that the resonance is located at an energy E_0 that is much greater than E and Γ, the denominator is practically a constant. Recalling that the neutron width is given by

$$\Gamma_n(E) = \Gamma_n(E_0)\sqrt{E/E_0},$$

we reduce the Breit–Wigner expression to

$$\sigma_\gamma(E) = \frac{\pi\hbar^2}{2mE} g(J) \sqrt{\frac{E}{E_0}} \frac{\Gamma_n(E_0)\Gamma_\gamma}{E_0^2}.$$

All terms on the right side of this equation are constant except $\sqrt{E}/E$. Thus, in the absence of nearby resonances, the single-level $l = 0$ Breit–Wigner cross section varies as $1/v$. The capture cross section near the zero of neutron energy is the sum of the contributions from the tails of all the individual Breit–Wigner resonances, both positive and negative. These contributions add incoherently because the compound nucleus can break up by emission of a variety of different gammas, each one of which represents a distinct exit channel. Interference between channels is therefore negligible.

The Region of Overlapping Resonances

As the neutron energy increases toward the intermediate energy range (1 kev $\geq E \geq 0.5$ Mev in heavy nuclei), the radiative cross sections decrease but are still appreciable. If it were possible to experimentally resolve the individual resonances, it would make sense to analyze them using individual Breit–Wigner formulas. Actually the required resolution of an electron volt or so at several kilo-electron-volts is not yet within the capability of our experimenters; so we attempt to estimate the average cross section over many resonances. That is, we reproduce analytically the broad resolution that is presently feasible experimentally.

Consider first the area under a single resonance given by

$$\sigma_i = 2 \int_{E_0 - \Gamma/2}^{E_0 + \Gamma/2} \sigma_\gamma(E) \, dE, \tag{9.11}$$

where σ_i is the integrated cross section. We shall assume that $\sigma_\gamma(E)$ is given by the Breit–Wigner $l = 0$ formula (Equation 9.9) and that λ, $\Gamma_n(E)$, and Γ are approximately constant over the energy range of the resonance. Thus

$$\sigma_i = 2\pi\lambda^2 g(J)\Gamma_n\Gamma_\gamma \int_{E_0 - \Gamma/2}^{E_0 + \Gamma/2} \frac{dE}{(E - E_0)^2 + (\Gamma/2)^2}$$

$$= 2\pi\lambda^2 g(J)\Gamma_n\Gamma_\gamma \frac{1}{\Gamma/2} \left[\arctan \frac{E - E_0}{\Gamma/2} \Big|_{E_0 - \Gamma/2}^{E_0 + \Gamma/2} \right. \tag{9.12}$$

or

$$\sigma_i = 2\pi^2\lambda^2 g(J) \frac{\Gamma_n\Gamma_\gamma}{\Gamma}. \tag{9.13}$$

Consider now the effective cross section for a neutron beam with an energy spread ΔE. If D is the mean level spacing, the number of levels in this range of energy is $\Delta E/D$ (this is also the number of resonances). The average capture cross section, $\langle \sigma_\gamma \rangle$, then is $\Delta E/D$ times the average of the integrated cross section divided by the energy interval ΔE. Thus if Equation 9.13 is used in conjunction with the above arguments,

$$\langle \sigma_\gamma \rangle = 2\pi^2 \lambda^2 \left\langle \frac{g(J)\Gamma_n\Gamma_\gamma}{\Gamma D} \right\rangle . \tag{9.14}$$

A further simplification is possible based on the observation that in most nuclei in the intermediate-energy region $\Gamma_n \gg \Gamma_\gamma$ and $\Gamma \simeq \Gamma_n$; so

$$\langle \sigma_\gamma \rangle = 2\pi^2 \lambda^2 \left\langle \frac{g(J)\Gamma_\gamma}{D} \right\rangle . \tag{9.15}$$

The above expression is the final result. Since Γ_γ and $g(J)$ are roughly independent of energy and the level spacing D decreases only slowly with energy over ranges of the order of 1 Mev or less, the main energy variation of $\langle \sigma_\gamma \rangle$ is in λ^2, which goes as $1/E$. An example of this almost $1/E$ variation is shown in Figure 9.10.

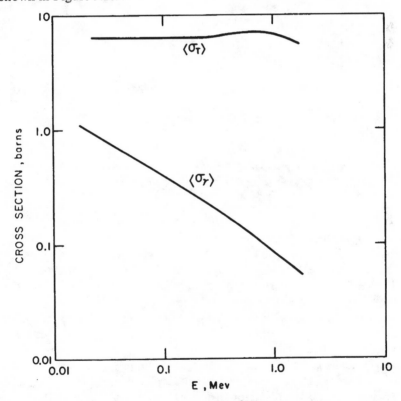

Figure 9.10—Average radiative-capture cross section, $\langle \sigma_\gamma \rangle$, of iodine in the intermediate-neutron-energy range. Also shown is the total cross section, averaged as is σ_γ by poor resolution. (From USAEC Report BNL-325.)

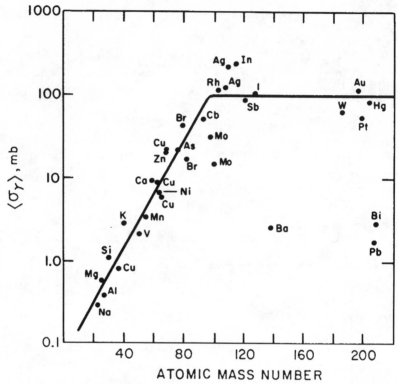

Figure 9.11—Variation of (n,γ) cross section with atomic mass number, A, at an effective energy of 1 Mev. The points lying appreciably below the smooth curve correspond to magic neutron numbers 50, 82, and 126.

Variation of the (n,γ) Cross Sections with Atomic Mass

Hughes, Spatz, and Goldstein* have made a systematic survey covering the (n,γ) cross section in 32 isotopes at an effective energy of about 1 Mev. Figure 9.11 summarizes their results, the most significant features of which are (1) a rapid increase of the (n,γ) reaction with atomic mass A (roughly exponential) from approximately 1 mb at $A = 35$ to approximately 200 mb at $A = 110$; (2) roughly constant (n,γ) cross sections of about 100 mb for $A \geq 120$; and (3) marked deviations from the norm for target nuclei containing neutron closed shells, e.g., those with 50, 82, or 126 neutrons. These magic-number nuclei have anomalously small (n,γ) cross sections as a result of their large level spacings and their small binding energies for an additional neutron.

*D. J. Hughes, W. D. B. Spatz, and N. Goldstein, *Phys. Rev.*, **75**: 1781 (1949).

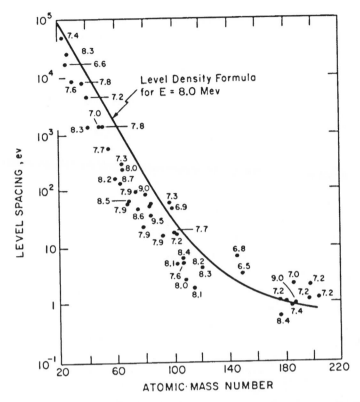

Figure 9.12—Average level spacings derived from neutron-capture cross sections at 1 Mev. The full line is the theoretical level spacing for 8-Mev excitation. The numbers refer to the excitation energies for specific nuclei in million electron volts.

The work of Hughes, Spatz, and Goldstein has been extended to many more isotopes with results that merely strengthen the conclusions outlined here. If Equation 9.15 is used in conjunction with these determinations of $\langle \sigma_\gamma \rangle$, it becomes an easy task to estimate the average level spacing D at a neutron energy of 1 Mev as a function of atomic mass. The results are shown in Figure 9.12. The abnormally large level spacings near $A = 150$ and $A = 210$ are associated with magic neutron numbers 82 and 126. It is from such data that the constants entering into the empirical level-density formula (Equation 6.4) are derived.

Although the analysis described here is based on $l = 0$ capture only, it may be expected to yield results that are better than an order of magnitude. As the neutron energy is raised, the capture cross section falls off less rapidly than $1/E$ as higher l interactions become important. But the onset of inelastic scattering tends to compensate for the higher l capture; so the analysis is not too deficient as long as it is confined to the energy range below several million electron volts.

The High-Energy Region

In the high-neutron-energy region, the behavior of $\sigma_\gamma(E)$ may be expected to differ from that described above for several reasons: (1) with the onset of inelastic-scattering and (n,p) and (n,α) reactions, the competition in the deexcitation of the compound nucleus becomes much less favorable to the radiation process, (2) as the neutron energy increases, the possibility of compound-nucleus formation by $l > 0$ neutrons becomes appreciable, (3) direct radiative capture may become the dominant mechanism.

The high-energy radiative-capture cross sections are, in any event, very small. The 14-Mev radiative-capture cross sections of intermediate and heavy elements all lie between 1 and 10 mb. These values, although so small that they can be considered zero in most practical reactor applications, are very large compared to the prediction of compound-nucleus theory (about 10^{-3} mb) and have led to the strong conviction that direct capture is responsible. If the gross forces between an incident neutron with a nucleus are describable by an optical potential, then the neutron may radiate in the region of changing potential at the edge of the nucleus. Depending on how much gamma energy it radiates, the neutron may either escape the nucleus or be captured in something resembling a compound-nucleus state but one in which much of the energy resides in shape oscillations of the charge and mass density. It can be shown that the radiative decay probability of such a "collective state" is much larger than that of a single-particle compound-nucleus state and may explain the unexpectedly large (n,γ) cross sections at 14 Mev.* There is no unequivocal experimental data leading to the belief that such effects occur at low and intermediate neutron energies.

9.4 Gamma-Ray Emission from Excited Nuclei

In this section we discuss the factors influencing the emission of gamma rays from an excited nucleus. Specifically, we seek an expression for the decay probability from one excited level, E_a, to a lower level, E_b. It is assumed that we know the spin and parity of the initial state $I_a^{\Pi_a}$ and the final state $I_b^{\Pi_b}$.

A rigorous description of the emission of radiation by a quantum-mechanical system requires the formalism of the quantum theory of

---

*A. M. Lane, Direct Radiative Capture, in *Nuclear Structure Study with Neutrons*, Reference 1 at the end of this chapter.

radiation and a detailed knowledge of the nuclear wave functions of the initial and final states between which the transition occurs. Except in the case of the lightest nuclei, like the deuteron, these wave functions are generally unknown. However, considerable information can be obtained from a study of the classical radiation field emitted by a distribution of time-varying charges and currents. In fact, the semiclassical theory of radiation borrows this representation of the radiation source (nucleus) as an oscillating electric and magnetic moment. The electric and magnetic fields and the electric charges, currents, and magnetization density are expanded in spherical harmonics of order $l = 1, 2, 3, \ldots$, and the names dipole, quadrupole, octupole, $\ldots$, 2^l-pole are given both to these nuclear moments and to the resulting radiation (see Section 5.2). Each multipole component of the field and source has ascribed to it a particular angular momentum l and z component m and an oscillating frequency ω related to the gamma energy by $E = E_a - E_b = \hbar\omega$.

On the basis of a long analysis, too long to be repeated here, it has been shown* that the transition probabilities per unit time for electric and magnetic multipole quanta are given by:

$$\lambda_{El}(m) = \frac{8\pi(l+1)}{l[(2l+1)!!]^2} \frac{1}{\hbar}\left(\frac{\omega}{c}\right)^{2l+1} |G_E(l,m)|^2, \tag{9.16}$$

$$\lambda_{Ml}(m) = \frac{8\pi(l+1)}{l[(2l+1)!!]^2} \frac{1}{\hbar}\left(\frac{\omega}{c}\right)^{2l+1} |G_M(l,m)|^2, \tag{9.17}$$

where $(2l+1)!! \equiv 1 \cdot 3 \cdot 5 \cdots (2l+1)$. The quantities $G_E(l,m)$ and $G_M(l,m)$ are the matrix elements of the multipole operator of order l,m for electric and magnetic radiation, respectively. The matrix element for a transition from state ψ_a to ψ_b in a system of particles is given by the integral (see Section 3.3)

$$G(l,m) = \int \bar{\psi}_b Q(l,m)\psi_a \, d\tau, \tag{9.18}$$

where $\bar{\psi}_b$ is the complex conjugate of ψ_b, $d\tau$ is the volume element, and $Q(l,m)$ is an operator whose specific form depends on the nature of the transition, electric or magnetic, and its multipolarity. These operators are well known, but the wave functions in Equation 9.18 are unknown. To obtain numerical values for the transition probabilities, one must adopt a model for the nucleus, one that is sufficiently simple to allow an estimate of the wave functions. The independent-particle model is the most elementary since a nuclear transition is considered to involve the change in

*See Chapter XII, J. M. Blatt and V. F. Weisskopf, *Theoretical Nuclear Physics*, Reference 2 at the end of this chapter.

the quantum state of only one nucleon, the "valence" proton or neutron. Under simplifying assumptions for the radial dependence of the wave function of this single nucleon, we obtain the following estimate for the transition probability of an electric 2^l pole

$$\lambda_{El} = (2.4 \times 10^{21}) f_l R^{2l} \left(\frac{E}{197}\right)^{2l+1} \text{sec}^{-1},$$

(9.19)

and, for a magnetic 2^l pole,

$$\lambda_{Ml} = (1.1 \times 10^{21}) f_l R^{2l-2} \left(\frac{E}{197}\right)^{2l+1} \text{sec}^{-1}.$$

(9.20)

In these equations R is the nuclear radius in fermis, E is the gamma energy in million electron volts, and f_l is given by

$$f_l = \frac{2(l+1)}{l[(2l+1)!!]^2} \left(\frac{3}{l+3}\right)^2$$

(9.21)

and has the following values:

l	f_l
1	2.5×10^{-1}
2	4.8×10^{-3}
3	6.3×10^{-5}
4	5.3×10^{-7}
5	3.1×10^{-9}

The transition probabilities λ_l were obtained by summing $\lambda_l(m)$ over all m_b substates of the final state b and averaging $\lambda_l(m)$ over all m_a substates of the initial state a.

The expressions for the transition probabilities displayed above are very approximate, valid within about one or two orders of magnitude. Nevertheless, significant information can be gleaned from their examination. First, we note that f_l is a rapidly decreasing function of l, an increase of one unit of angular momentum decreases f_l by a factor of about 100. In consequence, transitions will generally proceed by emission of a gamma of the lowest allowed l value. Next, we notice that for a given value of l, electric radiation is more probable than magnetic radiation in the ratio

$$\frac{\lambda_{El}}{\lambda_{Ml}} \simeq 2R^2,$$

(9.22)

which for a medium-mass nucleus is approximately 30; so electric radiation is favored over magnetic radiation all else being equal. We note, finally, the strong dependence of the transition probability on the energy of the gamma ray. Consequently, high-energy transitions, for example, between the excited state and the ground state, are favored over low-energy transitions, all else being equal.

It is instructive to examine the order of magnitude of the mean lifetime as a function of the type of radiation and its polarity. Substituting the approximation $R = 1.4A^{1/3}$ into Equations 9.19 and 9.20, we find that for $l = 1$

$$\lambda_{E1} = 1.5 \times 10^{14} A^{2/3} E^3 \text{ sec}^{-1}, \tag{9.23a}$$

$$\lambda_{M1} = 2.8 \times 10^{13} E^3 \text{ sec}^{-1}, \tag{9.23b}$$

and for $l = 2$

$$\lambda_{E2} = 1.6 \times 10^8 A^{4/3} E^5 \text{ sec}^{-1}, \tag{9.23c}$$

$$\lambda_{M2} = 1.2 \times 10^8 A^{2/3} E^5 \text{ sec}^{-1}, \tag{9.23d}$$

and, finally, for an extreme angular-momentum change, $l = 5$,

$$\lambda_{E5} = 1.6 \times 10^{-11} A^{10/3} E^{11} \text{ sec}^{-1}, \tag{9.23e}$$

$$\lambda_{M5} = 7.5 \times 10^{-11} A^{8/3} E^{11} \text{ sec}^{-1}. \tag{9.23f}$$

The reciprocals of these transition probabilities are the mean lifetimes for these various types of radiation. It can be seen that the mean life for the emission of 1-Mev electric dipole ($E1$) radiation from a medium-weight ($A \simeq 80$) nucleus will be of the order of 10^{-16} sec; magnetic dipole ($M1$) radiation will have a mean lifetime of the order of 10^{-14} sec; electric quadrupole ($E2$), about 10^{-11} sec; magnetic quadrupole, about 10^{-9} sec; and so on. When the change in angular momentum of the nucleus is small, the transition probability is large and the mean life is short. As the change in angular momentum becomes larger, the mean life increases dramatically. For a change of five units, the mean lifetimes for emission of $E5$ and $M5$ radiation (1 Mev, $A = 80$) are of the order of 10^5 and 10^7 sec, respectively.

Selection Rules for Gamma-Ray Emission

A photon emitted in radiative transition carries away a certain number of units of orbital angular momentum, and this angular momentum is designated by the same type of quantum numbers (l,m) as the orbital angular momentum of material particles. The major difference is that for photons l can have only nonzero values since there is no multipole radiation with $l = 0$. Thus, the angular momentum of a photon is $\hbar[l(l + 1)]^{1/2}$, and its possible projections on any arbitrary axis are $m\hbar$ with maximum projection $l\hbar$, $l = 1, 2, 3, \dots$.

If the photon is emitted by a nucleus in going from a total angular-momentum state $\mathbf{I}_a$ to a state $\mathbf{I}_b$, then, since total angular momentum must be conserved,

$$\mathbf{l} = \mathbf{I}_a - \mathbf{I}_b. \tag{9.24}$$

That is, the angular-momentum quantum numbers l and m of the radiation must obey the selection rules

$$|I_a - I_b| \leq l \leq I_a + I_b, \tag{9.25}$$

$$m = m_a - m_b, \tag{9.26}$$

where m_a and m_b are the z components of $\mathbf{I}_a$ and $\mathbf{I}_b$. Since l can never take on the value zero, it is clear from Equation 9.25 that radiative transitions of the $0 \to 0$ type are absolutely forbidden.

Now let us consider the selection rules that arise from parity conservation. From the parities of the operators that give rise to the transitions, we can show that the parity of a gamma emitted in an electric multipole transition is given by

$$\Pi_y = (-1)^l \quad \text{(electric)} \tag{9.27}$$

and in a magnetic multipole transition by

$$\Pi_y = (-1)^{l+1} \quad \text{(magnetic)}. \tag{9.28}$$

Parity conservation in the transition from nuclear state $I_a^{\Pi_a}$ to state $I_b^{\Pi_b}$ demands that

$$\Pi_a = \Pi_b \Pi_y. \tag{9.29}$$

Thus, for example, gamma transitions between two states of the same parity ($\Pi_a = \Pi_b$) must take place by emission of an even-parity gamma. These are, from Equations 9.27 and 9.28, gammas of the type $E2$, $E4$, $E6$, ... and $M1$, $M3$, $M5$, ... Gamma transitions between states of opposite parity ($\Pi_a = -\Pi_b$) must take place by emission of an odd-parity gamma of the type $E1$, $E3$, $E5$ ... and $M2$, $M4$, $M6$ ....

When the selection rule arising from the conservation of angular momentum is combined with that arising from the conservation of parity and account is also taken of the extreme dependence of the transition probabilities on l, we can guess quite accurately what sort of radiation will be emitted in a transition between levels of known spin and parity. For example, what radiation will most likely be emitted in the transition from level $5/2^+$ to level $3/2^-$? From Equation 9.25 we see that the possible values of l are 1, 2, 3, and 4. But conservation of parity requires that, since $\Pi_a = -\Pi_b$, the quantum number l must be odd for electric radiation and even for magnetic radiation. Thus, $E1$, $M2$, $E3$, and $M4$ are the only possible radiations. However, the probability of an $E1$ transition is several orders of magnitude greater than the probability of any of the other possible transitions. Thus we can state with certitude that the radiation will be almost entirely electric dipole, $E1$.

Usually, as in this example, the transition proceeds by the lowest allowed l value. Occasionally the transition probability for electric radiation of order $l + 1$ is comparable to that for magnetic radiation of order l, and both transitions occur with frequencies proportional to their relative transition probabilities.

At this point the reader may wish to turn back and reread the part of this chapter dealing with isomeric transitions. The reasons for the existence of isomeric levels and the magnitudes of their half-lives will now be quite obvious.

Collective Transitions

The above evaluation of transition probabilities was based on the independent-particle model, which postulates that the "valence" nucleon moves in a static field produced by all the other particles in the nucleus. Another nuclear model that allows the evaluation of transition probabilities is the collective model. In this model the core of the nucleus is treated as a system of particles bound together so that they have the degrees of freedom of a hydrodynamical system; i.e., the core can exhibit collective motions, such as vibration and rotation.*

The collective model assumes that the surface of an ellipsoidal nucleus may vary in time in a wavelike way so that the nucleus gives the appearance of rotation. This motion will be quantized and give rise to low-lying rotational energy levels, in analogy with those of a rigid rotator. The energies of these rotational levels are given by

$$E_{rot} = \frac{\hbar^2}{2I_m}[I(I + 1) - I_0(I_0 + 1)], \tag{9.30}$$

where I_m is the effective moment of inertia of the nucleus and I_0 and I are the total angular-momentum quantum numbers of the ground state and the excited state, respectively. The value of I^{π} is restricted to certain values depending on I_0, e.g., when $I_0 = 0$

$I = 2, 4, 6$ (all even parity),

when $I_0 \neq 0$ or $1/2$

$I = I_0 + 1, I_0 + 2, \ldots$ (all the same parity),

*A. Bohr and B. R. Mottelson, *Kgl. Danske. Videnskab. Selskab. Mat.-Fys. Medd.*, **27**: No. 16 (1953).

and when $I_0 = 1/2$ there is an additional term in Equation 9.30 equal to

$$\Delta E_{rot} = \frac{\hbar}{2I_m}(-1)^{I-j+1}\left[j + \frac{I}{2} + \frac{1}{2}\right]$$ (9.31)

where j is the total angular momentum of the last unpaired nucleon.

Since successive rotational levels have a spin difference of 1 or 2 and have the same parity, rotational transitions will be mainly of the electric quadrupole $E2$ or magnetic dipole $M1$ type. In an even-even nucleus, since $\Delta I = 2$ between successive levels and all levels have even parity, the excited levels decay by a cascade of pure $E2$ radiation.

The transition probability for collective transitions is proportional to the square of the intrinsic quadrupole moment of the nucleus; thus it is generally only highly deformed nuclei that exhibit appreciable decay through collective rearrangements.

The matrix elements for the electric and magnetic multipole transitions in the collective model contain contributions both from transitions by individual nucleons and from multipole moments generated by the collective motions of the nuclear core. In highly deformed nuclei, transition probabilities for radiation of the collective type may exceed that of a single-particle type by an order of magnitude or more.

9.5 Statistical Theory of Gamma-Ray Spectra

Given an ideal situation in which the energy, spin, and parity of the capturing state and of all lower-lying states in the compound nucleus are known, it would be possible to work out in detail the capture gamma spectrum. The transition probabilities, together with the selection rules on spin and parity, would completely determine the cascade dynamics. The calculations would be long and tedious but basically very elementary. We might, for example, begin the analysis by calculating the probabilities of decay from the capturing state to all accessible lower states. Then the same calculation could be carried out for the second highest state, the probabilities of decay now being multiplied by the nuclear level population of the second state, which is the probability that the second highest state was excited by decay from the upper state. We could proceed stepwise in this fashion from level to level all the way to the ground state. At each step in the calculation, we would use the following relations:

$$P(E_a \rightarrow E_b) = \frac{\Gamma_\gamma(E_a \rightarrow E_b)}{\Gamma_\gamma(E_a)} = \frac{\lambda_\gamma(E_a \rightarrow E_b)}{\lambda_\gamma(E_a)},$$ (9.32)

where

$$\Gamma_\gamma(E_a) = \sum_{b=0}^{a-1} \Gamma_\gamma(E_a \to E_b), \tag{9.33a}$$

$$\lambda_\gamma(E_a) = \sum_{b=0}^{a-1} \lambda_\gamma(E_a \to E_b), \tag{9.33b}$$

and

$$W(E_b) = \sum_{a=b+1}^{N} P(E_a \to E_b)W(E_a). \tag{9.34}$$

Equation 9.32 merely defines the probability of a transition between an upper level E_a and any lower level E_b in terms of either the partial widths or the transition probabilities. Equations 9.33 are definitions of the total width and decay probability, and Equation 9.34 gives the nuclear level population of any level E_b due to transitions from all upper levels. The population of the capturing state $W(E_c)$ may be set to unity, or 100, or any other desired normalization. The denominator of Equation 9.32 must be suitably modified if there are modes of deexcitation of the levels other than gamma emission.

A slight complication arises if one considers compound-nucleus excitations in which either the mean level spacing is so small that the states effectively form a continuum or the states are reasonably discrete but unresolved in energy, spin, and parity. In this event, the sums over discrete terms are replaced by integrals weighted by the level density $\rho(E)$; thus

$$P(E' \to E) = \frac{\Gamma_\gamma(E' \to E)\rho(E)}{\Gamma_\gamma(E')}, \tag{9.35}$$

where

$$\Gamma_\gamma(E') = \int_{E_r}^{E'} \Gamma_\gamma(E' \to E)\rho(E)\, dE + \sum_{b=1}^{N} \Gamma_\gamma(E' \to E_b) \tag{9.36}$$

and

$$W(E) = P(E_c \to E) + \int_{E}^{E_c} W(E')P(E' \to E)\, dE'. \tag{9.37}$$

The summation in Equation 9.36 accounts for the N resolved levels. The energies E_r and E_c are the lower bound of the unresolved region and the energy of the capturing state, respectively. The first term in Equation 9.37 accounts for direct transitions from the capturing state into the continuum. In the resolved energy region, the discrete level populations are given by

$$W(E_b) = P(E_c \rightarrow E_b) + \int_{E_r}^{E_c} W(E')P(E' \rightarrow E_b)\,dE'$$

$$+ \sum_{a=b+1}^{N} W(E_a)P(E_a \rightarrow E_b). \quad (9.38)$$

At this point the reader might well entertain the conviction that, although the above development is obviously correct, it is practically useless because the spins and parities of the levels in the continuum are unknown and are likely to remain so. Actually, calculations of capture gamma spectra can be carried out, and in fact have been successfully carried out, despite the absence of detailed information about the continuum. This is feasible because the details of the continuum have little effect on the capture gamma spectrum. The transition probabilities (Equations 9.23) strongly favor high-energy gamma transitions, i.e., transi-

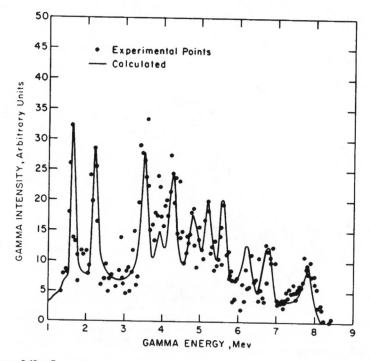

Figure 9.13—Comparison between experimental and calculated neutron-capture gamma-ray spectra for the 89-kev ($J^{\pi} = 3^{+}$) resonance in ^{27}Al. [From K. J. Yost, *Nucl. Sci. Eng.*, **32**: 62 (1968).]

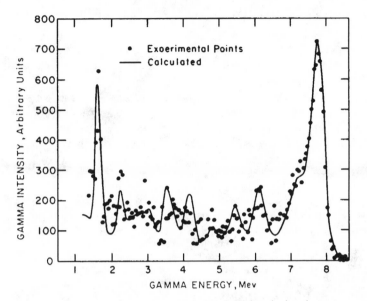

Figure 9.14—Comparison between experimental and calculated neutron-capture gamma-ray spectra for the 120-kev ($J^{\pi} = 2^{-}$) resonance in ^{27}Al. [From K. J. Yost, *Nucl. Sci. Eng.*, **32**: 62 (1968).]

tions to levels at or near the ground state. Thus the energies, spins, and parities of the low-lying states and of the capturing state are primarily responsible for the shape of the capture gamma spectrum. The continuum states are of secondary importance since transitions between continuum states are relatively unlikely.

The density of energy levels of spin J at energy E can be estimated from the semiempirical equations displayed in Chapter 6,

$$\rho(E,J) = \{(2J + 1) \exp\left[-J(J + 1)/2\sigma^2\right]\}[C \exp 2\sqrt{aE}], \quad (9.39)$$

and the parities can be assigned randomly with equal probability. Alternatively, the distribution of partial decay widths in the continuum may be determined by sampling from an appropriate Porter–Thomas distribution.

The first calculation of the type outlined above was reported by K. J. Yost* in 1968. Figures 9.13 and 9.14 are comparisons of his calculated capture gamma spectra and the experimental spectra. To effect meaningful comparisons, Yost smoothed the calculated intensities with the response matrix used to unfold the experimental pulse-height data. In obtaining

*K. J. Yost, *Nucl. Sci., Eng.*, **32**: 62 (1968).

these results. Yost employed only $E1$, $E2$, $M1$, and $M2$ transitions. Because of the lack of sufficiently accurate transition probabilities, he used empirically determined spin branching ratios but showed that the resulting spectra were relatively insensitive to the choice of any reasonable set of branching ratios.

Figures 9.13 and 9.14 clearly show the substantial changes in spectrum shape that can occur with shifts in the spin and parity of the capturing state; in this instance, from $J^{\Pi} = 3^+$ for the 89-kev ^{27}Al resonance to $J^{\Pi} = 2^-$ for the 120-kev ^{27}Al resonance. In particular, the very strong transition to the ground state at about 7.8 Mev from capture in the 120-kev resonance is strongly suppressed in capture in the 89-kev resonance.

SELECTED REFERENCES

1. M. N. de Mevergnies, P. VanAssche, and J. Vervier (eds.), *Nuclear Structure Study with Neutrons*, North Holland Publishing Company, Amsterdam, 1966. A collection of some twenty papers from an international conference dealing with a wide range of topics in neutron interaction theory and experiment. The discussions at the end of each article are particularly valuable in pointing out areas in which the theory is far from complete. The student will want to read at least the paper by A. M. Lane, "Direct Radiative Capture," the paper by J. Vervier, "Slow Neutron Capture Gamma Rays," and the paper by J. Julien, "Determination of Neutron Resonance Parameters and Their Significance for Nuclear Physics."

2. John M. Blatt and Victor F. Weisskopf, *Theoretical Nuclear Physics*, John Wiley & Sons, Inc., New York, 1952. By all means read Chapter XII, "Interaction of Nuclei with Electromagnetic Radiation," which contains a derivation of the expressions for multipole emission probabilities.

EXERCISES

1. Show that the following three commonly encountered forms of the $l = 0$ Breit–Wigner resonance formula are equivalent:

(a) $\sigma_\gamma(E) = \pi\lambda^2 g(J) \dfrac{\Gamma_n(E)\Gamma_\gamma}{(E - E_0)^2 + (\Gamma/2)^2}$,

(b) $\sigma_\gamma(E) = \pi\lambda_0^2 g(J) \sqrt{\dfrac{E_0}{E}} \dfrac{\Gamma_n(E_0)\Gamma_\gamma}{(E - E_0)^2 + (\Gamma/2)^2}$,

(c) $\sigma_\gamma(E) = \pi\lambda\lambda_0 g(J) \dfrac{\Gamma_n(E_0)\Gamma_\gamma}{(E - E_0)^2 + (\Gamma/2)^2}$,

where $\Gamma_n(E) = \Gamma_n(E_0)\sqrt{\dfrac{E}{E_0}}$,

$\lambda = \dfrac{\sqrt{2mE}}{\hbar}$,

$\lambda_0 = \dfrac{\sqrt{2mE_0}}{\hbar}$.

2. After capturing a thermal neutron, a particular compound nuclear species always decays to the ground state by emitting two successive gamma rays of energy E_1 and E_2. It is known that the probability of emission of the second gamma in the solid angle $d\Omega(\theta)$ relative to the direction of the first is given by $P(\theta)\, d\Omega$. (a) Derive an expression for the distribution in energy of the recoil nuclei. (b) Derive an expression for the average energy of the recoil nuclei.

3. What are the most probable multipole gamma transitions between the following states?

(a) $\dfrac{11^-}{2} \to \dfrac{3^+}{2}$ (b) $\dfrac{11^-}{2} \to \dfrac{5^+}{2}$ (c) $\dfrac{5^-}{2} \to \dfrac{1^-}{2}$ (d) $\dfrac{5^-}{2} \to \dfrac{1^+}{2}$

(e) $\dfrac{9^+}{2} \to \dfrac{3^-}{2}$ (f) $\dfrac{5^+}{2} \to \dfrac{1^+}{2}$ (g) $2^- \to 2^-$ (h) $0^+ \to 0^+$

4. A fictitious nuclide has an $I^\Pi = 1/2^+$ ground state and approximately evenly spaced (1 Mev apart) excited states with $I^\Pi = 9/2^+$, $1/2^+$, $3/2^-$, $5/2^+$, and $7/2^-$ in order of ascending energy. Draw an energy-level diagram for this nuclide, and indicate on this diagram the most likely gamma transition from each level (five lines in all). Write beside the lines their multipolarity ($E1$, $M2$, etc.). Write on the excited levels their approximate mean lifetimes.

5. Referring to Figure 9.2, state the most likely types of gamma radiation (electric or magnetic and l value) that will be emitted in the isomeric transitions $^{77m}\text{Ge} \to {}^{77}\text{Ge}$, $^{115m}\text{Cd} \to {}^{115}\text{Cd}$, and $^{115m}\text{In} \to {}^{115}\text{In}$. Estimate the half-life of each of these isomeric transitions.

6. Suppose the total decay rate of an excited state is given by $\lambda = \lambda_1 + \lambda_2$, where λ_1 and λ_2 are the probabilities per unit time of decay by processes 1 and 2, respectively. In terms of λ_1 and λ_2, what is the expression for (a) the half-life of this excited state and (b) the branching ratio for the processes 1 and 2?

7. Assume that after capturing a neutron of laboratory energy $\tilde{E}$ the compound nucleus decays by emitting a single gamma ray to its ground state. Perform an exhaustive kinematical analysis of this reaction. If the gamma ray is emitted at laboratory angle $\tilde{\theta}$, what are the laboratory energy and angle of the recoil nucleus?

10 CHARGED-PARTICLE EMISSION

The compound nucleus formed by the capture of a neutron may sometimes decay by emission of a charged particle. Among the possible reactions, (n,p) and (n,α) are most common; (n,d) reactions are less common primarily because their Q values are less favorable. These charged-particle-out (cpo) reactions are significant to nuclear engineers for many reasons; a few of which are:

1. Most cpo reactions lead to radioactive daughter nuclides that engender radiation problems in reactor systems. For example, the ^{16}N produced in the ^{16}O$(n,p)^{16}$N reaction decays with the emission of very energetic gamma rays, which constitute the major radiation source requiring shielding in the primary loops of pressurized-water reactors.

2. The cpo reactions ^{10}B$(n,\alpha)^7$Li and ^{6}Li$(n,\alpha)^3$H have large slow-neutron cross sections and can be employed both in detecting slow neutrons and in shielding against them.

3. The cpo reaction ^{6}Li$(n,\alpha)^3$H may furnish the major source of tritium fuel for future thermonuclear reactors.

4. The cpo reaction ^{14}N$(n,p)^{14}$C produces radioactive ^{14}C whenever nuclear explosions are conducted in the atmosphere.

10.1 Kinematics of Charged-Particle-Out Reactions

The kinematics of cpo reactions, which are completely straightforward, can be exemplified by the (n,α) reaction:

$$A + n \rightarrow (C^*) \rightarrow B + \alpha + Q.$$

The Q value of the reaction is given by

$$Q = 931 \frac{\text{Mev}}{\text{amu}} [M(A) + m_n - M(B) - M(^4\text{He})], \tag{10.1}$$

where the atomic mass of ^{4}He, not the mass of the alpha particle, is used to properly account for the masses of the orbital electrons.

If Q is positive, the reaction is exoergic and may be triggered by a neutron of any energy. If Q is negative, the reaction is endoergic, and the threshold energy is given by

$$\tilde{E}_t = \frac{M(A) + m_n}{M(A)} (-Q). \tag{10.2}$$

The threshold energy, it will be recalled, is the minimum neutron laboratory kinetic energy at which the reaction can possibly proceed. The center-of-mass threshold energy is simply the negative of the Q value,

$$E_t = -Q. \tag{10.3}$$

Thresholds: Actual and Apparent

It is a peculiarity of cpo reactions that they may not proceed with a detectable cross section until the energy of the incoming neutron is much greater than the threshold energy. For example, consider the reaction $^{27}\text{Al}(n,\alpha)^{24}\text{Na}$. Its Q value may be found from the following atomic masses:

$$M(^{27}\text{Al}) = 26.98153, \qquad M(^{24}\text{Na}) = 23.99066,$$

$$m_n = \frac{1.008665}{27.990195}, \qquad M(^4\text{He}) = \frac{4.00260}{27.99326}.$$

Thus $Q = 27.990195 - 27.99326 = -0.003065$ amu $= -2.85$ Mev. The center-of-mass threshold energy is 2.85 Mev; the laboratory threshold is $28/27 \times 2.85$ Mev $= 2.96$ Mev.

The cross section for this reaction is shown in Figure 10.1 plotted both linearly and logarithmically against energy. From the linear plot one would be tempted to place the threshold at about 5 Mev. This is the *apparent threshold*; it is also, in a sense, the *practical threshold* since the cross section becomes very small indeed below 5 Mev. It is not, however, the *actual threshold*, which, as calculated above, lies at 2.96 Mev. The logarithmic plot makes clear that the cross section is decreasing exponentially below 5 Mev. With sensitive apparatus the cross section should be detectable down to the actual threshold.

10.2 Semiclassical Theory of Alpha Emission

The threshold behavior of cpo-reaction cross sections can be explained by an elementary semiclassical analysis. Although far from rigorous, this analysis clearly illustrates the reason for the existence of apparent thresholds and for the exponential increase in the cross section with energy in the vicinity of an apparent threshold.

On the assumption that alpha-particle emission follows the formation of a compound nucleus, the cross section for the (n,α) reaction can be written in the familiar form

$$\sigma_\alpha = \sigma_c \frac{\Gamma_\alpha}{\Gamma}$$

or in the equivalent form

$$\sigma_\alpha = \sigma_c \frac{\lambda_\alpha}{\lambda}, \tag{10.4}$$

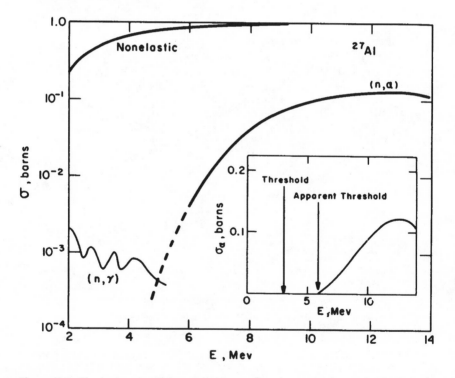

Figure 10.1—The (n,α) cross section of aluminum plotted on a semilog scale. The insert shows the same cross section on a linear scale.

where λ_α is the decay constant for alpha-particle emission by the compound nucleus and λ is the total decay constant. At high neutron energies, σ_c is a slowly varying function of energy; therefore the behavior of σ_α is strongly dependent on the variation of the decay constant λ_α with energy. The decay constant can be estimated by considering the transparency of the Coulomb barrier that holds the alpha particle within the nucleus.

Figure 10.2 is a highly simplified representation of the potential energy $V(r)$ of a system consisting of a charged particle and a residual nucleus that remains after emission of the charged particle. It is assumed that the potential energy is given by Coulomb's law at all distances greater than some channel radius R and by a constant V_0 at separation distances less than R:

$$V(r) = \begin{cases} \dfrac{zZe^2}{r} & \text{(for } r > R), \\ -V_0 & \text{(for } r < R). \end{cases} \tag{10.5}$$

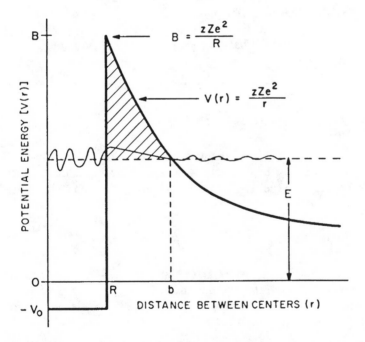

Figure 10.2—Simplified potential between a charged particle and the residual nucleus for the case $l = 0$. The major parameters determining the transparency of the barrier are E, B, R, and b.

The channel radius is a parameter that can be adjusted to optimize the agreement between theory and experiment. For example, the empirical formula

$$R = (1.4A^{1/3} + 1.2) \text{ fm}, \tag{10.6}$$

in which the first term is the radius of the residual nucleus and the second term is an effective radius for the alpha particle, has been found to produce surprisingly good agreement with experimental values of radioactive decay constants.*

The total energy of the system, E, is taken as the total kinetic energy of the alpha particle and the residual nucleus when the two are widely separated:

$$E = \tfrac{1}{2}mv_r^2,$$

where v_r is the relative velocity of the two bodies and m is the reduced mass of the system. In most cases the mass of the daughter nucleus is much larger than the mass of the alpha particle; therefore E may be taken as the kinetic energy of the emitted alpha particle.

The decay constant λ_α, which is defined as the probability of escape per second, will be assumed to be equal to the number of times the alpha particle strikes the potential barrier each second multiplied by the transparency of the barrier:

$$\lambda_\alpha = \text{frequency of hitting the barrier} \times \text{transparency}$$

or

$$\lambda_\alpha = f \times T.$$

For an order of magnitude estimate, we assume that the particle is rebounding back and forth with velocity v within a nuclear potential of diameter $2R$ and that the transparency is given by the approximate expression, Equation 3.142:

$$\lambda_\alpha \stackrel{\circ}{=} \frac{v}{2R} \exp\left\{ -\frac{2}{\hbar} \int_R^b \sqrt{2m[V(r) - E]} \right\} dr. \tag{10.7}$$

To order of magnitude, we expect the velocity of the alpha particle in the nucleus to be commensurate with a kinetic energy in the range of about 1 to 10 Mev; i.e., about 10^9 cm/sec. Again, to order of magnitude, the diameters of most nuclei are about 10^{-12} cm. Thus

$$\lambda_\alpha \stackrel{\circ}{=} 10^{21} e^{-G} \text{ sec}^{-1}, \tag{10.8}$$

*John M. Blatt and Victor F. Weisskopf, *Theoretical Nuclear Physics* (John Wiley & Sons, Inc., New York, 1952), p. 578.

where, by definition,

$$G \equiv \frac{2}{\hbar} \int_{R}^{b} \sqrt{2m[V(r) - E]} \, dr. \tag{10.9}$$

The transparency e^{-G} is widely known as the *Gamow factor*. It was first employed by Gamow and independently by Gurney and Condon in 1928 in the derivation of a successful theory of radioactive alpha decay.[*] It is clear that the argument that led to the preceding order of magnitude expressions is equally applicable to proton or deuteron emission.

The function G can be readily integrated (start by substituting $r = y^2$) with the result

$$G = \frac{R\sqrt{8mB}}{\hbar} \left(\frac{\cos^{-1}\sqrt{x}}{\sqrt{x}} - \sqrt{1 - x} \right), \tag{10.10}$$

where $B \equiv zZe^2/R$ is the maximum of the electrostatic potential barrier and $x \equiv E/B$ is the ratio of the energy of the emitted particle to the barrier height B. Inspection of this equation shows that the transparency e^{-G} is an extremely sensitive function of x, the ratio of the energy of the emitted particle to the barrier height. As an illustration of the effect of barrier penetration on the cross sections above the thresholds of cpo reactions, we will consider the reaction $^{27}\text{Al}(n,\alpha)^{24}\text{Na}$. The threshold energy for this reaction is 2.96 Mev. The parameter x is therefore given by

$$x = \frac{E}{B} = \frac{(\tilde{E}_n - 2.96 \text{ Mev})}{B},$$

where $\tilde{E}_n$ is the laboratory kinetic energy of the neutron. Using the rough approximation $R = 1.5A^{1/3}$ fm for the channel radius, we find that the electrostatic barrier height, B, is

$$B = \frac{zZe^2}{R} = 7.3 \text{ Mev},$$

and the dimensionless coefficient in the function G is

$$\frac{R\sqrt{8mB}}{\hbar} = 9.6.$$

The variation of the decay constant λ_α and its reciprocal τ_α, the mean life of the compound nucleus for alpha decay, as functions of neutron energy are shown in Table 10.1. It can be seen that the mean life for alpha

[*] G. Gamow, Z. *Physik*. **51**: 204 (1928); R. W. Gurney and E. U. Condon, *Nature*. **122**: 439 (1928).

decay is extremely long compared to other nuclear mean lives until the neutron energy is approximately 4 to 5 Mev. In this energy range the mean life for alpha emission (10^{-14} to 10^{-15} sec) is of the same order of magnitude as that for gamma emission. One would therefore expect the cross sections for these two modes of decay of the compound nucleus to be of the same order of magnitude. They indeed are, as can be seen in Figure 10.1. This agreement, however, must be considered fortuitous in view of the questionable nature of the assumptions and methods we have employed. In the following sections more rigorous, fully quantum-mechanical methods will be explored.

Table 10.1—Dependence of the Decay Constant λ_α on the Neutron Kinetic Energy in the Reaction $^{27}Al(n,\alpha)^{24}Na$

Neutron kinetic energy ($\bar{E}_n$), Mev	E/B	G	Decay constant ($\lambda_\alpha \triangleq 10^{21} e^{-G}$), sec^{-1}	Mean life ($\tau_\alpha = 1/\lambda_\alpha$), sec
3.03	0.01	131	10^{-40}	10^{40}
3.25	0.04	56	10^{-4}	10^{4}
3.69	0.10	29	10^{8}	10^{-8}
4.42	0.20	15.2	10^{14}	10^{-14}
5.88	0.40	6.0	10^{18}	10^{-18}
7.34	0.60	2.4	10^{20}	10^{-20}

10.3 Charged-Particle Penetrabilities

The basic quantities that must be known before one can begin a rigorous calculation of cross sections are the transmission coefficients or penetrabilities. In modern application these are obtained from a numerical solution of the Schrödinger equation including a realistic semiempirical complex potential well with diffuse edges, spin-orbit coupling, and so on. With charged particles a Coulomb potential is added to the strictly nuclear potentials. Therefore, in general, the penetrabilities are functions of the orbital angular-momentum quantum number l, the channel-spin quantum number j, and the total angular-momentum quantum number J. They are also functions of all the parameters that describe the well shape and, of course, the channel energy. It is found, however, that the major parameters determining the penetrabilities are the channel energy, E and the orbital quantum number, l, and, in the case of charged particles, the coulomb barrier energy, B. It is therefore instructive to examine the variation of the penetrabilities as a function of these parameters only. To this end, we consider the case of a particle incident on a simple spherical well

$$V(r) = \begin{cases} -V_0 & \text{(for } r < R), \\ 0 & \text{(for } r > R, \text{ neutrons)}, \\ \dfrac{zZe^2}{r} & \text{(for } r > R, \text{ charged particles)}. \end{cases}$$

The transmission coefficient $T_l(E)$ is defined as the probability that an incident particle of angular momentum l and channel energy E enters the nucleus (or, in view of reciprocity, leaves the nucleus). It is given by

$$T_l(E) = 1 - |U_l(E)|^2, \tag{10.11}$$

where U_l is the reflection factor, the ratio of the amplitudes of the outgoing and incoming waves. The general expression for the reflection factor is

$$U_l = \left[\frac{D_l - R\left[\dfrac{dW_l^{(-)}/dr}{W_l^{(-)}}\right]}{D_l - R\left[\dfrac{dW_l^{(+)}/dr}{W_l^{(+)}}\right]} \frac{W_l^{(-)}}{W_l^{(+)}} \right]_{r=R} \tag{6.23'}$$

The notation used here is identically that of Section 6.2, to which the reader is referred for definitions.

To obtain the most elementary form for the reflection factor, we shall assume that the nucleus is totally absorbing, i.e., black toward the incoming neutrons. This condition is mathematically expressed by employing as the wave function within the nucleus one that consists only of an incoming wave and has no outgoing component. Under the further assumption that the potential is constant within the nucleus, the inner wave function may be written as

$$W_l = e^{-iKr}, \tag{10.12}$$

where K is the wave number corresponding to the kinetic energy in the interior of the nucleus,

$$K = \sqrt{2m(E + V_0)}/\hbar. \tag{10.13}$$

The "logarithmic derivative" D_l is therefore

$$D_l \equiv R\left[\frac{dW_l/dr}{W_l}\right]_{r=R} = -iKR. \tag{10.14}$$

After this expression is substituted for D_l in Equation 6.23′ and the resulting expression is substituted for U_l in Equation 10.11, the following formula for T_l eventually emerges from the algebra

$$T_l = \left(\frac{4}{2 + \dfrac{X}{x} |Z_l|^2 + xX|Z_l'|^2} \right)_{r=R} ;$$

(10.15)

where

$$x = kR,$$

$$X = KR,$$

and

$$Z_l = \begin{cases} j_l(kr) + in_l(kr) & \text{(neutrons),} \\ F_l(\eta,kr) + iG_l(\eta,kr) & \text{(charged particles),} \end{cases}$$

where j_l and n_l are the spherical Bessel and Neumann functions, F_l and G_l are the regular and irregular Coulomb wave functions, and Z_l' is the derivative of Z_l with respect to kr. The parameter η in the Coulomb wave functions is defined as

$$\eta \equiv \frac{kzZe^2}{2E} = \frac{xB}{2E}.$$

(10.16)

To illustrate the behavior of these transmission coefficients, we have plotted $T_l(E)$ as a function of channel energy for neutrons in Figure 10.3 and for protons in Figure 10.4. Both sets of curves were drawn assuming a nuclear well depth $V_0 = 22$ Mev and a nuclear radius of 5 fm. For the proton curves the nuclear charge was taken as $Z = 19$.

Several features of these curves are worthy of attention. Both sets of curves show the effect of the centrifugal barrier in reducing the transmission coefficients for $l > 0$ particles. The charged-particle transmission coefficients become appreciable only as E approaches the Coulomb barrier energy B. In contrast, the neutron penetrabilities show no such "apparent threshold." It is also worth noting that the charged-particle penetrabilities remain less than unity at channel energies well above the Coulomb barrier. This strictly quantum-mechanical effect was not taken into account in the crude analysis of the penetrabilities based on the Gamow factor; as a result that analysis has very little validity as E approaches and exceeds the barrier height.

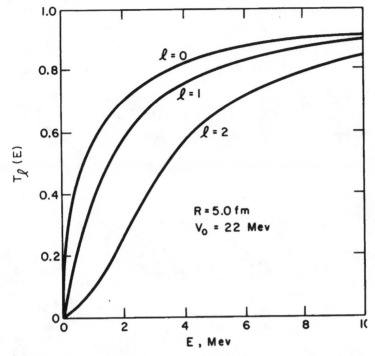

Figure 10.3—Black-nucleus transmission coefficients for neutrons as a function of channel energy for l = 0, 1, and 2.

Graphs of neutron transmission coefficients for values of l as high as 7 may be found in USAEC Report NYO-636.* Extensive tables of charged-particle transmission coefficients for values of l as high as 14 may be found in NYO-3077.† These graphs and tables, though based on the black-nucleus assumption and on the simple nuclear potential $V(r) = -V_0$, are quite useful in obtaining somewhat better than order of magnitude estimates of penetrabilities. To obtain penetrabilities based on more realistic optical-model potentials, numerical solutions of the Schrödinger equation, such as that embodied in the digital computer program ABACUS-2 ,‡ may be employed.

*B. T. Feld et al., *Final Report of the Fast-Neutron Data Project*, USAEC Report NYO-636, Nuclear Development Associates, January 1951. See also H. Feshbach and V. F. Weisskopf, *Phys. Rev.*, **76**: 1550 (1949).

†H. Feshbach, M. M. Shapiro, and V. F. Weisskopf, *Tables of Penetrabilities for Charged-Particle Reactions*, USAEC Report NYO-3077, Nuclear Development Associates, June 1953.

‡E. H. Auerbach, *ABACUS-2, Program Operation and Input Description*, USAEC Report BNL-6562, Brookhaven National Laboratory, November 1962.

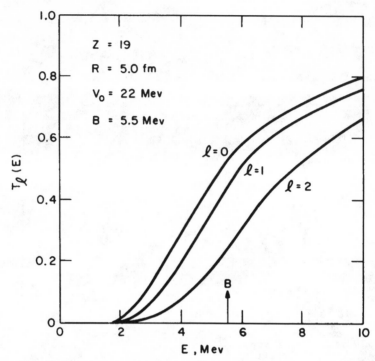

Figure 10.4—Black-nucleus transmission coefficients for protons as a function of channel energy for l = 0, 1, and 2. The Coulomb-barrier height at the nuclear surface is denoted by the arrow.

10.4 Hauser–Feshbach Calculations of Charged-Particle-Out Cross Sections

Under the assumption that the neutron and charged-particle penetrabilities are independent of the channel spin and the total angular momentum, the general Hauser–Feshbach expression for the reaction leading from channel α to channel α' can be written in the form

$$\sigma(\alpha|\alpha') = \pi\lambda^2 \sum_{ljJl'j'} \frac{2J + 1}{(2s + 1)(2I + 1)} \frac{T_{\alpha l}T_{\alpha'l'}}{\sum_{\alpha''l''} T_{\alpha''l''}}, \qquad (10.17)$$

where s, I, and J are the spin of the neutron, the spin of the ground state of the target nucleus, and the total angular-momentum quantum number, respectively; $T_{\alpha l}$ is the neutron penetrability at the incident channel energy, $T_{\alpha'l'}$ is the charged particle penetrability at the emergent channel energy, and $\sum_{\alpha''l''} T_{\alpha''l''}$ is the summation of the penetrabilities over all possible emergent channels and orbital angular momenta. The particles that may emerge include elastically and inelastically scattered neutrons, protons, alpha particles, deuterons, and gamma rays.

As an illustration of the Hauser–Feshbach procedure for determining cpo cross sections, we shall consider the calculations that were carried out by Troubetzkoy et al.* in 1961 to find the (n,p) and (n,α) cross sections of ^{40}Ca. (This target nucleus was previously used to illustrate the calculation of elastic and inelastic scattering cross sections in Chapters 7 and 8.) Although these cross sections can be more reliably calculated today because of our increased knowledge of energy-level spins and parities and the use of more realistic optical potentials, nevertheless these calculations clearly illustrate the general procedure as well as the behavior of cpo cross sections as functions of neutron energy.

At neutron energies in the range of about 1 to 5 Mev, it is assumed that the following four reactions have appreciable cross sections.

$$^{40}\text{Ca} + n \rightarrow (^{41}\text{Ca})^* \rightarrow \begin{bmatrix} ^{40}\text{K} + p \\ ^{37}\text{Ar} + \alpha \\ ^{40}\text{Ca} + n' \\ ^{40}\text{Ca} + n \end{bmatrix}$$

The Q values for the (n,p) and (n,α) reactions leading to the ground states of the residual nuclei are -0.537 and $+1.742$ Mev, respectively.

Knowledge of the low-lying level schemes of ^{40}K and ^{37}Ar is required before one can proceed with the calculation. The level schemes employed by Troubetzkoy et al. are shown in Figures 10.5 and 10.6. Assignments of spins and parities in parentheses were estimated in whole or in part. The level scheme of ^{40}Ca, which is also required to calculate the penetrabilities of the inelastically scattered neutrons, was shown in Chapter 8.

Let us focus our attention on the ^{40}Ca$(n,p)^{40}$K reaction. At a neutron energy E within the range of about 2.5 to 5 Mev, the residual nucleus, ^{40}K, which remains after emission of a proton, may be left in the ground state or in one of the four excited states extending up to 1.639 Mev (Figure 10.5). The total (n,p) cross section is the sum of the partial cross sections for the excitation of each of the levels. Let us therefore consider the cross section for the excitation of one of the levels, to be specific, the level at 0.885 Mev. The proton that is emitted in exciting this level has an energy $E' = E + Q - 0.885$ Mev $= E - 1.422$ Mev. The excited level will decay to the ground state by emitting a 0.885-Mev gamma ray.

* E. S. Troubetzkoy, M. H. Kalos, H. Lustig, J. H. Ray, and B. H. Trupin, *Fast-Neutron Cross Sections of Manganese, Calcium, Sulfur, and Sodium*, Report NDA-2133-4, Nuclear Development Associates, December 1961.

The first step in evaluating the Hauser–Feshbach expression, Equation 10.17, consists in doing the necessary angular-momentum bookkeeping to determine the possible values of l, J, and l' which enter into the summation. The steps in chronological order are

1. Determine the channel spins of the initial constellation. Since $I = 0$ and $s = 1/2$, the only possible channel spin is $j = 1/2$.

2. Determine the channel spins of the final constellation. Since $I' = 5$ and $s' = 1/2$, the two possible channel spins are $j' = 9/2$ and $j' = 11/2$.

3. For each orbital angular-momentum quantum number of the initial constellation $l = 0, 1, 2, \ldots$, determine the J^{Π} of the excited com-

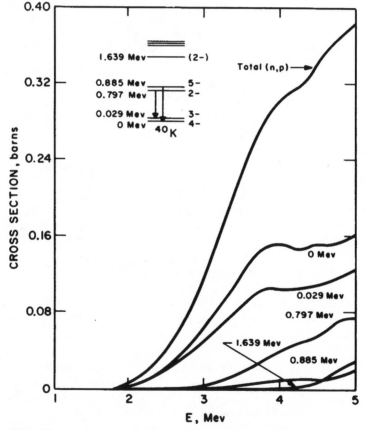

Figure 10.5—Calculated total (n,p) cross section and cross section for excitation of specific levels in ⁴⁰K in the reaction ⁴⁰Ca(n,p)⁴⁰K. (From E. S **Troubetzkoy et al., Report** NDA-2133-4, 1961.)

pound nuclear state. Since the channel spin has only the value 1/2, J is determined from $J = |l \pm j|$. The parity Π is determined from $\Pi = (-1)^l \Pi_A \Pi_n = (-1)^l$. The last step follows from the fact that both the parity of the neutron and the parity of the ground state of ^{40}Ca are positive.

4. Determine the possible angular-momentum quantum numbers of the final constellation corresponding to each J and j' using $|J - j'| \leq l' \leq (J + j')$. Some of these l' values may be rejected because they do not

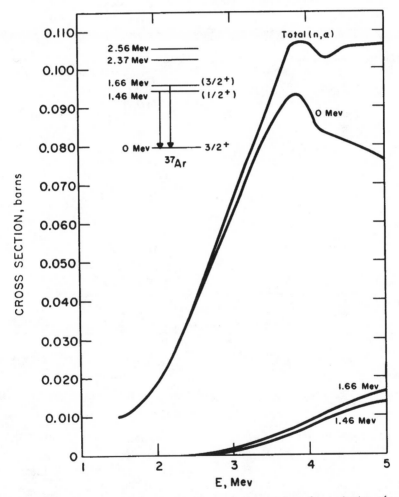

Figure 10.6—Calculated total (n,α) cross section and cross sections for excitation of levels in ^{37}Ar in the reaction ^{40}Ca$(n,\alpha)^{37}$Ar. (From E. S. Troubetzkoy et al., Report NDA-2133-4, 1961.)

conform to the parity conservation rule

$$\Pi = (-1)^l \Pi_B \Pi_p = -(-1)^{l'}.$$

The last step follows from the fact that the parity of the proton is $+1$ and the parity of the 0.885-Mev level in ^{40}K is -1.

The results of this bookkeeping are listed in Table 10.2 for values of l from 0 to 3 inclusive.

Table 10.2—Permitted Values of l' and J for an (n,p) Reaction from a 0^+ to 5^- Level for $l \le 3$

^{40}Ca $+ n$		$\rightarrow$	$(^{41}$Ca$)^*$	$\rightarrow$	$(^{40}$K$)^* + p$	
I^Π	j	l	J^Π	$(I^\Pi)'$	j'	l'
0^+	$1/2$	0	$1/2^+$	5^-	$9/2$	5
					$11/2$	5
		1	$1/2^-$		$9/2$	4
					$11/2$	6
		1	$3/2^-$		$9/2$	$4,6$
					$11/2$	$4,6$
		2	$3/2^+$		$9/2$	$3,5$
					$11/2$	$5,7$
		2	$5/2^+$		$9/2$	$3,5,7$
					$11/2$	$3,5,7$
		3	$5/2^-$		$9/2$	$2,4,6$
					$11/2$	$4,6,8$
		3	$7/2^-$		$9/2$	$2,4,6,8$
					$11/2$	$2,4,6,8$

Using this table of possible lJl' combinations in conjunction with Equation 10.17, we can immediately write the expression for the cross section. The terms corresponding to the first two J^Π values of Table 10.2 are

$$\frac{\sigma_p(0^+|5^-)}{\pi \lambda^2} = \left[\frac{2T_0^n T_5^p}{\sum\limits_{(1/2)^+} T_{\alpha''l''}} \right] + \left[\frac{T_1^n(T_4^p + T_6^p)}{\sum\limits_{(1/2)^-} T_{\alpha''l''}} \right] \qquad (10.18)$$

The superscripts n and p designate neutron and proton penetrabilities, respectively. The summations in the denominator are over all possible exit channels consistent with the designated values of J^Π. In these expressions

$$T_l^n = T_l^n(E),$$

$$T_{l'}^p = T_{l'}^p(E - 1.422 \text{ Mev}),$$

and the various $T_{\alpha''}$ are evaluated at the energy peculiar to the particular decay channel α''.

The results published by Troubetzkoy et al. for this cross section, as well as for the other (n,p) and (n,α) cross sections, are shown in Figures 10.5 and 10.6. These calculations are based on neutron, proton, and alpha-particle penetrabilities obtained from the simple black-nucleus model discussed in the previous section. Despite this great oversimplification of the potential, the cross sections were found to agree with the few available experimental values to within about 30 to 50%. It may be assumed that a modern calculation based on realistic optical-model potentials would produce considerably better agreement.

Under the assumption that statistical theory is valid, the angular distribution of the emitted particle, which is given, in general, by Equation 8.43, can be written as

$$\sigma(\alpha|\alpha',\theta) = \frac{\lambdabar^2}{4} \sum_{ljJl'j'L} \frac{(-1)^{j'-j}}{(2s+1)(2I+1)} \frac{T_{\alpha l}T_{\alpha'l'}}{\sum_{\alpha''l''} T_{\alpha''l''}}$$

$$\times\; Z(lJlJ;jL)Z(l'Jl'J;j'L)P_L(\cos\theta),$$

(10.19)

where it is again assumed that the penetrabilities are independent of the channel spin and the total angular momentum.

The summation index L takes on only even values less than or equal to the minimum of $(2l, 2l', 2J)$. Employing this rule in conjunction with Table 10.2, we see that the angular distribution of protons that have proceeded through the $J^\eta = (1/2)^+$ compound state to excite the 0.885-Mev level in ^{40}K is isotropic in the center-of-mass system. Those protons which proceed through the $(1/2)^-$ level also are isotropic (although $2J = 1$, L must be even). Those which proceed through the $(3/2)^-$ level will contain P_0 and P_2 terms and so on.

The strengths of the P_0, P_2, P_4, and higher components in the differential cross section depend on the relative sizes of the coefficients of these terms. Usually, the larger the angular momentum l or l', the smaller the associated penetrability T_l or $T_{l'}$. Since the penetrabilities always appear as products, it is almost impossible to estimate the angular distribution without a full-fledged evaluation of many of the terms in Equation 10.19.

The measured angular distributions of charged particles from cpo reactions are often not symmetrical about 90°. Two effects may contribute to this lack of agreement between the predictions of Hauser–Feshbach theory and experiment. First, the assumptions on which Hauser–Feshbach theory is based may be totally or partly invalid. Perhaps the density of levels in the compound nucleus is not great enough to ensure that a statistical average is obtained over all possible phases of the state. In this case the separate l and J contributions to the cross section will be additive in scattering amplitude rather than in cross section. Second, the assumption that a compound nucleus is formed may be partially invalid. Part of the interaction may take place directly without formation of a compound nucleus. The angular distribution in this case is highly forward peaked without a corresponding symmetrical peak in the backward direction.

10.5 Slow-Neutron-Induced Charged-Particle-Out Reactions

Every exoergic cpo reaction has a finite cross section regardless of how small the energy of the bombarding neutron, but, because of the Coulomb barrier, the cross sections of the vast majority of these exoergic reactions are vanishingly small at low neutron energies and cannot be experimentally detected until the neutron energy reaches the million-electron-volt region and above. The exceptional cases are listed in Table 10.3. None of the target nuclides in Table 10.3 has a Z greater than 19. As we might expect, slow-neutron-induced cpo reactions have appreciable cross sections only with target nuclides of low atomic number, hence low Coulomb-

Table 10.3—Charged-Particle-Out Reactions Induced by Slow Neutrons*

Reaction	Target Z	$\sigma_\alpha(0.025\ \mathrm{ev})$, barns	Q value, Mev
^{3}He$(n,p)^3$H	2	5327	0.76
^{14}N$(n,p)^{14}$C	7	1.8	0.63
^{33}S$(n,p)^{33}$P	16	0.002	0.53
^{35}Cl$(n,p)^{35}$S	17	0.4	0.62
^{40}K$(n,p)^{40}$Ar	19	4.0	0.93
^{6}Li$(n,\alpha)^3$H	3	950	4.8
^{10}B$(n,\alpha)^7$Li	5	3840	2.8
^{17}O$(n,\alpha)^{14}$C	8	0.24	1.8
^{32}S$(n,\alpha)^{29}$Si	16	0.007	1.5
^{33}S$(n,\alpha)^{30}$Si	16	0.2	3.5
^{36}Ar$(n,\alpha)^{33}$S	18	0.006	2.1

*Only reactions with naturally occurring target nuclides are tabulated.

barrier height. To the nuclear engineer the most interesting reactions in Table 10.3 are those involving the target nuclei ^{10}B and ^{6}Li.

^{10}B(n,α)^{7}Li

The ^{10}B(n,α)^{7}Li reaction is undoubtedly the cpo reaction of greatest interest to nuclear engineers. Because of its large thermal cross section, boron is often employed in the control of nuclear reactors and in shielding against slow neutrons. In addition, the cross section for this reaction is widely used as the standard against which other cross sections are measured in the neutron-energy range from thermal to about 10 kev.

The cross section has a value of 3837 $\pm$ 10 barns at a neutron energy of 0.025 ev and varies as $1/v$ to about 100 kev. As of 1967 it had been measured to within about $\pm 2\%$ over this energy range. Above 100 kev the cross section begins to deviate from $1/v$ because of the proximity of resonances near 140, 250, 420, and 490 kev. These resonances are attributed to energy levels in the compound nucleus ^{11}B at excitation energies of 11.60, 11.68, 11.85, and 11.94 Mev (Figure 10.7).

Because of the low Coulomb barrier (about 2.5 Mev) and the relatively high Q value (2.8 Mev), charged-particle emission is highly favored over radiative decay of the compound nucleus, the ratio of σ_α to σ_γ being about 8000 to 1 at low neutron energies.

The alpha particles are emitted into two channels, the first corresponding to the ground state of ^{7}Li, and the second to the first excited state of ^{7}Li located 0.48 Mev above the ground state (Figure 10.7). Emission to the excited state is favored by about 14 to 1 and is independent of neutron energy from thermal to about 100 kev:

$$\frac{\sigma_\alpha(^7\text{Li } 0.48 \text{ Mev})}{\sigma_\alpha(^7\text{Li ground state})} = 14, \qquad 0 \le E \le 100 \text{ kev}.$$

The magnitude and the $1/v$ variation of the ^{10}B(n,α)^{7}Li cross section can be qualitatively understood by recourse to the Breit–Wigner formula for the (n,α) reaction. The resonances responsible for this cross section are probably those with centers located at neutron energies between 140 and 2000 kev. Although the resonance parameters have not been completely determined for these levels, one can reproduce the cross section in the $1/v$ region by ascribing values to these parameters (e.g., $\Gamma_n \simeq 20$ kev and $\Gamma_\pi \simeq 200$ to 500 kev) which agree reasonably well with the measured cross sections.

The $1/v$ variation of the cross section over the entire range from thermal to about 100 kev is due to the large reaction width Γ_α. Because $\Gamma_\alpha \gg \Gamma_n(E)$ and $E_0 \gg E$, the resonance denominator in the Breit–Wigner expression is practically constant over this range and the $1/v$ variation arises from the numerator terms $\lambda^2\Gamma_n(E)$; i.e.,

$$\sigma_\alpha(E) = \pi\lambda^2 \frac{2J + 1}{2(2I + 1)} \frac{\Gamma_n(E)\Gamma_\alpha}{(E - E_0)^2 + (\Gamma/2)^2}$$

$$\propto \lambda^2\Gamma_n(E) \propto \frac{1}{E}\sqrt{E} \propto \frac{1}{v}.$$

(10.20)

It is instructive to derive this $1/v$ variation in another way, based on black-nucleus penetrabilities. If we assume that the reaction proceeds

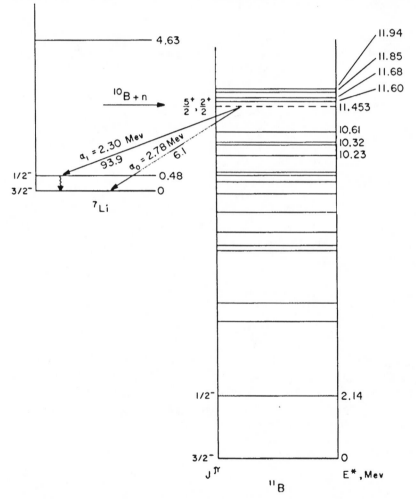

Figure 10.7—Energy levels of ^{11}B and ^{7}Li showing the two alpha-particle decay channels for the slow-neutron-induced reaction ^{10}B$(n,\alpha)^7$Li.

through a single $l = 0$ entrance channel, we can write Equation 10.17 in the form

$$\sigma_\alpha(E) = \pi \lambda^2 T_0^n(E) \sum_J g(J) \frac{T_{l'}^\alpha(E')}{\sum_{\alpha''l''} T_{\alpha''l''}}. \tag{10.21}$$

The neutron penetrabilities are obtained by substituting the black-nucleus boundary condition $D_l = -iKR$ into the reflection factor for neutrons, Equation 6.29,

$$U_l = \frac{D_l - a_l + ib_l}{D_l - a_l - ib_l} e^{2i\Delta_l}. \tag{6.29'}$$

Then

$$T_l(E) = 1 - |U_l|^2 = \frac{4b_l KR}{a_l^2 + (b_l + KR)^2}.$$

At low neutron energies, $k \ll K$; thus both a_l and b_l are small compared with KR. Hence

$$T_l^n(E) = \frac{4b_l}{KR} \quad \text{(for } k \ll K\text{)}. \tag{10.22}$$

At low neutron energies, $kR \ll 1$ and $b_l \simeq (kR)^{2l+1}$. Therefore all the $T_l^n(E)$ are much less than unity, actually several orders of magnitude less than unity. On the other hand, because the exit energy E' of the alpha particle is greater than the barrier energy B, the $T_{l'}^\alpha(E')$ are of order unity. Thus we come finally to the conclusion that the sum over the ratio of exit-channel penetrabilities in Equation 10.21 is for all practical purposes equal to unity; therefore

$$\sigma_\alpha(E) = \pi \lambda^2 T_0^n(E). \tag{10.23}$$

We could have arrived at this same equation by simply assuming that the cross section for the (n,α) reaction was equal to the cross section for formation of the compound nucleus by $l = 0$ neutrons.

Upon substituting the expression for T_0^n from Equation 10.20 into Equation 10.23, we find

$$\sigma_\alpha(E) = \pi \lambda^2 \frac{4k}{K} = \frac{4\pi \lambda}{K}. \tag{10.24}$$

This is the final result. Since K is practically constant for any neutron energy below 100 kev, the entire energy variation of the cross section is in λ, which varies as $1/\sqrt{E}$.

Having succeeded in deriving the $1/v$ variation of the low-energy (n,α) cross section, we note that we can go one step further and obtain the correct magnitude of the cross section by making a reasonable choice of K. In fact, the experimental curve

$$\sigma_\alpha(\text{barns}) = 611/\sqrt{E(\text{ev})}$$

is reproduced very closely by the natural choice $K = 10^{13}\text{cm}^{-1}$, corresponding to a well depth of about 20 Mev.

$^6\text{Li}(n,\alpha)^3\text{H}$

At present this reaction is of considerable interest to nuclear engineers and physicists primarily because of its usefulness in the measurement of neutron-energy spectra. The numbers and energies of high-energy neutrons incident on lithium-drifted-germanium solid-state detectors can be obtained from the pulse-height spectra of the emitted alpha particles and tritons. In the future this reaction may become the primary source of tritium generated by and for thermonuclear reactors.

The cross section for this reaction is 950 barns at 0.025 ev and varies as $1/v$ to about 30 kev. It has been measured to within about $\pm 2\%$ over this energy range. The cross section at higher energies is dominated by a single broad resonance centered at about 250 kev. There are hints of further unresolved resonances beyond 1 Mev.

Only one alpha channel is open, that to the ground state of ^3H. The final channel energy, $E' = E + Q = E + 4.8$ Mev, is shared between the alpha particle and the triton in such a way, of course, as to conserve linear momentum and total energy.

SELECTED REFERENCE

1. J. B. Marion and J. L. Fowler (eds.), *Fast Neutron Physics, Part II: Experiments and Theory*, Interscience Publishers, New York, 1963. A wealth of interesting titbits on cpo reactions are found in the twenty articles comprising this volume. The reader could profitably spend his time reading the following chapters, particularly Chapter V.N:

Chap. V.D., "Direct Reaction Theories," by N. Austern.
Chap. V.E., "Resonance Processes with Fast Neutrons," by H. B. Willard, L. C. Biedenharn, P. Huber, and E. Baumgartner.
Chap. V.N., "Neutron-Induced Reactions," by F. L. Ribe.

EXERCISES

1. What reasons can you suggest for the facts that: (a) electrons are never emitted in the decay of a compound nucleus, i.e., there are no known (n,e) reactions, and (b) there are no known slow-neutron-induced (n,d) reactions.

2. Show by actual calculation that the remarkable sensitivity of the half-lives of natural alpha emitters to the energies of the emitted alphas, as shown in the following table, can be explained on the basis of the semiclassical theory of alpha emission. (Assume for the purpose of calculation that the channel radii are given by Equation 10.6 and that $\lambda_\alpha = 10^{21} e^{-G}$ sec^{-1}).

Parent	Daughter	Alpha energy, Mev	Half-life
$^{238}_{92}$U	$^{234}_{90}$Th	4.25	4.51×10^9 y
$^{234}_{92}$U	$^{230}_{90}$Th	4.84	2.35×10^5 y
$^{228}_{90}$Th	$^{224}_{88}$Ra	5.52	1.91 y
$^{222}_{86}$Em	$^{218}_{84}$Po	5.59	3.83 d
$^{218}_{84}$Po	$^{214}_{82}$Pb	6.11	3.05 m
$^{214}_{84}$Po	$^{210}_{82}$Pb	7.83	1.6×10^{-4} s
$^{212}_{84}$Po	$^{208}_{82}$Pb	8.95	3.0×10^{-7} s

3. Use the data in the table in Exercise 2 in conjunction with the semiclassical theory of alpha emission to determine the channel radii in the decays ^{238}U $\rightarrow$ ^{234}Th $+ \alpha$ and ^{212}Po $\rightarrow$ ^{208}Pb $+ \alpha$. How well do these channel radii conform to the empirical expression, Equation 10.6?

4. Investigate the feasibility of building a detector to determine neutron energy by measuring the angle β between the reaction products in the reaction ^{6}Li$(n,\alpha)^3$H. What is the lowest neutron energy that could be measured if the angle β could be determined with a resolution of 1 degree? (Reference: M. G. Silk, British Report AERE-M-1850, 1967.)

5. Prepare an angular-momentum bookkeeping table similar to Table 10.2 for the reaction ^{40}Ca $(n,\alpha)^{37}$Ar leading to the 1.66-Mev level in ^{37}Ar. Assume this level has $I^\Pi = (3/2)^+$. Consider l values from 0 to 3 inclusive. (a) Using this table, write the expression for the (n,α) cross section in a form similar to Equation 10.18. At what energies are the various penetrabilities evaluated? (b) Write a similar expression for the cross section, assuming that you have available a set of neutron and alpha-particle penetrabilities that depend on the channel spin as well as on l.

11 THE FISSION REACTION

In this chapter emphasis is placed on those aspects of the fission reaction which are of most interest to the nuclear engineer, namely, the kinematics, cross sections, and products of the neutron-induced binary fission reaction in uranium, thorium, and plutonium. In addition, a brief description of fission theory is offered. More-extensive and detailed information can be found in the selected references at the end of the chapter.

11.1 History

Following the discovery of the neutron by Chadwick in 1932 and of artificial radioactivity by Curie and Juliot in 1934, Enrico Fermi and his associates in Rome began a series of experiments to see whether or not neutrons would transform natural elements into radioactive nuclei. Within a month after the discovery of artificial radioactivity, Fermi announced that various elements did indeed become radioactive when exposed to neutrons. About a month later he announced that neutron bombardment of uranium produced radioactivity that he speculated must be due to transuranic elements produced in (n,γ) reactions that were followed by β^- decays.

The radioactivity of the "transuranic elements," however, was most peculiar. There were too many decay chains, and the chains were much longer than those of other heavy elements in the uranium region. The mystery deepened and the speculation grew during the next five years until the radiochemists Hahn and Strassman in 1939 correctly identified barium ($Z = 56$) as one of the products of the reaction. They thus succeeded in proving beyond any doubt that, when uranium is bombarded with neutrons, radioelements are formed whose atomic numbers are roughly half those of the parents.

Lise Meitner and Otto Frisch immediately developed the concept of, and the name, *fission*. They guessed that Hahn's results must be explained by Bohr's idea that the nucleus was like a liquid drop, a drop that could, under certain conditions, elongate and divide itself. On the basis of the mass-defect curve, Meitner calculated that the energy available in such a breakup would be about 200 Mev. Since this was approximately the amount of energy that the electrostatic repulsion of the fission fragments would give them, it was clear that the process of fission was energetically possible. Frisch soon afterwards confirmed that large amounts of energy were indeed emitted in fission. He irradiated a thin uranium foil in an ionization chamber with neutrons and observed great bursts of ionization corresponding to energies of the order of 100 Mev.

It was soon speculated that some of the neutron-rich fission fragments might reach stability by neutron emission as well as by β^- decay. Such neutrons were soon observed, and measurements showed that on the average about two or three neutrons were emitted per fission. The possibility of a chain reaction was immediately recognized.

Differences in the fissionabilities of the various heavy nuclides were soon found. Niels Bohr was the first, in 1939, to attribute the slow-neutron fission of uranium to the isotope ^{235}U, which occurs only to the extent of 0.7% in natural uranium. It was later found that the abundant isotope ^{238}U could be made to fission under bombardment by high-energy neutrons.

It is now known that most heavy nuclides will undergo fission provided enough energy is supplied by the bombarding particle. Fission has been produced in the isotopes of many heavy ($Z \geq 70$) elements by bombardment with high-energy protons, deuterons, alpha particles, photons, and light ions, such as ^{12}C and ^{16}O. In addition, some heavy nuclides have been found to undergo spontaneous fission as an alternative to the more common alpha and beta modes of radioactive decay.

11.2 Nomenclature

It has become common practice in nuclear engineering to use the word *fissionable* to describe those elements which have a significant cross section (say, ≥ 1 barn) for fission by neutrons of energy less than 18 Mev. Thus ^{238}U is said, to be fissionable since its fission cross section is significant for neutrons having energies above 1 Mev. Elements that are fissionable by room-temperature thermal neutrons are said to be *fissile*. The five common fissionable nuclides are ^{232}Th, ^{238}U, ^{233}U, ^{235}U, and ^{239}Pu, but only the last three of these nuclides are also fissile.

11.3 Characteristics

The fission reaction has several characteristics that are not found in other nuclear reactions. The most striking of these is the large energy release. Another is the fact that, unlike most nuclear reactions, the fission reaction is not completed in one or two stages.

The actual splitting (*scission*) of the excited compound nucleus is only the first stage of a complex chain of events. Though the major portion of the energy released in fission appears in the form of kinetic energy of the fission fragments, a considerable amount (some 40 Mev) resides in excitation energy of the fission fragments.

These fragments give up their excess internal energy in a series of steps. They first emit prompt neutrons and gamma rays, probably within 10^{-14} sec after scission. The resulting nuclei are then either in the ground states of radioactive isotopes or in isomeric states. If in isomeric states, they decay primarily by gamma emission.

During a time of the order of. 10^{-11} sec after scission, the fragments give up all their kinetic energy and come to rest. At times greater than about 10^{-3} sec, those fission products that are radioactive begin chains of beta decays. Most of the products of these beta decays are in excited states that reach the ground state by gamma emission. A small fraction of the products of these beta decays emit neutrons to reach their ground states. These are the sources of the delayed neutrons that are emitted with apparent half-lives ranging from 0.2 sec to about 1 min.

11.4 Kinematics of the Fission Reaction

From a kinematic viewpoint the fission reaction consists initially of the absorption of a neutron by a target nucleus to form a compound nucleus and the subsequent scission of this nucleus into two fragments:

$$Z^A + n \rightarrow (Z^{A+1})^* \rightarrow Z_1^{A_1} + Z_2^{A_2}. \tag{11.1}$$

Conservation of protons and neutrons in the reaction is expressed by

$$Z_1 + Z_2 = Z \quad \text{and} \quad A_1 + A_2 = A + 1.$$

The Q value for the fission reaction is given by

$$Q = 931 \frac{\text{Mev}}{\text{amu}} [M_T + m_n - M_1 - M_2], \tag{11.2}$$

where M_T is the rest mass of the target nucleus and M_1 and M_2 are the rest masses of the two fragments. Scission occurs in a variety of ways; therefore the masses M_1 and M_2 vary from fission to fission. Thus the Q value for fission is not a single well-defined quantity but rather an average over the probability distributions of the fragment masses. Regardless of the mode of fission, however, the Q value is found to lie somewhere in the approximate range 190 to 210 Mev, with an average for the thermal-neutron fission of ^{235}U of 206 Mev.

The energy released in fission can be estimated from the binding-energy curve, Figure 5.1. As shown in this figure, the value of the average binding energy per nucleon is about 7.6 Mev in the vicinity of uranium and plutonium and about 8.5 Mev in the range of mass numbers from 40 to 150. Thus the average binding energy per particle is about 0.9 Mev greater in the fission products than in the compound nucleus. Since the compound nucleus of the fissionable elements contains some 234 to 242 nucleons, the total energy release per uranium or plutonium fission is on the average somewhat in excess of 200 Mev.

Scission of the compound nucleus into three or more fragments also occurs, but is rare. For example, in the thermal-neutron fission of ^{235}U, there is evidence that the frequency of ternary fission is roughly 1 in 400 and of quaternary fission is less than 1 in 3000 normal binary events.

Ternary fission almost always occurs with the emission of an alpha particle in coincidence with two heavy fragments. Occasionally, however, a triton is emitted rather than an alpha particle. Triton emission occurs in only about one in 10^4 fission events but is significant in fuel reprocessing because it produces hazardous radioactive tritium.

The nucleus rarely scissions into three or four roughly equal fragments. The upper limit for this process in the thermal fission of ^{235}U is one such event in 10^5 normal binary events.

The Barrier Against Fission

Purely from mass considerations, based on the binding-energy curve, fission becomes energetically possible, in fact, is an exoergic reaction, for nuclei with masses greater than about 90 amu. In view of this it might be expected that all heavy nuclei with $A > 90$ would undergo spontaneous fission. Actually spontaneous fission has never been observed for any nuclide lighter than ^{230}Th, whose half-life for fission is extremely long ($> 10^{17}$ years). The half-life for fission decreases rapidly with increasing atomic number until at element 102 it is only of the order of seconds. In the common fissionable isotopes of thorium, uranium, and plutonium, the half-lives for spontaneous fission are all greater than 10^{10} years.

The half-life for the spontaneous fission of ^{236}U from its ground state is about 2×10^{16} years. How is it then that the absorption of a slow neutron by ^{235}U may result in the essentially instantaneous fission of the compound nucleus $(^{236}U)^*$? Apparently the nucleus has a very small probability of splitting until it has first acquired a certain minimum energy, an *activation energy*. This energy is not available in the ground state of ^{236}U but is supplied in the form of the neutron separation energy when ^{235}U absorbs a thermal neutron to become $(^{236}$U$)^*$.

Figure 11.1 is a simplified drawing of the potential energy of the nucleus as a function of the distance between the centers of two virtual or actual fission fragments. At infinite separation the potential is zero. At closer distances the Coulomb potential between the two charged fragments rises as $1/r$. In the merged state the attractive nuclear forces eventually exceed the repulsive Coulomb force so that the coalesced fragments reside in a potential well surrounded by a potential barrier. This barrier

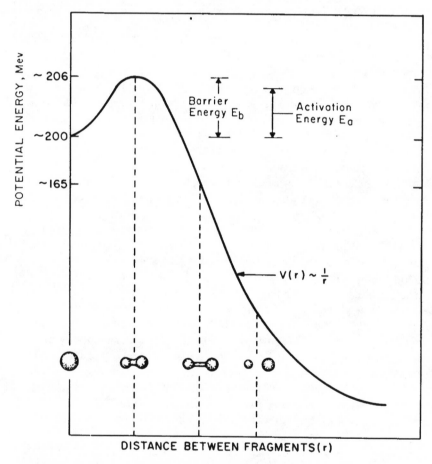

Figure 11.1—Schematic diagram of the potential energy between two fission fragments. The numbers are approximately those which apply to thorium, uranium, and plutonium.

qualitatively resembles that which encloses an alpha particle within a radioactive nucleus. The probability of penetrating the barrier is relatively small unless the nucleus is given an excitation energy E^* that is almost equal to the barrier energy E_b. The following empirical relation has been derived for the barrier energy:[†]

$$E_b = \frac{19.0 - 0.36Z^2}{A + \delta} \text{ Mev,} \tag{11.3}$$

where Z is the atomic number, A is the mass number, and δ is a parameter that is equal to zero for even-even, 0.7 for odd-odd, and 0.4 for even-odd and odd-even compound nuclei.

For fission to compete with gamma emission, its half-life must be comparable with that for gamma emission, i.e., about 10^{-14} sec. Given an excitation energy equal to E_b, the hypothetical time for fission would be about 10^{-21} sec. The two fragments can penetrate the potential barrier that binds them in a time of the order of 10^{-14} sec if they are given an energy within about 0.9 Mev of the top of the barrier. It follows therefore that the activation energy for fission, E_a, is about 0.9 Mev less than the barrier energy, E_b.

When a thermal neutron is captured by a target nucleus, the resulting compound nucleus acquires an excitation energy equal to the separation energy S_n of the neutron from the compound nucleus. If this excitation energy exceeds the activation energy, then the nucleus may fission. In this event the target nucleus is said to be fissile. If additional energy is required in the form of kinetic energy of the incoming neutron, the target nucleus will exhibit an apparent threshold for fission.

Since fission is exoergic for all elements with $A \gtrsim 90$, fission thresholds are not true thresholds. Owing to the barrier-penetration nature of the fission process, there is no neutron energy below which fission cannot be induced in the heavy elements. Hence the apparent threshold for fission must be arbitrarily defined. We shall define it as that energy below which the fission cross section is less than 0.1 barn.

Table 11.1 shows that whenever the neutron separation energy, S_n, exceeds the calculated activation energy, E_a, the target nuclide has no threshold. Conversely, when $E_a > S_n$, there is an apparent threshold, which in most cases is approximately equal to the difference between E_a and S_n. The agreement between experimentally observed and predicted apparent thresholds is surprisingly good.

† Earl K. Hyde, *The Nuclear Properties of the Heavy Elements*, Vol. III, Fission Phenomena, Reference 1 at the end of this chapter.

Table 11.1—Correlation of Apparent Thresholds for Fission with the Difference between the Activation Energy and the Neutron Separation Energy from the Compound Nucleus

Target nuclide	Activation energy (E_a),* Mev	Separation energy (S_n),† Mev	Difference $E_a - S_n$, Mev	Apparent threshold (E_t),‡ Mev
^{233}U	5.1	6.8	−1.7	0
^{235}U	5.2	6.5	−1.3	0
^{239}Pu	4.8	6.4	−1.6	0
^{232}Th	6.0	4.9	1.1	1.6
^{234}U	5.5	5.2	0.3	0.3
^{236}U	5.6	5.3	0.3	0.8
^{238}U	5.7	4.8	0.9	0.9
^{237}Np	5.7	5.5	0.2	0.35

*The activation energy is taken to be 0.9 Mev less than the barrier energy as determined by Equation 11.3.
†Neutron separation energy from the nucleus $A + 1$.
‡The apparent threshold is arbitrarily defined as the energy below which the fission cross section is less than 0.1 barn.

11.5 Fission Cross Sections at Low Energies

In the energy region below about 1 kev, the fission cross sections of the fissile nuclides have a complicated resonance structure that is similar to the (n,γ) cross sections of all the heavy elements. Since it can be said with almost complete certainty that the fission reaction proceeds through the formation and subsequent decay of a compound nucleus, the quest for a theoretical cross section for this reaction naturally begins with the Breit–Wigner isolated ($l = 0$) resonance equation

$$\sigma_f(E) = \pi\lambda^2 g(J) \frac{\Gamma_n\Gamma_f}{(E - E_0)^2 + (\Gamma/2)^2}, \tag{11.4}$$

where Γ_f is the width for fission of the compound nucleus and Γ is the total width, which for low-energy neutrons is the sum of $\Gamma_n + \Gamma_\gamma + \Gamma_f$.

If this equation accurately described the energy variation of the fission cross section, we should expect that (1) at very low energies and far from resonances, the cross section would vary as $1/v$ and (2) the shape of the cross section would be symmetric about the resonance energy. Actually the experimental cross sections for fission show neither the $1/v$ behavior nor complete symmetry. Both these departures from expectation have been the subjects of intense investigation.

Departure from 1/v of the Fission Cross Section at Very Low Energies
When the product $\sigma_f \sqrt{E}$ is plotted against E in the energy region from
0 to about 0.1 ev, the resulting curves are not horizontal straight lines,
as they would be if the cross sections varied inversely with the neutron
velocity. The departure from $1/v$ behavior is quite striking, as can be seen
in Figure 11.2. Similar plots show that $\sigma_\gamma(E)$ also fails to vary as $1/v$.

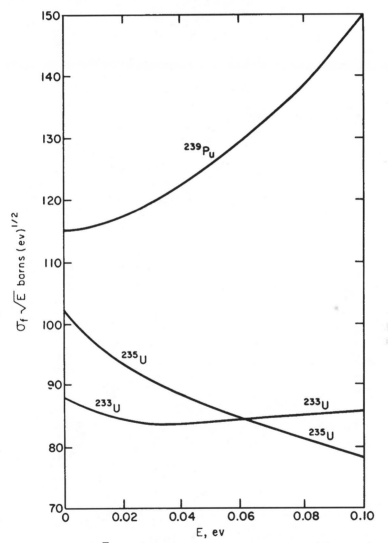

Figure 11.2—Plot of $\sigma_f \sqrt{E}$ vs. energy showing the non-1/v nature of the very-low-
energy fission cross section.

It is not at all surprising that the very-low-energy reaction cross sections of the fissile elements fail to vary as $1/v$ when it is recalled that a necessary condition for the $1/v$ variation is the absence of nearby resonances. Such a condition never obtains in the very heavy elements, whose energy levels are separated by only electron volts at excitation energies of the order of 6 to 7 Mev, the neutron separation energy from the compound nucleus.

The energies and partial reaction widths of the first five resonances in the fissile elements are tabulated in Table 11.2. Of particular note is the fact that the lowest-lying level in each of the isotopes is only of the order of 0.2 to 0.3 ev above zero and has fission and capture widths of the order of 1/20 ev. Thus the first resonance is, in fact, not far from the very-low-energy region. This accounts for at least part of the non-$1/v$ character of the very-low-energy cross sections.

It is not possible to derive $\sigma_f(E)$ at very low energies by simply summing the contributions from each of the measured resonances, assuming in the process that each resonance is describable by a single-level Breit–

Table 11.2—Resonance Parameters of the Lowest-Lying Levels* in the Fissile Nuclides ^{233}U, ^{235}U, and ^{239}Pu

Target nuclide and I^Π	E_0, ev	Γ_f, mv	Γ_γ, mv	$2g\Gamma_n$, mv
^{233}U	0.17	60	40	0.0002
$5/2^+$	1.55	600	50	0.17
	1.78	220	40	0.31
	2.30	46	40	0.18
	3.66	180	53	0.141
^{235}U	0.29	100	35	0.0032
$7/2^-$	1.135	115	42	0.0154
	2.040	10	37	0.0077
	2.84	160	40	0.008
	3.15	90	47	0.029
^{239}Pu	0.296	60	39	0.121
$1/2^+$	7.85	40	41	0.132
	10.95	130	40	2.62
	11.90	22	41	1.61
	14.30	60	50	0.68

*Abstracted from USAEC Report BNL-325, 2nd ed., supplement 2, vol. 3, Brookhaven National Laboratory, 1965.

Wigner equation. To obtain the proper magnitude and energy variation of $\sigma_f(E)$, we must assume the existence of a rather large negative energy resonance. This resonance is assumed to be due to an energy level in the compound nucleus that lies just a fraction of an electron volt below the excitation energy corresponding to the neutron separation energy. It is assumed to be the highest bound level in the compound nucleus (see Figure 6.1). This level is at a negative energy in the sense that it cannot be reached by scattering neutrons off a target nucleus.

Since nothing whatsoever is known experimentally about this hidden level, its properties $(E_0, \Gamma_f, \Gamma_\gamma,$ and $\Gamma_n)$ are tailored in such a way that when its Breit–Wigner contribution is added to that of the positive-energy resonances, the observed variations of σ_f and σ_γ with energy are reproduced. If the analysis based on one hidden level fails to reproduce the observed cross sections, several hidden levels are assumed.

Departure from Symmetry of the Fission Resonance Cross Sections
Upon closely examining the fission resonances, we find that they do not exhibit the symmetry about E_0 that would be expected if the single-level Breit–Wigner equation were valid. This is in marked contrast to the radiative-capture resonances, which usually do have symmetrical forms (see Figure 11.3).

Apart from the asymmetry, there is another distinctive difference between fission and radiative-capture resonances. There is a considerable variation in the fission width Γ_f from one resonance to another but hardly

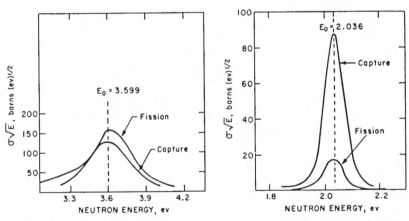

Figure 11.3—Fission and radiative-capture cross sections for the 2.036- and 3.599-ev resonances in ^{235}U. The capture curve is symmetrical about E_0, whereas the fission curve is not. [From F. J. Shore and V. L. Sailor, *Phys. Rev.*, **112**: 191 (1958).]

any variation in the radiation width Γ_γ. This can be seen in the data in Table 11.2.

The asymmetry of the fission resonances and the variation of the fission widths are both believed to be due to interference between levels. A better understanding of this can be obtained by recalling the reasons for the symmetry of the radiative-capture resonances and the uniformity of Γ_γ from one resonance to another.

The radiative-capture width Γ_γ is the sum of a very large number of partial widths, each being related to the probability of decay from a fixed compound-nucleus level to any one of hundreds of lower-lying levels. Because there are so many exit channels, the individual variations in the partial widths from one resonance to another are averaged out, and Γ_γ is very nearly the same for all resonances. Also because there are so many different exit channels, the interference between levels is practically non-existent, and the cross-section curves for radiative capture exhibit Breit–Wigner symmetry.

On the other hand, fission resonances show both asymmetry and variation in width. The converse of this explanation, then, leads us to conclude that the number of channels open for fission is very low and there is interference between levels. This interference excludes the rigorous application of the single-level Breit–Wigner equation for the analysis of fission cross sections in the thermal and resonance regions. Therefore multilevel Breit–Wigner equations, derived from R-matrix theory, are increasingly being employed and with good results. The success of the multilevel Breit–Wigner analysis based on one, or at most a few, exit channels is surprising. The large spectrum of fission products would suggest an equally large number of channels for fission since each pair of fragments might be thought of as a distinct exit channel. We shall defer the explanation of this seeming anomaly to a later section.

It is also possible that the asymmetry is due in part to the existence of small unresolved resonances that lie near the prominent ones. Recent theoretical calculations employing mock resonance parameters indicate that as many as half or more of the actual levels may be experimentally unobserved because of overlap.*

11.6 Fission Cross Sections at High Energies

In the neutron-energy region from about 1 to 20 Mev, fission cross sections exhibit the smooth variation that is typical of other reaction cross

---

*J. E. Lynn, Quasi-Resonances and the Channel Theory of Neutron-Induced Fission, in *International Conference on the Study of Nuclear Structure with Neutrons, July 19–23, 1965*, Antwerp, pp. 125–155, North Holland Publishing Company, Amsterdam, 1966.

sections in the continuum region. Figure 11.4 shows the high-energy cross sections for the common fissionable nuclides.

The high-energy cross sections of the fissile nuclides are simply continuations of the resonance cross sections into the region of overlapping resonances. The cross sections are now most simply expressed as the product of the cross section for formation of the compound nucleus σ_c and the branching ratio for decay by fission:

$$\sigma_f(E) = \sigma_c(E)\frac{\Gamma_f(E^*)}{\Gamma_T(E^*)}. \tag{11.5}$$

In this expression the branching ratio is to be interpreted as an average over an energy range determined by the spread in the incident-neutron beam.

The behavior of the fission cross section of ^{238}U is typical of that of nuclides which have pseudo fission thresholds in the million-electron-volt region. It rises rapidly to an appreciable fraction of the geometrical cross section, then remains constant for about 5 or 6 Mev, after which it rapidly rises to another plateau. Measurements on ^{238}U indicate that the fission cross section rises at neutron energies near 1, 6, and 13 Mev.

The first rise (at 1 Mev) is due simply to the vastly increased probability of fission as the excitation energy of the compound nucleus approaches the barrier energy. The cross section then remains constant until the excitation energy is high enough to permit the nucleus to expel one neutron without reducing the excitation of the residual nucleus below its

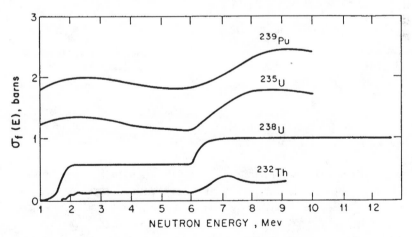

Figure 11.4—Fission cross sections in the million-electron-volt region for ^{232}Th, ^{235}U, ^{238}U, and ^{239}Pu.

fission threshold. Thus the second rise (at 6 Mev) is due to the onset of the (n,nf) reaction, sometimes called *second-chance fission*. The next rise (at 13 Mev) occurs when the (n,2nf) reaction becomes possible. Measurements on ^{238}U indicate that there may be a further threshold near 19 Mev for the (n,3nf) reaction. Other nuclides that exhibit much the same sort of fission cross section as ^{238}U are ^{232}Th, ^{234}U, ^{236}U, ^{237}Np, ^{240}Pu, and ^{241}Am. As can be seen in Figure 11.4, the fissile nuclides also exhibit an almost constant cross section from 1 to 6 Mev, then a rise to a new plateau as second-chance fission becomes energetically possible.

11.7 Distribution of Fission Energy

There is still considerable uncertainty about the exact distribution of the energy produced in fission among the various forms in which it appears. A representative set of values from recent measurements for the slow-neutron fission of ^{235}U is shown in Table 11.3. The uncertainty in the fragment kinetic energies is of the order of 3%. The uncertainty in the other energies is generally of the order of 10%. The total energy release is known to within about 5%.

Table 11.3—Energy Distribution per Slow-Neutron Fission of ^{235}U

Fragment kinetic energies before neutron emission	167.7 Mev
Kinetic energy of fission neutrons	4.8 Mev
Prompt gamma radiation ($t < 1$ μsec)	8.0 Mev
Fission-product gamma radiation ($t > 1$ μsec)	7.6 Mev
Fission-product beta radiation	7.7 Mev
Neutrino energy	10.0 Mev
Total energy release	205.8 Mev

Kinetic Energies of the Fission Fragments
About four-fifths of the energy released in fission appears in the form of kinetic energy of the fragments. The remainder appears at the instant of scission as excitation energy within the fragments. It has been found experimentally that the degree of excitation of the compound nucleus has little or no effect on the kinetic energies of the fragments. For example, fission of ^{235}U by thermal and by 90-Mev neutrons produces fission fragments with almost identical average kinetic energies. The excess excitation energy resides in the internal energies of the fragments and results in the boiling off of more neutrons. Further proof of this is the fact that fission of the compound nucleus produced by thermal-neutron capture

results ultimately in an average of almost one more neutron than does spontaneous fission of the same nuclide.

Because the energy liberated in fission is so great, the nucleus undergoing fission can be considered, for almost all practical purposes, to be at rest in the laboratory system at the instant of scission. The fission fragments must therefore recede from one another with speeds V_1 and V_2 in the laboratory system such that linear momentum is conserved:

$$M_1 V_1 = M_2 V_2. \tag{11.6}$$

The kinetic energies of the fragments are then in the ratio

$$\frac{E_1}{E_2} = \frac{M_1 V_1^2}{M_2 V_2^2} = \frac{M_2}{M_1}. \tag{11.7}$$

The kinetic energies are seen to be inversely proportional to the masses.* Hence from the measured kinetic-energy distribution of the fragments, the mass distribution can be obtained.

Mass Distribution of the Fission Products

The fission yield for a particular nuclide is defined as the probability, expressed as a percentage, of forming that nuclide, or the chain of which it is a member, in fission. Since each binary fission results in two nuclei, the total fission yield is 200%.

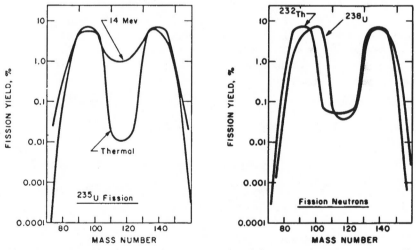

Figure 11.5—Mass distributions of ^{235}U, ^{238}U, and ^{232}Th fission products. [From S. Katcoff, *Nucleonics,* **18** (11): 201 (1960).]

---

*This is only approximately correct. For a full discussion, see the article by James Terrell, *Phys. Rev.,* **113**: 527–541 (1959).

When the fission yields are plotted against mass number, as in Figure 11.5, each curve displays two distinct peaks, corresponding to the emission of a light and a heavy fragment. The curves are not identical for the various heavy fissionable nuclides, but they are remarkably close to one another. Of great interest is the observation that symmetrical fission is extremely rare. For example, the slow-neutron-induced fission of the compound nucleus $(^{236}U)^*$ into two atoms of $A = 118$ is about 600 times less probable than its fission into fragments of approximate mass 95 and 140. The probability of symmetrical fission increases manyfold with higher energy neutrons; it is about 100 times more probable with 14-Mev neutrons than with slow neutrons. When the energy of the neutrons is raised to 90 Mev, only one peak is observed, corresponding to the fact that at these very high energies symmetrical fission becomes the favored mode of compound-nucleus splitting.

Angular Distributions of the Fission Fragments

It has been observed that fission fragments are not isotropically emitted when neutrons of energies in the million-electron-volt range or above are used to induce fission. At most energies above 1 Mev, the forward direction of emission is favored, but there are appreciable fluctuations in the magnitude of the forward peaking. In the high-energy fission of ^{238}U, for example, the ratio of fragment yield at 0° to the neutron beam to the yield at 90° fluctuates erratically between a minimum of 1.09 to a maximum of 1.70 in the neutron-energy range from 1.35 to 20 Mev.

In the neutron-induced fission of ^{232}Th, there is a striking reversal of the usual anisotropy at the resonance centered at 1.6 Mev. In this resonance, and apparently only in this resonance, the 0°/90° yield falls to the low value of 0.1. Immediately above this resonance at 1.8 Mev, the anisotropy has disappeared, and at still higher energies it resumes its usual form, favoring 0° emission by as much as a factor of 2 to 1.

The angular distribution of the fragments from the 1.6-Mev fission of ^{232}Th can be fit with an expression of the type

$$1 + a_2 P_2(\cos \theta) + a_4 P_4(\cos \theta) + a_6 P_6(\cos \theta),$$

where $P(\cos \theta)$ is a Legendre polynomial. This suggests that neutrons with l values up to 3 may be responsible for the fission.*

11.8 Theories of the Fission Reaction

If it is to be acceptable, any theory of the fission reaction must be able to account for at least the following facts:

1. Although the fission reaction is exoergic for heavy nuclides, only a few of the very heaviest nuclides actually fission spontaneously. Some

*R. L. Henkel and J. E. Brolley, *Phys. Rev.*, **94**: 640 (1954).

few fission with the addition of a slow neutron, but most require a high-energy neutron. There is an apparent barrier to fission.

2. The mass distribution of fission fragments is asymmetrical but tends toward symmetry at very high neutron energies.

3. In the favored form of fission, into fragments of unequal size, the fragments exhibit approximately equal internal energies.

4. Fission resonances exhibit interference, leading to the conclusion that there are but few channels open to fission.

5. The angular distribution of the fragments from fission induced by high-energy neutrons is anisotropic, favoring emission forward and backward along the beam.

The Liquid-Drop Model

Shortly after the announcement of fission in 1939, Niels Bohr and John Wheeler published a detailed model that satisfactorily explained the initial limited observations.* They suggested that fission could be understood on the basis of the *liquid-drop model* of the nucleus which had been proposed by Bohr some four years before.

Any nucleus that contains a reasonably large number of nucleons may be likened to a drop of liquid, the nucleons being analogous to the molecules in the liquid. The forces acting between the nucleons are of two kinds. One is the short-range charge-independent nucleon–nucleon force. The other is the Coulomb repulsive force between protons. The nuclear forces are known to be saturated so that the nucleons are bound only to their nearest neighbors. Since the nucleons on the surface are attracted only inwardly, a surface tension is set up which forces the drop to assume a stable spherical shape. The situation is thus analogous to a charged liquid drop where intermolecular forces give rise to surface tension and the stable spherical shape.

When energy is added to the nucleus (e.g., by the absorption of a neutron), this excitation energy may simply agitate the nucleons within a nearly spherical nucleus or it may set up oscillations that distort the shape of the drop. It is assumed that the volume of the drop remains constant; therefore any distortion from the spherical shape increases the surface area, and the surface tension tends to return the drop to its spherical shape. But the Coulomb forces, being repulsive, tend to increase the distortion. If the surface-tension forces are greater, the original spherical shape is eventually restored; but, if the Coulomb forces are greater, the deformation of the drop will increase with each oscillation, from spherical, through perhaps ellipsoidal, and eventually to a shape like a peanut with two lobes connected by a neck. At this point the force of repulsion between

*Niels Bohr and John A. Wheeler, The Mechanism of Nuclear Fission, *Phys. Rev.* 56 : 426 (1939).

the charges in the two lobes is stronger than the reduced cohesive force in the neck of the nucleus. The neck splits, and the electrostatic repulsion drives the two fragments apart.

The liquid-drop model explains why heavy nuclei can be fissioned more easily than light ones. The more protons in the nucleus, the greater the Coulomb repulsion between the ends of the distorted nucleus, hence the less energy required to get the process started.

The two fragments will eventually attain a total kinetic energy equal to the electrostatic potential energy at the moment of scission. As a rough estimate we can assume that this will be approximately equal to the electrostatic potential between two uniformly charged spheres whose centers are separated by distance R:

$$E = \frac{(Z_1 e)(Z_2 e)}{R}.$$

The total kinetic energy of the fragments, E, is known to be approximately 168 Mev. If we choose values of Z_1 and Z_2 corresponding to a probable mode of fission, say $Z_1 = 57$ and $Z_2 = 35$, then we obtain for the distance between the centers of the fragments at the moment of scission $R = 17 \times 10^{-13}$ cm. Since the sum of the radii of the two fragments is approximately 15×10^{-13} cm, the conclusion is that the neck is rather short and thick. The nucleus presumably looks more like a peanut than a dumbbell when it reaches the point beyond which it cannot return to a spherical shape and must scission.

Potential-Energy Maps

The droplet throughout its stages of deformation may be assumed, as a first approximation, to have an axis of symmetry. Its shape can then be expressed in terms of spherical polar coordinates R and θ by a function $R = R(\theta)$:

$$R = \frac{R_0}{\lambda}\left[1 + \sum_{n=1}^{\infty} \alpha_n P_n(\cos\theta)\right], \tag{11.8}$$

where R_0 is the radius of the undistorted spherical drop, P_n is the Legendre polynomial of order n, and λ is a scale factor adjusted to maintain constant volume.

The deformation parameters, α_n, account for all the shapes the nucleus may assume. All the Legendre polynomials produce shapes that are axially symmetric. The even polynomials produce shapes that are also symmetric toward reflection through the central plane perpendicular to the axis of symmetry. The odd polynomials do not have this latter symmetry.

The potential energy of the charged liquid-drop nucleus is a function of its shape, i.e., a function of the deformation parameters. The calculation of surface- and Coulombic-energy contributions to the potential energy $V(\alpha)$ is extremely tedious and is usually carried out on high-speed computers. For moderate distortions of the symmetric type, the parameters α_2 and α_4 contribute most significantly to the potential energy:

$$R(\theta) = \frac{R_0}{\lambda} [1 + \alpha_2 P_2(\cos \theta) + \alpha_4 P_4(\cos \theta)].$$

The resulting potential-energy contour map is shown in Figure 11.6. The normal spherical nucleus is in a potential-energy minimum at the origin. Energy must be added to distort it from the spherical shape, and the natural tendency is for it to return to the spherical shape. However, there is a path that leads through a potential-energy valley over a saddle point to a condition of scission. The potential energy at the saddle point, approximately 7 Mev in this example, is the minimum amount of energy that must be added to cause the charged drop to divide. The shape of the drop at the saddle point is that of an equal-lobed peanut, as can be verified by plotting Equation 11.8 with $\alpha_2 \simeq 0.7$ and $\alpha_4 \simeq 0.3$.

Asymmetric Mass Distribution in Fission

Potential-energy maps have been carried out to much higher orders in α_n, but to date there is no firm evidence from these studies that the pure liquid-drop model is capable of explaining the uneven mass split in nuclear fission. The difficulty stems from the fact that the liquid-drop model predicts that the nucleus consists of two equal lobes as it passes over the saddle point. One would therefore expect to get two fragments with almost equal internal energies in the rare symmetrical fission events, but the experimental evidence indicates that almost the opposite is true. When the masses are almost equal, the internal energies, as measured by the number of neutrons emitted from the fragments, are highly unequal. Conversely, when the nucleus splits in its most probable asymmetric mode, the internal energies are approximately equal (see Figure 11.7).

The most widely accepted explanation of these facts is based on a combination of the liquid-drop model and the shell model.* As the excited nucleus rapidly oscillates back and forth, its lobes tend to maintain as tightly bound a configuration as possible, i.e., configurations consisting of magic numbers of neutrons or protons. A rough picture of the ^{236}U nucleus just before scission might look like that shown in Figure 11.8. As the nucleus passes over the saddle point, the larger sphere most probably will contain 50 protons plus some neutrons, and the smaller,

---

*S. L. Whetstone, Jr., *Phys. Rev.*, **114**: 581 (1959).

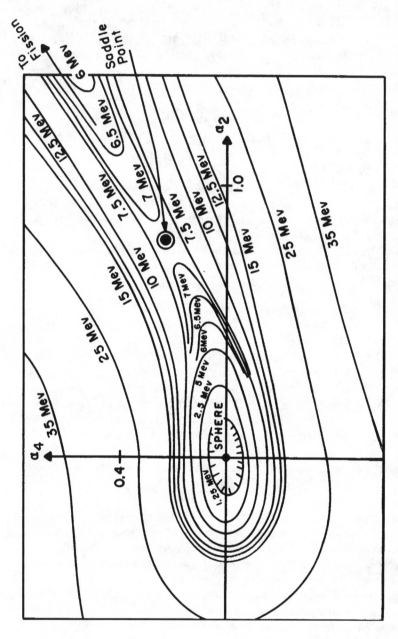

Figure 11.6—Potential energy of a heavy nucleus ($Z \simeq 90$) as a function of the deformation parameters α_2 and α_4. The potential energy is taken as zero for the undistorted spherical shape. [From E. Segrè (Ed.), *Experimental Nuclear Physics*, Vol. II, p. 133, John Wiley & Sons, Inc., New York, 1953.]

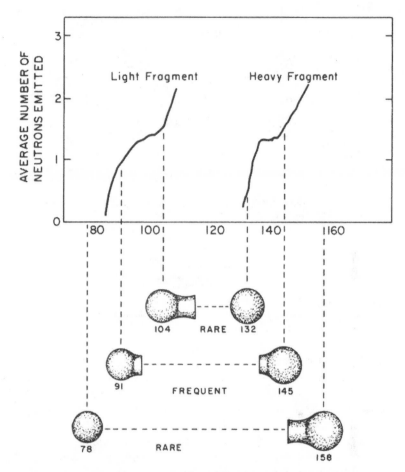

Figure 11.7—The average number of neutrons emitted from the light and heavy fragments as a function of the initial fragment mass. The most probable division shares the excitation energy equally between the fragments.

50 neutrons plus some protons. The rather thick neck contains about 26 nucleons.

It is reasonable to assume that the nucleus most probably will break in two at the thinnest portion of the neck, near its middle. This favors the observed asymmetric mass distribution, and it partitions the deformation energy of the neck equally between the two fragments. Since the two lobes would be expected to have little internal excitation energy before scission, the excitation energies of the fragments after scission will be nearly equal, and therefore the number of neutrons emitted from each

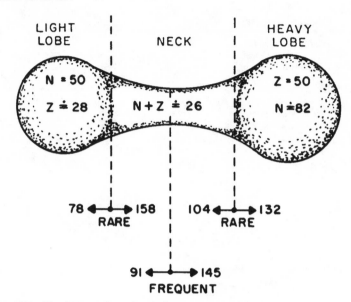

Figure 11.8—The ²³⁶U nucleus shortly before scission. The dashed lines indicate three of the many possible modes of scission. The most probable mode consists of division at the center of the neck of nucleons connecting the two magic lobes.

fragment will be, on the average, equal for the most probable mode of fission. Occasionally the neck will break close to one of the lobes. In these cases the fragment which carries off the neck will have much more excitation energy and therefore will boil off many more neutrons than the other. This effect is clearly evident in the data in Figure 11.7.

This model neatly explains the fission of such elements as uranium and plutonium into two unequal masses. It is not in conflict with the symmetric fission of slightly fissionable elements, such as lead and bismuth. In these lighter elements the shell effect apparently does not operate at all because the lobes cannot have a magic number of neutrons or protons for any reasonable division.

The Unified Model of Fission

In 1955 A. Bohr and B. Mottelson suggested a unified model to explain both the angular asymmetry in fission and the apparent fact that there are but a few channels open to fission. We quote Bohr:*

> When a heavy nucleus captures a neutron or absorbs a high energy photon, a compound nucleus is formed in which the excitation energy is

*A. Bohr, On the Theory of Nuclear Fission, in *Proceedings of the International Conference on the Peaceful Uses of Atomic Energy, Geneva, 1955,* Vol. 2, p. 151, United Nations, New York, 1956.

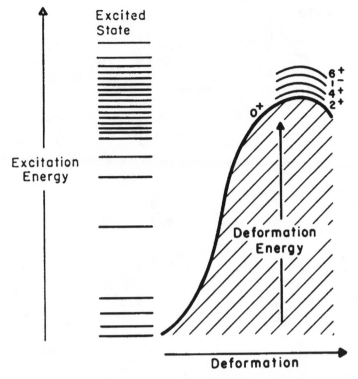

Figure 11.9—Schematic view of A. Bohr's suggestion that a nucleus about to fission may use up most of its excitation energy in deformation, leaving but a few quantum states available at the saddle point. (From Earl K. Hyde, *The Nuclear Properties of the Heavy Elements*, Vol. III, Fission Phenomena, p. 46, Prentice-Hall, Inc., Englewood Cliffs, N. J., 1964.)

distributed among a large number of degrees of freedom of the nucleus. The complex state of motion thereby initiated may be described in terms of collective nuclear vibrations and rotations coupled to the motion of individual nucleons.

The compound nucleus lives for a relatively very long period, usually of the order of a million times longer than the fundamental nuclear periods, after which it decays by emission of radiation or of neutrons, or by fission. The latter process occurs if a sufficient amount of energy becomes concentrated on potential energy of deformation to enable the nucleus to pass over the saddle point shape, at which the repulsive Coulomb forces balance the cohesive nuclear interactions.

For excitation energies not too far above the fission threshold, the nucleus, in passing over the saddle point, is 'cold,' since the major part of its energy content is bound in potential energy of deformation. The quantum states available to the nucleus at the saddle point of the 'fission channels' are then widely separated and represent relatively simple types of motion of the nucleus. These channels are expected to form a similar spectrum as the observed low-energy excitations of the nuclear ground state. [See Figure 11.9.]

This ordered character of the motion of the nucleus at saddle point gives rise to a number of regularities in the fission phenomena. Thus, a fission process passing through a single channel may exhibit a marked anisotropy in the angular distribution of the fragments, depending on the angular-momentum quantum numbers of the channel. Moreover, the wide spacing of the channels implies that the fission threshold may depend significantly on the spin and parity of the compound nucleus.

Bohr goes on to explain why low-lying states of even-even compound nuclei may be assumed to consist of a series of even parity rotational levels (0^+, 2^+, 4^+, ..., etc.) and a series of negative parity levels (1^-, 3^-, 5^-, ..., etc.) and how the occurrence of asymmetry in fission may be related to the existence of the negative parity states. To pursue these interesting details, however, would take us well beyond the intended scope of this book. The curious reader is advised to look into the selected references and the original literature.

SELECTED REFERENCES

1. Earl K. Hyde, *The Nuclear Properties of the Heavy Elements*, Vol. III, *Fission Phenomena*, Prentice Hall, Inc., Englewood Cliffs, N. J., 1964. This book gives a detailed account of almost everything that is known experimentally and theoretically about nuclear fission. It is exhaustively referenced and is an excellent place to start to learn about fission.

2. Lawrence Wilets, *Theories of Nuclear Fission*, Oxford University Press, London, 1964. This is a concise 132-page monograph on the various models of nuclear fission. Experimental results are discussed only in reference to a particular theory. Major emphasis is on the models that describe the nucleus up to the point of scission; relatively little space is devoted to cross-section theory and the products of fission.

EXERCISES

1. Show that the half-life for fission is of the order of 10^{-14} sec when the excitation energy of the compound nucleus is 0.9 Mev below the barrier energy. [*Hint:* Pattern your solution after that used in the alpha-decay problem.]

2. The half-lives for the spontaneous fission of ^{238}U and ^{248}Cf are of the order of 10^{16} years and 10^4 years, respectively. Show that these are the expected orders of magnitude on the basis of barrier penetration.

EFFECTS OF TARGET STRUCTURE AND ATOMIC MOTION **12**

A few general remarks may help to clarify the relation of the material in this chapter to that in the previous chapters. Previous chapters (particularly Chapters 6 through 10) were devoted to the analysis of cross sections for isolated neutron–nucleus interactions. The nucleus was assumed to be free; i.e., not bound in a molecule, a liquid, or a solid. Furthermore, because the cross-section analysis was carried out in the R-system, the motion of the nucleus was of no consequence. But, in the usual targets of interest (solids, liquids, and molecular gases), the nuclei are not isolated, not free, and not motionless; as a result, one can distinguish three effects on the measured cross sections.

1. Because the atoms in the targets are moving, the measured cross sections are an average of individual neutron–nucleus cross sections. This is the source of the "nuclear Doppler" effect.

2. Because the nuclei are in a regular spatial array in crystalline solids and, to a lesser degree, in molecular gases and liquids, the scattered neutron waves from various nuclei can add coherently and produce a measured differential cross section that is radically different from the individual neutron–nucleus cross sections.

3. Because the atoms are bound by chemical forces in molecules, liquids, and solids, a new type of inelastic scattering, involving changes in the energy level of the target material, occurs.

This chapter is devoted to the analysis of these cooperative effects, so called because they all result from the cooperation of more than one scattering center. The first of these cooperative effects, the Doppler effect, has significance for neutrons of all energies; the latter two affect only the low energy (generally less than 1 ev) neutrons.

12.1 The Doppler Effect

If the target nuclei were at rest in the laboratory frame, the cross section for an interaction would depend only on the neutron's kinetic energy. However, in an actual target the atoms are never completely at rest; they are always in thermal motion, vibrating around equilibrium positions in solids or moving more or less randomly in gases and liquids. Some of the atoms have a velocity component toward the incoming neutrons; others have a velocity component in the opposite direction. So the encounters between neutrons and nuclei are characterized by a range of relative velocities, hence a range of relative energies, and therefore a range of cross sections. The cross section that is measured in the laboratory (and reported, for example, in USAEC Report BNL-325) is actually a weighted average of the cross sections of all these individual encounters.

It has been stressed in previous chapters that individual neutron–nucleus cross sections depend on the total energy in the center-of-mass system. This energy, E, is given by

$$E = \tfrac{1}{2}mv^2, \tag{12.1}$$

where m is the reduced mass of the neutron–nucleus system and v is their relative speed of approach; that is,

$$m \equiv \frac{m_n M}{m_n + M} \tag{12.2}$$

and

$$v \equiv |\tilde{\mathbf{v}} - \tilde{\mathbf{V}}|, \tag{12.3}$$

where m_n and M are the masses of the neutron and nucleus, and $\tilde{\mathbf{v}}$ and $\tilde{\mathbf{V}}$ are the laboratory velocities of the neutron and the nucleus.

Definition of Effective Cross Section

The cross section that is measured in the laboratory, $\bar{\sigma}$, the *effective cross section*, is related to the cross section in the relative coordinate system by the equation

$$\tilde{v}\bar{\sigma}(\tilde{v}) = \int |\tilde{\mathbf{v}} - \tilde{\mathbf{V}}|\sigma(|\tilde{\mathbf{v}} - \tilde{\mathbf{V}}|)P(\tilde{\mathbf{V}})\,d\tilde{\mathbf{V}}, \tag{12.4}$$

where $\sigma(|\tilde{\mathbf{v}} - \tilde{\mathbf{V}}|)$ is the cross section at relative speed $|\tilde{\mathbf{v}} - \tilde{\mathbf{V}}|$ and $P(\tilde{\mathbf{V}})d\tilde{\mathbf{V}}$ is the fraction of the nuclei with velocities in the range $d\tilde{\mathbf{V}}$ about $\tilde{\mathbf{V}}$. To verify the validity of Equation 12.4, multiply both sides by the number densities of neutrons in the beam and nuclei in the target, and note that both sides are equal to the interaction rate per unit volume, I, which is independent of the coordinate system. The quantity $\bar{\sigma}(\tilde{v})$ is then seen to be given by the fundamental operational definition of a cross section; that is, $\bar{\sigma}(\tilde{v}) = I/[n][A]\tilde{v} = I/\phi_n[A]$.

Effect of Target Motion on 1/v Cross Sections

Before proceeding to a more general evaluation of Equation 12.4, let us consider one particularly simple, but significant, special case. Suppose the cross section is inversely proportional to the relative velocity, i.e.,

$$\sigma(|\tilde{\mathbf{v}} - \tilde{\mathbf{V}}|) = \frac{c}{|\tilde{\mathbf{v}} - \tilde{\mathbf{V}}|}$$

or

$$\sigma(v) = \frac{c}{v}, \tag{12.5}$$

where c is a constant. When this expression is inserted in Equation 12.4, the relative velocities cancel, and the result is

$$\bar{\sigma}(\tilde{v}) = \frac{c}{\tilde{v}} \int P(\tilde{\mathbf{V}})d\tilde{\mathbf{V}} = \frac{c}{\tilde{v}}, \tag{12.6}$$

where the last step follows from the normalization of the probability density function $P(\tilde{\mathbf{V}})$. From the above two equations, therefore,

$$\bar{\sigma}(\tilde{v}) = \sigma(v). \tag{12.7}$$

Stated in words, the important conclusion is: *A cross section that varies as 1/v is unaffected by the motion of the target atoms.* Thus the low-energy 1/v tails of reaction resonances are independent of the temperature of the target. We shall make use of this fact in discussing the overall effect of temperature on the shapes of Breit–Wigner reaction resonances.

Target Velocity Distribution

To proceed with the evaluation of Equation 12.4, we must choose a velocity distribution that is characteristic of the target atoms. Let us assume for the moment that the target atoms have a Maxwellian distribution in speeds, as they would if the target were a gas. The distribution of velocities is isotropic in the laboratory system; so the fraction of atoms with speeds in the range $d\tilde{V}$ about $\tilde{V}$ which are moving in solid angle $d\tilde{\Omega}$ is given by the well-known expression

$$P(\tilde{V})d\tilde{V}\,d\tilde{\Omega} = \left(\frac{M}{2\pi kT}\right)^{3/2} \exp\left(-\frac{M\tilde{V}^2}{2kT}\right)\,\tilde{V}^2\,d\tilde{V}\,d\tilde{\Omega}, \tag{12.8}$$

where k is Boltzmann's constant ($k = 8.617 \times 10^{-5}$ ev/°K) and T is the absolute temperature of the gas. The Maxwellian distribution function, which is certainly valid for the atoms in a gaseous target, has been shown to be approximately correct for the atoms in solid targets provided the actual temperature T is replaced by an effective temperature. This subject will be considered again in a later section. Assume there is a monoenergetic, monodirectional beam of neutrons incident on a target along the z-axis. Since the cross section is a function of the relative motion, we must seek an expression for the fraction of target atoms having velocity components along the z-axis in the range $d\tilde{V}_z$ about $\tilde{V}_z$. This distribution function will be called $Q(\tilde{V}_z)$, where, referring to Figure 12.1,

$$\tilde{V}_z = \tilde{V} \cos \tilde{\Theta} \equiv \tilde{V}\tilde{\mu}. \tag{12.9}$$

An atom with speed $\tilde{V}$ will contribute to $Q(\tilde{V}_z)$ if, and only if, it is moving in such a direction that $\tilde{V}\tilde{\mu} = \tilde{V}_z$; hence the fraction of atoms that contribute to $Q(\tilde{V}_z)d\tilde{V}_z$ is

$$Q(\tilde{V}_z)\,d\tilde{V}_z = \int_0^1 P(\tilde{V}_z/\tilde{\mu})\frac{d\tilde{V}_z}{\tilde{\mu}}\,2\pi\,d\tilde{\mu}, \tag{12.10}$$

where $P(\tilde{V}_z/\tilde{\mu})$ is given by Equation 12.8 with $\tilde{V}$ replaced by $\tilde{V}_z/\tilde{\mu}$. The straightforward integration over $\tilde{\mu}$ yields

$$Q(\tilde{V}_z)\,d\tilde{V}_z = \left(\frac{M}{2\pi kT}\right)^{1/2}\exp\left(-\frac{M\tilde{V}_z^2}{2kT}\right)d\tilde{V}_z. \tag{12.11}$$

The probability density function $Q(\tilde{V}_z)$ is normalized over the entire range $(-\infty \le \tilde{V}_z \le +\infty)$. Note that the integral in Equation 12.10 is extended only over the half space of positive projections of $\tilde{V}$ on the z-axis. If it were extended over the entire range, the result would be zero because the angular distribution of atom velocities is isotropic. For the same reason, the distribution function for negative $\tilde{V}_z$ is identical to $Q(\tilde{V}_z)$, i.e., $Q(-\tilde{V}_z) = Q(\tilde{V}_z)$.

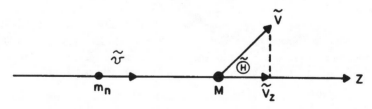

Figure 12.1—The relation between the atom's speed $\tilde{V}$, its z component $\tilde{V}_z$, and the neutron's speed $\tilde{v}$.

Effect of Target Motion on Constant Cross Sections

Let us next consider what is apparently a trivial case. Suppose the cross section is constant; i.e., $\sigma(v) = \sigma_0$. It might be guessed that this cross section would be unaffected by target motion. Actually, the effective cross section in this case,

$$\bar{\sigma}(\tilde{v}) = \frac{\sigma_0}{\tilde{v}} \int |\tilde{\mathbf{v}} - \tilde{\mathbf{V}}| P(\tilde{\mathbf{V}}) \, d\tilde{\mathbf{V}}, \tag{12.12}$$

is a function of target motion because of the quantity $|\tilde{\mathbf{v}} - \tilde{\mathbf{V}}|$ within the integral. If we employ a Maxwellian distribution of velocities for $P(\tilde{\mathbf{V}})$, we find that $\bar{\sigma}$ deviates more and more from σ_0 as (1) the mass of the atom decreases, (2) $\tilde{v}$ decreases, and (3) the temperature increases. All these are obviously related to the value of $|\tilde{\mathbf{v}} - \tilde{\mathbf{V}}|$ in Equation 12.12. When it is a good approximation to set $|\tilde{\mathbf{v}} - \tilde{\mathbf{V}}| = \tilde{v}$, Equation 12.12 reduces immediately to $\bar{\sigma} = \sigma_0$. Otherwise, we find that the effective cross section is somewhat different from the theoretical cross section and is not independent of $\tilde{v}$.

With a Maxwellian velocity distribution for $P(\tilde{\mathbf{V}})$, the integration of Equation 12.12 yields*

$$\bar{\sigma}(\tilde{v}) = \frac{\sigma_0}{x_0} \left[\left(x_0 + \frac{1}{2Ax_0} \right) \text{erf} \left(x_0 \sqrt{A} \right) + \frac{1}{\sqrt{\pi A}} \exp \left(-Ax_0^2 \right) \right]. \tag{12.13}$$

In this equation, A is the mass of the target atom expressed in units of the neutron mass and x_0 is the ratio of the neutron velocity $\tilde{v}$ to the most probable velocity of neutrons having a Maxwellian distribution characterized by the temperature T of the scatterer, that is,

$$A \equiv \frac{M}{m_n}, \tag{12.14}$$

$$x_0 \equiv \frac{\tilde{v}}{\sqrt{2kT/m_n}}.$$

The error function is defined as

$$\text{erf}(a) = \frac{2}{\sqrt{\pi}} \int_0^a e^{-u^2} \, du. \tag{12.15}$$

---

*For details, see p. 33 of the book by V. F. Turchin, *Slow Neutrons*, Reference 2 at the end of this chapter.

For values of x_0 larger than about 4, $\bar{\sigma}$ is almost equal to σ_0, and there is practically no effect due to thermal motion. At the other extreme, for values of $x_0 \ll 1$, the error function behaves as

$$\text{erf}(x_0\sqrt{A}) \simeq \frac{2}{\sqrt{\pi}}x_0\sqrt{A},$$

and Equation 12.13 reduces to

$$\bar{\sigma}(\tilde{v}) = \frac{\sigma_0}{x_0}\frac{2}{\sqrt{\pi A}} = \sigma_0\frac{\frac{2}{\sqrt{\pi}}\sqrt{\frac{2kT}{M}}}{\tilde{v}} = \sigma_0\frac{\langle\tilde{V}\rangle}{\tilde{v}}, \qquad (12.16)$$

where $\langle\tilde{V}\rangle$ is the average thermal velocity of the target atoms. At this extreme any cross section that is constant for fixed targets behaves inversely with the neutron velocity due to thermal motion. This means, for example, that the cross section for neutrons scattered by an atomic-hydrogen gas target will be constant (at 20.4 barns/atom) in the electron-volt range but will behave as $1/\tilde{v}$ at lower neutron energies, diverging toward infinity as $\tilde{v} \to 0$. The higher the temperature of the target, the higher the neutron energy at which the thermal effect would become apparent.

Let us summarize this discussion by stating that: A neutron cross section that is constant in the R-system will have the same constant value in the laboratory provided $\tilde{v} \gg \sqrt{2kT/m_n}$, but it will behave as $1/\tilde{v}$ as $\tilde{v} \to 0$.

Effect of Target Motion on a Breit–Wigner Reaction Resonance

Target motion produces its most dramatic effect on the shapes of Breit–Wigner resonances. Consider, for example, the 6.7-ev resonance in ^{238}U. If the experimental method used in the measurement had infinite precision, the peak value of the resonance would be measured as $\bar{\sigma} = 8000$ barns. Correcting for room-temperature atomic motion, the peak value in the relative coordinate system is $\sigma = 22,000$ barns, a substantial increase. At target temperatures above room temperature, the mean velocity of the target atoms is still higher, resulting in a more pronounced Doppler effect. The decrease in the peak value with temperature is coupled with a broadening of the width of the resonance, which is called "Doppler broadening."

The Breit–Wigner single level, $l = 0$, reaction equation may be written in the form

$$\sigma(E) = \frac{\sigma_0}{1 + \left[\dfrac{E - E_0}{\Gamma/2}\right]^2}, \qquad (12.17)$$

where E is the available energy in the center-of-mass system, E_0 is the available energy at exact resonance, and σ_0 is given by

$$\sigma_0 = 4\pi\lambda_0^2 g(J)\frac{\Gamma_n(E_0)\Gamma_i}{\Gamma^2}\sqrt{\frac{E_0}{E}}, \tag{12.18}$$

where $\Gamma_n(E_0)$ is the neutron width at the resonance energy E_0, Γ_i is the width for the outgoing particle, and Γ is the total width. Note that σ_0 is not a constant; it varies as $E^{-1/2}$.

To find the effective cross section in the vicinity of a resonance, we must express the integrand in the fundamental equation, Equation 12.4, in terms of a single variable. It is convenient to choose the relative neutron-nucleus kinetic energy E for this purpose because the Breit–Wigner equation is expressed in this variable. Since the velocity distribution function is expressed in terms of $\tilde{V}_z$, the first task is to relate $\tilde{V}_z$ and E. Rigorously,

$$E = \tfrac{1}{2}m|\tilde{\mathbf{v}} - \tilde{\mathbf{V}}|^2 = \tfrac{1}{2}m(\tilde{v}^2 + \tilde{V}^2 - 2\tilde{v}\tilde{V}_z). \tag{12.19}$$

This expression can be simplified by restricting the analysis: (1) to heavy nuclei, $M \gg m$, allowing the replacement of the reduced mass m by the neutron mass m_n and (2) to neutron energies considerably in excess of the thermal energy so that $\tilde{V}^2$ may be neglected in comparison to $\tilde{v}^2$. (This latter approximation should not be made when deducing the position E_0 of the resonance from the experimentally observed cross section.) Thus, the simplified expression is

$$E = \tfrac{1}{2}m_n\tilde{v}^2 - m_n\tilde{v}\tilde{V}_z = \tilde{E} - \sqrt{2m_n\tilde{E}}\,\tilde{V}_z,$$

or, solving for $\tilde{V}_z$,

$$\tilde{V}_z = \frac{E - \tilde{E}}{\sqrt{2m\tilde{E}}}, \tag{12.20}$$

where the subscript on m has been dropped since, under the first assumption above, $m_n \simeq m$. Substituting this expression for $\tilde{V}_z$ into the distribution function, Equation 12.11, we get

$$Q(\tilde{V}_z)\,d\tilde{V}_z = \left(\frac{M}{2\pi kT}\right)^{1/2}\exp\left[\frac{-M(E - \tilde{E})^2}{(2m\tilde{E})(2kT)}\right]\frac{dE}{\sqrt{2m\tilde{E}}}. \tag{12.21}$$

One further defining equation

$$\Delta \equiv \sqrt{\frac{4m\tilde{E}kT}{M}}, \tag{12.22}$$

when substituted into Equation 12.21, yields the final result for the distribution function in terms of the relative kinetic energy

$$Q(E)dE = \frac{1}{\Delta\sqrt{\pi}} \exp\left[-\left(\frac{E - \tilde{E}}{\Delta}\right)^2\right] dE. \tag{12.23}$$

The quantity Δ introduced above is called the "Doppler width"; as will be soon made clear, it is the most important parameter determining the effect of thermal atomic motion on effective resonance cross sections. Note that Δ depends directly on temperature and on neutron kinetic energy $\tilde{E}$. Thus the hotter the target material and the higher the energy of the resonance, the greater the Doppler width.

Two of the three factors in the integral of Equation 12.4 have now been expressed in terms of the relative energy E. The remaining factor $|\tilde{v} - \tilde{V}|$ will be approximated by $\tilde{v}$ since as explained above the assumption is that the neutron's velocity is much greater than the velocity of the target nucleus. After substituting the single-level formula (Equation 12.17), the probability density function (Equation 12.23), and $|\tilde{v} - \tilde{V}| = \tilde{v}$ into Equation 12.4, we get

$$\bar{\sigma}(\tilde{E}) = \frac{1}{\Delta\sqrt{\pi}} \int_0^\infty \frac{\sigma_0}{1 + \left(\dfrac{E - \tilde{E}}{\Gamma/2}\right)^2} \exp\left[-\left(\frac{E - \tilde{E}}{\Delta}\right)^2\right] dE. \tag{12.24}$$

If we now assume the resonance to be narrow and lying essentially within a small energy interval, we may neglect the slight energy variation of σ_0 and assign a value to σ_0 that corresponds to the cross section at E_0. This approximation seems even less drastic when one recalls the previously discovered fact that a $1/v$ cross section, like σ_0, is unaffected by the motion of target atoms. Thus

$$\bar{\sigma}(\tilde{E}) = \frac{\sigma_0}{\Delta\sqrt{\pi}} \int_0^\infty \frac{\exp\left[-\left(\dfrac{E - \tilde{E}}{\Delta}\right)^2\right]}{1 + \left(\dfrac{E - \tilde{E}}{\Gamma/2}\right)^2} dE. \tag{12.25}$$

Equation 12.25, or the more rigorous Equation 12.24, is the Doppler-broadened line shape. The manipulation from here on is strictly for mathematical convenience. Defining the deviation from exact resonance in units of the resonance half-width as

$$x \equiv \frac{\tilde{E} - E_0}{\Gamma/2},$$

$$y \equiv \frac{E - E_0}{\Gamma/2}, \tag{12.26}$$

Table 12.1—The ψ-Function*

ξ	0	0.5	1	2	4	8	10	20	40
0.05	0.04309	0.04308	0.04306	0.04298	0.04267	0.04145	0.04055	0.03380	0.01639
0.10	0.08384	0.08379	0.08364	0.08305	0.08073	0.07208	0.06623	0.03291	0.00262
0.15	0.12239	0.12223	0.12176	0.11989	0.11268	0.08805	0.07328	0.01695	0.00080
0.20	0.15889	0.15854	0.15748	0.15331	0.13777	0.09027	0.06614	0.00713	0.00070
0.25	0.19347	0.19281	0.19086	0.18324	0.15584	0.08277	0.05253	0.00394	0.00067
0.30	0.22624	0.22516	0.22197	0.20968	0.16729	0.07042	0.03880	0.00314	0.00065
0.35	0.25731	0.25569	0.25091	0.23271	0.17288	0.05724	0.02815	0.00289	0.00064
0.40	0.28679	0.28450	0.27776	0.25245	0.17359	0.04566	0.02109	0.00277	0.00064
0.45	0.31477	0.31168	0.30261	0.26909	0.17502	0.03670	0.01687	0.00270	0.00064
0.50	0.34135	0.33733	0.32557	0.28286	0.16469	0.03025	0.01446	0.00266	0.00063

*From T. D. Beynon and I. S. Grant, Evaluation of the Doppler-Broadened Single-Level and Interference Functions, *Nucl. Sci. Eng.*, **17**: 545 (1963).

we can then write Equation 12.25 as

$$\bar{\sigma}(\tilde{E}) = \frac{\sigma_0 \Gamma}{2\sqrt{\pi}\Delta} \int_{-2E_0/\Gamma}^{\infty} \frac{\exp\left[-(\Gamma/2\Delta)^2(x-y)^2\right]}{1+y^2}\, dy. \tag{12.27}$$

For almost all resonances of interest, $2E_0/\Gamma$ is a very large number, and, since the integrand of the above equation is small for large values of y, the interval $(-2E_0/\Gamma, \infty)$ may be extended with negligible error to $(-\infty, \infty)$. Equation 12.27 is commonly written in the form

$$\sigma(\tilde{E}) = \sigma_0 \psi(\xi,x), \tag{12.28}$$

where $\psi(\xi,x)$ is defined by the integral

$$\psi(\xi,x) \equiv \frac{\xi}{2\sqrt{\pi}} \int_{-\infty}^{\infty} \frac{\exp\left[-\frac{1}{4}\xi^2(x-y)^2\right]}{1+y^2}\, dy, \tag{12.29}$$

where the parameter ξ is defined as the ratio of natural width to the Doppler width

$$\xi \equiv \frac{\Gamma}{\Delta}. \tag{12.30}$$

The function $\psi(\xi,x)$ is called the "Doppler-broadened line shape" and sometimes the "Voight profile" in honor of the man who first investigated it.* A short tabulation of $\psi(\xi,x)$ is presented in Table 12.1. More extensive tabulations can be found in M. E. Rose et al., USAEC Report WAPD-SR-

*W. Voight, *S. B. bayer. Akad. Wiss.*, p. 603 (1912).

506 (1954), in D. W. Posener, *Aust. J. Phys.*, **12**: 184 (1959), and elsewhere in the technical literature.

Let us now examine a few extreme cases of the Doppler-broadened reaction line shape as given by Equation 12.28.

Example 12.1   Natural Width Much Greater Than Doppler Width,

$\Gamma/\Delta \gg 1$

There is a theorem in the theory of distributions which states that if $S(r)$ is a nonnegative function normalized such that

$$\int_{-\infty}^{\infty} S(r)dr = 1,$$

then

$$\lim_{K \to \infty} KS(Kr)dr = \delta(r),$$

where $\delta(r)$ is the Dirac delta function. Let

$$S(x - y) = \frac{1}{\sqrt{\pi}} \exp[-(x - y)^2],$$

which is nonnegative. Furthermore,

$$\frac{1}{\sqrt{\pi}} \int_{-\infty}^{\infty} \exp[-(x - y)^2]dy = \frac{1}{\sqrt{\pi}} \int_{-\infty}^{\infty} \exp[-r^2]dr = 1,$$

where the change of variables $r = y - x$ has been made. Thus the function $S(x - y)$ satisfies the conditions of the theorem allowing us to write

$$\lim_{\xi \to \infty} \frac{\xi}{2\sqrt{\pi}} \exp\left[-\left(\frac{\xi}{2}\right)^2 (x - y)^2\right] = \delta(x - y).$$

Hence, by comparing the function $S(x - y)$ and the Voight profile, we have

$$\bar{\sigma}(\tilde{E}) = \sigma_0 \int_{-\infty}^{\infty} \frac{\delta(x - y)}{1 + y^2} dy = \frac{\sigma_0}{1 + x^2}$$

or

$$\bar{\sigma}(\tilde{E}) = \frac{\sigma_0}{1 + \left(\dfrac{\tilde{E} - E_0}{\Gamma/2}\right)^2},$$

which is exactly the unbroadened line shape with E replaced by $\tilde{E}$. This lack of Doppler effect was to be expected since small Δ implies low target temperature, hence negligible target motion.

Example 12.2 *Doppler Width Much Greater Than Natural Width,*
$\Gamma/\Delta \ll 1$

The actual line profile can be approximated near the line center by setting $y = 0$ in the exponential of Equation 12.29 so that

$$\bar{\sigma}(\tilde{E}) \simeq \sigma_0 \frac{\xi}{2\sqrt{\pi}} \int_{-\infty}^{\infty} \frac{\exp(-\frac{1}{4}\xi^2 x^2)}{1 + y^2} \, dy,$$

which can be readily integrated to

$$\bar{\sigma}(\tilde{E}) \simeq \sigma_0 \frac{\sqrt{\pi}\,\xi}{2} \exp(-\tfrac{1}{4}\xi^2 x^2). \tag{12.31}$$

This is the "pure Doppler line shape." At the exact center of the resonance, $x = 0$, the measured peak cross section is given by

$$\bar{\sigma}(\tilde{E}_0) \simeq \sigma_0 \frac{\sqrt{\pi}}{2} \frac{\Gamma}{\Delta}, \tag{12.32}$$

which shows that when conditions are such that the Doppler width is much greater than the natural width, the measured peak value of the resonance in the laboratory is less than the value in the R-system by the factor $(\sqrt{\pi}/2)(\Gamma/\Delta)$.

Example 12.3 *Invariance of the Line Area*

In addition to decreasing the observed peak value of a resonance, the Doppler effect broadens the resonance in such a way that the area under the resonance peak remains constant (provided the temperature range is no greater than a few thousand degrees). The invariance of the area is proved as follows: The area under the natural line shape is

$$\int_0^{\infty} \frac{\sigma_0}{1 + \left(\dfrac{E - E_0}{\Gamma/2}\right)^2} \, dE = \frac{2\sigma_0}{\Gamma} \int_{-\infty}^{\infty} \frac{dy}{1 + y^2} = \frac{2\pi\sigma_0}{\Gamma}. \tag{12.33}$$

The area under the broadened peak is

$$\int_0^{\infty} \sigma_0 \psi(\xi,x) d\tilde{E} = \frac{\sigma_0}{\Delta\sqrt{\pi}} \int_{-\infty}^{\infty} dy \int_{-\infty}^{\infty} dx \, \frac{\exp[-(\Gamma/2\Delta)^2(x - y)^2]}{1 + y^2}.$$

Integrating first over x, then over y,

$$\frac{\sigma_0}{\Delta\sqrt{\pi}} \left(\frac{\Delta}{\Gamma} 2\sqrt{\pi}\right) \int_{-\infty}^{\infty} \frac{dy}{1 + y^2} = \frac{2\pi\sigma_0}{\Gamma}, \tag{12.34}$$

showing that the area under the Doppler-broadened resonance is the same as the area under the natural line shape, provided the approximations inherent in the above proof are met. This result implies, for example, that the total activation induced in an optically thin target by a beam of neutrons having a constant velocity spectrum is independent of the temperature of the absorber. The same is not true for an optically thick absorber because of changes in the self-shielding of the absorber material. In fact, the Doppler effect is exploited in some lumped-fuel reactors to provide a prompt negative temperature coefficient, as will be discussed later.

Figure 12.2 shows the effect of temperature on the radiative-capture cross section of two isolated s-wave capture resonances. Both resonances are assumed to have $\Gamma_n(E_0) = \Gamma_\gamma = 0.1$ ev, $g = 1/2$, and total width $\Gamma = 0.2$ ev. One is centered at 0.5 ev; the other, at 5.0 ev. An extremely high temperature ($kT = 1.0$ ev) was chosen to demonstrate clearly the effect of broadening in this small figure. Perhaps the most noteworthy features of this figure are the following: (1) The higher the energy of the resonance, the greater the Doppler decrease in the peak height. This follows from the fact that the Doppler width is proportional to the square root of the resonance energy. (2) There is no Doppler effect on the $1/v$ tails of the resonance, as expected.

The "proof" that the line area is invariant, which is represented by Equations 12.33 and 12.34, obviously lacks rigor, being based on the assumptions that σ_0 is truly constant and that $\psi(\xi,x)$ accurately represents the Doppler-broadened line shape. A direct rigorous analysis shows that the area is a function, though a weak one, of temperature. Specifically, it can be shown to depend on the dimensionless ratio $\gamma = kT/AE_0$, where A is the mass of the target atom expressed in neutron mass units.[*] For values of γ less than 10^{-2} such as exist in nuclear reactors, the line area is within 0.5% constant. However, at extremely high temperatures ($10^5 - 10^{7}°$K), typical of stellar interiors and nuclear explosions, the Doppler area may be as much as a factor of 3 greater than the natural line area.

Effect of Target Motion on a Breit–Wigner Scattering Resonance

The effect of target motion on scattering resonances can usually be ignored simply because the natural widths of scattering resonances are usually much larger than any conceivable Doppler width. The exception to this rule is to be found in the narrow scattering resonances in the heavy elements.

[*] E. H. Canfield. Temperature Dependence of Effective Cross Sections and Areas under Resonances, *Trans. Am. Nucl. Soc.*, **11** (1): 185 (1968).

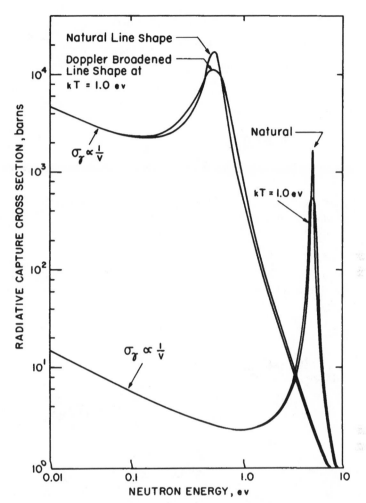

Figure 12.2—Natural ($kT = 0$) and Doppler-broadened ($kT = 1.0$ ev) radiative-capture resonances centered at 0.5 and 5.0 ev with parameters $\Gamma_n(E_0) = \Gamma_\gamma = 0.1$ ev, $\Gamma = 0.2$ ev, and $g = 1/2$.

The procedure for determining the Doppler effect on scattering resonances is essentially identical to that used for reaction resonances. Assuming the resonances are $l = 0$ type, one begins with the single-level, $l = 0$, Breit–Wigner equation for a scattering resonance (Equation 7.23),

$$\sigma_n(E) = 4\pi R^2 + 4\pi \lambda^2 g \left[\left| \frac{\Gamma_n/2}{E - E_0 + i\Gamma/2} + \frac{R}{\lambda} \right|^2 - \left(\frac{R}{\lambda} \right)^2 \right]$$

which can be written

$$\sigma_n(E) = \sigma_p + \sigma_0(1 + y^2)^{-1} + (\sigma_0\sigma_p g)^{1/2}2y(1 + y^2)^{-1}, \tag{12.35}$$

where, by definition,

$$\sigma_0 \equiv 4\pi\lambda^2 g(\Gamma_n/\Gamma)^2, \tag{12.36a}$$

$$\sigma_p \equiv 4\pi R^2, \tag{12.36b}$$

and, as before,

$$y \equiv \frac{E - E_0}{(\Gamma/2)} \tag{12.36c}$$

The procedure from this point on is identical to that used to determine the effective cross section for a reaction resonance. The result is

$$\bar{\sigma}(\tilde{E}) = \sigma_p + \sigma_0\psi(\xi,x) + (\sigma_0\sigma_p g)^{1/2}\chi(\xi,x), \tag{12.37}$$

where $\chi(\xi,x)$, the "Doppler interference function," which arises from the integration over the third term in Equation 12.35, is given by

$$\chi(\xi,x) \equiv \frac{\xi}{2\sqrt{\pi}} \int_{-\infty}^{\infty} \frac{\exp[-\tfrac{1}{4}\xi^2(x - y)^2]}{1 + y^2} 2y \, dy. \tag{12.38}$$

A short tabulation of $\chi(\xi,x)$ is given in Table 12.2. More extensive tables can be found in the literature.

Table 12.2—The χ-Function[*]

ξ	0	0.5	1	2	4	8	10	20	40
0.05	0	0.00120	0.00239	0.00478	0.00951	0.01865	0.02297	0.04076	0.05221
0.10	0	0.00458	0.00915	0.01821	0.03573	0.06626	0.07833	0.10132	0.05957
0.15	0	0.00986	0.01968	0.03894	0.07470	0.12690	0.14096	0.12219	0.05341
0.20	0	0.01680	0.03344	0.06567	0.12219	0.18538	0.19091	0.11754	0.05170
0.25	0	0.02515	0.04994	0.09714	0.17413	0.23168	0.22043	0.11052	0.05103
0.30	0	0.03470	0.06873	0.13219	0.22694	0.26227	0.23199	0.10650	0.05069
0.35	0	0.04529	0.08940	0.16976	0.27773	0.27850	0.23236	0.10437	0.05049
0.40	0	0.05674	0.11160	0.20890	0.32442	0.28419	0.22782	0.10316	0.05037
0.45	0	0.06890	0.13498	0.24880	0.36563	0.28351	0.22223	0.10238	0.05028
0.50	0	0.08165	0.15927	0.28875	0.40075	0.27979	0.21729	0.10185	0.05022

[*] From T. D. Beynon and I. S. Grant, Evaluation of the Doppler-Broadened Single-Level and Interference Functions, *Nucl. Sci. Eng.*, **17**: 545 (1963).

Equivalent Temperatures of Solids

The preceding Doppler-effect equations were derived on the assumption that the target nuclei were moving with a Maxwellian velocity distribution. The validity of this assumption when applied to solids was considered by Lamb.* He showed that $P(\bar{V})$ is indeed Maxwellian for any atom bound in the solid by Hooke's force law, but the temperature parameter in the Maxwellian, T^*, is not, in general, equal to the thermodynamic temperature, T. For high T the velocity distribution is Maxwellian with the effective temperature $T^* = T$. For low T and atoms bound by Hooke's law, the distribution is Maxwellian but with $T^* > T$. Atoms bound other than by Hooke's law will exhibit a low-temperature distribution that is not Maxwellian.

Two simple models of the solid state have been employed to determine T^*. In the Einstein model the atoms are assumed to be bound to lattice sites in such a way that they vibrate independently with frequencies that are integral multiples of the fundamental frequency v_E. The hydrogen in zirconium hydride, for example, behaves approximately this way. The effective temperature is given by

$$T^* = \tfrac{1}{2}\theta_E \coth(\theta_E/2T), \tag{12.39}$$

where the parameter θ_E is the temperature corresponding to the lowest vibrational mode, i.e.,

$$k\theta_E = hv_E. \tag{12.40}$$

At low temperatures most solids behave approximately like the Debye model. In this model the atoms vibrate with a continuous frequency distribution $F(v) = v^2$ from $v = 0$ to some maximum frequency v_D. The effective temperature in this case is given by

$$T^* = \tfrac{3}{2}\theta_D \int_0^1 v^3 \coth(v\theta_D/2T)\, dv, \tag{12.41}$$

where the Debye temperature, θ_D, is the temperature corresponding to the maximum vibrational frequency v_D, i.e.,

$$k\theta_D = hv_D. \tag{12.42}$$

Debye temperatures of solids commonly found in reactors are listed in Table 12.3. A more complete list may be found in the *American Institute of Physics Handbook*, McGraw-Hill Book Company, Inc.

Lamb showed that, provided the condition

$$\Delta^* + \Gamma \gg 2k\theta_D \tag{12.43}$$

*W. E. Lamb, Jr., *Phys. Rev.*, 55: 190 (1939).

Table 12.3—Debye Temperatures of Selected Solids

Solid	$\theta_D(°K)$	Solid	$\theta_D(°K)$
Graphite	~1000	Na	160
Al	375	Ni	413
Be	1160	Pb	96
Cd	165	Sn	195
Fe	355	Ta	230
Hf	213	U	200
In	109	W	270
Mn	410	Zr	265
Mo	360	UO_2	160

is met, the atoms in a crystal at temperature T give the same absorption line shape as they would in a gas at a temperature T^* such that kT^* equals the average energy per vibrational degree of freedom of the crystal. In the above Δ^* is the Doppler width (Equation 12.22) at effective temperature T^*.

The general trend of the relation between T^* and T, as derived from Equation 12.41, is shown in Table 12.4. The ratio T^*/T is shown as a function of the ratio of thermodynamic temperature to Debye temperature. The significant fact that emerges from Tables 12.3 and 12.4 is that the effective temperature T^* is practically the same as the thermodynamic temperature T for most solids at room temperature. The notable exceptions to this rule are graphite and beryllium because of their unusually high Debye temperatures.

Table 12.4—T^*/T as Function of T/θ_D

T/θ_D	T^*/T
0.10	3.77
0.25	1.68
0.50	1.19
0.75	1.09
1.00	1.05
2.00	1.01
≥ 3.00	1.00

Energy of the Resonance Center

The Breit–Wigner reaction equation in the coordinate system in which the compound nucleus is at rest (the R-system) contains a resonance denominator of the form $(E - E_0)^2 + (\Gamma/2)^2$. In the laboratory system the recoil energy of the nucleus $(m_n/M)\tilde{E}$ shifts the peak of the resonance from E_0 to a higher energy. If the nucleus is completely free, the resonance peak will occur at laboratory energy

$$\tilde{E}_0 = E_0 + (m_n/M)E_0. \tag{12.44}$$

If, on the other hand, the nucleus is firmly bound to a solid lattice so that it is not free to recoil, the resonance peak will occur at $\tilde{E}_0 = E_0$. In the intermediate case in which the nucleus is neither completely free nor solidly bound, the peak may be expected to occur at some intermediate energy between these two extremes.

All nuclei are effectively free toward neutrons with energies much greater than the binding energies in molecules ($\sim 1 - 10$ ev) or solids ($\sim 10 - 50$ ev). Thus, a resonance that occurs in the thousand-electron-volt region or above will be shifted as though the nucleus were free, regardless of the physical or chemical form of the target.

Significance of the Doppler Effect in Nuclear Reactors

The mechanisms responsible for the inherent quenching of power excursions in nuclear reactors are many and varied. In highly enriched plate-type thermal reactors, fuel expansion and moderator heating are usually the most significant. But, in low-enrichment UO_2-fueled fast and thermal reactors, the time required for heat transfer delays the effect of fuel and moderator expansion so that these mechanisms are relatively ineffective in preventing catastrophic power surges. It is the Doppler effect that is primarily responsible for the prompt negative temperature coefficient in these reactors.

As the temperature of the reactor core rises, the absorption resonances broaden, and, because the fuel is lumped, the probability of parasitic absorption increases, and the reactivity of the reactor decreases. The magnitude of this effect and, indeed, its sign depend on the degree of enrichment because the fission resonances in the fissile elements (e.g. ^{235}U and ^{239}Pu) are broadening as well as the parasitic resonances in the other materials (e.g., ^{238}U and ^{232}Th). Therefore, it is essential that the reactor designer carefully consider the Doppler effect in order to exploit it to provide adequate reactor stability and inherent safety.

12.2 Interactions in the Chemical Region

The remainder of this chapter is concerned with the interactions of neutrons having energies less than, or of the order of, 1 ev. This energy region is commonly called the "chemical region" because the energy of the neutron is of the same order as the chemical binding energies of atoms in molecules or crystals. In this region two new phenomena come into play:

1. Because the kinetic energy of the neutron is of the order of, or less than, the binding energy of the atoms in the target molecules or solid, the atoms can no longer be considered to be free. Energy interchange between the neutron and the nucleus is quantized and can occur only through the excitation of discrete vibrational and rotational energy levels.

2. Because the neutron wavelength in the chemical region is of the order of the interatomic spacing in molecules or crystals, the waves scattered from various nuclei in the same molecule or crystal interfere with one another and produce essentially all the diffraction and reflection effects commonly associated with X-rays.

Both these phenomena are associated with the structure of the target. If the target were a monatomic gas, neither would occur, and the cross section for the gas would simply be a linear sum of the cross sections of the individual nuclei weighted by their abundances in the gas. Because the targets commonly encountered by neutrons in nuclear reactors are either solids (e.g., graphite, beryllium, zirconium hydride, uranium metal, and uranium oxide) or liquid (e.g., water, heavy water, and liquid sodium), both these structure-dependent phenomena do occur in reactors. Because many reactors are particularly sensitive to the interactions of neutrons in the thermal region, it is essential that the nuclear engineer understand these low-energy effects.

In addition to the two new phenomena enumerated above, the Doppler effect due to the thermal agitation of the target atoms also exists. At low energies the Doppler effect makes constant (scattering) cross sections behave like $1/\bar{v}$ in the laboratory but leaves unaffected $1/v$ (absorption) cross sections. All three phenomena are present in varying degrees depending on the structure and temperature of the target, and with some targets it is practically impossible to quantitatively separate the various contributions to a measured cross section.

Interactions in the chemical region may be divided into the usual three categories: elastic scattering, inelastic scattering, and reactions. An elastic collision is one that leaves the target molecule or crystal in its initial state. An inelastic collision is one that results in an energy exchange between the neutron and the molecule or crystal; the neutron may either lose or gain energy depending on whether the target undergoes a transition from a lower to a higher energy state, or vice versa. A reaction is an interaction in which the neutron disappears. It should be noted that the mechanism of inelastic scattering in the chemical region is quite distinct from what it is in the high-energy region. At high energies inelastic scattering is a nuclear phenomenon; the incident neutron is absorbed by the target nucleus to form a compound nucleus, which then decays by emitting a neutron of lower energy. In the chemical region, on the other hand, inelastic scattering is associated with the bonds between the struck nucleus and the molecule or crystal of which the nucleus is a part. The neutron may be pictured as colliding with a nucleus that is vibrating on the end of a spring, exchanging some energy, and moving off in another direction. No compound nucleus is formed. The same particles exist in the initial and the final constellation, only their energy states have changed. The interaction may be symbolically expressed in the form

$$n + T_i \rightarrow T_f + n, \tag{12.45}$$

where T_i is the target molecule or crystal in its initial state i, and T_f is the target in its final state f. If $i \neq f$, the interaction is called inelastic scattering; if $i = f$, the interaction is called elastic scattering.

Reactions such as (n,γ), (n,p) and (n,f) take place in the chemical region exactly as in any other region. The structure of the target does not affect reaction cross sections except through the motion of the target atoms, the previously discussed Doppler effect. The reason reaction cross sections are unaffected by chemical binding is rather obvious. The binding energies of the atoms in the target molecules or crystals are only of the order of electron volts; reactions, however, take place through the decay of a compound nucleus, a process that involves millions of electron volts.

Although reaction cross sections are unaffected by chemical binding, there is one reaction in which the final constellation may be affected. The daughter nucleus that remains after gamma decay from a low-lying isomeric state may be unable to leave its site in a solid because the recoil energy it receives may be less than the energy with which it is bound to the lattice. This is the source of the well-known and extremely useful Mossbauer effect.

12.3 Scattering by a Single Spinless Nucleus

A brief review will be given to establish the formalism that will be used in the consideration of neutron interactions at very low energies. It will be recalled that the total wave function in the region where the potential is zero can be written in the form

$$\psi(r) = e^{ikz} + f(0)\frac{e^{ikr}}{r} \qquad \text{(for } r > R\text{),} \qquad (12.46)$$

where $f(0)$ is the amplitude of the scattered wave and R is the radius beyond which the potential is zero. In general, $f(0)$ is a function of both the scattering angle 0 and the available energy E of the incident neutron. It was shown in Chapter 4 that $f(\theta)$ was given by a sum over the contributions from the various partial waves:

$$f(\theta) = \frac{1}{2ki} \sum_{l=0}^{\infty} (2l + 1)(e^{2i\delta_l} - 1)P_l(\cos 0). \qquad (12.47)$$

A great simplification is possible in the chemical region because in the low-energy region only the $l = 0$ waves contribute to the cross section, and the differential scattering cross section from a single scattering center is therefore isotropic in the R-system. The scattering amplitude at low energies is simply

$$f = \frac{1}{2ki}(e^{2i\delta} - 1), \qquad (12.48)$$

where $\delta \equiv \delta_0$ is the s-wave phase shift, a complex number in general.

In terms of the scattering amplitude, the elastic-scattering cross section, the reaction cross section, and the total cross section are given, respectively, by

$$\sigma_s = 4\pi|f|^2, \qquad (12.49a)$$

$$\sigma_r = \frac{4\pi}{k}\operatorname{Im}f - 4\pi|f|^2, \qquad (12.49b)$$

$$\sigma_T = \sigma_s + \sigma_r = \frac{4\pi}{k}\operatorname{Im}f, \qquad (12.49c)$$

where Im f is the imaginary part of the scattering amplitude f. Equations 12.49 can be readily verified by substituting the reflection factor $U \equiv e^{2i\delta} = 2kif + 1$ into Equations 6.14, 6.16, and 6.17. Note that Equation 12.49c implies that the scattering amplitude must always contain an imaginary part.

The theoretical calculation of f or its equivalents δ and U was the subject of previous chapters. We shall not be concerned with it in this chapter. Instead we shall take the viewpoint that f has been determined either theoretically or experimentally.

Observed neutron cross sections at low energies tell us a great deal about the energy variation of the scattering amplitude f in the chemical region. Experimentally, the scattering cross section is constant (in the

R-system), and the reaction cross section varies as $1/v$ in the absence of near-by resonances. These facts in conjunction with Equations 12.49 imply that the scattering amplitude f is independent of energy. The $1/v$ nature of low-energy reaction cross sections obtains because of the factor $1/k$ in Equation 12.49b, not because of Im f, which is constant except in the vicinity of a resonance.

Equations 12.49 can also be used in conjunction with experimental data to calculate the values of $|f|$, Im f, and $|$Re $f|$. Note that the algebraic sign of the real part of f cannot be determined from the equations above. In almost all nuclides the imaginary part of f has a much smaller magnitude than the real part. Consider an extreme case to prove this point: The element boron has an elastic-scattering cross section of about 5 barns in the chemical region and a total cross section of 764 barns at 0.025 ev. From Equations 12.49a and 12.49c, therefore,

$$|f| = \sqrt{\sigma_s/4\pi} = 6.3 \text{ fm}$$

and

$$\text{Im} f = \frac{\sigma_T k}{4\pi} = \frac{\sigma_T \sqrt{2mE}}{4\pi\hbar} = 0.2 \text{ fm.}$$

The absolute value of the real part of f is given by

$$|\text{Re } f| = (|f|^2 - |\text{Im } f|^2)^{1/2} \simeq |f| = 6.3 \text{ fm,}$$

and it is seen that even in a strong absorber such as boron the imaginary part of f is only a small fraction of the real part. It is much smaller in most elements; values of the order of 10^{-4} to 10^{-6} are usual for the ratio of Im $f/|$Re $f|$.

In summary, both the real and imaginary parts of the scattering amplitude f are energy independent in the low-energy region (sans near-by resonances), and Im f is positive and much smaller than $|$Re $f|$.

The Scattering Length
For all but a few elements, the algebraic sign of the real part of f is negative. It has become common, therefore, to describe interactions in the low-energy region in terms of the scattering length a, which is related to the scattering amplitude in the low-energy region by the equation

$$a = - \lim_{k \to 0} (\text{Re} f) \simeq - \lim_{k \to 0} f. \tag{12.50}$$

The physical significance of the scattering length is related to the behavior, as the energy approaches zero, of the reduced wave function outside the nucleus, $W(r) = r\psi(r)$. From Equation 12.46

$$W(r)\big|_{\substack{k \to 0 \\ r > R}} = r + \lim_{k \to 0} f = r - a. \tag{12.51}$$

Thus the scattering length a is seen to be the point on the r axis at which the reduced outside wave function goes to zero as $k \to 0$. The wave function is depicted in Figure 12.3 for the common case of a positive scattering length. Replacing r by the nuclear radius R in Equation 12.51, we find that the scattering length is given by

$$a = R[1 - \psi(R)], \tag{12.52}$$

where $\psi(R)$, the radial wave function of the nuclear surface, is determined by the details of the potential. If the nucleus were an impenetrable sphere, $\psi(R) = 0$, and the scattering length would be simply the radius of the sphere. It will be left as an exercise for the student to show that for a more realistic potential (e.g., a nuclear square well) $\psi(R)$ will usually be less than unity; so a is positive but there are some particular combinations of well depth V_0 and radius R that result in $\psi(R) > 1$, hence a negative scattering length.

In subsequent sections the scattering length will be employed in place of the scattering amplitude. The fundamental equation for the total wave function outside the potential region, Equation 12.46, will be

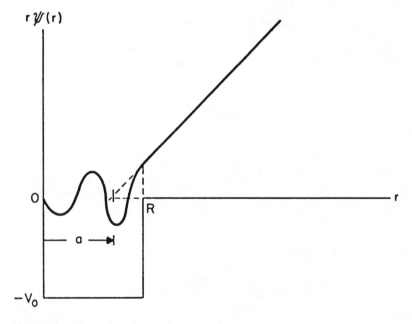

Figure 12.3—The reduced wave function $r\psi(r)$ of s-wave neutrons interacting with a square potential well in the limit as the energy approaches zero.

expressed in the form

$$\psi(r) = e^{ikz} - a\frac{e^{ikr}}{r}.$$ (12.53)

The scattering length is the energy-independent (below 1 ev, at least), real, and usually positive part of the scattering amplitude. The scattering cross section of a single nucleus is given by $4\pi a^2$.

12.4 The Fermi Pseudopotential

To analyze the interaction between a neutron beam and a complex target like a molecule or a crystal, we find it is convenient to have an analytical form for the two-body neutron–nucleus potential. Realistic potential models like the optical model would be extremely cumbersome to use in problems involving many scattering centers. Furthermore, it would be rather a waste of effort to employ such a potential when the answer to the two-body problem is in fact already known in the form of the experimentally determined scattering length.

Fermi* was the first to call attention to the fact that it is possible to use the first Born approximation to calculate the scattering amplitude from individual nuclei in a molecule or crystal provided the neutron energy is very low. He suggested choosing the simplest potential for the interaction, namely, a delta function centered at the position of the nucleus, the origin in the case of a single nucleus:

$$V(\mathbf{r}) = A\delta(\mathbf{r}).$$ (12.54)

The constant A is to be adjusted so that the first Born approximation is forced to give the proper value for the free-nucleus scattering length.

The first Born approximation (Equation 4.142) for the scattering length is given by

$$a = -f(0) = \frac{m}{2\pi\hbar^2} \int_{\text{all space}} e^{ik\mathbf{r}'\cdot(\mathbf{u}_0-\mathbf{u})}V(\mathbf{r}')d\tau',$$ (12.55)

where m is the reduced mass of the system. Upon substituting $A\delta(\mathbf{r}')$ for $V(\mathbf{r}')$ under the integral sign and making use of the defining equation of the delta function of vectorial argument, namely,

$$\int_{\text{all space}} F(\mathbf{r}')\delta(\mathbf{r}' - \mathbf{r}_0)d\tau' = F(\mathbf{r}_0),$$ (12.56)

we get $A = 2\pi\hbar^2 a/m$. Therefore, the Fermi pseudopotential, which used in conjunction with the first Born approximation gives the correct scattering

*E. Fermi, *Ric. Sci.*, 1: 13 (1936).

length for a single free nucleus, is given by

$$V(\mathbf{r}) = \frac{2\pi\hbar^2}{m} a\, \delta(\mathbf{r}).$$ (12.57)

This potential and its use in the Born approximation have been shown to be valid as long as the neutron wavelength is much greater than the scattering length. In the chemical region this condition is always met, since neutron wavelengths are of the order of 10^{-9} cm or longer and scattering lengths are of the order of nuclear radii, 10^{-12} cm.

The Fermi pseudopotential is not unique, as can be most easily verified by writing Equation 12.55 in the form $a = -\lim_{k\to 0} f(\theta)$. The exponential goes to unity, and we find that any potential whose integral is given by

$$\int_{\text{all space}} V(\mathbf{r}')d\tau' = \frac{2\pi\hbar^2}{m} a$$ (12.58)

correctly reproduces the scattering length. The delta function happens to be the most easily handled of all such potentials because it replaces the actual volume of the nucleus by a point located at its center and therefore eliminates the spatial integration.

In the general case of many scattering centers, the Fermi pseudopotential for the jth scattering center is given by

$$V(\mathbf{r} - \mathbf{R}_j) = \frac{2\pi\hbar^2}{m_j} a_j\, \delta(\mathbf{r} - \mathbf{R}_j),$$ (12.59)

where $\mathbf{r}$ and $\mathbf{R}_j$ are the position vectors of the neutron and the jth nucleus, respectively, and m_j and a_j are the reduced mass and the scattering length of the jth nucleus, respectively.

12.5 Reduced-Mass Effect

The reduced-mass effect arises because the scattering length, a, is proportional to the reduced mass of the neutron–nucleus system; see, for example, Equation 12.55. The scattering cross section, which is given by $4\pi a^2$, is thus proportional to the square of the reduced mass of the system,

$$\sigma_s \propto m^2.$$ (12.60)

If the atoms in the target are free, the scattering cross section is given by $\sigma_s = \sigma_{\text{free}}$, the "free-atom scattering cross section" which is constant in the absence of near-by resonances and equal to

$$\sigma_{\text{free}} = 4\pi a^2.$$ (12.61)

The measured scattering cross section will be constant and equal to σ_{free}

at all energies from epithermal down to $E = 0$ provided (1) the atoms in the target are free and (2) appropriate corrections are made for the Doppler effect.

Suppose now the atoms are bound in a molecule or crystal. At neutron energies above about 1 ev the atoms are effectively free, and one measures σ_{free} again. But, as the neutron energy is lowered below the binding energy of the atoms in the molecule or crystal, the atoms are unable to rebound freely, and the effective mass of the scattering center rises. The scattering cross section, being proportional to the square of the reduced mass, behaves as follows:

$$\sigma_s(E) = \sigma_{free} \left(\frac{m_{eff}}{m_{free}} \right)^2,$$
(12.62)

where m_{free} is the reduced mass of the free atom and neutron system and m_{eff} is the effective reduced mass of the bound system,

$$m_{free} = \frac{m_n M}{m_n + M},$$

$$m_{eff} = \frac{m_n M_{eff}}{m_n + M_{eff}}.$$
(12.63)

In these equations m_n is the neutron mass, M is the actual mass of the target atom, and M_{eff} is the effective mass of the target atom. The important point is that the effective mass is a function of neutron energy, varying between the atomic mass M at high energies, where the atom acts as if it were free, and the molecular or crystal mass $\overline{M}$ at very low neutron energies, where the molecule or crystal acts as a rigid system that must recoil as a whole. In the low-energy limit, Equation 12.62 becomes

$$\lim_{E \to 0} \sigma_s(E) = \sigma_{free} \left(\frac{\overline{m}}{m_{free}} \right)^2 \equiv \sigma_b,$$
(12.64)

where $\overline{m}$ is the reduced mass of the neutron–molecule or neutron–crystal system,

$$\overline{m} = \frac{m_n \overline{M}}{m_n + \overline{M}}.$$
(12.65)

The low-energy limit of the scattering cross section is called the "bound-atom cross section" and is designated σ_b, as in Equation 12.64.

The ratio of the bound- and free-atom cross sections depends, as can be seen, on the mass of the scattering atom. For heavy atoms the reduced mass is practically equal to the neutron mass, and the ratio is unity. For the lightest atom, ^{1}H, the effect is most pronounced. Consider four cases, atomic hydrogen gas, molecular hydrogen gas, water, and paraffin ($C_{22}H_{46}$). The molecular masses $\overline{M}$ are respectively 1, 2, 10, and 310 neutron masses. Using Equation 12.64,

Scatterer	σ_b/σ_{free}
H	1
H_2	$(4/3)^2 = 1.78$
H_2O	$(20/11)^2 = 3.31$
$C_{22}H_{46}$	$(620/311)^2 = 3.98$

In an infinite-mass scatterer, such as a crystal, the bound cross section of hydrogen is exactly 4 times the free cross section. Thus the bound cross section of hydrogen will vary between 1.78 and 4 times the free cross section, depending on the mass of the molecule or crystal.

The effect of molecular binding on the hydrogen cross section in H_2 and H_2O is shown in Figure 12.4. Unfortunately, these curves do not show only the reduced-mass effect. Because they are total cross sections, they include other effects, such as inelastic scattering with excitation of vibrational and rotational levels and coherent scattering by the two protons in the molecule. The curve labelled "H_2 gas (corrected)" is actually the raw data obtained by Melkonian* roughly corrected for thermal translational motion and with the $1/\tilde{v}$ capture cross section subtracted. If all the other effects were properly accounted for, the scattering cross section for bound H_2 would be expected to rise to 1.78×20.4 barns per hydrogen atom as $E \rightarrow 0$. Note that, as $E \rightarrow 1$ ev, all the cross sections converge to the constant free-atom scattering cross section of hydrogen.

Some confusion has arisen in the nuclear industry because the curve in the "Barn Book" (USAEC Report BNL-325, 2nd ed., 1958) labeled as the total cross section of ^{1}H is actually the total cross section per hydrogen atom in gaseous H_2, uncorrected for thermal motion, the reduced-mass effect, and the inelastic and coherent effects mentioned previously. These raw data must not be assumed to be the cross section per hydrogen atom regardless of the structure in which the hydrogen is bound. Serious errors can result from this misapplication, particularly if the hydrogenous substance is a neutron moderator or shield.

12.6 Coherent Scattering

Neutrons with energies in the chemical region have de Broglie wavelengths of the order of the interatomic spacings in molecules and solids, i.e., a few

*E. Melkonian, *Phys. Rev.*, 76: 1744 (1949).

Angstrom units, as can be seen by expressing the wavelength in the form

$$\lambda = \frac{h}{\sqrt{2m_n E}} = \frac{0.287 \text{ Å}}{\sqrt{E(\text{ev})}}.$$

(12.66)

When λ is of the order of the interatomic spacing, coherent effects caused by interference among the waves scattered from the various nuclei of the same crystal or molecule become detectable. The interference is particularly noticeable in crystalline targets but also can be observed with amorphous solids, liquids, and even molecular gases. In this section we discuss the basic reason for coherent scattering and departures from coherence, and we define the nomenclature used to describe this effect. In subsequent sections, specific attention will be directed to molecules and solids.

The difficulty students sometimes experience on being initiated into the subject of coherent scattering will not afflict the reader if he will keep

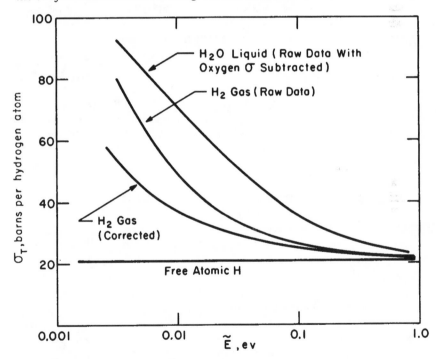

Figure 12.4—The total slow-neutron cross section per proton bound in H_2 gas and in water. The water curve has been corrected by subtracting the oxygen cross section, 3.73 barns/atom, but has not been corrected for thermal motion. Both the raw data and the corrected H_2 curves are shown. The scattering cross section of atomic hydrogen is 20.4 barns. The absorption cross section is 0.33 barns at 0.025 ev and varies as $1/\bar{v}$.

the following facts in mind throughout the remainder of this chapter.

1. Interference effects can occur only between waves that describe neutrons having the same energy and the same spin state.

2. Under the stimulation of the incident-neutron wave $(\exp ikz)$, each nucleus in a target independently emits a scattered wave. In the relative coordinate system attached to the ith nucleus, the scattered wave from this nucleus has the form of a spherical outgoing wave (s-wave)

$$\psi_{s,i} = - a_i \frac{e^{ik_i r_i}}{r_i}, \tag{12.67}$$

where a_i is the scattering length of the ith nucleus, k_i is the wave number of the scattered neutron in the R-system attached to the ith nucleus, and r_i is the distance measured from the ith nucleus.

3. If N of the nuclei scatter neutrons so that these neutrons have the same energy and spin in the laboratory system, then these scattered waves add coherently. The total scattered wave at laboratory point r is simply the sum of the individual waves from all the nuclei, i.e.,

$$\psi_s = \sum_{i=1}^{N} - a_i \frac{e^{ik'|\mathbf{r} - \mathbf{R}_i|}}{|\mathbf{r} - \mathbf{R}_i|}, \tag{12.68}$$

where $\mathbf{R}_i$ is the laboratory position of the ith scattering center at the instant of scattering and $k' = (2m_n E')^{1/2}/\hbar$, E' being the common laboratory energy of the scattered neutrons. The wave number k' of the scattered wave may differ from that of the incident wave k in the event of inelastic scattering. Nevertheless coherence will result if more than one nucleus scatters with the same wave number k'.

4. Perhaps most important of all is the simple fact that coherent scattering affects only the angular distribution of the scattered neutrons; it does not affect the total scattering cross section. If the scattered waves are found to add destructively in certain directions, then it will be found that they add constructively in other directions in just such magnitude as to ensure that the integral over all space of the scattered current remains constant. In other words, neutrons are conserved in scattering.

5. Because scattering lengths are so small (of the order of 10^{-12} cm), scattering by a single nucleus leaves the amplitude of the incident wave, which is normalized to unity before scattering, practically unchanged. However, extinction of the incident wave becomes a significant phenomenon when the target contains many (of the order of 10^{12}) scattering centers.

The existence and the form of coherent elastic scattering can be illustrated quite simply by analyzing the most primitive system capable of interference effects. Consider a plane wave of neutrons incident on a system of two rigidly fixed, identical, spinless nuclei (Figure 12.5). The incident neutrons will be represented by the unit amplitude wave function

$$\psi_{inc} = e^{ikz}. \tag{12.69}$$

Each of the nuclei produces a scattered wave that is spherically symmetrical with respect to its own center. Furthermore, because the nuclei are rigidly fixed in space, the wave number k of the scattered neutrons is identical to that of the incident neutrons, i.e., there is no energy exchange, and the laboratory and relative coordinate systems coincide.

Now let us calculate the total scattered wave that reaches a detector at point $(r,0,\phi)$ where r is much greater than the interatomic distance d.

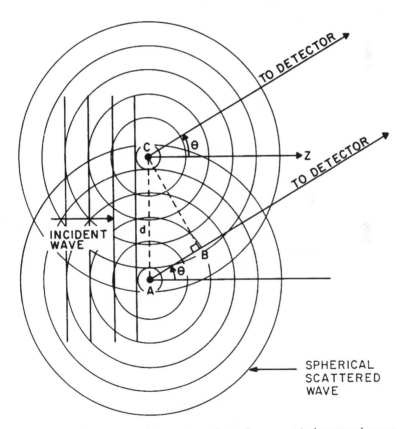

Figure 12.5—Two nuclei at A and C produce spherically symmetrical scattered waves under the influence of an incident-plane wave. The plane transmitted wave that continues to the right is omitted for the sake of clarity.

The scattering angle θ is measured with respect to the direction of the incident current, i.e., the z-axis. The azimuthal angle ϕ is measured in the plane that contains both nuclei and is perpendicular to the z-axis. The reference line for ϕ is the line segment between the two nuclei, i.e., AC in Figure 12.5. If the origin of the L-system is fixed at the center of the first nucleus, then the scattered wave from this nucleus has the form

$$\psi_{s,1}(r,\theta,\phi) = -a\frac{e^{ikr}}{r}, \tag{12.70}$$

where a is the scattering length of the nucleus and r is the distance to the detector. (Because the nuclei are assumed to be rigidly fixed, a is actually the bound-atom scattering length given by $[1 + (m_n/M)]a_{free}$). The scattered-wave function from nucleus 2 at the detector is given by

$$\psi_{s,2}(r,\theta,\phi) = -a\frac{e^{ik(r + \overline{AB})}}{r + \overline{AB}} = -a\frac{e^{ik(r + d\sin\theta\cos\phi)}}{r}, \tag{12.71}$$

where the distance $\overline{AB}$ has been replaced by $d\sin\theta\cos\phi$ in the exponent and by zero in the denominator. This is the usual inconsistency in treatment and is justified by the fact that the exponential is sensitive to small changes in its argument whereas the inverse-r term is practically unaffected by a slight change in r.

The total scattered-wave function at (r,θ,ϕ) is given by

$$\psi_s(r,0,\phi) \equiv \psi_{s,1} + \psi_{s,2} = -\frac{e^{ikr}}{r}[a + a\exp(ikd\sin\theta\cos\phi)]. \tag{12.72}$$

The differential scattering cross section per nucleus is given by one-half the absolute square of the scattered wave (the factor one-half is introduced because there are two nuclei).

$$\sigma_s(\theta,\phi) = \tfrac{1}{2}|a + a\exp(ikd\sin\theta\cos\phi)|^2$$
$$= a^2[1 + \cos(kd\sin\theta\cos\phi)]. \tag{12.73}$$

It is important to recognize that the above cross section is the differential scattering cross section per bound nucleus only in the presence of a second identical nucleus a distance d away. It is not the differential scattering cross section of an isolated bound nucleus, which, as we know, is isotropic and given by $\sigma(\theta) = a^2$ per steradian.

Let us now examine this differential cross section to see how it varies as a function of energy and angle. First consider the extreme cases: At very low energies $kd \ll 1$; so $\cos(kd\sin\theta\cos\phi) \simeq 1 - kd\sin\theta\cos\phi \simeq 1$.

In this case

$$\sigma_s(0,\phi) \simeq 2a^2 \qquad \text{(for } d/\lambda \ll 1\text{),} \qquad\qquad (12.74a)$$

and

$$\sigma_s = \int_{4\pi} \sigma(\theta,\phi) \, d\Omega = 8\pi a^2 \qquad \text{(for } d/\lambda \ll 1\text{).} \qquad (12.74b)$$

The very-low-energy scattering cross section assigned to each nucleus is seen to be isotropic and to have twice the value it would have if the nucleus were isolated.

At very high energies, $kd \gg 1$; so the term $\cos(kd \sin\theta \cos\phi)$ oscillates extremely rapidly between $+1$ and -1 as θ and ϕ take on their range of values. A detector that subtends a finite solid angle $\Delta\Omega$ would average this variation and see an average of $\cos(kd \sin\theta \cos\phi)$ equal to zero; thus, in effect,

$$\sigma_s(0,\phi) \simeq a^2 \qquad \text{(for } d/\lambda \gg 1\text{),} \qquad\qquad (12.75a)$$

and

$$\sigma_s = \int_{4\pi} \sigma(0,\phi) \, d\Omega = 4\pi a^2 \qquad \text{(for } d/\lambda \gg 1\text{).} \qquad (12.75b)$$

The very-high-energy cross section assigned to each nucleus is seen to be isotropic and equal to the value it would have if the nucleus were isolated.

In the intermediate-energy range, when $kd \simeq 1$, the differential cross section is, in general, a function of both θ and ϕ. To simplify the discussion, we will consider only the special case $\phi = 0$ corresponding to scattering in the plane formed by the velocity vector of the incident neutrons and the interatomic line. The differential cross section in this plane is given by

$$\sigma_s(0,0) = a^2[1 + \cos(kd \sin\theta)] \qquad\qquad (12.76)$$

or, in terms of the neutron wavelength, by

$$\sigma(0,0) = a^2\left[1 + \cos\left(2\pi \frac{d}{\lambda} \sin\theta\right)\right]. \qquad (12.77)$$

Plots of $\sigma(0,0)/2a^2$ for various values of d/λ are shown in Figure 12.6. It can be seen that for values of $d/\lambda \geq 1/2$ the cross section varies between $2a^2$ and zero, with the number of minima being equal to the integer part of $2d/\lambda$. For values of $d/\lambda < 1/2$, there is only one minimum, and this minimum is nonzero. Furthermore, the angular distribution becomes isotropic at neutron energies so low that $\lambda \gg d$.

Maxima in the coherent scattering cross section occur whenever the argument of the cosine in Equation 12.77 is an integral multiple of 2π, i.e., when

$$2\pi \frac{d}{\lambda} \sin \theta = n2\pi,$$

where n is an integer. Rearranging, we find

$$n\lambda = d \sin \theta, \tag{12.78}$$

which is a statement of Bragg's law for the situation depicted in Figure 12.5. A more common statement of Bragg's law may be obtained by imposing the condition that the neutrons are incident on the scattering centers at angle θ and are detected at angle θ. The path difference in this event, $2d \sin \theta$, must equal an integral number of wavelengths for maximum constructive interference, hence

$$n\lambda = 2d \sin \theta. \tag{12.79}$$

In a crystalline solid there are many values of d, each one corresponding to the distance between a particular plane of atoms in the crystal. Whenever the neutron energy is such that the Bragg law obtains, a peak is seen in the cross section. This is the reason for the characteristic jagged appearance of the low-energy cross sections of crystalline substances. Each of the peaks corresponds to Bragg scattering from a particular set of crystal planes.

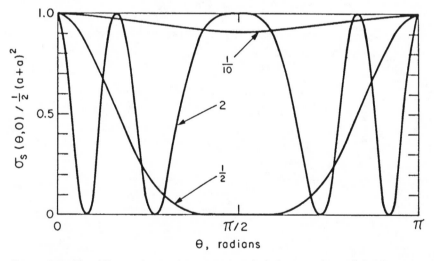

Figure 12.6—The differential cross section per nucleus for two identical, fixed $I = 0$ nuclei as a function of polar scattering angle θ at azimuthal angle $\phi = 0$. The curves are labelled by the ratio d/λ.

If the neutron wavelength is greater than $2d$, Equation 12.79 cannot be satisfied by any value of θ. Hence, coherent elastic scattering cannot occur. The wavelength that is equal to twice the minimum interplane distance in a crystal is called the "Bragg cutoff." Neutrons with wavelengths longer than the Bragg cutoff cannot undergo coherent elastic scattering. The effect of the Bragg cutoff on cross sections can be clearly seen in the low-energy cross sections of solids in USAEC Report BNL-325.

12.7 Sources of Incoherence

There are three fundamental sources of incoherence in low-energy neutron scattering from molecules and crystals. Any target that contains (1) nonidentical nuclei, or (2) nonzero nuclear spins, or (3) nonconstant internuclear spacing will produce some degree of incoherence. In other words, any violation of the conditions that were deliberately imposed in the preceding section to produce pure coherent scattering will result in some incoherence.

Before getting down to details, let us first note that the distinction between coherent and incoherent cross sections is made primarily on an experimental rather than a theoretical basis. When an experiment, such as the scattering of low-energy neutrons off a solid, is performed and it is found that the differential cross section consists of peaks and valleys superimposed on a constant background, the cross section is artificially divided into two components. The part that shows angular dependence is identified with the "coherent cross section," and the constant isotropic background is identified with the "incoherent cross section." The actual division between these two, which is not quite this simple, is explained shortly.

Nonidentical Nuclei—Isotopic Incoherence

Incoherent scattering will result if the nuclei in the scatterer are characterized by more than one scattering length. Targets containing several elements (e.g., NaCl) or one element with several isotopes (e.g., ^{28}Si, ^{29}Si, ^{30}Si) will exhibit this form of incoherence, which is commonly, though inexactly, called isotopic incoherence.

To see how isotopic incoherence arises, consider the two fixed, spinless nuclei again, but assign them different scattering lengths, say, a_1 and a_2, respectively. The analysis is identical to the previous case down to Equation 12.73, which becomes

$$\sigma_s(\theta,\phi) = \tfrac{1}{2}|a_1 + a_2 \exp(ikd \sin\theta \cos\phi)|^2 \qquad (12.80)$$
$$= \tfrac{1}{2}[a_1^2 + a_2^2 + 2a_1a_2 \cos(kd \sin\theta \cos\phi)].$$

As the cosine ranges from $+1$ to -1, the cross section ranges from $(a_1 + a_2)^2/2$ to $(a_1 - a_2)^2/2$. The minimum is not zero, as it was for the identical nuclei, but rather $(a_1 - a_2)^2/2$. This minimum would appear in an experiment as a constant, isotropic background, as in Figure 12.7, and would be identified as resulting partly from incoherent scattering.

The actual splitting up of the measured cross section into coherent and incoherent components proceeds as follows: Consider the general case of N fixed nuclei characterized by scattering lengths $a_n(n = 1, N)$. Set up an origin of a fixed-coordinate system on one of these nuclei, and consider the phase difference between the wave scattered from this nucleus and the nth nucleus which is located at position $\mathbf{R}_n$. As shown in Figure 12.8, this phase difference can be written as

$$k\overline{AB} + k'\overline{BC} = kR_n \cos \theta_1 + k'R_n \cos \theta_2$$

$$= \mathbf{k}\cdot\mathbf{R}_n + \mathbf{k}'\cdot\mathbf{R}_n = (\mathbf{k} + \mathbf{k}')\cdot\mathbf{R}_n = \boldsymbol{\kappa}\cdot\mathbf{R}_n,$$

where use has been made of the definition

$$\boldsymbol{\kappa} \equiv \mathbf{k} + \mathbf{k}'. \tag{12.81}$$

The differential scattering cross section per nucleus may now be written in the convenient and general form

$$\sigma_s(\theta,\phi) = \frac{1}{N} \left| \sum_{n=1}^{N} a_n e^{i\boldsymbol{\kappa}\cdot\mathbf{R}_n} \right|^2 \tag{12.82}$$

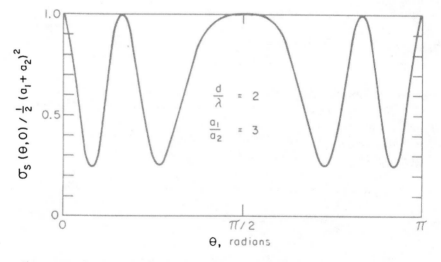

Figure 12.7—Illustrating the effect of isotopic incoherence on the differential scattering cross section per nucleus of two fixed spinless nuclei.

We shall now show that this expression can be factored into two terms, one of which is independent of scattering angle, i.e., independent of $\boldsymbol{\kappa} \cdot \mathbf{R}_n$. To this end, the absolute square in Equation 12.82 is expressed as a double summation

$$\sigma_s(\theta,\phi) = \frac{1}{N} \sum_{n,m} a_n a_m G_n \bar{G}_m, \tag{12.83}$$

where the phase factors have been written, for convenience, in the form

$$G_n \equiv e^{i\boldsymbol{\kappa} \cdot \mathbf{R}_n}, \tag{12.84}$$

and the bar indicates, as usual, a complex conjugate. The indices n and m in Equation 12.83 both range from 1 to N. When the identity substitutions

$$a_n = (a_n - \langle a \rangle) + \langle a \rangle,$$

$$a_m = (a_m - \langle a \rangle) + \langle a \rangle,$$

where $\langle a \rangle$ is the average value of all the scattering lengths; i.e.,

$$\langle a \rangle \equiv \frac{1}{N} \sum_{n=1}^{N} a_n, \tag{12.85}$$

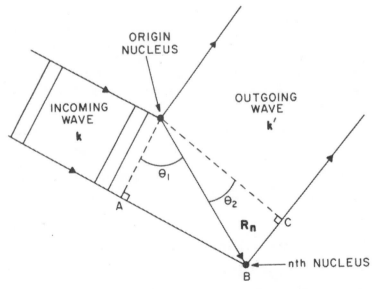

Figure 12.8—Illustrating the fact that the phase difference between the scattered waves from the origin nucleus and the nth nucleus is given by $k\,\overline{AB} + k'\,\overline{BC}$.

are made, the cross section assumes the form

$$\sigma_s(\theta,\phi) = \frac{1}{N} \sum_{n,\,m} [(a_n - \langle a \rangle)(a_m - \langle a \rangle) + \langle a \rangle(a_m - \langle a \rangle)$$
$$+ (a_n - \langle a \rangle)\langle a \rangle + \langle a \rangle\langle a \rangle] G_n \bar{G}_m.$$

When the right-hand side is summed, the second and third terms vanish because the indices m and n have the same range. For the same reason, only the factors with $m = n$ in the first term survive. Hence

$$\sigma_s(\theta,\phi) = \frac{1}{N} \sum_n (a_n - \langle a \rangle)^2 + \frac{1}{N} \sum_{n,m} \langle a \rangle^2 G_n \bar{G}_m.$$

When the first term is summed, there results

$$\sigma_s(\theta,\phi) = \underbrace{\langle a^2 \rangle - \langle a \rangle^2}_{\text{incoherent}} + \underbrace{\frac{1}{N} \langle a \rangle^2 \left| \sum_n e^{i\boldsymbol{\kappa}\cdot\mathbf{R}_n} \right|^2}_{\text{coherent}} \qquad (12.86)$$

where the average of the squared scattering length is defined as

$$\langle a^2 \rangle \equiv \frac{1}{N} \sum_{n=1}^{N} a_n^2. \qquad (12.87)$$

Equation 12.86 is the basis of the definitions of coherent and incoherent scattering. As indicated, the incoherent cross section is defined as that part of the scattering cross section which is independent of angle; the coherent cross section is that part which is dependent on angle. We note that, if all the scattering lengths were identical, then $\langle a^2 \rangle = \langle a \rangle^2$, isotopic incoherence would disappear and the resulting cross section would be completely coherent, as expected.

The scattering cross section integrated over all solid angles is commonly expressed as a sum of two cross sections, corresponding to coherent and incoherent scattering,

$$\sigma_s = \sigma_{\text{coh}} + \sigma_{\text{inc}}. \qquad (12.88)$$

The total scattering cross section is written as

$$\sigma_s = 4\pi\langle a^2 \rangle, \qquad (12.89a)$$

and the coherent cross section as

$$\sigma_{\text{coh}} = 4\pi\langle a \rangle^2. \qquad (12.89b)$$

Finally, the incoherent cross section is determined by subtraction:

$$\sigma_{\text{inc}} = \sigma_s - \sigma_{\text{coh}} = 4\pi[\langle a^2 \rangle - \langle a \rangle^2]. \qquad (12.89c)$$

The squares of the scattering length for total scattering, coherent scattering, and incoherent scattering are defined by means of

Equations 12.89

$$a_s^2 = \langle a^2 \rangle, \tag{12.90a}$$

$$a_{coh}^2 = \langle a \rangle^2, \tag{12.90b}$$

$$a_{inc}^2 = \langle a^2 \rangle - \langle a \rangle^2. \tag{12.90c}$$

The expressions for these cross sections can readily be specialized to the case of a mixture of distinct nuclides. If f_i is the fraction of the nuclei having a scattering length a_i, then

$$\sigma_s = 4\pi \sum_i f_i(a_i)^2, \tag{12.91a}$$

$$\sigma_{coh} = 4\pi \left(\sum_i f_i a_i \right)^2, \tag{12.91b}$$

$$\sigma_{inc} = \sigma_s - \sigma_{coh} = 4\pi \left[\sum_i f_i(a_i)^2 - \left(\sum_i f_i a_i \right)^2 \right] \tag{12.91c}$$

The distinction that has just been made between coherent and incoherent cross sections is unnecessary and somewhat arbitrary, but, since it is universally employed in the interpretation of experimental results, it has considerable practical importance. The definition of the total scattering cross section per nucleus, σ_s, has a clear and unambiguous physical meaning. The arbitrariness lies in the definition of the coherent cross section, σ_{coh}, which represents the cross section that, if attributed to each nucleus in an ideal molecule or crystal, gives the "same amount of interference" observed with the real molecule or crystal. The incoherent cross section is then defined simply as the differences between σ_s and σ_{coh}.

It must be recognized that, under conditions in which the outgoing wave functions from the various scattering centers are characterized by identical energies and spins, the waves add coherently. The resulting cross section is, in the most fundamental sense, completely coherent, regardless of its shape. Trying to ascribe to each nucleus a part of this essentially cooperative phenomenon is analogous to trying to ascribe a quantitative measure of "beauty" to each square centimeter of a Picasso so that the sum of all equals the whole. (Only the most severe critic of modern art would dare suggest a solution to this problem.)

Nonzero Nuclear Spins—Spin Incoherence

A second source of incoherent scattering is spin incoherence, which arises because the potential, hence the scattering length, between a nucleus and a neutron is a function of the relative orientations of their spin vectors. If a nucleus has spin I other than zero, the channel spin j is either $I + 1/2$ or $I - 1/2$, corresponding to parallel and antiparallel orientations of the neutron and nucleus spins. The two resulting scattering lengths will be denoted a_+ and a_-. Without going into details, which are very similar to those for isotopic incoherence, we find that for identical, fixed, $I > 0$ nuclei the scattering cross sections per nucleus are given by

$$\sigma_s = 4\pi(p_+ a_+^2 + p_- a_-^2), \tag{12.92a}$$

$$\sigma_{coh} = 4\pi(p_+ a_+ + p_- a_-)^2, \tag{12.92b}$$

$$\sigma_{inc} = \sigma_s - \sigma_{coh} = 4\pi p_- p_+ (a_+ - a_-)^2, \tag{12.92c}$$

where the quantities p_+ and p_- are the relative probabilities of spin-parallel and spin-antiparallel states. The fractions p_+ and p_- are simply the statistical factor $g(J) = (2J + 1)/(2s + 1)(2I + 1)$ with the total angular momentum J being replaced by the channel spin j since $l = 0$. Hence

$$\begin{aligned} p_+ &= \frac{I + 1}{2I + 1}, \\[2ex] p_- &= \frac{I}{I + 1}. \end{aligned} \tag{12.93}$$

Equations 12.92 are exactly what one would expect from the most primitive of arguments. They are in fact identical in form to those which obtain in the case of two distinct isotopes. This can be verified by replacing $f_1 a_1$ and $f_2 a_2$ by the analogous quantities $p_+ a_+$ and $p_- a_-$ in Equations 12.91 and noting that Equations 12.92 are reproduced.

For most nuclides both a_+ and a_- are positive quantities. Only about 3% of all nuclides have negative scattering lengths. In a couple of nuclides, it happens by chance that the spin-parallel scattering length has opposite sign to that of the spin-antiparallel scattering length and furthermore is of just the right magnitude so that the coherent scattering $(p_+ a_+ + p_- a_-)$ almost vanishes. Ordinary hydrogen is the most famous example of this condition. For ^{1}H, $a_+ = 5.3$ fm, $a_- = -24$ fm, and, of course, $I = 1/2$; therefore

$$\sigma_s = 4\pi(\tfrac{3}{4}a_+^2 + \tfrac{1}{4}a_-^2) = 20.36 \text{ barns},$$

$$\sigma_{coh} = 4\pi(\tfrac{3}{4}a_+ + \tfrac{1}{4}a_-)^2 = 2.0 \text{ barns}.$$

The ^{1}H coherent cross section is seen to be an order of magnitude less than the free-atom hydrogen cross section of 20.36 barns. This is the basis for the common approximation that hydrogen, whether free or bound in a crystal, such as zirconium hydride, scatters incoherently.

The source of the algebraic sign of the scattering length can be traced back to the Breit–Wigner expression for the elastic-scattering cross section,

$$\sigma_s = 4\pi(1 - g)R^2 + 4\pi g \left| R + \frac{\lambdabar_0 \Gamma_n(E_0)/2}{E - E_0 + i\Gamma/2} \right|^2.$$

If only potential scattering is operative in the thermal region, then $\Gamma_n = 0$ and $a_+ = a_- = R$; thus $a_{\text{coh}} = R$ and $a_{\text{inc}} = 0$. If both potential and compound elastic scattering are operative, then either a_+ or a_- (or both a_+ and a_- for $I = 0$) is given by

$$a_{\text{coh}} = R + \frac{g\lambdabar_0 \Gamma_n(E_o)/2}{E - E_0 + i\Gamma/2}.$$

Taking the limit of this equation as $E \to 0$ and making use of the previously discussed fact that the imaginary part of the scattering length is much smaller than the real part, we get

$$\lim_{E \to 0} a_{\text{coh}} \simeq R + \frac{g\lambdabar_0 \Gamma_n(E_0)/2}{-E_0},$$

where the reasonable assumption has been made that $E_0^2 \gg \Gamma^2/4$. Now we see that the algebraic sign of a_{coh} depends critically on the magnitude and sign of the resonance energy E_0. If E_0 is negative (a bound level), then a_{coh} is positive. This is the situation in neutron–proton scattering with spins parallel, since the bound state of the deuteron has $I = 1$. If E_0 is positive (a virtual level) and so small that the resonance term above is greater than R, then a_{coh} will be negative. This is the situation in spin-antiparallel neutron–proton scattering.

The basic reason why so few scattering lengths are negative can now be understood. A negative scattering length requires that the compound nucleus have a scattering resonance close to but above the separation energy of the neutron; that is, E_0 must be positive and so small that $|g\lambdabar_0 \Gamma_n(E_0)/2E_0| > R$. This is a rare occurrence because the level spacing is generally a factor of 10^2 to 10^3 greater than the level width.

Experimental Data

Table 12.5 contains a selected fraction of the available data on slow-neutron scattering amplitudes and cross sections. The third and fourth columns list the coherent-scattering amplitudes and corresponding coherent-scattering cross sections for bound nuclei. Most of these were obtained by measuring diffraction by crystals of known structure or by measuring the critical angle for total neutron reflection. The fifth column lists the total bound elastic-scattering cross sections. These were obtained primarily from transmission experiments with neutrons in the range of 1 to 10 ev. After the absorption cross section is subtracted from the total cross section, the remainder is the free-atom scattering cross section, which is then multiplied by $[(A + 1)/A]^2$ to obtain the bound-atom scattering cross section that is listed.

Table 12.5—Selected Values of a_{coh}, σ_{coh}, and σ_s Obtained from Neutron Diffraction, Reflection, and Transmission Studies[*]

Z	Nuclide or element	Bound-atom values[†]		
		a_{coh}, fm	σ_{coh}, barns	σ_s, barns
1	^{1}H	−3.67	1.78	80
	^{2}H	6.77	5.76	7.4
3	^{6}Li	1.8	0.4	
	^{7}Li	−2.1	0.55	~1
4	^{9}Be	7.8	7.7	7.5
6	^{12}C	6.4	5.2	5.2
7	^{14}N	9.14	10.5	10.5
8	O	5.8	4.2	4.2
12	Mg	5.2	3.4	3.5
13	^{27}Al	3.5	1.5	1.5
26	Fe	9.2	10.6	11.7
	^{54}Fe	4.2	2.2	2.5
	^{56}Fe	10.0	12.6	13
	^{57}Fe	2.3	0.64	2
27	^{59}Co	2.8	1.0	~5
28	Ni	10.3	13.4	17.3
	^{58}Ni	14.7	27.0	27.0
	^{60}Ni	2.8	1.0	1.0
	^{62}Ni	−8.5	9.1	9
29	Cu	7.6	7.3	7.8
30	Zn	5.7	4.1	4.2
40	Zr	7.0	6.2	6.4
50	Sn	6.1	4.6	4.9
55	^{133}Cs	4.9	3.0	~7
73	^{181}Ta	7.0	6.1	7.0
79	^{197}Au	7.7	7.5	~9
82	Pb	9.34	10.9	10.9
90	^{232}Th	9.8	12.1	13

[*]These are basically the values reported by C. G. Shull and E. O. Wollan in a series of articles in the *Physical Review* between 1948 and 1951 updated by the recommended values in USAEC Report BNL-325, 2nd ed., supplement 2, 1966.

[†]These are bound-atom scattering lengths and cross sections, i.e., $a_{coh} = [(A + 1)/A]a_{coh}$ (free) and $\sigma_s = [(A + 1)/A]^2\sigma_s$ (free).

The difference between σ_s and σ_{coh} is the incoherent cross section, from which a_{inc}^2 can be obtained. The incoherent cross section of even–even (zero spin) nuclides should be zero and, indeed, is within experimental error, as can be seen by inspection of columns four and five of Table 12.5.

Isotopic and Spin Incoherence

In the more general case when both nonzero spins and nonidentical nuclides are present in the same target, the general equation for the differential cross section, Equation 12.86, still applies. Only the definition of the average scattering length $\langle a \rangle$ and the average squared scattering length $\langle a^2 \rangle$ must be modified to include the average over spin states. If a_n^+ and a_n^- are the spin-parallel and spin-antiparallel scattering lengths of nucleus n, which has spin I_n, then

$$\langle a \rangle = \frac{1}{N} \sum_{n=1}^{N} \left(\frac{I_n + 1}{2I_n + 1} a_n^+ + \frac{I_n}{I_n + 1} a_n^- \right), \tag{12.94a}$$

$$\langle a^2 \rangle = \frac{1}{N} \sum_{n=1}^{N} \left[\frac{I_n + 1}{2I_n + 1} (a_n^+)^2 + \frac{I_n}{I_n + 1} (a_n^-)^2 \right]. \tag{12.94b}$$

Nonconstant Internuclear Spacing—Thermal Incoherence

The third and final source of incoherence stems from violation of the constancy of the internuclear spacing. There are several obvious conditions under which the internuclear spacing is not constant. For example, the atoms in monatomic gases and amorphous solids are randomly distributed in space; so coherence is effectively wiped out. Of much more interest is the situation in crystalline solids. In these, coherence still exists but is diminished because of defects, such as interstitials and vacancies, and the orientation of the crystallites with respect to one another. But, even in a perfect single crystal, coherence is less than perfect because of the thermal motion of the atoms about their equilibrium positions. As the temperature of such a crystal is increased, the coherent scattering decreases.

The effect of nuclear motion can be taken formally into account by modifying the general expression for the differential cross section, Equation 12.82, to include the possibility of nonfixed nuclei,

$$\sigma_s(\theta, \phi) = \frac{1}{N} \left| \sum_{n=1}^{N} a_n \int e^{i \kappa \cdot \mathbf{R}_n} P_n(\mathbf{R}_n) d\mathbf{R}_n \right|^2, \tag{12.95}$$

where $P_n(\mathbf{R}_n) d\mathbf{R}_n$ is the probability of finding the nth nucleus in volume element $d\mathbf{R}_n$ about point $\mathbf{R}_n$. The quantity P_n is, in other words, the squared wave function $\bar{\psi}_n \psi_n$ of the nth nucleus. It depends on the details of the potential well binding the atom to its lattice site. The integral is rigorously taken over all space, but, since in actuality the probability of finding a nucleus more than a few angstroms from its equilibrium position is vanishingly small, the integral may be pictured as extending only over a volume of the order of a few cubic angstroms. The general development of the theory of slow-neutron scattering is resumed in Section 12.9 after the following brief detour into the sources of inelastic effects.

12.8 Sources of Inelastic Effects

Almost all reactor materials are in the form of molecular liquids or crystalline solids, both of which contain energy levels because they consist of groups of atoms bound together by interatomic forces and subject to the rules of quantum mechanics. These levels are generally discrete and are separated by energies of the order of 0.1 ev in the energy range below about 1 ev, but they tend to overlap and become effectively continuous above approximately 1 ev. They finally become truly continuous above the binding energy of the atom in the molecule or solid, typically 5 ev to 25 ev.

The interaction of a slow neutron with these materials is, of course, conditioned by the distribution of the energy levels. The neutron can lose or gain energy only in discrete amounts, as determined by the intervals between energy levels in the target. If the energy level of the target is unaffected by the neutron, the interaction is called elastic scattering; otherwise it is called inelastic scattering.

To understand neutron interactions with liquids and solids we must examine the energy-level structure in these materials.

Energy Levels in Molecules

The energy levels in molecules are of three basic types: electronic, vibrational, and rotational. The vibrational and rotational levels, whose energies are determined by the motions of the atomic nuclei, are most strongly affected by neutron collision.

The sources of vibrational and rotational levels are most easily understood by consideration of the most primitive of molecules, the diatomic molecule. Diatomic molecules, such as H_2, O_2, and HCl, can rotate as a whole about an axis passing through the center of mass perpendicular to the internuclear axis, and the atoms in the molecule can vibrate relative to one another along the internuclear axis.

The most elementary possible model to explain the form of the vibrations in a diatomic molecule is that each atom moves along the internuclear axis in simple harmonic motion. In effect, the molecule is pictured as a pair of masses bound together by an elastic spring, a simple harmonic vibrator. The classical analysis of this problem is well known to the reader. The two-body problem is replaced by the equivalent one-body problem of a mass point having reduced mass $m = m_1 m_2/(m_1 + m_2)$ which is acted on by a force proportional to its displacement from an equilibrium position, Hooke's force law,

$$F = -k(x - x_0), \tag{12.96}$$

where k is called the force or spring constant. Newton's second law of motion,

$$-k(x - x_0) = m \frac{d^2x}{dt^2},$$

then yields the solution

$$x = x_0 \sin (2\pi v_v t + \phi),$$

where the classical vibrational frequency v_v is given by

$$v_v = \frac{1}{2\pi} \sqrt{\frac{k}{m}}. \tag{12.97}$$

The quantum-mechanical analysis consists of solving the Schrödinger equation for the reduced-mass particle that is acted on by a potential corresponding to Hooke's force law,

$$\frac{d^2\psi}{dx^2} + \frac{2m}{\hbar^2} (E - \tfrac{1}{2}kx^2)\psi = 0.$$

It is found that there are eigensolutions only for the discrete set of values of total energy given by

$$E_v = hv_v(n + \tfrac{1}{2}), \tag{12.98}$$

where v_v is the classical vibrational frequency and the vibrational quantum number, n, can take on only integer values 0, 1, 2, ... The vibrational eigenfunctions, ψ_v, are found to be the Hermite polynomials displayed in Chapter 3. For our purposes, it should be particularly noted that the energy-level spectrum of the simple harmonic vibrator consists of a series of equidistant levels, ranging upward from a minimum of $hv_v/2$, each separated from the next by an energy of hv_v.

Now let us consider the rotation of the molecule. The simplest model now consists of ignoring vibration and considering the molecule to be a rigid rotator, i.e., a pair of mass points bound together by a weightless rigid rod. If r_1 and r_2 are the distances of masses m_1 and m_2 from the center of mass of the system, then the classical energy of rotation is given by

$$E_r = \tfrac{1}{2}I\omega^2,$$

where $I = m_1 r_1 + m_2 r_2$ is the moment of inertia of the system, and ω is its angular velocity. This two-body problem is readily reduced to an equivalent one-body problem: Upon defining $r = r_1 + r_2$ and noting that $r_1 = [m_2/(m_1 + m_2)]r$ and $r_2 = [m_1/(m_1 + m_2)]r$, we find $I = [(m_1 m_2)/(m_1 + m_2)]r^2$ or $I = mr^2$; that is, the moment of inertia of the molecule is identical to the moment of inertia of a reduced-mass point located a distance r from the axis of rotation.

The quantum-mechanical analysis consists in solving the three-dimensional Schrödinger equation for the reduced-mass point. The potential is set equal to zero since no potential energy is associated with a rigid rotator. Energy eigenvalues appear because of the quantization of the orbital angular momentum. The rotational-energy eigenvalues are given by

$$E_r = \frac{\hbar^2}{2mr^2} J(J + 1) = \frac{\hbar^2}{2I} J(J + 1), \tag{12.99}$$

where, in accordance with common usage, the symbol J has been employed for the rotational quantum number instead of the symbol l and can assume any of the integer values 0, 1, 2, ... The rotational eigenfunctions, ψ_r, are the surface harmonic functions discussed in Chapter 4.

The rotational levels, unlike the vibrational levels, are not evenly spaced but are spread out approximately as J. In addition, the lowest energy level of the rotator is zero. More significant is the difference in magnitude of the energies associated with rotation and vibration. The energy interval between adjacent vibrational levels is usually of order 10^{-1} ev; between rotational levels, only about 10^{-2} ev. So the vibration-rotation energy-level diagram consists of a series of evenly spaced vibration levels, each of which has a satellite band of rotation levels. (Figure 12.9.)

Actual molecules are not as simple as the ideal picture we have constructed of them. Their vibrational motion is not purely harmonic, and their rotation is not purely rigid. Both these models are elementary first approximations to the truth.

The possible modes of motion in triatomic and larger molecules are more complicated. In general, if the molecule contains N atoms, there are $3N$ degrees of freedom. Six of these describe translations and rotations of the molecule as a rigid body. The remaining $3N - 6$ degrees of freedom (or $3N - 5$ in the special case of a linear configuration of atoms) represent oscillatory modes of motion. All modes other than the translational modes are quantized and give rise to inelastic slow-neutron scattering.

Energy Levels in Solids

So far as neutron scattering is concerned, the most significant attribute of any solid is its frequency spectrum $f(v)$, the number of frequencies per unit frequency range. As mentioned previously, two highly simplified

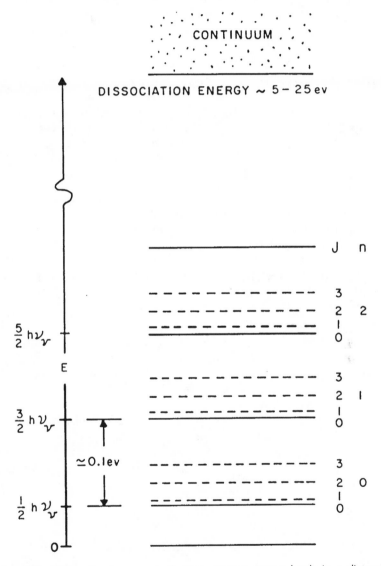

Figure 12.9—Schematic diagram of vibration–rotation energy levels in a diatomic molecule. Three rotational levels are shown for each of the first three vibrational levels.

models have been used to approximate $f(v)$. Historically, the oldest is due to Einstein who assumed that $f(v)$ was a series of delta functions at a fundamental frequency and its overtones. This was followed by the Debye model, in which $f(v)$ is proportional to v^2 from $v = 0$ to some maximum frequency v_D. Both these models, though crude, have some validity and still are used to some extent today. The Einstein model, for example, yields surprisingly good results for the scattering of neutrons off the metal hydrides ZrH_x and MgH_2. However, in general, the frequency spectra of most crystals are far more complex than that predicted by either the Einstein or the Debye model (see Figure 12.12).

The quanta of lattice vibrations in crystals are called phonons. The energy of a phonon of frequency v is hv, and the energy spectrum of phonons in a solid is $f(hv)$. The cross section for inelastic scattering in a solid may be regarded as arising from the interactions between the neutron and the crystal that involve the creation or destruction of one or more phonons. If no phonon is created or destroyed, the interaction is called elastic.

Energy Levels in Liquids

The liquids of primary interest to nuclear engineers are the reactor moderators and coolants: the molecular liquids, H_2O and D_2O, and the metallic liquids, Na and NaK.

Theories of the liquid state are not nearly so advanced as those of the gaseous and solid states. Neutron scattering is being employed primarily as a tool to check simplified models so as to develop theories of the liquid state. It is becoming increasingly clear that the liquid state is not completely disordered, like the gaseous state, but rather bears a resemblance to the polycrystalline solid state. There seem to be strong indications that groups of molecules bind together to form temporary globules with regular crystalline-type structures and that these globules diffuse, break up, and rearrange themselves in a complex dynamical dance. The quantized rotation and vibration levels that exist in isolated molecules are smeared out by the near proximity of other molecules; thus the frequency spectrum of a liquid is continuous like that of a solid (see Figure 12.15). Inelastic scattering from solid sodium has been found to be very similar to that from liquid sodium, indicating that the solid-type frequency spectrum undergoes but little change when a metal passes from the solid to the liquid state.

12.9 Theory of Slow-Neutron Scattering in Solids and Liquids

The theoretical analysis of slow-neutron scattering in solids and liquids has been based almost without exception on the space–time formulation of Van Hove.* The objective is to find the amplitude of the total neutron wave scattered from a system of moving nuclei when a monodirectional, monoenergetic neutron beam is incident.

The incident-neutron wave function is written as

$$\psi_{inc}(\mathbf{r},t) = e^{i(\mathbf{k_0} \cdot \mathbf{r} - \omega_0 t)}, \tag{12.100}$$

where $\hbar \mathbf{k_0}$ and $\hbar \omega_0$ are the neutron momentum and energy, respectively. Each nucleus, n, is assumed to provide a potential of the form of the Fermi pseudopotential.

$$V_n(\mathbf{r},t) = \frac{2\pi \hbar^2}{m} a_n \, \delta[\mathbf{r} - \mathbf{R}_n(t)], \tag{12.101}$$

where $\mathbf{R}_n(t)$ is the position of nucleus n at time t, and a_n and m are the bound scattering length and reduced mass of nucleus n, respectively. (It is assumed that the nuclei are equally massive.)

The total wave function at any time t due to interactions that have occurred at time t_0 is given by the solution to the time-dependent Schrödinger equation with $\mathbf{R}_n(t)$ replaced by $\mathbf{R}_n(t_0)$

$$i\hbar \frac{\partial}{\partial t} \psi(\mathbf{r},t) = \left\{ -\frac{\hbar^2}{2m} \nabla^2 + \sum_n \frac{2\pi \hbar^2}{m} a_n \, \delta[\mathbf{r} - \mathbf{R}_n(t_0)] \right\} \psi(\mathbf{r},t). \tag{12.102}$$

Making the usual first Born approximation that the source of the scattered wave, ψ_s, is the incident wave, ψ_{inc}, rather than the rigorously correct total wave ψ, we have,

$$\left(\frac{\hbar^2}{2m} \nabla^2 + i\hbar \frac{\partial}{\partial t} \right) \psi_s(\mathbf{r},t) = \frac{2\pi \hbar^2}{m} \sum_n a_n \, \delta[\mathbf{r} - \mathbf{R}_n(t_0)] \psi_{inc}(\mathbf{r},t_0). \tag{12.103}$$

The Green's function for this equation can be determined by replacing the summation on the right-hand side of Equation 12.103 by a delta-function source in space and time. The equation

$$\left(\nabla^2 + \frac{2mi}{\hbar} \frac{\partial}{\partial t} \right) \psi_0(\mathbf{r} - \mathbf{r}', t - t_0) = 4\pi \delta(\mathbf{r} - \mathbf{r}') \delta(t - t_0) \tag{12.104}$$

has the solution

$$\psi_0(\mathbf{r} - \mathbf{r}', t - t_0) = \sqrt{\frac{m}{2\pi \hbar}} \frac{i}{(t - t_0)^{3/2}} \exp\left[\frac{im|\mathbf{r} - \mathbf{r}'|^2}{2\hbar(t - t_0)} \right]. \tag{12.105}$$

(The nuclear engineer will recognize this Green's function. Except for constants and factors of i, it has the form of the Fermi age solution for the slowing down of neutrons from a point isotropic source.)

*L. Van Hove, *Phys. Rev.*, 95: 249 (1954).

The total scattered wave is the superposition of the Green's function solutions representing waves originating throughout the scatterer at all earlier times. That is, integration of the kernel ψ_0 must be carried out over both space and time. Thus

$$\psi_s(\mathbf{r}',t) = \int_{\mathbf{r},t_0} d\mathbf{r}\, dt_0\, \psi_0(\mathbf{r} - \mathbf{r}',t - t_0) \sum_n a_n \delta[\mathbf{r} - \mathbf{R}_n(t_0)]\psi_{inc}(\mathbf{r},t_0).$$

$$(12.106)$$

Substituting Equations 12.100 and 12.105 for ψ_{inc} and ψ_0, respectively, into Equation 12.106, we find that the scattered wave is given by

$$\psi_s(\mathbf{r}',t) = i\sqrt{\frac{m}{2\pi\hbar}} \int_{-\infty}^{t} dt_0 (t - t_0)^{-3/2} \sum_n a_n \int d\mathbf{r}\, \delta[\mathbf{r} - \mathbf{R}_n(t_0)]$$

$$\times \exp\left[\frac{im|\mathbf{r} - \mathbf{r}'|^2}{2\hbar(t - t_0)}\right] \exp[i(\mathbf{k}_0\cdot\mathbf{r} - \omega_0 t)],$$

$$(12.107)$$

where, it is to be noted, all spatial integrals are over the volume of the target. Equation 12.107 is the complete solution but in a very inconvenient form. The scattered wave is given as a function of the time in terms of the history of the positions of the scattering centers. To compare with experiment, we need the momentum ($\hbar k$) and energy ($\hbar\omega$) spectra of the scattered neutrons. They can be found by carrying out a Fourier transformation in the variables ω and t. The scattered wave function is expressed in the form

$$\psi_s(\mathbf{r}',t) = \sum_{\omega'} f(\mathbf{r}',\omega')e^{i\omega't},$$

$$(12.108a)$$

and the Fourier coefficients are given by

$$f(\mathbf{r}',\omega') = \frac{1}{T} \int_0^T dt\, e^{i\omega't}\psi_s(\mathbf{r}',t),$$

$$(12.108b)$$

where T, the period of the Fourier transformation, is a time that is long relative to the transit time of the neutron between scatterer and detector.

After considerable mathematical manipulation, which will not be repeated here,* the amplitude of the scattered wave is reduced to the form

$$f(\mathbf{k},\omega) = \frac{1}{Tr'} \int_0^T dt_0\, e^{i\omega't_0} \sum_n a_n \int d\mathbf{r}\, \delta[\mathbf{r} - \mathbf{R}_n(t_0)]e^{i\boldsymbol{\kappa}\cdot\mathbf{r}},$$

$$(12.109)$$

where $\boldsymbol{\kappa} = \mathbf{k}_0 - \mathbf{k}$ is the "momentum" change and $\omega' = \omega_0 - \omega$ is the "energy" change of the scattered neutron. The scattered intensity in a

---

* See the works by Egelstaff and by Turchin, References 2 and 3 at the end of this chapter.

small energy interval, ΔE, is given by the absolute square of the scattering amplitude at frequency ω times the number of frequencies in ΔE. The number of terms in ΔE is $T\Delta E/2\pi\hbar$, hence

I_s = Scattered intensity in ΔE

$$= \frac{T\Delta E}{2\pi\hbar}|f(\mathbf{k},\omega)|^2$$

$$= \frac{\Delta E}{h}\frac{1}{T(r')^2}\int_0^T dt_0\int_0^T dt\, e^{-i\omega'(t_0-t)}\sum_{m,n}\bar{a}_m a_n \tag{12.110}$$

$$\times \int d\mathbf{r}''\int d\mathbf{r}\, e^{i\mathbf{\kappa}\cdot(\mathbf{r}-\mathbf{r}'')}\,\delta[\mathbf{r}''-\mathbf{R}_m(t)]\delta[\mathbf{r}-\mathbf{R}_n(t_0)],$$

where the bar indicates a complex conjugate. Introducing the variable $\tau \equiv t_0 - t$ and assuming that the time T greatly exceeds the maximum time during which the motions of any two atoms are correlated, so that the limits of integration over τ are effectively infinite, we find

$$I_s = \frac{\Delta E}{h}\frac{1}{(r')^2 T}\int_0^T dt_0\int_{-\infty}^\infty d\tau\, e^{-i\omega'\tau}\sum_{m,n}\bar{a}_m a_n$$

$$\times \int d\mathbf{r}''\int d\mathbf{r}\, e^{i\mathbf{\kappa}\cdot(\mathbf{r}-\mathbf{r}'')}\,\delta[\mathbf{r}''-\mathbf{R}_m(\tau-t_0)]\delta[\mathbf{r}-\mathbf{R}_n(t_0)]. \tag{12.111}$$

The integration with respect to t_0 is carried out formally by replacing the delta-function product with its time-average value,

$$I_s = \frac{\Delta E}{h}\frac{1}{(r')^2}\int_{-\infty}^\infty d\tau\, e^{-i\omega'\tau}\sum_{m,n}\bar{a}_m a_n$$

$$\times \int d\mathbf{r}''\int d\mathbf{r}\, e^{i\mathbf{\kappa}\cdot(\mathbf{r}-\mathbf{r}'')}\langle\delta[\mathbf{r}''-\mathbf{R}_m(0)]\delta[\mathbf{r}-\mathbf{R}_n(\tau)]\rangle. \tag{12.112}$$

The differential cross section is given by r' times the scattered flux per unit energy interval, $I_s v/\Delta E$, divided by the incident flux, which is normalized to v_0. Thus,

$$\sigma(E_0|E,\theta) = \frac{1}{h}\frac{v}{v_0}\int_{-\infty}^\infty d\tau\, e^{-i\omega'\tau}\sum_{m,n}\bar{a}_m a_n$$

$$\times \int d\mathbf{r}'\int d\mathbf{r}\, e^{i\mathbf{\kappa}\cdot\mathbf{r}}\langle\delta[\mathbf{r}'+\mathbf{R}_m(0)-\mathbf{r}]\delta[\mathbf{r}-\mathbf{R}_n(\tau)]\rangle, \tag{12.113}$$

where the change of dummy variables $\mathbf{r}' = \mathbf{r} - \mathbf{r}''$ has been made within the spatial integrals.

The formidable-looking Equation 12.113 can be reduced to a more appealing form by introducing the definitions of the space–time correlation functions. This will also lead to the splitting of the differential cross section into coherent and incoherent parts. The space–time correlation function of the system is defined by Van Hove as

$$G(\mathbf{r}',\tau) \equiv \frac{1}{N} \sum_{m,n=1}^{N} d\mathbf{r} \langle \delta[\mathbf{r}' + \mathbf{R}_m(0) - \mathbf{r}]\delta[\mathbf{r} - \mathbf{R}_n(\tau)]\rangle. \tag{12.114}$$

This function can be further split into two parts, corresponding to diagonal terms in the summation $m = n$, and off-diagonal terms $m \neq n$. The diagonal terms produce the self-correlation function G_s, which is a measure of the positions of the same atom as a function of time; the off-diagonal terms produce the distant correlation function G_d, which is a measure of the relative positions of two different atoms as a function of time. The definitions are

$$G_s(\mathbf{r}',\tau) \equiv \frac{1}{N} \sum_{n=1}^{N} \int d\mathbf{r} \langle \delta[\mathbf{r}' + \mathbf{R}_n(0) - \mathbf{r}]\delta[\mathbf{r} - \mathbf{R}_n(\tau)]\rangle, \tag{12.115}$$

$$G_d(\mathbf{r}',\tau) \equiv \frac{1}{N} \sum_{n \neq m=1}^{N} d\mathbf{r} \langle \delta[\mathbf{r}' + \mathbf{R}_m(0) - \mathbf{r}]\delta[\mathbf{r} - \mathbf{R}_n(\tau)]\rangle. \tag{12.116}$$

Introducing these definitions into Equation 12.113, the differential cross section can now be written as a sum of coherent and incoherent cross sections

$$\sigma(E_0|E,\theta) = \sigma_{\text{coh}}(E_0|E,\theta) + \sigma_{\text{inc}}(E_0|E,\theta), \tag{12.117}$$

where

$$\sigma_{\text{coh}}(E_0|E,\theta) = \frac{\langle a \rangle^2}{h} \frac{v}{v_0} \int \int d\mathbf{r}' \, d\tau \, e^{i(\mathbf{\kappa \cdot r}' - \omega'\tau)} G(\mathbf{r}',\tau) \tag{12.118}$$

and

$$\sigma_{\text{inc}}(E_0|E,\theta) = \frac{[\langle a^2 \rangle - \langle a \rangle^2]}{h} \frac{v}{v_0} \int \int d\mathbf{r}' \, d\tau \, e^{i(\mathbf{\kappa \cdot r}' - \omega'\tau)} G_s(\mathbf{r}',\tau). \tag{12.119}$$

The average scattering length $\langle a \rangle$ and squared scattering length $\langle a^2 \rangle$ are defined by Equations 12.85 and 12.87, respectively. As should have been expected, the incoherent cross section depends only on the self-correlation function, whereas the coherent cross section depends on both self- and distant correlation.

When both isotopic and spin incoherence are absent, $\langle a^2 \rangle = \langle a \rangle^2 = a^2$ and the differential cross section becomes

$$\sigma(E_0|E,\theta) = \frac{a^2}{h} \frac{v}{v_0} \int d\mathbf{r}'\, d\tau\, e^{i(\kappa \cdot \mathbf{r}' - \omega'\tau)} G(\mathbf{r}',\tau)$$

$$= \frac{\sigma_b}{4\pi} \sqrt{\frac{E}{E_0}}\, \widetilde{S}(\kappa,\omega'), \tag{12.120}$$

where use has been made of the fact that the bound-atom cross section σ_b equals $4\pi a^2$, and h^{-1} times the integral over space and time has been defined as the function $\widetilde{S}(\kappa,\omega')$. This function has been shown to obey the detailed balance condition between the energy-loss and energy-gain cross sections, namely,

$$\widetilde{S}(-\kappa,-\omega') = e^{-\hbar\omega'/kT} \widetilde{S}(\kappa,\omega'), \tag{12.121}$$

where k is Boltzmann's constant and T is the absolute temperature of the scatterer. It has also been shown that the function $S'(\kappa,\omega')$, defined by

$$S'(\kappa,\omega') \equiv e^{-\hbar\omega'/2kT} \widetilde{S}(\kappa,\omega'), \tag{12.122}$$

is even in both κ and ω'. It has become common practice to define a dimensionless function $S(\kappa,\omega')$ as $S'(\kappa,\omega')$ times kT:

$$S(\kappa,\omega') \equiv (kT)S'(\kappa,\omega'). \tag{12.123}$$

The function $S(\kappa,\omega')$ is commonly called "the scattering law." In terms of the scattering law, the double differential cross section, Equation 12.120, can be written as

$$\sigma(E_0|E,\theta) = \frac{\sigma_b}{4\pi kT} \sqrt{\frac{E}{E_0}}\, e^{-\hbar\omega'/2kT} S(\kappa,\omega'), \tag{12.124}$$

in which, it should be recalled, $\hbar\omega' = E_0 - E$ and $\kappa = \mathbf{k}_0 - \mathbf{k}$ are the energy and momentum changes, respectively.

12.10 Measured Scattering Laws in Solids and Liquids

The Van Hove theory of slow-neutron scattering in solids and liquids, though elegant and rigorous, is not capable today of yielding calculated double differential cross sections of sufficient accuracy to preclude the need for their measurement. The main stumbling block to application of the theory is the difficulty attendant on the calculation of the space–time correlation function, which depends on details of the dynamic structure of the scatterer, some of which are unknown. Partial success has been achieved in the analysis of simple solids, but liquids have to the present time essentially defied theoretical analysis. Much research is being aimed at turning the problem around so as to determine the dynamic structure of liquids from measured neutron-scattering laws.

Most of the experimental data on double differential cross sections is presented in the form of plots of the scattering law. The double differential cross section is commonly expressed in the form

$$\sigma(E_0|E,\theta) = \frac{\sigma_b}{4\pi kT}\sqrt{\frac{E}{E_0}}\,e^{-\beta/2}S(\alpha,\beta), \tag{12.125}$$

where

E_0, E = the incident and final neutron energies (in the laboratory system

$\alpha = [E_0 + E - 2(E_0 E)^{1/2}\cos\theta]/AkT$

$\beta = (E - E_0)/kT$

A = mass of principal scattering atom in neutron mass units

k = Boltzmann constant (8.62×10^{-5} ev/°K)

T = temperature of scatterer (°K)

σ_b = bound-atom cross section of principal scattering atom

Equation 12.125 is merely a restatement of Equation 12.124, with the momentum exchange κ and the energy exchange ω' being replaced by the dimensionless momentum exchange α and energy exchange β. In the definitions of A and σ_b, the words "principal scattering atom" refer to situations in which the target is composed of more than one atomic species. The principal scattering atom is then chosen as the one with the largest macroscopic scattering cross section. For example, the principal scattering atom in H_2O is hydrogen, and in D_2O it is deuterium. When used to describe polyatomic scatters, Equation 12.125 is not a completely rigorous statement but, rather, should be considered as a convenient semitheoretical expression for fitting experimental data.

Two considerable practical advantages flow from writing the double differential cross section in the form of Equation 12.125. The number of independent variables is reduced from three (E_0, E, θ) to two (α, β). Thus plots or tables of the so-called scattering law $S(\alpha, \beta)$ over a range of α and β are sufficient to describe the differential cross section $\sigma(E_0|E,\theta)$ over a range in the three variables E_0, E, and θ. The second advantage is far more significant to the experimentalist. He can determine $S(\alpha, \beta)$, hence the double differential cross section in the entire chemical region, by measurements with incident neutrons of only a few energies. The amount of data required is thus cut down by an order of magnitude or more.

From the theoretical standpoint, this formulation is particularly inviting because the frequency distribution of the phonons in solids and liquids can, in some cases, be obtained directly from the experimental scattering law. It has been shown,* for example, that for an incoherent scatterer with a Gaussian self-correlation function the phonon frequency distribution $f(\beta)$ is given by

$$f(\beta) = 2\beta \sinh(\beta/2) \lim_{\alpha \to 0} \left[\frac{S(\alpha,\beta)}{\alpha} \right], \tag{12.126}$$

where $S(\alpha,\beta)$ is the measured scattering law. The frequency distribution $f(\beta)$ is often called the "spectral density function."

To illustrate the characteristic form of the double differential scattering cross section and the scattering law derived from it, we have chosen two sets of measurements on moderators of considerable interest to nuclear engineers: reactor-grade graphite and heavy water. Other scattering laws of interest, such as those of zirconium hydride and ordinary water, have been measured and are available in the technical literature and in the Brookhaven cross-section compilations.

Reactor-Grade Graphite

In the measurements of Whittemore,† neutrons of nine initial energies ranging from 0.129 to 0.611 ev were scattered from a 0.335-cm-thick 6- by 12-in slab of reactor-grade graphite that was maintained at room temperature ($\sim 300°$K). The energies of the neutrons scattered through angles of 30°, 60°, 90°, and 120° were determined by time-of-flight apparatus. The ranges of generalized momentum and energy transfer covered were $0.14 \leq \alpha \leq 5$ and $0 \leq \beta \leq 13$.

One of the set of 36 measured double differential cross sections is shown in Figure 12.10. The measured cross sections corrected for multiple scattering in the sample were reduced to the form of a scattering law using Equation 12.125 in the form

$$S(\alpha,\beta) = \frac{4\pi kT}{\sigma_b} \sqrt{\frac{E_0}{E}} \, e^{\beta/2} [\sigma(E_0|E,\theta)]_{\text{measured}}, \tag{12.127}$$

where $kT = 0.026$ ev, and $\sigma_b = 5.63$ barns for carbon in graphite.

---

*P. A. Egelstaff, *Nucl. Sci. Eng.*, **12**: 250 (1962); P. A. Egelstaff and P. Schofield, *Nucl. Sci. Eng.*. **12**: 260 (1962).
†W. L. Whittemore, *Nucl. Sci. Eng.*, **33**: 31 (1968).

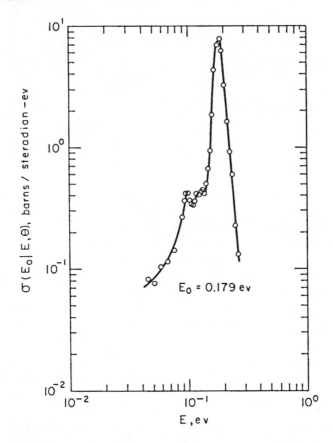

Figure 12.10—Distribution of neutron energies from scattering by graphite at $\theta = 30°$. Incident neutron energy is 0.179 ev. [From W. L. Whittemore, *Nucl. Sci. Eng.*, **33**: 31 (1968).]

The experimentally determined scattering law is shown in Figure 12.11. The theoretically derived scattering law, which was based on the theoretical frequency distribution shown in Figure 12.12, is seen to be in reasonable agreement with the experimental scattering law except at small values of the energy transfer β. The discrepancy at $\beta = 1$ is believed to be caused mainly by the effect of instrumental resolution broadening. The deviations from the otherwise smooth curves at $\beta = 5$ and 7 are believed to be due to coherent effects that were not taken into account in the theory.

The experimentally determined scattering law was employed to determine the frequency distribution of phonons in the graphite (Equation

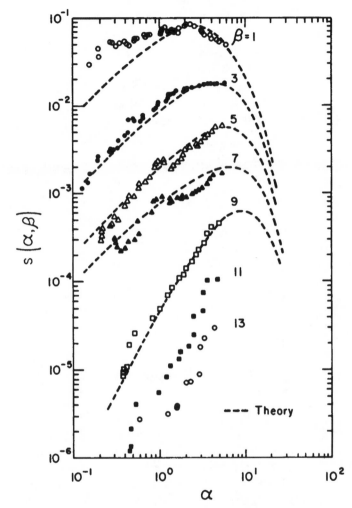

Figure 12.11—Scattering-law data for graphite at 300°K, corrected for multiple scattering and compared with theoretical calculations. [From W. L. Whittemore, *Nucl. Sci. Eng.*, **33**: 31 (1968).]

12.126). The experimental distribution, shown in Figure 12.12, is seen to be in rather poor agreement with the theoretical frequency distribution.

Heavy Water

In the measurements of Harling,* essentially monoenergetic neutrons with initial energies of 0.101 and 0.213 ev were scattered from a 0.158-cm-

*O. K. Harling, *Nucl. Sci. Eng.* **33**: 41 (1968).

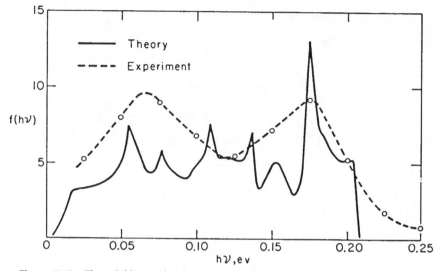

Figure 12.12—The solid line is the theoretical phonon-energy distribution employed in the calculation of the double differential cross section for graphite. The dashed line is the phonon-energy distribution obtained from analysis of the experimental cross section. [From W. H. Whittemore, *Nucl. Sci. Eng.*, **33**: 31 (1968).]

thick D_2O target that was maintained at room temperature (299°K). The energies of the neutrons scattered through ten angles from 15° to 155.5° were simultaneously measured with time-of-flight apparatus. These parameters allowed the generalized momentum- and energy-transfer ranges of $0.1 \leq \alpha \leq 16$ and $0.5 \leq \beta \leq 7$ to be investigated.

Figure 12.13 shows one of the two sets of double differential cross sections which were obtained. The reduction of these cross sections to the scattering law was accomplished through the use of Equation 12.127 with $\sigma_b = 7.6$ barns and $A = 2$, the bound-atom cross section and the mass of the principal scatterer (the deuteron), respectively. The resulting scattering law is shown in Figure 12.14.

The experimental spectral density function, determined from the scattering law through the use of Equation 12.126, is shown in Figure 12.15. The peak near $\beta = 2$ is believed to result from the hindered rotation band. However, it is possible that this peak is spurious since the raw data used by Harling to derive the scattering law and the spectral density function were not corrected for multiple scattering of neutrons within the D_2O target. More recent measurements by Whittemore indicate clearly that multiple scattering significantly alters the scattering law in D_2O.*

*W. L. Whittemore, *Nuc. Sci. Eng.*, **33**: 195 (1968).

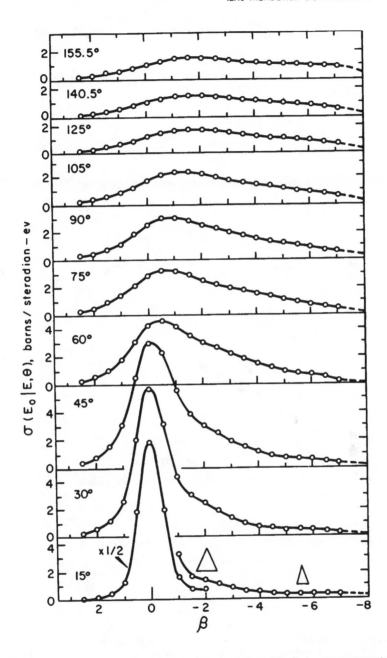

Figure 12.13—Measured double differential cross section of D_2O at room temperature. Incident neutron energy is 0.213 ev. [From O. K. Harling, *Nucl. Sci. Eng.*, **33**: 41 (1968).]

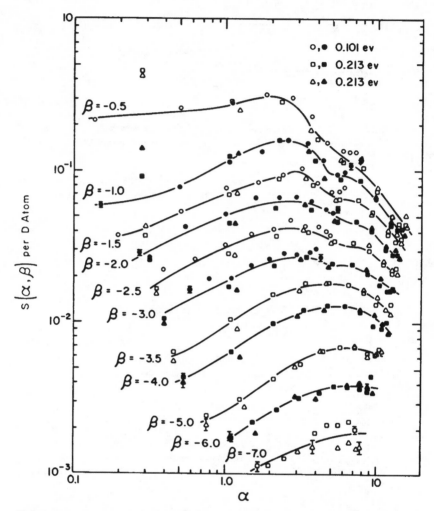

Figure 12.14—Scattering-law data for D_2O at 299°K. The curves were drawn by eye through the points. [From O. K. Harling, *Nucl. Sci. Eng.*, **33**: 41 (1968).]

SELECTED REFERENCES

1. Edoardo Amaldi, "The Production and Slowing Down of Neutrons," a 659-page "article" in the *Encyclopedia of Physics*, Vol. XXXVIII/2, Springer-Verlag, 1959. This is an excellent reference for the person who wishes a conversant knowledge of low-energy-neutron interactions without extensive mathematical details. About 100 pages of this long article are devoted to the topics of this chapter: Doppler effect, coherent scattering, and chemical binding effects. Extensive references to the pre-1959

hν , e v

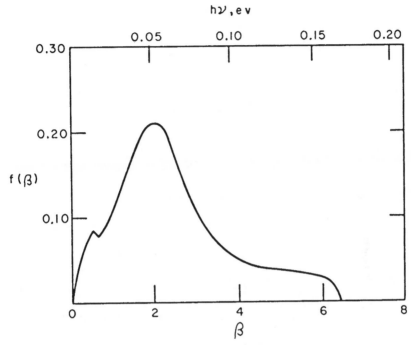

Figure 12.15—Experimentally determined spectral density function of the deuteron in D₂O at 299°K. [From O. K. Harling, *Nucl. Sci. Eng.*, **33**: 41 (1968).]

literature are given. Amaldi provides an extremely lucid presentation; a pleasure to read.

    2. V. F. Turchin, *Slow Neutrons*, 1963; translated from Russian by Israel Program for Scientific Translations, Jerusalem, 1965; distributed by Daniel Davey & Co., Inc., 257 Park Avenue South, New York. This work is more of a textbook than either Egelstaff or Amaldi. Turchin develops the relevant equations from first principles, leaving out practically no steps. All the material of this chapter is included in Turchin and much more, both theoretical and experimental. For a self-study program, it is suggested that the student spend a weekend with the Amaldi work, followed by a month with the Turchin book, and then go on to special topics in Egelstaff's book.

    3. P. A. Egelstaff (Ed.), *Thermal Neutron Scattering*, Academic Press Inc., New York, 1965. This is an up-to-date, thoroughly referenced treatment of the theoretical and experimental basis of low-energy-neutron scattering from solids, liquids, and molecular gases. It is highly recommended as the next step after having read Amaldi and studied Turchin.

    4. Lawrence Dresner, *Resonance Absorption in Nuclear Reactors*, Pergamon Press, New York, 1960. For those nuclear engineers interested in the effect of Doppler broadening on effective resonance integrals, this is a good place to start. The book includes a review of all significant work prior to 1959.

5. G. E. Bacon, *Neutron Diffraction*, 2nd ed., Oxford University Press, London, 1962. This book, written almost entirely from the experimental point of view, explains how low-energy-neutron diffraction is employed in the study of structures, particularly crystalline solid and magnetic solid structures.

EXERCISES

1. Verify the equivalence of the following sets of equations for s-wave neutrons

$$\sigma_n = 4\pi|f|^2 \qquad\qquad \sigma_n = \pi\lambda^2|1 - U|^2$$

$$\sigma_r = (4\pi/k)\,\text{Im}\,f - 4\pi|f|^2 \qquad \sigma_r = \pi\lambda^2(1 - |U|^2)$$

$$\sigma_T = (4\pi/k)\,\text{Im}\,f \qquad\qquad \sigma_T = 2\pi\lambda^2(1 - \text{Re}\,U)$$

where $f = (e^{2i\sigma} - 1)/2ki$ and $U = e^{2i\sigma}$.

2. Assume that the nuclear potential can be represented by a simple square well $V(r) = -V_0$ in the region $0 \le r \le R$, where the nuclear radius is given by $R = 1.5A^{1/3} \times 10^{-13}$ cm. Show that under this assumption about 97% of all nuclei will have positive scattering lengths and only about 3% will have negative scattering lengths.

3. Show that the Fermi pseudopotential in conjunction with the first Born approximation is valid provided $|ka| \ll 1$.

4. Calculate the differential scattering cross section per nucleus of a system of two identical nuclei under the assumption that the first nucleus is fixed and the second nucleus is vibrating along a line from the first nucleus such that it has equal probability of lying in any unit length between d_1 and d_2 from the first nucleus and has zero probability elsewhere.

APPENDIX
CONSTANTS AND CONVERSION FACTORS

1 Universal Physical Constants

Avogadro's number

$$N = 6.0225 \times 10^{23} \text{ mole}^{-1}$$

Boltzmann's constant

$$k = 1.3805 \times 10^{-16} \text{ erg/°K} = 8.617 \times 10^{-5} \text{ ev/°K}$$

Electronic charge

$$e = 4.8030 \times 10^{-10} \text{ esu} = 1.60210 \times 10^{-19} \text{ coulomb}$$

Planck's constant

$$h = 6.626 \times 10^{-27} \text{ erg-sec} = 4.135 \times 10^{-15} \text{ ev-sec}$$

$$\hbar = h/2\pi \text{ where } \pi = 3.14159$$

Speed of light in vacuum

$$c = 2.997925 \times 10^{10} \text{ cm/sec}$$

Universal gravitational constant

$$G = 6.670 \times 10^{-8} \text{ dyne-cm}^2/\text{g}^2$$

2 Masses*

Electron rest mass

$$m_e = 9.1091 \times 10^{-28} \text{ g} = 5.48597 \times 10^{-4} \text{ amu} = 0.511006 \text{ Mev}$$

*The atomic mass unit is defined as 1/12 of the mass of the ^{12}C atom. In this scale, 1 amu = 1.66043×10^{-24} g.

Neutron rest mass

$$m_n = 1.67482 \times 10^{-24} \text{ g} = 1.008665 \text{ amu} = 939.550 \text{ Mev}$$

Proton rest mass

$$m_p = 1.67252 \times 10^{-24} \text{ g} = 1.007277 \text{ amu} = 938.256 \text{ Mev}$$

3 Conversion Factors

Length

   1 angstrom (Å) $= 10^{-8}$ cm

   1 fermi (fm) $= 1$ femtometer $= 10^{-13}$ cm

Area

   1 barn $= 10^{-24}$ cm^2

Mass

   1 amu (^{12}C scale) $= 1.66043 \times 10^{-24}$ g

   1 amu (^{12}C scale) $= 1.000318$ amu (^{16}O scale)

Energy

   1 ev $= 1.60210 \times 10^{-12}$ erg

   1 joule $= 10^7$ erg $= 1$ watt-sec

   1 Mev $= 1.60210 \times 10^{-13}$ watt-sec

   1 calorie $= 4.1856$ joule $= 4.1856 \times 10^7$ erg

   1 erg $= 1$ dyne-cm $= 1$ g cm^2/sec^2

   1 Mev $= 10^6$ ev

   1 mev $= 10^{-3}$ ev

4 Derived Constants

Classical electron radius

$$r_e = e^2/m_e c^2 = 2.8178 \times 10^{-13} \text{ cm}$$

Coulomb constant (squared electron charge)

$$e^2 = 2.307 \times 10^{-19} \text{ erg-cm} = 1.44 \times 10^{-7} \text{ ev-cm}$$

Coulomb potential

$$V = zZe^2/r$$

$$V(\text{Mev}) = 1.44\, zZ/r(\text{fm})$$

De Broglie wave number and wave length of the neutron

$$k = 1/\lambda = \sqrt{2mE}/\hbar$$

$$k(\text{cm}^{-1}) = 2.19 \times 10^9 \sqrt{E(\text{ev})} \quad \text{if } m = 1 \text{ amu}$$

$$\lambda^2 = \hbar^2/2mE$$

$$\lambda^2(\text{barns}) = 2.1 \times 10^5/E(\text{ev}) \quad \text{if } m = 1 \text{ amu}$$

Fine-structure constant

$$\alpha = 2\pi e^2/hc = 1/137.027$$

5 Prefixes

Deca	= 10	Deci	= 10^{-1}
Hecto	= 10^2	Centi	= 10^{-2}
Kilo	= 10^3	Milli	= 10^{-3}
Mega	= 10^6	Micro	= 10^{-6}
Giga	= 10^9	Nano	= 10^{-9}
Tera	= 10^{12}	Pico	= 10^{-12}

6 Mathematical Symbols

Equal to	$=$
Identically equal to	$\equiv$
Approximately equal to	$\simeq$
Order of magnitude equal to	$\stackrel{\circ}{=}$
Asymptotically equal to	
Tends to (or yields)	
Proportional to	$\propto$
Average value of y	$\langle y \rangle$
Absolute value of y	$|y|$
Complex conjugate of y	$\bar{y}$
Real part of y	Re y
Imaginary part of y	Im y

INDEX